# International Series in Operations Research & Management Science

Volume 280

More information about this series at http://www.springer.com/series/6161

Pierre-Olivier Pineau • Simon Sigué
Sihem Taboubi
Editors

# Games in Management Science

## Essays in Honor of Georges Zaccour

Springer

*Editors*
Pierre-Olivier Pineau
Chair in Energy Sector Management
HEC Montréal Energy Sector Management
Montréal, QC, Canada

Simon Sigué
Faculty of Business
Athabasca University
Athabasca, AB, Canada

Sihem Taboubi
Department of Marketing
GERAD and HEC Montréal
Montréal, QC, Canada

ISSN 0884-8289 ISSN 2214-7934 (electronic)
International Series in Operations Research & Management Science
ISBN 978-3-030-19106-1 ISBN 978-3-030-19107-8 (eBook)
https://doi.org/10.1007/978-3-030-19107-8



This Springer imprint is published by the registered company Springer Nature Switzerland AG.
The registered company address is: Gewerbestrasse 11, 6330 Cham, Switzerland

# Preface

This volume is written to mark the 60th birthday of Professor Georges Zaccour. The contributors, who are either former Ph.D. students or successful research collaborators and friends of Professor Zaccour, want to take this opportunity to acknowledge his contributions and to honor his many scientific achievements.

It would be unrealistic and inadvisable to attempt to give a full account of Georges Zaccour's scientific achievements in a book preface. We hope, however, that the few points below will help the reader appreciate the extent of his contribution to science.

All things considered, Georges Zaccour is one of the most prolific scholars of his generation in the fields of operations research and management science. His work covers theoretical developments as well as applications of dynamic optimization and dynamic games in various fields including economics, energy, the environment, marketing, and supply chain management. His work is published regularly in top-ranked journals and is funded by the Natural Sciences and Engineering Research Council of Canada and the Social Sciences and Humanities Research Council of Canada. His publications include 4 books, 14 edited volumes, over 150 refereed journal articles, and 24 chapters in edited books. His Google Scholar profile reflects the impacts of his contributions in his research fields (close to 6000 citations and an h indicator of 39). Among many scientific distinctions and awards, Georges Zaccour is a fellow of the Royal Society of Canada and a recipient of the 2018 Isaacs Award for his outstanding contribution to the theory and application of dynamic games. A complete list of his publications is included in this book.

Service to and leadership in the scientific community have always been a top priority for Georges Zaccour. Among other things, he is the editor in chief of *Dynamic Games and Applications,* and the associate editor of *International Game Theory Review, Environmental Modeling & Assessment, Computational Management Science*, and *INFOR,* and he sits on the editorial boards of several other prestigious journals in various fields. He is the holder of the Chair in Game Theory and Management at HEC Montréal and a former president of the International Society of Dynamic Games (ISDG, 2002–2006). He is also a former director of the Group for Research in Decision Analysis (GERAD, 2001–2005), a world-renowned research center in operations research and management science that brings together several universities in Montréal.

But perhaps Georges Zaccour's most significant contribution to science is his dedication to, passion for, and unparalleled skill at training graduate students and his ability to help them develop into thriving researchers. Thanks to his outstanding work in this area and his capacity to create a blossoming research environment, the future of research in the field of differential games and their applications in management science has never been as promising as it is today. Over the years, Professor Zaccour has created a scientific family of more than 35 Ph.D. and post-doc graduates, who are established in seven different countries across four continents. This family will soon grow, with the additional 15 students currently working under his supervision.

Professor Georges Zaccour with his Ph.D. students. From left to right: Pierre-Olivier Pineau, Sihem Taboubi, Simon Sigué. 9th ISDG Symposium, Adelaide, South Australia, December 18–21, 2000.

We, the editors of this volume and the organizers of the 11th Workshop on Dynamic Games and Management Science (Montréal, October, 24–25, 2019), being held in his honor, were among Professor Zaccour's first Ph.D. students (2000, 1999, and 2002, respectively). We have the privilege of publicly acknowledging the extraordinary support and mentorship he has provided us, first during our training at HEC Montréal, and later in our respective careers and personal lives. What we have become, we owe to him. It goes without saying that we are very proud to be part of Professor Zaccour's multigenerational scientific family, which includes not only his extensive network of distinguished collaborators and friends worldwide, but also, as of this volume, his daughter, Suzanne Zaccour. Suzanne, the youngest contributor to this collection, is pursuing a Ph.D. with the Faculty of Law at Oxford University. With Michèle Breton, she has co-authored a chapter that opens doors to new applications of management science in the field of law.

## Contents of the Book

This book collects 21 chapters reviewed according to international journal standards. The chapters cover theoretical developments in game theory and present a broad spectrum of their applications in management science. These applications include such areas as cyber defense, energy, and environmental management, healthcare management, marketing, and supply chain management.

We have divided the volume into three parts. Part I, composed of six chapters, is dedicated to *Marketing and Supply Chain Games*. In Chap. 1, S. Jørgensen and S. P. Sigué investigate a differential game that takes place in a duopolistic market where firms control their advertising and pricing decisions. A novel aspect of their study is that they introduce price as an additional instrument used by competing firms in the commonly used Lanchester model, which originally solely captured the effects of advertising competition on the evolution of the firms' market shares. In Chap. 2, G. Martín-Herrán and S. Taboubi study the pricing of optional contingent products (i.e., a set of products that includes one that is useless without the other) and compare prices and profits when these products are sold by the same company or by two separate firms linked by the interdependency of their product demand. Their study provides an interesting "logical experimental" perspective on marketing modeling and game theory, two areas of expertise of Professor Zaccour. In Chap. 3, S. Karray conducts one of the first studies to examine the efficiency of rebate programs in marketing channels where competition is considered at both the manufacturer and retailer levels. Rebate programs are price discounts that consumers acquire by purchasing a product at an initial period and that they then use on the next purchasing occasion. The author uses a two-period game, where manufacturers decide in the first period whether or not to implement a rebate program, while retailers react by fixing the rebate level and the price to consumers. Retailers also set the retail prices in period 2 by taking into account the decisions made in the first period. Backward induction is used to solve the games under three scenarios, in which the rebate program can be implemented or not by one or both manufacturers in the distribution channel. In Chap. 4, L. Lambertini deals with the issue of channel coordination through the use of two-part tariffs. These are price mechanisms that manufacturers can use in order to allow a decentralized channel to replicate the results of a vertically integrated one. The author extends the study of Zaccour (2008), on the efficiency of such mechanisms in a dynamic setting, by introducing competition both upstream and downstream. In Chap. 5, O. Rubel studies contractual agreements in the event of major crisis events from a channel's perspective. He uses a stochastic differential game in a bilateral monopoly to examine the impact of a product recall on the pricing strategies and profits of both firms. The author investigates whether vendor agreements, which are signed before any unit is sold, could aggravate the double-marginalization problem in the channel, for two recall cost structures. In Chap. 6, P. De Giovanni and T. S. Genc investigate a closed-loop supply chain where members can implement either a traditional wholesale pricing contract or a revenue-sharing contract. Unlike previous studies, where the return rate is solely affected by the manufacturer's green efforts, the authors introduce the retail prices set by the retailer into the dynamics, and demonstrate that this variable plays a key role in identifying the best contract for achieving coordination and for reaching environmental objectives.

Part II contains six chapters dedicated to *Resources Games*, which also includes environmental and climate topics. In Chap. 7, O. Bahn and A. Haurie formulate a steady-state game model with two types of production economy ("dirty" and "clean") and two types of emission-reduction technology (investment in carbon capture and sequestration, and a technology of direct air capture). Their model presents the results of negotiations among different coalitions of countries in managing a net-zero Greenhouse Gas (GHG) emissions regime. The authors use it to compute and compare various environmental and economic indicators (i.e., GHG emissions, capital stocks, labor allocation, consumption) obtained under a Nash game and a cooperative game. In Chap. 8, H. Dawid, R. F. Hartl, and P. M. Kort build a general dynamic model for a firm that uses as an input in its production process an energy that can be delivered either conventionally or by building a stock of green energy capital. The authors analyze variations of this model and examine the impacts on the results. A key element is their explicit modeling of the positive side effect of investing in green energy. This is done to capture the fact that firms investing in green energy are positioning themselves as "green firms," which helps enhance the impact of their advertising efforts on their goodwill and on demand. In Chap. 9, D. Tasneem and H. Benchekroun review the experimental literature that analyzes the behavior of agents in dynamic common-pool resource games (such as fisheries, forestry, and water). The authors propose a classification of this literature into three groups: studies that compare behavior to predictions of cooperative and noncooperative theoretical benchmarks in the presence of dynamic externalities; studies that aim to find behavioral support for the use of specific types of strategies; and studies that examine behaviors in a continuous time setting. In Chap. 10, D. Claude and M. Tidball study vertical externalities and strategic delegation, a topic that links two fields of research to which Professor Zaccour has made significant contributions: interactions in marketing channels, and environmental and resource economics. The authors consider a vertical market structure with an upstream monopoly, which fixes the input price, and a downstream quantity competition, where firms generate pollution emissions when they process the intermediate product into a final good. They compute and compare the results of their model under two scenarios, depending on whether the upstream monopolist is able to pre-commit to a fixed input price or whether it relies on a flexible pricing scheme. In Chap. 11, J. de Frutos and G. Martín-Herrán use a linear-quadratic transboundary pollution game to illustrate how nonlinear incentive strategies can sustain an agreement over time. The authors use a less restrictive definition of incentive equilibria with respect to the existing literature, in the sense that they look for an incentive strategy equilibrium that allows the pollution stock under a noncooperative mode of play to be close enough, but not necessarily identical, to its value under cooperation. The authors compare the incentive equilibrium strategies, their credibility, and the players' payoff under open-loop and Markovian strategies. In Chap. 12, F. J. André and L. M. Castro survey the literature on emission trading under market power. They develop a unifying two-period model that allows them to replicate some of the main results in this literature and to analyze the relationship between permit prices and the degree of competition in the output and emissions markets for different market structures (depending on which is the mainstream market).

Finally, Part III deals with *Social Games*. It combines nine chapters on various topics related to health, security, social norms, the law, etc., including some theoretical developments in game theory. In Chap. 13, M. Breton and S. Zaccour launch a new and original conversation between the law and game theory on the personhood status of environmental entities. By granting personhood status to a river suffering from a firm's polluting activities, the authors allow the river to become an active downstream player in a cooperative game. They show that cooperation with environmental entities having a personhood status may be preferable to alternative solutions such as laissez-faire, government regulation, and noncooperative or cooperative solutions involving interested parties.

In Chap. 14, N. Van Long introduces the concept of a feedback Kant-Nash equilibrium in a discrete time model of resource exploitation. The author revisits the well-known dynamic model of the tragedy of the commons, where he considers a subset of Kantian agents guided by a moral norm. This norm allows them to explain their actions according to different rules than those used by rational agents (i.e., Nashians). One of the main results obtained is that, even without external punishment for violations of social norms, if a sufficiently large fraction of the population consists of Kantian agents, the tragedy of the commons can be substantially attenuated.

In Chap. 15, E. Billette de Villemeur and P. Pineau investigate the double prisoner's dilemma resulting from the fact that some individuals continue to consume increasing amounts of oil, despite their high price, while at the same time, environmental militants oppose production in the oil industry. The authors examine the impacts on the individual and collective outcomes of two sets of choices: being an environmental militant or not, and adopting for a frugal level of energy consumption or not. One of their main results indicates that the highest collective outcome is obtained when frugal behavior is adopted but militancy is avoided. This result allows them to conclude that effective environmental action should avoid opposing oil supply sources, while encouraging consumers to become more frugal.

In Chap. 16, F. Cabo, A. Garcia-Gonzales, and M. Molpeceres-Abella study compliance with social norms that are optimally established by a benevolent central planner as an evolutionary stable equilibrium. Compliance is analyzed in a two-population evolutionary game, where individuals from one population play against and imitate agents within their own, but also the other, population. The paper distinguishes two types of agents, namely, the standard pro-self agents (Sanchos), whose payoffs are defined by a prisoner's dilemma game dominated by a noncompliance strategy; and the pro-social Quixotes, who still have an incentive to free-ride, although they prefer compliance over mutual defection, as in a snowdrift game. The authors analyze the conditions under which the interaction with the population of selfish Sanchos increases or decreases the compliance rate among altruistic Quixotes. In Chap. 17, F. Ngendakuriyo and P. V. Reddy analyze a differential game that takes place between an active civil society and a government. They examine the conditions under which a country can switch from an initial situation of endogenous corruption to a society with no, or little, corruption. The authors extend the model initially examined in Ngendakuriyo and Zaccour (2013) by introducing the effects of social inertia in the society, which induces positive (negative) feedback, depending on the social perception of the prevailing institutional quality. Their study demonstrates that an increase in optimism (pessimism) in the society leads it to invest less (more) effort to fight corruption, whereas a corrupt government invests more (less) effort in a repression policy.

In Chap. 18, A. Sokri discusses how game theory can be applied to cyber defense problems. The author provides an extensive review of the literature in this area and classifies it into three major categories: resource allocation, network security, and cooperation models. Finally, he suggests replacing one assumption used in the existing literature on security games, in order to capture more realistic features in these problems. The assumption states that defenders and attackers are able to accurately evaluate their own payoffs and those of their opponents. The author proposes using new approach that introduces uncertainty into the model. He ends by discussing the main challenges associated with the applicability of game-theoretic methods in cyberspace and the avenues for future research. In Chap. 19, A. Buratto, L. Grosset, and B. Viscolani investigate the recent problem related to the increase in the number of unvaccinated people and study the effectiveness of healthcare management policies on this issue. The authors formulate and solve a differential game that takes place between the healthcare system, whose aim is to minimize the number of unvaccinated people, and a pharmaceutical firm, which produces and sells a given type of vaccine. To pursue their objectives, the two players run appropriate vaccination advertising campaigns. The authors find that the pharmaceutical firm's communication policy helps the healthcare system decrease the number of unvaccinated people. In Chap. 20, S. Debia analyzes the interaction between international trade and pollution mitigation. While the traditional game-theoretic literature on pollution mitigation found its analysis on homogeneous goods, this study assumes a setup of intra-industry trade between two countries, each producing a differentiated product. The author builds a bi-matrix noncooperative game about the decision to cooperate (or not) in a nonbinding environmental agreement (such as Paris COP 21) and shows that both countries cooperate if their products are complements. This contradicts the traditional prisoner's dilemma result for two-player noncooperative games with positive externalities (such as pollution mitigation). Finally, in Chap. 21, E. Parilina and A. Sedakov examine the stability problem of coalition structures. The authors use a dynamic competition model in discrete time, where they consider that firms choose their outputs in each time period. In their model, the market price is based on the firms' decision and on the price in the previous time period. They use two approaches to determine the firms' profits in the game: one approach where profits are not redistributed within the coalition and one where profits are redistributed using a solution from cooperative game theory. For each case, the authors study the stability of the coalition structure and verify its dynamic stability.

Montréal, QC, Canada — Pierre-Olivier Pineau
Athabasca, AB, Canada — Simon Sigué
Montréal, QC, Canada — Sihem Taboubi

# Contributions by Georges Zaccour


**Books**

Basar, T., & Zaccour, G. (2018). *Handbook of dynamic game theory* (Vol. 1, 670 p). Springer.

Basar, T., & Zaccour, G. (2018). *Handbook of dynamic game theory* (Vol. 2, 635 p). Springer.

Haurie, A., Brawczyk, J. B., & Zaccour, G. (2012). *Games and dynamic games* (465 p). NOW World Scientific Publishing Co.

Jørgensen, S., & Zaccour, G. (2004). Differential games in marketing, International series in quantitative marketing. Kluwer Academic Publishers.

**Edited Volumes**

Krivan, V., & Zaccour, G. (Eds.). (2013). Advances in dynamic games: theory, applications and numerical methods. Annals of the international society of dynamic games (Vol. 13, 350 pp). Springer.

Loulou, R., Waaub, J.-Ph., & Zaccour, G. (Eds.). (2005). *Energy and environment*. Springer.

Haurie, A., & Zaccour, G. (Ed.). (2005). *Dynamic games: theory and applications*. Springer.

Zaccour, G. (Ed.). (2002). Optimal control and differential game: Essays in honor of Steffen Jørgensen. Advances in computational management science series. Kluwer Academic Publishers.

Zaccour, G. (Ed.). (2002). Decision and control in management science: Essays in honor of Alain Haurie. Advances in computational management science series. Kluwer Academic Publishers.

Zaccour, G. (Ed.). (1998). Deregulation of electric utilities. Topics in regulatory economics and policy series. Kluwer Academic Publishers.

Jørgensen, S., & Zaccour, G. (Eds.). (1996). Dynamic competitive analysis in marketing. Lecture notes in economics and mathematical systems (Vol. 444). Springer-Verlag.

Carraro, C., Haurie, A., & Zaccour, G. (Eds.). (1994, December). Environmental management in a transition to market economy: A challenge to governments and business. Éditions Technip.

Breton, M., & Zaccour, G. (Eds.). (1994). *Industrie pétrolière: Outils d'aide à la décision et études de cas*. Éditions Technip.

Breton, M., & Zaccour, G. (Eds.). (1994). Proceedings of the Sixth International Symposium on Dynamic Games and Applications, Symposium, Montréal.

Lapointe, A., & Zaccour, G. (Eds.). (1993). Ajustements structurels et gestion du secteur énergétique en Afrique. Éditions Technip.

Poirier, A., & Zaccour, G. (Eds.). (1991). *Maritime and pipeline transportation of oil and gas*. Éditions Technip.

Breton, M., & Zaccour, G. (Eds.). (1991). *Advances in operations research in the oil and gas industry*. Éditions Technip.

Haurie, A., Poirier, A., Smeers, Y., & Zaccour, G. (Eds.). (1988). *Modélisation et analyse des marchés du gaz naturel*. Actes du colloque tenu à Montréal, HEC.

**Papers in Refereed Journals**

Pazoki, M., & Zaccour, G. (2019, April). A mechanism to promote product recovery and remanufacturing. *European Journal of Operational Research, 274*(2), 601–614.

Breton, M., Dahmouni, I., & Zaccour, G. (2019, March). Equilibria in a two-species fishery. *Mathematical Biosciences, 309*, 78–91.

Sacco, A., & Zaccour, G. (2019, January). Impact of social externalities on the formation of an international environmental agreement: an exploratory analysis. *International Transactions on Operational Research, 26*(1), 64–79.

Vardar, B., & Zaccour, G. (2018, December). The strategic impact of adaptation in a transboundary pollution dynamic game. *Environmental Modeling & Assessment, 23*(6), 653–669.

Crettez, B., Hayek, N., & Zaccour, G. (2018, November). Brand imitation: A dynamic-game approach. *International Journal of Production Economics, 205*, 139–155.

Ben Rhouma, T., & Zaccour, G. (2018, June). Optimal marketing strategies for the acquisition and retention of service subscribers. *Management Science, 64*(6), 2609–2627.

Pnevmatikos, N., Vardar, B., & Zaccour, G. (2018, June). When should a retailer invest in brand advertising? *European Journal of Operational Research, 267*(2), 754–764.

Zhang, Q., Zhang, J., Zaccour, G., & Tang, W. (2018, May). Strategic technology licensing in a supply chain. *European Journal of Operational Research, 267*(1), 162–175.

Crettez, B., Hayek, N., & Zaccour, G. (2018, March). Existence and uniqueness of optimal dynamic pricing and advertising controls without concavity. *Operations Research Letters, 46*(2), 199–204.

Oubraham, A., & Zaccour, G. (2018, March). A survey of applications of viability theory to the sustainable exploitation of renewable resources. *Ecological Economics, 15*, 346–367.

Zaman, H., & Zaccour, G. (2018). Vehicle scrappage incentives to accelerate the replacement decision of heterogeneous consumers. *Omega.*

Kort, P. M., Taboubi, S., & Zaccour, G. (2018). Pricing decisions in marketing channels in the presence of optional contingent products. *Central European Journal of Operations Research.*

Sebri, M., & Zaccour, G. (2017, November). Cross-country differences in private-label success: An exploratory approach. *Journal of Business Research, 80*, 116–126.

Parilina, E., & Zaccour, G. (2017, October). Node-consistent shapley value for games played over event trees with random terminal time. *Journal of Optimization Theory and Applications, 175*(1), 236–254.

Sebri, M., & Zaccour, G. (2017, September). Estimating umbrella-branding spillovers: A retailer perspective. *European Journal of Marketing, 51*(9/10), 1695–1712.

Karray, S., Martín-Herrán, G., & Zaccour, G. (2017, May). Assessing the profitability of cooperative advertising programs in competing channels. *International Journal of Production Economics, 187*, 142–158.

Groot, N., Zaccour, G., & De Schutter, B. (2017, April). Hierarchical game theory for system-optimal control: Applications of reverse Stackelberg games in regulating marketing channels and traffic routing. *IEEE Control Systems Magazine, 37*(2), 129–152.

Parilina, E., Sedakov, A., & Zaccour, G. (2017, April). Price of anarchy in a linear-state stochastic dynamic game. *European Journal of Operational Research, 258*(2), 790–800.

Masoudi, N., & Zaccour, G. (2017, April). Adapting to climate change: Is cooperation good for the environment? *Economics Letters, 153*, 1–5.

Samano Sanchez, M., Santugini, M., & Zaccour, G. (2017, April). Dynamics in research joint ventures and R&D collaborations. *Journal of Economic Dynamics and Control, 77*, 70–92.

Reddy, P. V., & Zaccour, G. (2017, February). Feedback Nash equilibria in linear-quadratic difference games with constraints. *IEEE Transactions on Automatic Control, 62*(7), 590–604.

Ngendakuriyo, F., & Zaccour, G. (2017, January). Should civil society organizations cooperate or compete in fighting a corrupt government? *Mathematical Social Sciences, 85*, 30–36.

Reddy, P. V., Wrzaczek, S., & Zaccour, G. (2016, December). Quality effects in different advertising models – An impulse control approach. *European Journal of Operational Research, 255*(3), 984–995.

Nguyên Hoang, L., Soumis, F., & Zaccour, G. (2016, December). Measuring unfairness feeling in allocation problems. *Omega, 65*, 138–147.

Nguyên Hoang, L., Soumis, F., & Zaccour, G. (2016, November). A heuristic optimization of Bayesian incentive-compatible cake-cutting. *Computers & Operations Research, 75*, 76–82.

Kanani Kuchesfehani, E., & Zaccour, G. (2016, September). Incentive equilibrium strategies in dynamic games played over event trees. *Automatica, 71*, 50–56.

Masoudi, N., Santugini, M., & Zaccour, G. (2016, July). A dynamic game of emissions pollution with uncertainty and learning. *Environmental and Resource Economics, 64*(3), 349–372.

Reddy, P. V., & Zaccour, G. (2016, July). A friendly computable characteristic function. *Mathematical Social Sciences, 82*, 18–25.

Buratto, A., Grosset, L., & Zaccour, G. (2016, July). Strategic pricing and advertising in the presence of a counterfeiter. *IMA Journal of Management Mathematics, 27*(3), 397–418.

Ben-Brahim, M., Zaccour, G., & Ben Abdelaziz, F. (2016, July). Strategic investments in R&D and efficiency in the presence of free riders. *RAIRO-Operations Research, 50*(3), 611–625.

Parilina, E., & Zaccour, G. (2016, June). Strategic support of node-consistent cooperative outcomes in dynamic games played over event trees. *International Game Theory Review, 18*(2), 1640002.

De Giovanni, P., Reddy, P. V., & Zaccour, G. (2016, March). Incentive strategies for an optimal recovery program in a closed-loop supply chain. *European Journal of Operational Research, 249*(2), 605–617.

Parilina, E., & Zaccour, G. (2015, September). Approximated cooperative equilibria for games played over event trees. *Operations Research Letters, 43*(5), 507–513.

Andres-Domenech, P., Martín-Herrán, G., & Zaccour, G. (2015, September). Cooperation for sustainable forest management: An empirical differential game approach. *Ecological Economics, 117*, 118–128.

Reddy, P. V., & Zaccour, G. (2015, September). Open-loop Nash equilibria in a class of linear-quadratic difference games with constraints. *IEEE Transactions on Automatic Control, 60*(9), 2559–2564.

Lambertini, L., & Zaccour, G. (2015, July). Inverted-U aggregate investment curves in a dynamic game of advertising. *Economics Letters, 132*, 34–38.

Parilina, E., & Zaccour, G. (2015). Node-consistent core for games played over event trees. *Automatica, 55*, 304–311.

Dantas, D., Taboubi, S., & Zaccour, G. (2014, October). Which business model for ebook pricing. *Economics Letters, 25*(1), 126–129.

Jørgensen, S., & Zaccour, G. (2014, August). A survey of game-theoretic models of cooperative advertising. *European Journal of Operational Research, 237*(1), 1–14.

Erdlenbruch, K., Tidball, M., & Zaccour, G. (2014). Quantity-quality management of a groundwater resource by a water agency. *Environmental Science & Policy, 44*, 201–214.

Kanani Kuchesfehani, E., & Zaccour, G. S-adapted equilibria in games played over event trees with coupled constraints. *Journal of Optimization Theory and Applications*. DOI 10.1007/s10957-014-0623-6.

Jansens, G., & Zaccour, G. (2014). Strategic price subsidies for new technologies. *Automatica, 50*, 1999–2006.

Andres-Domenech, P., Saint-Pierre, P., Smala Fanokoa, P., & Zaccour, G. (2014, August). Sustainability of the dry forest in Androy: A viability analysis. *Ecological Economics, 104*, 33–49.

De Giovanni, P., & Zaccour, G. (2014, January). A two-period game of closed-loop supply chain. *European Journal of Operational Research, 232*(1), 22–40.

Masoudi, N., & Zaccour, G. (2014, March). Emissions control policies under uncertainty and rational learning in a linear-state dynamic model. *Automatica, 50*, 719–726.

Eyland, T., & Zaccour, G. (2014). Carbon tariffs and cooperative outcomes. *Energy Policy, 65*, 718–728.

Amrouche, N., Ben Rhouma, T., & Zaccour, G. (2014), Branding decisions for retailers' private labels. *Journal of Marketing Channels, 21*, 100–115.

Ben Youssef, S., & Zaccour, G. (2014). Absorptive capacity, R&D spillovers, emissions taxes and R&D subsidies. *Strategic Behavior and the Environment, 4*(1), 41–58.

Martín-Herrán, G., Sigué, S. P., & Zaccour, G. (2014, January). Downstream horizontal integration and multiunit dealership. *International Transactions in Operational Research, 21*(1), 81–101.

Genc, T. S., & Zaccour, G. (2013, September). Capacity investments in a stochastic dynamic game: Equilibrium characterization. *Operations Research Letters, 41*(5), 482–485.

Masoudi, N., & Zaccour, G. (2013). A differential game of international pollution control with evolving environmental costs. *Environment and Development Economics, 18*(6), 680–700.

Ngendakuriyo, F., & Zaccour, G. (2013). Fighting corruption: To precommit or not? *Economics Letters, 120*(2), 149–154.

Reddy, P. V., Shevkoplyas, E., & Zaccour, G. (2013). Time-consistent shapley value for games played over event trees. *Automatica, 49*(6), 1521–1527.

Ben Youssef, S., Breton, M., & Zaccour, G. (2013, April). Cooperating and noncooperating firms in inventive and absorptive research. *Journal of Optimization Theory and Applications, 157*(1), 229–251.

Eyland, T., & Zaccour, G. (2012). Strategic effects of a border tax adjustment. *International Game Theory Review, 14*(3), 22.

Ingene, C., Taboubi, S., & Zacour, G. (2012). Game-theoretic coordination mechanisms in distribution channels: integration and extensions for models without competition. *Journal of Retailing, 88*(4), 476–496.

Ben Rhouma, T., & Zaccour, G. (2012). An empirical investigation of late bidding in online auctions. *Economics Letters, 117*(3), 715–717.

Martín-Herrán, G., Taboubi, S., & Zaccour, G. (2012). Dual role of price and myopia in a marketing channel. *European Journal of Operational Research, 219*(2), 284–295.

Pineau, P.-O., Rasata, H., & Zaccour, G. (2011). A dynamic oligopolistic electricity market model with interdependent segments. *Energy Journal, 32*(4), 183–217.

Domenech, P.-A., Saint-Pierre, P., & Zaccour, G. (2011, Decmber). Forest conservation and $CO_2$ emissions: A viable approach. *Environmental Modeling & Assessment, 16*, 6, 519–539.

Martín-Herrán, G., Sigué, S. P., & Zaccour, G. (2011, September). Strategic interactions in traditional franchise systems: Are franchisors always better? *European Journal of Operational Research, 213*(3), 526–537.

Pineau, P.-O., Rasata, H., & Zaccour, G. (2011, August). Impact of some parameters on investments in oligopolistic electricity markets. *European Journal of Operational Research, 213*, 180–195.

André, F., Sokri, A., & Zaccour, G. (2011, July). Public disclosure programs vs. traditional approaches for environmental regulation: green goodwill and the policies of the firm. *European Journal of Operational Research, 212*, 199–212.

Smala Fanokoa, P., Telahigue, I., & Zaccour, G. (2011). Buying cooperation in an Asymmetric Environmental Differential Game. *Journal of Economic Dynamics and Control, 35*, 935–946, 2011.

Kort, P. M., & Zaccour, G. (2011). When a firm should open its source code: A strategic analysis. *Production and Operations Management, 20*(6), 877–888.

Legras, S., & Zaccour, G. (2011). Temporal flexibility of permit trading when pollutants are correlated. *Automatica*, 47(5), 909–919.

Nasiri, F., & Zaccour, G. (2010, November). Renewable portfolio standard policy: A game-theoretic Analysis. *INFOR*, *48*(4), 251–260,

Jørgensen, S., Martín-Herrán, G., & Zaccour, G. (2010). The Leitmann-Schmitendorf advertising differential game. *Applied Mathematics and Computation, 217*, 1110–1116.

Jørgensen, S., Martín-Herrán, G., & Zaccour, G. (2010). Dynamic games in the economics and management of pollution. *Environmental Modeling & Assessment, 15*(6), 433–467.

Martín-Herrán, G., Sigué, S.-P., & Zaccour, G. (2010). The dilemma of pull and push price promotions. *Journal of Retailing, 86*(1), 51–68.

Breton, M., Sbragia, L., & Zaccour, G. (2010). Dynamic models for international environmental agreements. *Environmental and Resource Economics, 45*, 25–48.

Claude, D., & Zaccour, G. (2009). Investment in tourism market and reputation. *Journal of Public Economic Theory, 11*(5), 797–817.

Nasiri, F., & Zaccour, G. (2009). An exploratory game-theoretic analysis of biomass electricity generation supply chain. *Energy Policy, 37*(11), 4514–4522.

Buratto, A., & Zaccour, G. (2009). Coordination of advertising strategies in a fashion licensing contract. *Journal of Optimization Theory and Applications, 142*, 31–53.

Tidball, M., & Zaccour, G. (2018). A differential environmental game with coupling constraints. *Optimal Control Applications and Methods, 30*, 197–207.

Bahn, O., Breton, M., Sbragia, L., & Zaccour, G. (2009). Stability of international environmental agreements: An illustration with asymmetrical countries. *International Transactions in Operational Research*, 1–18.

Amrouche, N., & Zaccour, G. (2009, April). A shelf-space-dependent wholesale price when manufacturer and retailer brands compete. *OR Spectrum, 31*(2), 361–383.

Martín-Herrán, G., & Zaccour, G. (2009, April). Credible linear incentive equilibrium strategies in linear-quadratic differential games. *Annals of the International Society of Dynamic Games, 10*, 261–291.

Jorgensen, S., Kort, P., & Zaccour, G. (2009, March). Optimal pricing and advertising policies for an entertainment event. *Journal of Economic Dynamics and Control, 33*(3), 583–596.

Viscolani, B., & Zaccour, G. (2009, January). Advertising strategies in a differential game with negative competitor's interference. *Journal of Optimization Theory and Applications, 140*(1), 153–170.

Amrouche, N., Martín-Herrán, G., & Zaccour, G. (2008, November). Feedback Stackelberg equilibrium strategies when the private label competes with the national brand. *Annals of Operations Research, 164*(1), 79–95.

Krajewska, M. A., Kopfer, H., Laporte, G., Ropke, S., & Zaccour, G. (2008, November). Horizontal cooperation among freight carriers: request allocation and profit sharing. *Journal of Operational Research Society, 59*(11), 1483–1491.

Zaccour, G. (2008). Time consistency in cooperative differential games: A tutorial. *INFOR, 46*(1), 81–92.

Amrouche, N., Martín-Herrán, G., & Zaccour, G. (2008, June). Pricing and advertising of private and national brands in a dynamic marketing channel. *Journal of Optimization Theory and Applications, 137*(3), 465–483.

Zaccour, G. (2008, May). On the coordination of dynamic marketing channels and two-part tariffs. *Automatica, 44*(5), 1233–1239.

Ben Abdelaziz, F., Ben Brahim, M., & Zaccour, G. (2008, January). R&D equilibrium strategies with surfers. *Journal of Optimization Theory and Applications, 136*(1), 1–13.

Martín-Herrán, G., Rubel, O., & Zaccour, G. (2008, February). Competing for consumer's attention. *Automatica, 44*, 361–370.

Breton, M., Sokri, A., & Zaccour, G. (2008). Incentive equilibrium in an overlapping-generations environmental game. *European Journal of Operational Research, 185*, 687–699.

Karray, S., & Zaccour, G. (2007). Effectiveness of coop advertising programs in competitive distribution channels. *International Game Theory Review, 9*(2), 151–167.

Pineau, P.-O., & Zaccour, G. (2007). An oligopolistic electricity model with interdependent market segments. *The Energy Journal, 28*(3), 165–185.

Rubel, O., & Zaccour, G. (2007). A differential game of a dual distribution channel. *Annals of the International Society of Dynamic Games, 9*, 547–568.

Jørgensen, S., & Zaccour, G. (2007, April). Developments in differential game theory and numerical methods: Economic and management applications. *Computational Management Science, 4*(2), 159–182.

Amrouche, N., & Zaccour, G. (2007, July). Shelf-space allocation of national and private brands. *European Journal of Operational Research, 180*(2), 648–663.

Karray, S., & Zaccour, G. (2006, September). Could co-op advertising be a manufacturer's counterstrategy to store brands? *Journal of Business Research, 59*(9), 1008–1015.

Jorgensen, S., Kort, P., & Zaccour, G. (2006, August). Advertising an event. *Automatica, 42*(8), 1349–1355.

Breton, M., Martín-Herrán, G., & Zaccour, G. (2006, July). Equilibrium investment strategies in foreign environmental projects. *Journal of Optimization Theory and Applications, 130*(1), 23–40.

Breton, M., Jarrar, R., & Zaccour, G. (2006, May). A note on feedback Stackelberg equilibria in a Lanchester model with empirical application. *Management Science, 52*(5), 804–811.

Beltov, T., Jørgensen, S., & Zaccour, G. (2006). Optimal retail price promotions. *Anales de Estudios Económicos y Empresariales, XVI*, 9–36.

Fredj, K., Martín-Herrán, G., & Zaccour, G. (2006, April). Incentive mechanisms to enforce sustainable forest exploitation. *Environmental Modelling & Assessment, 11*(2), 145–156.

Martín-Herrán, G., Taboubi, S., & Zaccour, G. (2006, February). The impact of manufacturers' wholesale prices on retailer's shelf-space and pricing decisions. *Decision Sciences, 37*(1), 71–90.

Breton, M., Vencatachellum, D., & Zaccour, G. (2006, February). Dynamic R&D with strategic behavior. *Computers & Operations Research, 33*(2), 426–437.

Breton, M., Zaccour, G., & Zahaf, M. (2006, January). A game-theoretic formulation of joint implementation of environmental projects. *European Journal of Operational Research, 168*(1), 221–239.

Breton, M., Fredj, K., & Zaccour, G. (2006, January). International cooperation, coalitions stability and free-riding in a game of pollution control. *The Manchester School, 74*(1), 103–121.

Jørgensen, S., Taboubi, S., & Zaccour, G. (2006). Incentives for retailer promotion in a marketing channel. *Annals of the International Society of Dynamic Games, 8*, 365–378.

Jørgensen, S., Martín-Herrán, G., & Zaccour, G. (2005, December). Sustainability of cooperation overtime in linear-quadratic differential games. *International Game Theory Review, 7*(4).

Breton, M., Zaccour, G., & Zahaf, M. (2005, October). A differential game of joint implementation of environmental projects. *Automatica, 41*(10), 1737–1749.

Martín-Herrán, G., & Zaccour, G. (2005, August). Credibility of incentive equilibrium strategies in linear-state differential games. *Journal of Optimization Theory and Applications, 126*(2), 367–389.

Tidball, M., & Zaccour, G. (2005, June). An environmental game with coupling constraints. *Environmental Modelling & Assessment, 10*(2), 153–158.

Martín-Herrán, G., Taboubi, S., & Zaccour, G. (2005, June). A time-consistent open-loop Stackelberg equilibrium of shelf-space allocation. *Automatica, 41*(6), 971–982.

Jarrar, R., Martín-Herrán, G., Zaccour, G. (2004, July). Markov perfect equilibrium advertising strategies of Lanchester duopoly model: A technical note. *Management Science, 50*(7), 995–1000.

Breton, M., Turki, A., & Zaccour, G. (2004). Efficiency of Bertrand and Cournot under precommitment. *Game Theory and Applications, 10*, 31–38.

Gagné, R., Nguimbus, R., & Zaccour, G. (2004, spring). The effects of location and non-location factors on gasoline station performance. *Energy Studies Review, 12*(2), 153–169.

Haurie, A., & Zaccour, G. (2004). S-adapted equilibria in games played over event trees: An overview. *Annals of the International Society of Dynamic Games, 7*, 367–400.

Fredj, K., Martín-Herrán, G., & Zaccour, G. (2004, February). Slowing deforestation pace through subsidies: A differential game. *Automatica, 40*(2), 301–309.

Breton, M., Turki, A., & Zaccour, G. (2004, October). Dynamic model of R&D, spillovers and efficiency of Bertrand and Cournot equilibria. *Journal of Optimization Theory and Applications, 123*(1), 1–25.

Jørgensen, S., & Zaccour, G. (2003). A differential game of retailer promotions. *Automatica, 39*(7), 1145–1155.

Jørgensen, S., Martín-Herrán, G., & Zaccour, G. (2003, October). Agreeability and time-consistency in linear-state differential games. *Journal of Optimization Theory and Applications, 119*(1), 49–632003.

Jørgensen, S., Taboubi, S., & Zaccour, G. (2003, October). Retail promotions with negative brand image effects: Is cooperation possible? *European Journal of Operational Research, 150*(2), 395–405.

Zaccour, G. (2003, April). Computation of characteristic function values for linear-state differential games. *Journal of Optimization Theory and Applications, 117*(1), 183–194.

Petrosjan, L., & Zaccour, G. (2003). Time-consistent shapley value of pollution cost reduction. *Journal of Economic Dynamics and Control, 27*(3), 381–398.

Jørgensen, S., & Zaccour, G. (2002). Channel coordination over time: Incentive equilibria and credibility. *Journal of Economic Dynamics and Control, 27*(5), 801–822.

Jørgensen, S, & Taboubi, S., & Zaccour, G. (2001). Cooperative advertising in a marketing channel. *Journal of Optimization Theory and Applications, 110*, 145–158.

Jørgensen, S., & Zaccour, G. (2001). Time consistent side payments in a dynamic game of downstream pollution. *Journal of Economic Dynamics and Control, 25*(2), 1973–1987.

Jørgensen, S., Sigué, S. P., & Zaccour, G. (2001). Stackelberg leadership in a marketing channel. *International Game Theory Review, 3*, 1–14.

Breton, M., & Zaccour, G. (2001). Equilibria in an asymmetric duopoly facing a security constraint. *Energy Economics, 23*(4), 457.

Jørgensen, S., & Zaccour, G. (2001). Incentive equilibrium strategies and welfare allocation in a dynamic game of pollution. *Automatica, 37*, 29–36.

Gagné, R., Kooli, N., & Zaccour, G. (2000). Impacts des prix promotionnels sur les ventes. *Revue internationale de gestion, 24*, 34–39.

Jørgensen, S., Sigué, S. P., & Zaccour, G. (2000). Dynamic cooperative advertising in a channel. *Journal of Retailing, 76*(Winter), 71–92.

Jørgensen, S., & Zaccour, G. (2000). Optimal output strategies in a two-stage game with entry, learning-by-doing and spillovers. *Game Theory and Applications, 5*, 65–72.

Petrosjan, L., & Zaccour, G. (2000). A multistage supergame of downstream pollution. *Annals of the International Society of Dynamic Games, 5*, 387–404.

Jørgensen, S., & Zaccour, G. (1999). Equilibrium pricing and advertising strategies in a marketing channel. *Journal of Optimization Theory and Applications, 102*, 111–125.

Jørgensen, S., Kort, P., & Zaccour, G. (1999). Production, inventories, and pricing under cost and demand learning effects. *European Journal of Operational Research, 117*, 382–395.

Jørgensen, S., & Zaccour, G. (1999). Price subsidies and guaranteed buys of a new technology. *European Journal of Operational Research, 114*, 338–345.

Krawczyk, J., & Zaccour, G. (1999). Management of pollution from decentralised agents. *International Journal of Environment and Pollution, 12*(2/3), 343–347.

Gagné, R., Sigué, S. P., & Zaccour, G. (1998, décembre). Droit d'entrée et taux de redevance dans les franchises d'exploitation au Québec. *L'Actualité Économique, 74*(4), 651–668.

Breton, M., Chauny, F., & Zaccour, G. (1997). A leader-follower dynamic game of new product diffusion. *Journal of Optimization Theory and Applications, 92*(1), 77–98.

Zaccour, G. (1996). A differential game model for optimal price subsidy of new technologies. *Game Theory and Applications, 2*, 103–114.

Tolwinski, B., & Zaccour, G. (1995, March). Learning from each other in a duopoly: Numerical approach. *Journal of Optimization Theory and Applications, 84*(3), 617–633.

Haurie, A., & Zaccour, G. (1995). Differential game models of global environmental management. *Annals of the International Society of Dynamic Games, 2*, 3–24.

Haurie, A., Smeers, Y., & Zaccour, G. (1992, August). Toward a contract portfolio management model for a gas producing firm. *INFOR, 30*(3), 257–273.

Amesse, F., & Zaccour, G. (1991). Les gestionnaires et l'exportation, Différences de perception et de caractéristiques entre gestionnaires de firmes exportatrices et non exportatrices. *Revue Canadienne des Sciences de l'Administration, 8*(3), 148–160.

Haurie, A., Zaccour, G., & Smeers, Y. (1990, August). Stochastic equilibrium programming for dynamic oligopolistic markets. *Journal of Optimization Theory and Applications, 66*(2), 243–253.

Zaccour, G., & Kestemont, M. P. (1989). Classification hiérarchique des données: une application à la segmentation des marchés internationaux. *STATISTICA, 28*, 1.

Zaccour, G. (1988). Valeur de Shapley et partage équitable des ressources. *L'Actualité Économique, 64*(2), 96–121.

Leroux, F., & Zaccour, G. (1984). Structures du marché et intensité de la publicité. *L'Actualité Économique, 60*(3), 341–354.

**Book Chapters**

Petrosjan, L. A., & Zaccour, G. (2018). Cooperative differential games with transferable payoffs. In T. Basar & G. Zaccour (Eds.), *Handbook of dynamic game theory* (Vol. 1, pp. 595–632). Springer.

Carlson, D. A., Haurie, A., & Zaccour, G. (2018). Infinite horizon concave games with coupled constraints. In T. Basar & G. Zaccour (Eds.), *Handbook of dynamic game theory* (Vol. 1, pp. 111–156). Springer.

Basar, T., Haurie, A., & Zaccour, G. (2018). Nonzero-sum differential games. In T. Basar & G. Zaccour (Eds.), *Handbook of dynamic game theory* (Vol. 1, pp. 61–110), Springer.

Zaccour, G. (2017). Sustainability of cooperation in dynamic games played over event trees. In R. Melnik, R. Makarov, & J. Belair (Eds.), *Recent progress and modern challenges in applied mathematics, modeling and computational science. Fields institute communications series* (pp. 419–437).

De Giovanni, P., & Zaccour, G. (2013). Cost-revenue sharing in a closed-loop supply chain. In P. Cardaliaguet & R. Cressman (Eds.), *Advances in dynamic games – Theory, applications, and numerical methods for differential and stochastic games – Annals of the international society of dynamic games* (pp. 395–421). Birkhauser.

Zaccour, G. (2013, March). In E. H. Kessler (Ed.), *Game theory, encyclopedia of management theory* (pp. 297–304). Sage Publications.

Martín-Herrán, G., Sokri, A., & Zaccour, G. (2007). Dynamic oligopoly competition with public environmental information disclosure. In L. A. Petrosjan & N. A. Zenkevitch (Eds.), *Contributions to game theory and management* (pp. 326–345).

Karray, S., & Zaccour, G. (2005). A differential game of advertising for national and store brands. In A. Haurie & G. Zaccour (Eds.), *Dynamic games: Theory and applications* (pp. 213–230), Springer.

Taboubi, S., & Zaccour, G. (2002). Impact of retailer's myopia on channel's strategies. In G. Zaccour (Ed.), *Optimal control and differential game: Essays in honor of Steffen Jørgensen* (pp. 179–192). Kluwer Academic Publishers.

Jørgensen, S., & Zaccour, G. (2002). Time consistency in cooperative differential game. In G. Zaccour (Ed.), *Decision and control in management science: In honor of Professor Alain Haurie* (pp. 349–366). Kluwer Academic Publishers.

Jørgensen, S., & Zaccour, G. (1996). A three-stage differential game of channel cooperation. In S. Jørgensen & G. Zaccour (Eds.), *Dynamic competitive analysis in marketing* (pp. 158–174). Berlin: Springer-Verlag.

Krawczyk, J., & Zaccour, G. (1996). Pollution management trough levies and subsidies. In L. Vlacic et al. (Eds.), *Modelling and control of national and regional economies* (pp. 241–246). Pergamon, an Imprint of Elsevier Science.

Haurie, A., Smeers, Y., & Zaccour, G. (1995). Gas contract portfolio management: Experiments with a stochastic programming approach. In J.-B. Lesourd, J. Percebois, & F. Valette (Eds.), *Models for energy policy*. Routledge.

Toumi, T., & Zaccour, G. (1995). Évaluation d'un projet de pipeline en Tunisie. In A. Lapointe & H. Taghvaï (Eds.), *L'industrie des hydrocarbures: défis et opportunités* (pp. 277–293). Paris: Éditions Technip.

El Sallaly, A., & Zaccour, G. (1993). Gestion optimale d'un parc de camions-citernes: le cas d'ESSO-Égypte. In M. Breton & G. Zaccour (Eds.), *Industrie pétrolière: outils d'aide à la décision et études de cas*. Paris: Éditions Technip.

Zaccour, G. (1993). Problèmes et politiques d'exploitation du gaz naturel dans les pays en voie de Développement. In A. Lapointe & G. Zaccour (Eds.), *Ajustements structurels et gestion du secteur énergétique en Afrique* (pp. 13–26). Paris: Éditions Technip.

Lord, L., Amesse, F., & Zaccour, G. (1992). Le choix séquentiel des marchés d'exportation des PME québécoises. In Volume édité par J. E. Denis, Dans La gestion de l'exportation (pp. 33–42). Ottawa: Centre de droit et de politique commerciale.

Haurie, A., Smeers, Y., & Zaccour, G. (1991). Gas contract portfolio management: A stochastic programming approach. In M. Breton & G. Zaccour (Eds.), *Advances in operations research in the oil and gas industry* (pp. 133–150). Paris: Éditions Technip.

Zaccour, G., & Poirier, A. (1991). Maritime transport of crude oil: Background and outlook. In A. Poirier & G. Zaccour (Eds.), *Maritime and pipeline transportation of oil and gas* (pp. 3–22). Paris: Éditions Technip.

Breton, M., & Zaccour, G. (1991). Sharing the cost of a pipeline: A game theoretic approach. In A. Poirier & G. Zaccour (Eds.), *Maritime and pipeline transportation of oil and gas* (pp. 291–306). Éditions Technip.

Haurie, A., & Zaccour, G. (1991). A game programming approach to efficient management of interconnected power networks. In R. Hämäläinen & H. K. Ehtamo (Eds.), *Differential games: Developments in modelling and computation* (pp. 275–284). Springer Verlag.

Breton, M., Haurie, A., & Zaccour, G. (1990). Methods in the analysis of multistage commodity markets. In C. T. Leondes (Ed.), *Advances in large scale systems dynamics* (pp. 75–106). Academic Press.

Flåm, S., & Zaccour, G. (1989). Nash-Cournot equilibria in the european gas market: A case where open-loop and feedback solutions coincide. In P. Bernhard & T. Basar (Eds.), *Differential games and applications* (pp. 145–156). Springer-Verlag.

Haurie, A., Smeers, Y., & Zaccour, G. (1989). Stochastic equilibrium programming for dynamic oligopolistic markets. In S. W. Wallace (Ed.), *Algorithms and model formulations in mathematical programming* (pp. 158–160).

**Conference Proceedings**

Breton, M., Fredj, K., & Zaccour, G. (2002, juillet). Characteristic functions, coalitions stability and free-riding in a game of pollution control. In *Proceedings of the Tenth International Symposium on Dynamic Games and Applications* (Vol. I, pp. 129–138).

Jørgensen, S., Taboubi, S., & Zaccour, G. (2000). Cooperative advertising under decreasing marginal effects of promotions on goodwill. In *Proceedings of the 9th International Symposium on Dynamic Games and Applications*, Adelaide, Australia (pp. 302–304).

Jørgensen, S., & Zaccour, G. (2000). Advertising allowances and pass-through in a randomly terminating channel relationship. In *Proceedings of the 9th International Symposium on Dynamic Games and Applications*, Adelaide, Australia (pp. 305–308).

Breton, M., Yezza, A., & Zaccour, G. (1996, décembre). Feedback Stackelberg equilibria in a dynamic game of advertising competition: A numerical analysis. In *Proceedings of the 7th International Symposium on Dynamic Games and Applications*, Japan (Vol. 1, pp. 47–58).

Jørgensen, S., & Zaccour, G. (1996, décembre). Dynamic pricing and advertising strategies and sustainable channel cooperation. In *Proceedings of the 7th International Symposium on Dynamic Games and Applications*, Japan (Vol. 1, pp. 389–405).

Krawczyk, J., & Zaccour, G. (1995). Pollution management trough levies and subsidies. In *Proceedings of the Symposium on Modelling and Control of National and Regional Economies*, Gold Coast, Australie, 2–5 July 1995.

Flåm, S., & Zaccour, G. (1992). Stochastic games; event-adapted equilibria and their computation. In *Proceedings of the Fifth International Symposium on Dynamic Games and Applications*, Grimentz, 15–17 juillet 1992.

Amesse, F., & Zaccour, G. (1989, juin). Une analyse discriminante des gestionnaires de firmes exportatrices et non exportatrices. In *Proceedings of the Administrative Science Association of Canada Conference* (Vol. 10).

Haurie, A., Zaccour, G., Legrand, J., & Smeers, Y. (1988). Un modèle de Nash Cournot Stochastique et dynamique pour le marché européen du gaz. In the *Actes du colloque Modélisation et analyse des marchés du gaz naturel*, HEC, Montréal.

Amesse, F., & Zaccour, G. (1987). Size as a determinant of export sales: Effect studied in thirty-two industrial sectors. In *Proceedings of the Administrative Sciences Association of Canada Conference*, Toronto (Vol. 8), 1–3 juin 1987.

Haurie, A., & Zaccour, G. (1986). A differential game model of power exchange between interconnected utilities. In *Proceedings of the 25th Conference on Decision and Control*, Athènes, Grèce (pp. 262–266).

# Acknowledgments

This project would not have been possible without the authors and reviewers of these papers. We are grateful to all of them for their response at each stage of the project, for their valuable comments and suggestions, and most of all, for their ability to keep this book project "top secret" over the last 3 years.

We also want to thank GERAD and the Chair in Energy Sector Management for their financial support of the 11th Workshop on Dynamic Games and Management Science (October 24–25, 2019). Special thanks go to Marilyne Lavoie and Marie Perrault for their help organizing this event, and to Olfa Taboubi for volunteering to design the visuals and poster. (Thanks also to Leon Petrosjan for sharing his photo albums of various meetings of our scientific community. Even though the card game between our plenary speakers depicted in the poster never actually happened—at least it hasn't yet—the material he provided allowed our designer to create the image.)

The editors would also like to thank Matthew Amboy and Faith Su for coordinating this project with Springer and Josée Lafrenière for the revision of some sections in this volume.

Finally, Sihem Taboubi acknowledges the Research and Knowledge Transfer Office at HEC Montréal for supporting the publication of this volume (CCS 32 153 300 56 R2363).

Poster of 11th Workshop on Dynamic Games and Management Science (Montreal, Canada, October 24–25, 2019)

# Contents

# Contributors

**Francisco J. André** Department of Economic Analysis and ICEI, Universidad Complutense de Madrid, Madrid, Spain

**Olivier Bahn** GERAD and Department of Decision Sciences, HEC Montréal, Montréal, QC, Canada

**Hassan Benchekroun** McGill University, Montréal, QC, Canada

**Michèle Breton** HEC Montréal, Montréal, QC, Canada

**Alessandra Buratto** Department of Mathematics Tullio Levi-Civita, University of Padova, Padova, Italy

**Francisco Cabo** IMUVA, Universidad de Valladolid, Valladolid, Spain

**Denis Claude** LEDI, Université de Bourgogne-Franche Comté, Dijon, France

**Herbert Dawid** Department of Business Administration and Economics and Center for Mathematical Economics, Bielefeld University, Bielefeld, Germany

**Sébastien Debia** HEC Montréal, Montréal, QC, Canada

**Luis M. de Castro** ICEI, Universidad Complutense de Madrid, Madrid, Spain

**Javier de Frutos** IMUVA and Dpto. Matemática Aplicada, Facultad de Ciencias, Universidad de Valladolid, Valladolid, Spain

**Pietro De Giovanni** Department of Operations Management, ESSEC Business School, Paris, France

**Etienne Billette de Villemeur** Université de Lille, Lille, France

**Ana Garcıa-Gonzalez** IMUVa, Universidad de Valladolid, Valladolid, Spain

**Talat S. Genc** Department of Economics, University of Guelph, Guelph, ON, Canada

**Luca Grosset** Department of Mathematics Tullio Levi-Civita, University of Padova, Padova, Italy

**Richard F. Hartl** Department of Business Administration, Production and Operations Management, University of Vienna, Vienna, Austria

**Alain Haurie** ORDECSYS, Chêne-Bougeries, Switzerland
University of Geneva, Geneva, Switzerland
GERAD, HEC Montréal, Montréal, QC, Canada

**Steffen Jørgensen** Department of Business and Economics, University of Southern Denmark, Odense, Denmark

**Salma Karray** Faculty of Business and IT, University of Ontario Institute of Technology, Oshawa, ON, Canada

**Peter M. Kort** Department of Econometrics and Operations Research & CentER, Tilburg University, Tilburg, The Netherlands
Department of Economics, University of Antwerp, Antwerp, Belgium

**Luca Lambertini** Department of Economics, University of Bologna, Bologna, Italy

**Guiomar Martín-Herrán** IMUVA and Dpto. Economía Aplicada, Facultad de Ciencias Económicas y Empresariales, Universidad de Valladolid, Valladolid, Spain

**Mercedes Molpeceres-Abella** Universidad de Valladolid, Valladolid, Spain

**Fabien Ngendakuriyo** Ministry of Finance Québec, Montréal, Québec, Canada

**Elena Parilina** Saint Petersburg State University, Saint Petersburg, Russia

**Pierre-Olivier Pineau** Chair in Energy Sector Management, HEC Montréal Energy Sector Management, Montréal, QC, Canada

**Olivier Rubel** Graduate School of Management, University of California, Davis, CA, USA

**Artem Sedakov** Saint Petersburg State University, Saint Petersburg, Russia

**Simon Sigué** Faculty of Business, Athabasca University, Athabasca, AB, Canada

**Abderrahmane Sokri** Government of Canada, Department of National Defence, Ottawa, ON, Canada

**Sihem Taboubi** Department of Marketing, GERAD and HEC Montréal, Montréal, QC, Canada

**Dina Tasneem** American University of Sharjah, Sharjah, United Arab Emirates

**Mabel Tidball** CEE-M, Université de Montpellier, Inra, Cnrs, SupAgro, Montpellier, France

**Ngo Van Long** McGill University, Montréal, QC, Canada

**Bruno Viscolani** Department of Mathematics Tullio Levi-Civita, University of Padova, Padova, Italy

**Puduru Viswanada Reddy** Department of Electrical Engineering, Indian Institute of Technology Madras, Chennai, India

**Suzanne Zaccour** Oxford University, Oxford, UK

# About the Editors

**Pierre-Olivier Pineau** is a full professor in the Department of Decision Sciences at HEC Montréal and has held the Chair of Energy Sector Management since December 2013. He is a specialist in energy policy, particularly in the electricity sector. His research focuses on the integration of electricity markets in Canada and around the world, as well as optimal approaches to balancing production and energy consumption. He has worked in Finland, Peru, and Cameroon on energy policy issues. He is a member of the Board of Directors of the Canadian Energy Economics Association. He is a member of the CIRODD (Interuniversity Research Center on the Operationalization of Sustainable Development) and a Fellow at CIRANO (Center for Interuniversity Research and Analysis of Organizations). Prior to joining HEC Montréal, he was a professor at the University of Victoria (BC) from 2001 to 2006.

**Simon Sigué** is a professor of marketing at Athabasca University, Canada. His research covers such topics as franchising, marketing channel, customer relationship management, international marketing and trade, and corruption. It has been published in various economics, management science, and marketing journals, including *European Journal of Operational Research*, *International Journal of Production Economics*, *Journal of Economic Dynamics & Control*, *Journal of Retailing*, *Journal of Service Research*, and *Omega: The International Journal of Management Science*. He is also the President of the Academy of African Business and Development and past editor of *Journal of African Business*.

**Sihem Taboubi** holds a Ph.D. in administration and an M.Sc. in marketing from HEC Montreal. She is a professor of marketing at HEC Montreal and a member of the Group of Research in Decision Analysis (GERAD) and the International Society of Dynamic Games (ISDG). Her main research covers applications of game theory in marketing and supply chains. More particularly, she uses optimal control, static games, and differential games to examine pricing, advertising, and promotion decisions in distribution channels. Her research is published in various prestigious journals in the fields of Management Science and Marketing. Professor Taboubi serves as a referee for several journals focusing on the applications of game theory in these research areas.

# A Lanchester-Type Dynamic Game of Advertising and Pricing

**Steffen Jørgensen and Simon Sigué**

**Abstract** The paper studies a differential game played by two competing firms over a finite time horizon. As the game progresses, the firms observe the position of the game, i.e., the current time and the current market shares. Each firm uses pricing and advertising in order to influence market shares. We suggest a generalization of the Lanchester market share dynamics such that the rates at which firms attract market share from each other are determined not only by their advertising efforts but also by the consumer prices charged in the market. A full characterization of Nash equilibrium price and advertising strategies is obtained.

**Keywords** Price and advertising competition · Duopoly · Differential game · Markovian Nash equilibrium

## 1 Introduction and Literature Review

The current research is devoted to the analysis of a dynamic noncooperative game of advertising and pricing played by the firms in a duopolistic market. We construct a differential game model which describes the evolution of market shares in a duopolistic market. This model extends the well-known Lanchester model of advertising competition by incorporating price competition. To the best of our knowledge, this development is new.

The survey paper by Huang et al. (2012) provides a good account of dynamic advertising research and may serve as an introduction to the problem to be dealt with in the current research. Huang et al. organized their presentation according to four different types of dynamics: *Nerlove-Arrow advertising goodwill models,*

S. Jørgensen
Department of Business and Economics, University of Southern Denmark, Odense, Denmark

S. Sigué (✉)
Faculty of Business, Athabasca University, Athabasca, AB, Canada
e-mail: simon.sigue@fb.athabascau.ca



P.-O. Pineau et al. (eds.), *Games in Management Science*,
International Series in Operations Research & Management Science 280,
https://doi.org/10.1007/978-3-030-19107-8_1

*Vidale-Wolfe sales response models, new-product diffusion models,* and *Lanchester market share models.* For the first three types of models we provide some examples of dynamic games in which competitors decide on their prices and advertising efforts.

- **Nerlove-Arrow Advertising Goodwill Models**

Nair and Narasimhan (2006) studied duopolistic competition with investment in quality, price, and advertising effort as the decision variables of a firm. A firm's sales depend on its own price and advertising goodwill. The right-hand sides of the Nerlove-Arrow goodwill dynamics depend on quality investment and advertising effort of both firms.

Erickson (2009) considered a differential game in the marketing and operations interface. Evolution of advertising goodwill follows the Nerlove-Arrow model while sales are a linear function of a firm's price and advertising goodwill. The competitor's price and advertising, however, do not appear in the dynamics of a specific firm.

Fruchter (2009) assumed that advertising goodwill signals quality and replaced advertising with quality. Price and advertising appear linearly on the right-hand side of the equations specifying the quality dynamics. The competitor's price and advertising are, however, absent in the dynamics for the quality variable of a specific firm.

- **Vidale-Wolfe Sales Response Models**

Fruchter and Messinger (2003) studied a situation in which a "fringe" firm enters the market of an incumbent firm. Prices and advertising efforts are a firm's decision variables and the right-hand side of fringe sales dynamics depend on all four marketing decision variables. The authors consider two cases: (1) Fringe firms are price-takers and advertise. (2) Fringe firms are not price-takers and advertise. In the first case, the only one price to be determined is that of the price-leader.

Krishnamoorthy et al. (2010) studied optimal pricing and advertising in a durable-good duopoly. The right-hand sides of the sales dynamics depend on a firm's own advertising and price.

- **New Product Diffusion Models**

Teng and Thompson (1984) studied price and advertising competition in a dynamic oligopoly. The model is a combination of the Bass new product diffusion model and the Vidale-Wolfe model. Due to the complex sales dynamics, the authors assumed that—in the case of a duopoly—one of the firms is a price leader. Hence one price only is to be determined.

- **Lanchester Market Share Models**

The simple Lanchester advertising model considers advertising as an offensive marketing tool, the only purpose of which is to attract customers from a rival firm (e.g., Erickson 1985; Jarrar et al. 2004). Some later extensions of the Lanchester model suppose that firms also use defensive advertising. The purpose of this type of advertising is to protect a firm's customer base (e.g., Erickson 1993; Martín-Herrán et al. 2012). Prices are exogenously given in these models and market shares are exclusively affected by advertising efforts (Huang et al. 2012). Jørgensen and Sigué (2015) studied a Lanchester model with three types of advertising: offensive, defensive, and generic. While the first two types of advertising are aimed at customers who are already in the market, the third type aims at influencing potential buyers with the aim of expanding demand for the product category. We have not been able to find differential game models having both prices and advertising efforts in the market share dynamics.[1]

*Remark 1* In most of the literature, the market structure in Lanchester models is a duopoly. See Kress et al. (2018) for a Lanchester model of a triopoly.

The model suggested in this paper extends the simple Lanchester model by incorporating offensive advertising efforts as well as retail prices in the market share dynamics. The retail price charged by a firm can play the role as a defensive as well as an offensive marketing instrument that is used to (1) protect a firm's own customer base and (2) attract customers from the rival firm.

Clearly, in many real-life markets firms compete on both price and advertising (as well as other marketing instruments) and we believe it makes good sense to study a scenario that reflects this type of competition. It will be shown that there is an inverse relationship between a firm's price and its advertising effort (and hence its advertising cost). Thus, when advertising is very costly, a firm will reduce its advertising effort (and its cost) and reduce its retail price. Conversely, when advertising is less costly, a firm will increase its advertising effort and increase its price.

The remainder of the paper is organized as follows. Section 2 presents a Lanchester-type differential game model of duopolistic competition on advertising efforts as well as prices. The novelty is that the rate at which a firm attracts market share from its rival *(also known as the attraction rate)* depends not only on advertising effort but also on prices. Using standard techniques of differential game theory, Sect. 3 derives a Markovian Nash equilibrium of the game and characterizes explicitly the equilibrium advertising effort rates, prices, and optimal profits. Section 4 concludes.

## 2 A Differential Game of Pricing and Advertising

In this section we construct a two-player differential game in which firms compete on price and advertising. Market shares evolve according to two ordinary differential equations that determine the current market share of a firm as a function of the advertising and pricing decisions taken by both firms.

[1]This observation was also made in the survey by Huang et al. (2012).

Time $t$ is continuous and the planning period (being the same for both firms) starts at $t = 0$ and ends at $t = T$. The horizon date $T$ is fixed and finite. Firm $i \in \{1, 2\}$ controls its rate of advertising efforts $a_i(t)$ and retail price $p_i(t)$. We suppose that advertising efforts and prices have short-term effects on market shares only, a hypothesis that is used in dynamic game literature dealing with sales response, Lanchester, and new product diffusion models. Clearly, the assumption is plausible in some, but certainly not all markets.

*Remark 2* A stream of literature builds on the hypothesis that advertising efforts have carry-over effects, that is, the effect of current advertising persists in (at least part of) the future. One way to model this is to assume that as time progresses, advertising efforts accumulate into a *stock of* "advertising goodwill." An example of this type of modelling is the Nerlove-Arrow model mentioned above.

Let $X_i(t), i \in \{1, 2\}$, represent the market share of firm $i$ such that $X_i(t) \in [0, 1]$ and $X_1(t) + X_2(t) = 1$. Using a dot to represent time-derivatives, the latter implies $\dot{X}_1(t) = -\dot{X}_2(t)$. The evolution of market shares is described by an extension of the Lanchester advertising model. In the basic Lanchester model, advertising is the only marketing instrument of a firm and market shares typically evolve according to the dynamics

$$\dot{X}_1(t) = \varphi_1 a_1(t) X_2(t) - \varphi_2 a_2(t) X_1(t)$$
$$\dot{X}_2(t) = \varphi_2 a_2(t) X_1(t) - \varphi_1 a_1(t) X_2(t)$$

in which $\varphi_i$ is a positive parameter. The above equations state that the rate of change of a firm's market share is determined by two factors that work in opposite directions. Consider, for example, the first equation: The first term on the right-hand side is the gain in market share of firm 1, coming from market share attracted from firm 2 and due to the advertising effort of firm 1. The second term is the loss of market share of firm 1, caused by the advertising effort of firm 2. The term $\varphi_i a_i(t)$ will be referred to as an *attraction rate*, for the simple reason that it is the rate at which a firm attracts market share from its rival.

*Remark 3* The Lanchester dynamics can be expressed in terms of sales rates, say, $S_1(t)$ and $S_2(t)$, instead of market shares. To do so, we need to assume that the market is mature. This means that industry sales remain constant, i.e., $S_1(t) + S_2(t) = m$ where $m > 0$ is the fixed market potential. Let $p_i(t)$ be the price charged by firm $i$. The revenue rate of firm $i$ is then $p_i(t) S_i(t)$. Clearly, $m X_i(t) = S_i(t)$ which means that the revenue rate can be expressed as $p_i(t) m X_i(t)$. The value of the constant $m$ is of no significance to our analysis and we put it equal to one. Then the revenue rate is $p_i(t) X_i(t)$.[2]

[2]In quite many games played with the Lanchester dynamics it is assumed that the revenue rate is $\pi_i X_i(t)$ where $\pi_i > 0$ is the constant revenue per unit of market share. This formulation often simplifies the analysis considerably.

Researchers working with marketing applications of differential games have experienced that games involving both advertising and pricing often cause analytical difficulties. We analyzed quite many games using alternative specifications of the attraction rates. Our original modelling assumption was that attraction rates are functions of product prices and advertising efforts of the firms, in general

$$\begin{aligned}\dot{X}_1(t) &= f_1(a_1(t), a_2(t), p_1(t), p_2(t))\sqrt{X_2(t)} \\ &\quad - f_2(a_1(t), a_2(t), p_1(t), p_2(t))\sqrt{X_1(t)} \\ \dot{X}_2(t) &= f_2(a_1(t), a_2(t), p_1(t), p_2(t))\sqrt{X_1(t)} \\ &\quad - f_1(a_1(t), a_2(t), p_1(t), p_2(t))\sqrt{X_2(t)}\end{aligned}$$

in which certain constraints have to be placed on functions $f_i$. The assumption that attraction rates depend on prices and advertising rates turned out to be too ambitious. We examined a series of functional forms of $f_i$ to model the influence of prices and advertising rates on market shares but all failed to provide usable results. It turned out, however, that a quite straightforward modification of the Lanchester dynamics is analytically tractable:

$$\begin{aligned}\dot{X}_1(t) &= a_1(t)\frac{p_2(t)}{p_1(t)}\sqrt{X_2(t)} - a_2(t)\frac{p_1(t)}{p_2(t)}\sqrt{X_1(t)} \\ \dot{X}_2(t) &= a_2(t)\frac{p_1(t)}{p_2(t)}\sqrt{X_1(t)} - a_1(t)\frac{p_2(t)}{p_1(t)}\sqrt{X_2(t)}.\end{aligned} \tag{1}$$

As in the basic Lanchester model, advertising efforts enter linearly in the attraction rates. On the right-hand sides, market shares appear nonlinearly.[3] The implication is that the marginal impact of current market share is diminishing. The novelty in the dynamics are the relative price terms in the attraction rates. These terms state that the ability of a firm to attract customers from its rival depends on its own pricing decisions as well as the rival's pricing decisions. Firm 1 tries to attack market share of firm 2 through advertising and pricing; The latter can defend its market share (customer base) by reducing her retail price.[4] Note that while advertising of a firm is purely offensive, i.e., intended to steal market share from the competitor, price is both an offensive and a defensive marketing tool that can be used (1) to protect a firm's own customer base and (2) to attack the customer base of the rival.

Firms play a noncooperative game in which firms are supposed to use Markovian strategies. A Markovian strategy makes a firm's actions depend on the position of

---

[3]The idea of letting market shares enter the right-hand sides of the market share dynamics as $\sqrt{X_i(t)}$ most likely originated in Sorger (1989) and has gained some popularity in the literature.

[4]Erickson (1993) did not consider pricing but included defensive advertising in the attraction rates. The purpose of defensive advertising is to defend a firm's customer base against offensive advertising done by the rival firm. In our model this is accomplished by pricing.

the game, that is, the current time $t$ as well as the current state $(X_1, X_2)$. Note that when a firm knows its own market share it will know the position of the game. In what follows we characterize a Markovian Nash equilibrium and determine the equilibrium advertising and pricing strategies as well as the resulting market shares and profits.

Disregarding production costs—which are not important for our problem—the objective functionals of the firms are

$$J_1(p_1, a_1) = \int_0^T \left[p_1(t) X_1(t) - \frac{c_1}{2} a_1^2(t)\right] dt + \sigma_1 X_1(T)$$

$$J_2(p_2, a_2) = \int_0^T \left[p_2(t) X_2(t) - \frac{c_2}{2} a_2^2(t)\right] dt + \sigma_2 X_2(T)$$

in which $c_i > 0$ and $\sigma_i > 0$, $i \in \{1, 2\}$ are parameters.

*Remark 4* The terms $p_i(t) X_i(t)$ should more correctly be $p_1(t) m X_1(t)$ where $m$ is the constant total market size. To simplify notation we have put $m$ equal to one. See also Remark 1 above.

Quadratic advertising costs quite have often been assumed in the literature and have the implication that advertising efforts exhibit decreasing marginal returns to advertising. The terms $\sigma_i X_i(T)$ are salvage values, used to truncate the horizon. Parameters $\sigma_i$ represent the value to a firm of a unit of market share at time $T$. Note that since salvage values are part of the objectives, the objective of a firm is to maximize salvage value as well as its overall profits. For three reasons we do not discount future profits. (1) The planning period normally is rather short in advertising and pricing planning, (2) interest rates are low, and (3) we believe that introducing discounting would not change, qualitatively speaking, our results.

The optimality conditions to be employed are based on Hamilton-Jacobi-Bellman (HJB) equations. Denote the value functions by $V_1(X_1, X_2, t)$ and $V_2(X_1, X_2, t)$, respectively. Omitting some technicalities, we need the existence of continuously differentiable value functions that satisfy the HJB equations for all $(X_1, X_2, t)$ such that $t \in [0, T]$, $X_i \in [0, 1]$ and $V_i(X_1, X_2, T) = \sigma_i X_i(T)$ for all $(X_1, X_2)$. These conditions are sufficient for a Markovian Nash equilibrium.[5]

Omitting the arguments of the value functions, the HJB equations are

$$-\frac{\partial V_1}{\partial t} = \max_{a_1 \geq 0, p_1 > 0} \left\{p_1 X_1 - \frac{c_1}{2} a_1^2 + \frac{\partial V_1}{\partial X_1} \dot{X}_1 + \frac{\partial V_1}{\partial X_2} \dot{X}_2\right\} \tag{2}$$

$$-\frac{\partial V_2}{\partial t} = \max_{a_2 \geq 0, p_2 > 0} \left\{p_2 X_2 - \frac{c_2}{2} a_2^2 + \frac{\partial V_2}{\partial X_2} \dot{X}_2 + \frac{\partial V_2}{\partial X_1} \dot{X}_1\right\}$$

[5]Rigorous statements of the optimality conditions can be found in, e.g., Dockner et al. (2000), Haurie et al. (2012).

in which $\dot{X}_i$ is given by (1). The chances of finding an explicit solution to these partial differential equations are small and we shall use another approach to determine the value functions. The idea is to guess what the value functions would be and if the guess turns out to be right (i.e., the guessed value functions satisfy the HJB equations) we are done.[6] Our guess is that value functions are linear in market shares:

$$V_1 = \gamma_1(t)X_1 + \eta_1(t)X_2,\ V_2 = \gamma_2(t)X_2 + \eta_2(t)X_1$$

in which $\gamma_i(t)$ and $\eta_i(t)$ are time-functions that will be determined later on.[7] It follows that

$$\frac{\partial V_1}{\partial X_1} = \gamma_1(t), \frac{\partial V_1}{\partial X_2} = \eta_1(t); \quad \frac{\partial V_2}{\partial X_2} = \gamma_2(t), \frac{\partial V_2}{\partial X_1} = \eta_2(t)$$

$$-\frac{\partial V_1}{\partial t} = -\dot{\gamma}_1(t)X_1 - \dot{\eta}_1(t)X_2; \quad -\frac{\partial V_2}{\partial t} = -\dot{\gamma}_2(t)X_2 - \dot{\eta}_2(t)X_1.$$

The coefficients $\gamma_i(t)$ and $\eta_i(t)$ of the value functions have an interpretation as shadow prices of market shares, that is, the marginal increase in value caused by a marginal increase in market share. We expect $\gamma_i(t) > 0$ and $\eta_i(t) < 0$.

*Remark 5* In the sequel we omit the arguments of strategies, market shares, and value function parameters whenever it is safe to do so.

To satisfy the terminal conditions $V_i(X_1, X_2, T) = \sigma_i X_i$ it must hold for all pairs $(X_1, X_2)$ that

$$V_1(X_1, X_2, T) = \gamma_1(T)X_1 + \eta_1(T)X_2 = \sigma_1 X_1$$
$$V_2(X_1, X_2, T) = \gamma_2(T)X_2 + \eta_2(T)X_1 = \sigma_2 X_2$$

and hence we need to have

$$\gamma_1(T) = \sigma_1 > 0, \eta_1(T) = 0 \text{ and } \gamma_2(T) = \sigma_2 > 0, \eta_2(T) = 0. \tag{3}$$

The conditions $\eta_i(T) = 0$ are intuitive. From the point of view of firm $i$, the $\eta_i$-parameter is the shadow price of a unit of market share of the rival firm $j$. Clearly, at the horizon date $T$ the market share of firm $j$ is irrelevant for the value of firm $i$.

[6] If the guess turned out to be wrong, we can make another guess.

[7] The above value functions also appear in Sorger (1989, p. 66).

The terms in curly brackets on the right-hand sides of (2) can be written as

$$p_1 X_1 - \frac{c_1}{2} a_1^2 + \gamma_1 \left( a_1 \frac{p_2}{p_1} \sqrt{X_2} - a_2 \frac{p_1}{p_2} \sqrt{X_1} \right)$$
$$+ \eta_1 \left( a_2 \frac{p_1}{p_2} \sqrt{X_1} - a_1 \frac{p_2}{p_1} \sqrt{X_2} \right)$$

and

$$p_2 X_2 - \frac{c_2}{2} a_2^2 + \gamma_2 \left( a_2 \frac{p_1}{p_2} \sqrt{X_1} - a_1 \frac{p_2}{p_1} \sqrt{X_2} \right)$$
$$+ \eta_2 \left( a_1 \frac{p_2}{p_1} \sqrt{X_2} - a_2 \frac{p_1}{p_2} \sqrt{X_1} \right)$$

and performing the maximizations indicated in (2) yields the Markovian price and advertising strategies

$$\hat{p}_1(X_1, X_2, t) = \frac{(\gamma_1 - \eta_1)^3 X_2^2}{c_2 (\gamma_2 - \eta_2) X_1^2} + \frac{(\gamma_2 - \eta_2)^2 X_1}{c_1 X_2} \tag{4}$$
$$\hat{p}_2(X_1, X_2, t) = \frac{(\gamma_2 - \eta_2)^3 X_1^2}{c_1 (\gamma_1 - \eta_1) X_2^2} + \frac{(\gamma_1 - \eta_1)^2 X_2}{c_2 X_1}$$
$$\hat{a}_1(X_1, X_2, t) = \frac{(\gamma_2 - \eta_2) X_1}{c_1 \sqrt{X_2}}; \ \hat{a}_2(X_1, X_2, t) = \frac{(\gamma_1 - \eta_1) X_2}{c_2 \sqrt{X_1}}.$$

We expect to find that $\gamma_i > \eta_i$, that is, the value of a marginal increase in own market share exceeds the value of a marginal increase of the rival's market share. This is intuitive. Clearly, price and advertising of a firm decrease if advertising becomes more costly. The reason is that if advertising becomes more costly, it makes sense to reduce advertising effort, and at the same time lower the price in order to counterbalance the effects of the reduced advertising effort.

Inserting the price and advertising strategies from (4) on the right-hand sides of (2) yields

$$\hat{p}_1 X_1 - \frac{c_1}{2} \hat{a}_1^2 + \frac{\partial V_1}{\partial X_1} \dot{X}_1 + \frac{\partial V_1}{\partial X_2} \dot{X}_2 = \frac{3 (\gamma_2 - \eta_2)^2}{2 c_1} \frac{X_1^2}{X_2} > 0$$
$$\hat{p}_2 X_2 - \frac{c_2}{2} \hat{a}_1^2 + \frac{\partial V_2}{\partial X_2} \dot{X}_2 + \frac{\partial V_2}{\partial X_1} \dot{X}_1 = \frac{3 (\gamma_1 - \eta_1)^2}{2 c_2} \frac{X_2^2}{X_1} > 0.$$

Using this result, and (2), shows that value function parameters $\gamma_i$ and $\eta_i$ must be chosen such that the following equations are satisfied:

$$-\dot{\gamma}_1 X_1 - \dot{\eta}_1 X_2 = \frac{3\left(\gamma_2 - \eta_2\right)^2}{2c_1}\frac{X_1^2}{X_2}$$

$$-\dot{\gamma}_2 X_2 - \dot{\eta}_2 X_1 = \frac{3\left(\gamma_1 - \eta_1\right)^2}{2c_2}\frac{X_2^2}{X_1}.$$

Writing these equations as

$$\frac{3\left(\gamma_2 - \eta_2\right)^2}{2c_1} X_1^2 + \dot{\gamma}_1 X_1 X_2 + \dot{\eta}_1 X_2^2 = 0$$

$$\frac{3\left(\gamma_1 - \eta_1\right)^2}{2c_2} X_2^2 + \dot{\gamma}_2 X_2 X_1 + \dot{\eta}_2 X_1^2 = 0$$

it follows that

$$\left(\frac{3\left(\gamma_2 - \eta_2\right)^2}{2c_1} - \dot{\eta}_2\right) X_1^2 + \left(\dot{\gamma}_1 - \dot{\gamma}_2\right) X_1 X_2 + \left(\dot{\eta}_1 - \frac{3\left(\gamma_1 - \eta_1\right)^2}{2c_2}\right) X_2^2 = 0.$$

To satisfy this equation for all $X_1, X_2$, the parameters $\gamma_i$ and $\eta_i$ must be chosen such that

$$\dot{\eta}_2 = \frac{3\left(\gamma_2 - \eta_2\right)^2}{2c_1}, \; \dot{\gamma}_1 = \dot{\gamma}_2, \; \dot{\eta}_1 = \frac{3\left(\gamma_1 - \eta_1\right)^2}{2c_2}. \tag{5}$$

*Remark 6* We shall consider solutions in which $\gamma_1$ and $\gamma_2$ are constant. Then $\dot{\gamma}_1 = 0$, $\dot{\gamma}_2 = 0$. On the other hand, if we assumed that $\gamma_1$ and $\gamma_2$ are time-dependent, we would find that solving $\dot{\gamma}_1(t) = \dot{\gamma}_2(t)$ yields $\gamma_1(t) = \gamma_2(t) + C$ which satisfies the transversality conditions $\gamma_1(T) = \sigma_1$, $\gamma_2(T) = \sigma_2$ if we set $C = \sigma_1 - \sigma_2$. However, the equations in (5) cannot be solved as one equation is missing. This problem does not arise if we look for a solution in which $\gamma_1$ and $\gamma_2$ are constant. The interpretation of this situation is that the value of having an "extra unit of market share" is the same, no matter at which time this happens.

Now we have four differential equations

$$\dot{\eta}_2 = \frac{3\left(\gamma_2 - \eta_2\right)^2}{2c_1}; \; \dot{\eta}_1 = \frac{3\left(\gamma_1 - \eta_1\right)^2}{2c_2}; \; \dot{\gamma}_1 = 0; \; \dot{\gamma}_2 = 0$$

that have solutions

1. $\eta_1(t) = C_{21}, \; \eta_2(t) = -\dfrac{2c_1 - 3C_{15}t + 2c_1 C_{15} C_{19}}{3t - 2c_1 C_{19}}$

   $\gamma_1(t) = C_{21}, \; \gamma_2(t) = C_{15}$

$$2.\ \eta_1(t) = -\frac{2c_2 - 3C_{27}t + 2c_2C_{27}C_{30}}{3t - 2c_2C_{30}},\ \eta_2(t) = C_{15}$$

$$\gamma_1(t) = C_{27},\ \gamma_2(t) = C_{15}$$

$$3.\ \eta_1(t) = C_{27},\ \eta_2(t) = C_{15}$$

$$\gamma_1(t) = C_{27},\ \gamma_2(t) = C_{15}$$

$$4.\ \eta_1(t) = -\frac{2c_2 - 3C_{21}t + 2c_2C_{21}C_{25}}{3t - 2c_2C_{25}},\ \eta_2(t) = -\frac{2c_1 - 3C_{15}t + 2c_1C_{15}C_{19}}{3t - 2c_1C_{19}}$$

$$\gamma_1(t) = C_{21},\ \gamma_2(t) = C_{15}$$

in which $C_{ij}$ are arbitrary constants.

Recall the boundary conditions on shadow prices from (3):

$$\gamma_1(T) = \sigma_1 > 0, \eta_1(T) = 0, \gamma_2(T) = \sigma_2 > 0, \eta_2(T) = 0.$$

These conditions must be satisfied. Solution 1 does not satisfy the requirement since we must have $\eta_1(T) = C_{21} = 0$ and then $\gamma_1(T) = C_{21} = \sigma_1 > 0$ is not satisfied. Similar arguments apply to solutions 2 and 3. We are left with solution 4 and choose $C_{21}$ and $C_{15}$ such that $C_{21} = \gamma_1(T) = \sigma_1$ and $C_{15} = \gamma_2(T) = \sigma_2$. Furthermore, we must determine $C_{19}$ and $C_{25}$ to have $\eta_i(T) = 0, i = 1, 2$. The terminal values $\eta_i(T)$ are

$$\eta_1(T) = -\frac{2c_2 - 3\sigma_1 T + 2c_2\sigma_1 C_{25}}{3T - 2c_2C_{25}};\ \eta_2(T) = -\frac{2c_1 - 3\sigma_2 T + 2c_1\sigma_2 C_{19}}{3T - 2c_1C_{19}}.$$

Solving these equations yields

$$C_{25} = -\frac{2c_2 - 3T\sigma_1}{2\sigma_1 c_2};\quad C_{19} = -\frac{2c_1 - 3T\sigma_2}{2\sigma_2 c_1}$$

and hence

$$\eta_1(t) = -\frac{3\sigma_1^2(T-t)}{2c_2 + 3\sigma_1(t-T)};\quad \eta_2(t) = -\frac{3\sigma_2^2(T-t)}{2c_1 + 3\sigma_2(t-T)}.$$

These expressions show that $\dot{\eta}_i(t) > 0$ and we conclude that $\eta_i(t) < 0$ for all $t < T$ (because $\eta_i(T) = 0$). This result is expected. For firm $i$ it has a negative value if firm $j$ gets an extra unit of the market share, for the simple reason that if the rival's market share is increased by one unit, market share of firm $i$ must decrease by one unit.

What remains is to determine the time paths of equilibrium market shares. The market share dynamics are given by (1) and equilibrium advertising efforts and prices are given by (4). Using these formulas, the equilibrium attraction rate of firm 1 is

$$\hat{a}_1 \frac{\hat{p}_2}{\hat{p}_1} = \frac{X_1^2}{X_2^{\frac{3}{2}} c_1 (\gamma_1 - \eta_1)} (\gamma_2 - \eta_2)^2$$
$$= \frac{2}{\sigma_1} \sigma_2^2 \frac{X_1^2}{X_2^{\frac{3}{2}}} \frac{c_1}{c_2} \frac{2c_2 - 3T\sigma_1 + 3t\sigma_1}{(2c_1 - 3T\sigma_2 + 3t\sigma_2)^2}.$$

*Remark 7* Since $X_2 = 1 - X_1$ it suffices to determine the equilibrium market share of firm 1. Inserting the equilibrium price and advertising strategies and the shadow prices into (1) yields

$$\dot{X}_1 = g_1(t) \frac{X_1^2}{1 - X_1} - g_2(t) \frac{(1 - X_1)^2}{X_1} \tag{6}$$

where

$$g_1(t) = \frac{\sigma_2 + \frac{3\sigma_2^2(T-t)}{2c_1 + 3\sigma_2(t-T)}}{c_1} \frac{\sigma_2 c_1 (2c_2 + 3\sigma_1 (t - T))}{\sigma_1 c_2 (2c_1 + 3\sigma_2 (t - T))}$$
$$g_2(t) = \frac{\sigma_1 + \frac{3\sigma_1^2(T-t)}{2c_2 + 3\sigma_1(t-T)}}{c_2} \frac{\sigma_1 c_2 (2c_1 + 3\sigma_2 (t - T))}{\sigma_2 c_1 (2c_2 + 3\sigma_1 (t - T))}.$$

The differential equation in (6) is highly nonlinear and does not have a closed-form solution.

Total profits in equilibrium are given by

$$V_1(x_{10}, x_{20}, 0) = \sigma_1 x_{10} - \frac{3\sigma_1^2 T}{2c_2 - 3\sigma_1 T} x_{20}$$
$$V_2(x_{10}, x_{20}, 0) = \sigma_2 x_{20} - \frac{3\sigma_2^2 T}{2c_1 - 3\sigma_2 T} x_{10}$$

and to see which firm will earn the largest profits we have the following evaluation:

$$V_1(x_{10}, x_{20}, 0) \begin{pmatrix} > \\ < \end{pmatrix} V_2(x_{10}, x_{20}, 0)$$
$$\Leftrightarrow \sigma_1 x_{10} - \frac{3\sigma_1^2 T}{2c_2 - 3\sigma_1 T} x_{20} \begin{pmatrix} > \\ < \end{pmatrix} \sigma_2 x_{20} - \frac{3\sigma_2^2 T}{2c_1 - 3\sigma_2 T} x_{10}.$$

Confirming intuition, the evaluation shows that a firm having a "large" initial market share $x_{i0}$ and a "high" unit salvage value $\sigma_i$ stands to earn the highest profits.

Comparative statics with respect to selected parameters show that

$$\frac{\partial}{\partial x_{10}}\left(\sigma_1 x_{10} - \frac{3\sigma_1^2 T}{2c_2 - 3\sigma_1 T} x_{20}\right) = \sigma_1 > 0$$

$$\frac{\partial}{\partial c_2}\left(\sigma_1 x_{10} - \frac{3\sigma_1^2 T}{2c_2 - 3\sigma_1 T} x_{20}\right) = \frac{6T\sigma_1^2 x_{20}}{(2c_2 - 3T\sigma_1)^2} > 0$$

$$\frac{\partial}{\partial T}\left(\sigma_1 x_{10} - \frac{3\sigma_1^2 T}{2c_2 - 3\sigma_1 T} x_{20}\right) = -\frac{6\sigma_1^2 c_2 x_{20}}{(2c_2 - 3T\sigma_1)^2} < 0$$

which means that equilibrium profits of, say, firm 1 would increase if its initial market share or the rival's advertising cost were increased, and decreases if the planning period were increased. The two first observations are intuitive: A larger initial market share and a higher advertising cost of the rival clearly are favorable for a firm.

Finally, the equilibrium value functions are

$$V_1(X_1, X_2, t) = \sigma_1 X_1 - \frac{3\sigma_1^2 (T-t)}{2c_2 + 3\sigma_1 (t-T)} X_2$$

$$V_2(X_1, X_2, t) = \sigma_2 X_2 - \frac{3\sigma_2^2 (T-t)}{2c_1 + 3\sigma_2 (t-T)} X_1$$

which shows, as expected, that a firm's value increases if own market share increases and decreases if the rival's market share increases.

## 3 Concluding Remarks

The current research has analyzed a dynamic noncooperative game of advertising and pricing, played by the firms in a duopolistic market. We have suggested a differential game model that extends the Lanchester model of advertising competition by taking into account price competition. Advertising effort is purely offensive, in the sense that its single aim is to steal customers from the competitor while pricing is both offensive and defensive, aimed at attracting new customers from the rival firm and mitigating the effects of pricing and advertising efforts of the rival.

The main novelty of our work lies in the market share dynamics that depend on both advertising efforts and price levels. With this starting point we constructed a differential game model and characterized explicitly Markovian equilibrium advertising efforts and prices.

In particular, and consistent with previous works that focus exclusively on offensive advertising, we found that a firm's offensive advertising efforts increase with its own market share (see, e.g., Jarrar et al. 2004). This finding differs from the one of Erickson (1993) who considered both offensive and defensive advertising. In this scenario, a firm's offensive advertising decreases with its market share.

The impact of a firm's market share on its optimal pricing strategy is not obvious, most likely due to the dual role (offensive, defensive) of pricing in our model. We found an inverse relationship between a firm's price and offensive advertising efforts (and costs). The implication is that when advertising is very (less) costly, a firm will advertise less (more) and charge a smaller (larger) price.

Our results demonstrate that it is possible to include prices in the Lanchester market share dynamics and still obtain closed-form results for the equilibrium marketing instruments (prices and advertising efforts), market share dynamics, and value functions. We leave it to future research to construct and analyze more complex specifications of the market share dynamics, for instance by including both offensive and defensive advertising.

## References

Dockner, E. J., Jørgensen, S., Long, N. V., & Sorger, G. (2000). *Differential games in economics and management science*. Cambridge: Cambridge University Press.

Erickson, G. M. (1985). A model of advertising competition. *Journal of Marketing Research, 22*, 297–304.

Erickson, G. M. (1993). Offensive and defensive marketing: Closed-loop duopoly strategies. *Marketing Letters, 4*, 285–295.

Erickson, G. M. (2009). A differential game model of the marketing-operations interface. *European Journal of Operational Research, 197*(1), 374–388.

Fruchter, G. (2009). Signaling quality: Dynamic price-advertising model. *Journal of Optimization Theory and Applications, 3*(4), 301–314.

Fruchter, G., & Messinger, P. E. (2003). Optimal management of fringe entry over time. *Journal of Economic Dynamics & Control, 28*(3), 445–466.

Haurie, A., Krawczyk, J. B., & Zaccour, G. (2012). *Games and dynamic games*. Singapore: World Scientific.

Huang, J., Leng, M., & Liang, L. (2012). Recent developments in dynamic advertising research. *European Journal of Operational Research, 220*, 591–609.

Jarrar, R., Martín-Herrán, G., & Zaccour, G. (2004). Markov perfect equilibrium advertising strategies of a Lanchester duopoly model: A technical note. *Management Science, 50*(7), 995–1000.

Jørgensen, S., & Sigué, S. P. (2015). Defensive, offensive, and generic advertising in a Lanchester model with market expansion. *Dynamic Games and Applications, 4*(5), 523–539.

Kress, M., Caulkins, J. P., Feichtinger, G., Grass, D., & Seidl, A. (2018). Lanchester model for three-way combat. *European Journal of Operational Research, 264*, 46–54.

Krishnamoorthy, A., Prasad, A., & Sethi, S. P. (2010). Optimal pricing and advertising in a durable-good duopoly. *European Journal of Operational Research, 200*(2), 486–497.

Martín-Herrán, G., McQuitty, S., & Sigué, S. P. (2012). Offensive versus defensive marketing: What is the optimal spending allocation. *International Journal of Research in Marketing, 29*, 210–219.

Nair, A., & Narasimhan, R. (2006). Dynamics of competing with quality- and advertising-based goodwill. *European Journal of Operational Research, 175*(1), 462–474.
Sorger, G. (1989). Competitive dynamic advertising: A modification of the Case game. *Journal of Economic Dynamics & Control, 13*, 55–80.
Teng, J. T., & Thompson, G. L. (1984). Optimal pricing and advertising policies for new product oligopoly models. *Marketing Science, 3*(2), 148–168.

# On the Modelling of Price Effects in the Diffusion of Optional Contingent Products

**Guiomar Martín-Herrán and Sihem Taboubi**

**Abstract** In this chapter, we study the pricing strategies of firms in a multi-product diffusion model where we use a new formalization of the price effects. More particularly, we introduce the impact of prices on one of the factors that affect the diffusion of new products: the innovation coefficient. By doing so, we relax one of the hypotheses in the existing literature stating that this rate is constant. In order to assess the impact of this functional form on the pricing policies of firms selling optional contingent products, we use our model to study two scenarios already investigated in the multiplicative form model suggested by Mahajan and Muller (M&M).

We follow a "logical experimentation" perspective by computing and comparing the results of three models: (1) The M&M model, (2) a modified version of M&M where the planning horizon is infinite, and (3) our model, where the new formalization of the innovation effect is introduced. This perspective allows us to attribute the differences in results to either the length of the planning horizon or to our model's formalization. Besides its contribution to the literature on pricing and diffusion, this paper highlights the sensitivity of results to the hypothesis used in product diffusion modelling and could explain the mixed results obtained in the empirical validations of diffusion models.

**Keywords** Marketing modelling · Diffusion models · Dynamic pricing · Contingent products

G. Martín-Herrán
IMUVA and Dpto. Economía Aplicada, Facultad de Ciencias Económicas y Empresariales, Universidad de Valladolid, Valladolid, Spain
e-mail: guiomar@eco.uva.es

S. Taboubi (✉)
Department of Marketing, GERAD and HEC Montréal, Montréal, QC, Canada
e-mail: Sihem.taboubi@hec.ca



P.-O. Pineau et al. (eds.), *Games in Management Science*,
International Series in Operations Research & Management Science 280,
https://doi.org/10.1007/978-3-030-19107-8_2

# 1 Introduction

This chapter contributes to the marketing modelling literature on products' diffusion and pricing. More particularly, we extend the existing literature that incorporated the price effects in the Bass diffusion model (Bass 1969) by providing a new functional form that captures the negative impact of prices on the innovation rate, a key parameter in the diffusion process. By using this formalization, we investigate the pricing strategies of firms in the case where contingent products are sold in the market. Contingent products are products positively inter-related in their demands. That is, an increase in the demand of one of them contributes the demand increase of the other one. The marketing literature distinguishes between two types of product contingency: The optional contingent products and the captive contingent ones (Peterson and Mahajan 1978; Kotler 1988, pp. 516–517). The latter case captures situations where none of the products can be used without the other, while the former case describes situations where one of the products (i.e., the contingent) is useless without the other, often denoted by "the base" (i.e., or primary) product.

In all cases, the interdependencies in products' demands highlight the importance of studying the influence of each product's price on its diffusion and the diffusion of the other product. Indeed, if a firm decides to fix a high (low) price at the launch of a new primary product in order to skim (penetrate) the market, we expect this price to have an impact on the adoption of the primary and the contingent products. These effects depend necessarily on the sensitivity of consumers to prices, which could differ depending on whether they belong to the innovators or the imitators' segments, as classified in the Bass model. Another important question to be addressed when examining the pricing strategies for interdependent products is related to the firms controlling the prices of these products: How pricing strategies compare whenever both products are sold via a single firm versus separate firms?. It has been often observed in practice and reported in the marketing literature (Kotler 1988) that a single firm controlling the prices of two products with optional-contingent interdependencies uses one of the products as a springboard to increase the adoption of the second one (Kort et al. 2018). It is not clear in this case how independent firms controlling each one of these products separately will fix their pricing strategies for the product under their control.

The pricing of contingent products by a single versus two independent firms is a topic that has been previously investigated in the study of Mahajan and Muller (1991) (M&M). The authors used an extension of the Bass model where prices affect the base and the contingent product's demands while the diffusion parameters (i.e., innovation and imitation) are considered constant. Hence, their model does not capture the fact that prices could influence endogenously one or both of these diffusion parameters, as suggested in the literature on products diffusion and captured by the functional form that we propose in this chapter.

To the best of our knowledge, our study is one of the first to examine the pricing strategies of firms in the complex case where the innovation parameter is price-dependent, and imitation is taken into account. Indeed, the existing literature that introduced endogenously the price effect on the diffusion parameters considered that diffusion is driven only by innovation. The imitation effect, which implies the modelling of the interaction between the state variable representing the cumulative number of adopters of the product, the number of non-adopters[1] and the control variable (i.e., price) was systematically omitted. With this simplifying assumption, the diffusion process is represented by a dynamic function which is multiplicatively separable in the state and the price, a feature that makes the model analytically tractable. The new formulation that we suggest in this study provides solutions for the more complex case where both innovation and imitation drive diffusion, and innovation is a price-dependent parameter.

Besides extending the diffusion literature to capture additional features in the diffusion process, our objectives are to assess the impact of using this new functional form and modifying the length of the planning horizon on the previous results obtained in the literature regarding the pricing policies of firms selling contingent products. To achieve these objectives, we compute prices, cumulative sales and profits for the base and the contingent products under the two scenarios investigated in M&M: A scenario where both products are sold by the same monopolistic firm, and a scenario where two distinct firms sell each one of the products. Our results are then compared to those obtained in a modified version of the M&M model[2] where we consider an infinite planning horizon. This modified version of the M&M model is used as a benchmark to assess the impact on results of two modelling hypotheses in the diffusion literature: The formalization of the parameters' effect and the length of the planning horizon.

The chapter is organized as follows. Section 2 revisits the Bass model, while Sect. 3 presents the M&M model and its original results under a finite-time horizon. Section 4 presents our model, and Sect. 5 extends the results obtained by M&M to an infinite-time horizon and compares the results obtained with this modified version to those obtained under our formulation of the price effects. Section 6 concludes.

## 2 The Bass Model

The seminal paper of Bass (1969) is the first study that introduced an analytical model of diffusion in the marketing literature. According to this model, the diffusion of a new product in the market can be modelled as a dynamic process describing the evolution in the number of adopters. According to the Bass model, buyers can be either innovators or imitators. The innovators are the portion of the market potential

[1]The number of non-adopters depends also on the state variable. It captures the remaining potential market at a particular time period. It is computed as the difference, in each time period, between the total market potential and the cumulative number of adopters at that time period.

[2]In the M&M model, the authors study additional scenarios involving the two cases of contingency described above (i.e., optional and captive). In order to focus on the main objective of our study, we restrict our analysis to investigate only the case of optional contingent products.

that adopt the new product after being exposed to external factors. The remaining portion of the market is composed of imitators who adopt the new product because they are sensitive to the word-of-mouth conveyed by those who already adopted the innovation. Denoting by $x(t)$ the cumulative number of adopters by time $t \in [t_0, T]$, the diffusion process in Bass (1969) is captured by the following differential equation

$$\dot{x}(t) = \left(a + \frac{b}{M}x(t)\right)(M - x(t)),$$

where $\dot{x}(t)$ denotes the adoption rate of the new product at time $t$, and $M$ is a positive parameter representing the market potential of the new product.[3] In the normalized Bass model, this equation can be rewritten as

$$\frac{\dot{x}(t)}{1 - x(t)} = a + bx(t)$$

which indicates that $(a + bx(t))$ is a hazard function representing the likelihood of adoption of a new product at time $t$ given that it has not yet been adopted. This function, called also the conversion function, depends on the positive parameters $a$ and $b$, which correspond to the innovation and the imitation rates, as well as the the cumulative number of adopters. In the early version of the Bass model, the parameters $a$ and $b$, and the market potential $M$, are considered constant, and the model is solved by considering a finite time-horizon.

By solving the differential equation given in the Bass model, it is possible not only to estimate the number of adopters at each time $t$, but also to compute the timing and the magnitude of the peak sales (or adoptions).

Despite its performance to forecast the number of adopters for various durable products, the Bass model has been criticized by many researchers in marketing management because of some of its underlying hypotheses. Indeed, various authors raised some concerns related to the use of constant parameters capturing the innovation and the imitation rates, on top of considering that the market potential of the new product does not change. Furthermore, the original version of the Bass model did not account for the impacts of the marketing efforts deployed by the firm launching the new product on its diffusion, nor the impacts of their competitors' strategies or the other products available in the market.[4]

Hence, over the years, this model provided a basic framework for many extensions in the product diffusion modelling literature. The main objective behind these

---

[3]The state dynamics of the cumulative number of adopters when the innovation rate parameter $a$ equals zero corresponds to the well-known logistic equation usually described by $\dot{P} = rP(1 - P/k)$, where $P$ is the population size, $r$ a constant that defines the growth rate, and $k$ is the carrying capacity.

[4]For a detailed analysis of these limits, see Mahajan et al. (1990).

developments was to enhance the model forecasting performance and to transform it from a time-series model into an optimization tool for decision-making.

Introducing the effect of different decision variables, such as prices and advertising, is one of the developments investigated in the marketing science literature on product diffusion. Considering the interactions between multiple products is another important one. The multiple-products' diffusion models encompass cases where (1) products are sold by competing firms in the market; (2) one of the products is a new generation of another one; and (3) products are complementary or contingent. The extension of diffusion models to handle these various cases is justified by the competitive environment of firms and the fact that most of them often face situations where they have to manage various products interrelated in their demands (i.e., a product mix).

As mentioned in the introduction, our paper is an extension of the Bass (1969) model that takes into account the two development axes described above, in addition of exploring the impact of extending the length of the planning horizon. We investigate the product diffusion process by introducing prices as decision variables in an innovative manner, and by considering the interactions between two products. More particularly, we investigate the case of optional contingent products. This paper goes in the same vein than the M&M study who examined the same issue with a different model's formalization of the price effects on diffusion. In the M&M model, prices affect the diffusion dynamics in a multiplicative separable manner, as suggested initially by Robinson and Lakhani (1975) and generalized through the GBM[5] by Bass et al. (1994). This formulation captures the fact that prices have an effect on the probability of purchase of non-adopters, rather than on the model's key parameters, that is: The market potential and the innovation and imitation rates. Hence, prices can shift the probability of purchase upward or downward, and consequently, speed or slow down the adoption, but do not influence in an endogenous manner the number of innovators or imitators, nor the total market potential.

Because the multiplicative model formalization introduces the price effect in the diffusion process without affecting the model's parameters, the GBM model remains analytically tractable, even when multiple firms selling competing or contingent products are considered in the market. This mathematical convenience allowed this functional form to be widely used in the diffusion literature (Jørgensen and Zaccour 2004, p. 73). The general result obtained with this formalization states that, when the imitation effect is high, a firm should use a low price for the launch of its new

[5]I.e., GMB states for the General Bass Model. The GBM is an extension of the Bass model that incorporates, in a multiplicative way, the effect of the marketing variables. According to this model, diffusion is described by the following differential equation:

$$\dot{x}(t) = (a + bx(t))(M - x(t))\,g(V),$$

with $g(V)$ representing a function capturing the impact of firms' decision variables (e.g. advertising, price, etc.)

product in order to attract innovators. The price can be increased later when the product is established and then decreased again.

When the price effects in the diffusion model are introduced by varying the market potential, a skimming strategy where prices are fixed at a high level at the introductory phase is optimal when the imitation effect is low, while a penetration strategy is recommended when the imitation effect is high (Horsky 1990).

As mentioned in the introductory section, our formalization of the price impact on diffusion is introduced in our model through the innovation rate. We model this parameter as a price-dependent function which drives an increase in the innovation rate whenever a firm decides to lower its price (i.e., a penetration strategy) or a decrease in the opposite case (i.e., a skimming strategy).

The choice of this formalization can be justified from a managerial point of view: Firms often target innovators when they choose their marketing strategies since innovators are the early adopters of the new product, and they have the power of influencing the buying decisions of imitators (Jørgensen and Zaccour 2004, p. 49). The study of Mesak (1996) provides an empirical validation of this hypothesis with real data. Indeed, the author tested various diffusion models incorporating price, advertising and distribution in order to identify which one fits better the data on the diffusion of cable TV. One of its main findings is that, for this innovation, the model that performs better is the one where price affects only the innovation rate, advertising affects the diffusion rate,[6] and distribution affects the market potential.

Most of the studies that considered the case where the diffusion model's parameters are endogenous focused on firms' advertising strategies while only few of them used a similar formulation for pricing decisions. The advertising effect, considered as positive, has been introduced either through the innovation rate only (Horsky and Simon 1983) or through both the innovation and the imitation rates (Dockner and Jørgensen 1988; Teng and Thompson 1983). In Thompson and Teng (1984), the authors introduced both the price and the advertising effects in the diffusion process. They considered that advertising affects both the innovation and the imitation rates, while prices have a multiplicative effect that shifts all the diffusion function as in the BGM. Additional extensions of the Bass model incorporated also the impact of economic conditions and marketing decision variables (e.g., distribution) on the market potential (Mahajan et al. 1979; Jones and Ritz 1991; Kalish 1985).[7]

## 3 The Mahajan and Muller (1991) Model

The first adaptation of the Bass model to the multi-product context, and more particularly to the case of contingent products' demands is presented in Peterson and Mahajan (1978). Mahajan and Muller (1991) provided an additional extension

[6]The effect here is found to be multiplicative, as in the GBM, meaning that advertising affects equally the innovation and the imitation rates.

[7]See Peres et al. (2010) for a more recent review of diffusion models.

of this model by introducing the price effects according to a multiplicative-form model, as in the GBM.

The authors considered a market where two products $i = \{1, 2\}$ are sold at prices $p_i(t)$: Product 1 is a base (or primary) product that could be bought independently by consumers, while product 2 is a contingent product purchased only by consumers who already bought the base product. By letting $x_1(t)$ and $x_2(t)$ denote the cumulative sales of the base and the contingent products respectively, and $M$ represent the market potential of these products. The instantaneous sales for both products are described by the following dynamic equations:

$$\dot{x}_1(t) = (a_1 + b_1 x_1(t))(M - x_1(t))\, e^{-\epsilon_1 p_1(t)}, \tag{1}$$

$$\dot{x}_2(t) = (a_2 + b_2 x_2(t))(x_1(t) - x_2(t))\, e^{-\epsilon_2 p_2(t)}, \tag{2}$$

where $a_i$ and $b_i$ represent the product $i$'s innovation and imitation parameters, respectively, and $\epsilon_i$ denote the price sensitivity parameter of product $i$.

The first differential equation is the standard diffusion equation used in the GBM, where $g(V)$ is a price-dependent function. It describes the adoption process of the base product (i.e., product 1) through the effects of innovation ($a_1$), imitation ($b_1$), and retail price. The main feature of Eq. (1) is that it is independent from Eq. (2), meaning that the diffusion of the base product is not affected by the diffusion of the contingent product. This characteristic captures the one-way demand interdependencies between both products in the case of an optional contingency relationship.

The second equation describes a similar process for the contingent product (i.e., product 2), but features a dependence of the contingent-product's diffusion on the diffusion of the base product. We can clearly see that the market potential for the contingent product is determined by the current cumulative sales volume of the base product. As in Eq. (1), Eq. (2) indicates that, at each time $t$, the number of potential buyers of the contingent product corresponds to the total number of buyers of the base product who did not previously buy the contingent product.[8] Here again, the adoption of the contingent product is subject to the impact of the constant parameters of innovation and imitation, and the product's retail price.

By considering a finite time horizon and a production cost ($c_i$), the authors computed and compared price trajectories, sales, and profits for both products under the scenarios where they are controlled by the same[9] or by different firms in order to test the following propositions[10]:

[8]Hence, the maximum number of consumers who could buy the contingent product should not exceed the maximum number of consumers who already bought the base product. (i.e. the market potential $M$).

[9]We use the superscript $m$ to denote the case of an integrated monopolist (i.e., a single firm selling both products).

[10]Remark: In the case of captive contingency, the price effect is introduced by considering that each one of the product diffusion processes is affected by not only its own price, but also by the price of the other product.

- **Proposition A**: When both products are sold by the same firm (i.e., an integrated monopolist), the firm sets a lower price on the base product and a higher price on the contingent one, with respect to the case where both products are sold by independent firms. This translates into the following inequalities:

$$p_1^m(t) < p_1(t) \quad \text{and} \quad p_2^m(t) > p_2(t).$$

- **Proposition B**: When both products are sold by the same firm, most of the profit is earned from the contingent product rather than the base one, and the markup for this product is set at a higher level. Hence, by using the notation $z_i$ to designate the total profit generated by product $i$ in the integrated monopolist scenario, this proposition translates into the following inequalities:

$$z_1/(z_1+z_2) < z_2/(z_1+z_2) \quad \text{and} \quad (p_1-c_1) < (p_2-c_2).$$

To test these propositions, the authors solved both scenarios by using in the first case an optimal control problem, where the unique decision maker (i.e., the monopolistic firm) chooses the two control variables $(p_1, p_2)$. His objective is to maximize his total discounted profit while taking into account the dynamics of the two state variables $x_1, x_2$. The optimality conditions are a system of four differential equations with two initial conditions for the state variables $x_1$ and $x_2$ and two final conditions for the corresponding costate variables.

In the case of the independent producers, on the one hand, the primary producer solves a control problem, where he decides its control variable $p_1$ with the objective of maximizing his discounted cumulative profits while taking into account the state variable $x_1$. The optimality conditions are a system of two differential equations with one initial condition for variable $x_1$ and one final condition for the corresponding costate variable. On the other hand, the contingent-product producer solves a control problem where he decides his control variable $p_2$ in order to maximize his objective functional taking into account the state variables $x_1$ and $x_2$. The optimality conditions in this case are a system of three differential equations with two initial conditions for variables $x_1$ and $x_2$ and one final condition for one costate variable. Putting together the optimality conditions for both firms (primary and contingent products' producers), these conditions constitute a system of four differential equations with two initial conditions for variables $x_1$ and $x_2$ and two final conditions for the corresponding costate variables. Both problems are solved by considering a finite-horizon and the resolution is made numerically.

The numerical simulations are made by considering the following parameters' values:

$$a_1 = a_2 = 0.015, \quad b_1 = b_2 = 0.4/60{,}000, \quad c_1 = c_2 = 60, \quad r = 0.01, \tag{3}$$

$$\varepsilon_1 = \varepsilon_2 = 0.01, \quad M = 60{,}000, \quad x_{10} = 6000, \quad x_{20} = 0. \tag{4}$$

The authors analyzed the steady state of the different scenarios as well as the optimal paths converging to the steady state. In the two scenarios, both $x_1$ and $x_2$ take the value $M$ at the steady state. That is, along all the optimal paths both $x_1(t)$ and $x_2(t)$ converge towards $M$ (See note 8).

The results obtained by M&M provided only a partial validation of proposition A and failed to support proposition B. Indeed, the authors were not able to prove, for any instant of time $t$, that the integrated monopolist producer will fix a lower price for the primary product and a higher price for the contingent one, while the independent producers will do the opposite (i.e., proposition A). The analytical results proved that $p_1^m < p_1$ for the same level of penetration ( i.e., $x_1 = x_1^m$ ) only when the time horizon $T$ is large. Furthermore, they found an opposite result with respect to the pricing of the contingent product. Indeed, their results indicate that the integrated monopolist fixes a lower price for the contingent product, when compared to the price that an independent firm would fix for that product. This result is obtained for the same level of penetration, when $T$ is large, and the discount rate $(r)$ is small.

Nevertheless, the authors proved that the diffusion of both the base and the contingent products are faster when both products are controlled by the same firm.

Proposition B was tested by running numerous simulations. The results obtained did not support anyone of the hypotheses in this proposition. Indeed, the authors found that, in many solutions (for certain time periods), the markup for the primary product was higher than for the contingent product, but it was also lower in other cases.[11] Furthermore, the base product is found to contribute more to total profits with respect to the contingent one.

## 4 Our Model

In our model, we examine the same issue on the pricing of contingent products and investigate the same scenarios as in the M&M's model. The main difference is that in M&M, the innovation rate is a constant parameter, while in our model, we specify a functional form that allows the innovation rate to vary through the impact of price decisions. Indeed, the functional form that we use takes explicitly into account the negative effect of product's price on early-adopters of the new product (i.e., innovators). The innovation rate is then given by the following equation

$$a_i\left(p_i\left(t\right)\right) = \alpha_i - \beta_i p_i\left(t\right), \qquad (5)$$

[11]This result indicates that $p_1^m$ can be greater or lower than $p_2^m$ for some time periods because, as mentioned above, M&M consider the symmetric scenario with respect to the parameters, including the symmetry in production costs (i.e., $c_1 = c_2$).

where $\alpha_i$ and $\beta_i$ are positive parameters. Parameter $\alpha_i$ can be interpreted as a price-independent innovation rate (as in the original Bass model) that captures the fact that a portion of innovators will adopt the new product regardless of its price. Parameter $\beta_i$ reflects price-sensitivity of innovators. Notice that when $\beta_1 = 0$, the original Bass model is obtained, which ignores this price effect on innovators.

The dynamics of the cumulative adoption of the primary product, $x_1$, and the contingent product, $x_2$, can be written as follows[12]:

$$\dot{x}_1(t) = (\alpha_1 - \beta_1 p_1(t) + b_1 x_1(t))(M - x_1(t)), \quad x_1(0) = x_{10}, \tag{6}$$

$$\dot{x}_2(t) = (\alpha_2 - \beta_2 p_2(t) + b_2 x_2(t))(x_1(t) - x_2(t)), \quad x_2(0) = x_{20}. \tag{7}$$

Because the model is not analytically tractable when a finite-time horizon is considered, we studied the problem with infinite horizon. Therefore, the objective in the case of the integrated monopolist is to choose the prices, $p_1$ and $p_2$, in order to maximize the following functional:

$$\int_0^\infty e^{-rt} \left[ (p_1(t) - c_1)\dot{x}_1(t) + (p_2(t) - c_2)\dot{x}_2(t) \right] dt$$

taking into account (6) and (7).

In the second scenario where two independent producers control the pricing decisions of the primary and the contingent product, we have two objective functionals: the objective of the primary-product producer and the objective of the contingent-product producer.

The objective for the primary-product's producer is to choose the price $p_1$ in order to maximize the following functional:

$$\int_0^\infty e^{-rt} (p_1(t) - c_1)\dot{x}_1(t)\, dt$$

taking into account (6).

The contingent-product's producer chooses the price $p_2$ in order to maximize the following objective functional

$$\int_0^\infty e^{-rt} (p_2(t) - c_2)\dot{x}_2(t)\, dt$$

taking into account (6) and (7).

Hence, the problems that we examine are solved according to the same procedure as the one used in the M&M model and described in the previous section. In

[12]Initially M&M assume $x_{20} = 0$, but in their numerical simulations it seems that they consider other initial values for variable $x_2$ positive but lower than the initial value for the variable $x_1$.

Appendix, we provide a detailed description of the different steps used in the resolution of both scenarios.

As mentioned in the introduction, our study allows us to attain two objectives: The primary objective is to assess the impact of using a new functional form capturing the price effects on diffusion. A second objective is to investigate the sensitivity of results to the length of the planning horizon. In order to remove any confusion between the original M&M model solved under a finite horizon and its counterpart when the planning horizon is infinite, we designate by "Modified M&M model" all the results obtained when the M&M model is solved under an infinite planning horizon. This modified version of the M&M model allowed us to assess the impact of the length of the planning horizon. Indeed, by comparing the results obtained in the original study of M&M where the time horizon is finite with those obtained with the modified M&M model, and by using the same parameters' values, we fulfill the secondary objective. The latter is an important step to fulfill the primary objective. Indeed, by comparing the results of our model, where the planning horizon is infinite, with the modified version of the M&M model, and by using the same parameters' values, we removed any explanation of the difference in results between the M&M results and the results obtained with our model, which is not related to the model's functional form.

## 5 Results

In this section, we present the results obtained with the three models described above, that is, the original M&M model, its modified version, and our suggested model, where a price-dependent innovation parameter is introduced in the diffusion process. We compare and contrast these results in order to assess the impact of using a different planning horizon and a new model's formalizations of the price-effects. In all cases, we start by reporting in a bullet-point form the "technical" results about the number of equilibria and the analytical tractability of the models; then present the main managerial findings related to the comparisons of pricing strategies, products' diffusion, and firms' profits under the two investigated. All the results are compared to the results obtained under the original version of the M&M model.

### *5.1 Results Under the Modified M&M Model*

When the M&M model is solved under an infinite-time horizon, we find that, for both scenarios investigated:

- The dynamical system of the optimality condition only presents a unique steady state, and there is a unique optimal path converging to this steady state.

- There are two possible cases depending on whether the Jacobian matrix of the dynamical system evaluated at the steady state presents two different negative eigenvalues or presents only one double negative eigenvalue. However, when we follow the symmetry hypothesis in M&M to analyze the particular case where all parameters are identical for both products, the Jacobian matrix of the dynamical system evaluated at the steady state in this particular case presents only one double negative eigenvalue.
- The steady-state values of the two scenarios can be analytically compared. The result of the comparison at the steady state (denoted by the subscript $ss$) is as follows:

$$x^m_{1ss} = x_{1ss}; \quad x^m_{2ss} = x_{2ss}; \quad p^m_{1ss} < p_{1ss}; \quad p^m_{2ss} = p_{2ss}.$$

As in the original M&M model, we test propositions A and B by running numerical simulations where we use the parameters' values given in (3) and (4). The state and price trajectories under both scenarios of the modified M&M model can be illustrated in Figs. 1 and 2.[13]

Figure 1 indicates that, under both scenarios, the optimal cumulative adoption paths for both products (i.e., $x_1(t)$ and $x_2(t)$) increase monotonously towards the market potential level $M$ with $x_1(t)$ always higher than $x_2(t)$ for all $t$. As already obtained in M&M in a finite horizon, our results indicate that, under an infinite time horizon, the diffusion of the base and the contingent products are faster when they are sold by the same firm.

Figure 2 shows that under the two scenarios, both the optimal price paths $p_1(t)$ and $p_2(t)$ increase monotonously towards their steady state values. This result differs from its counterpart in the original M&M model since the authors found that the price of the primary product could drop when a finite planning horizon is considered.

Furthermore, we can see from this figure that in the integrated monopolist case, $p^m_2(t) > p^m_1(t)$ for all $t$, while in the independent producers scenario $p_1(t) > p_2(t)$ for all $t$. Figure 2 shows also that $p_1(t) > p^m_1(t)$ for all $t$, and $p_2(t) < p^m_2(t)$ for all $t$, except for a very short initial period of time. This indicates that the result on the price comparisons for the contingent product provided in Proposition A is always supported when an infinite time horizon is considered.

[13]Similar figures have been computed for the following cases:

- $x_{20} = 3000, 6000$;
- $M = 40{,}000, 70{,}000, 80{,}000$;
- $\varepsilon_1 = \varepsilon_2 = 0.02$.

showed qualitatively similar results.

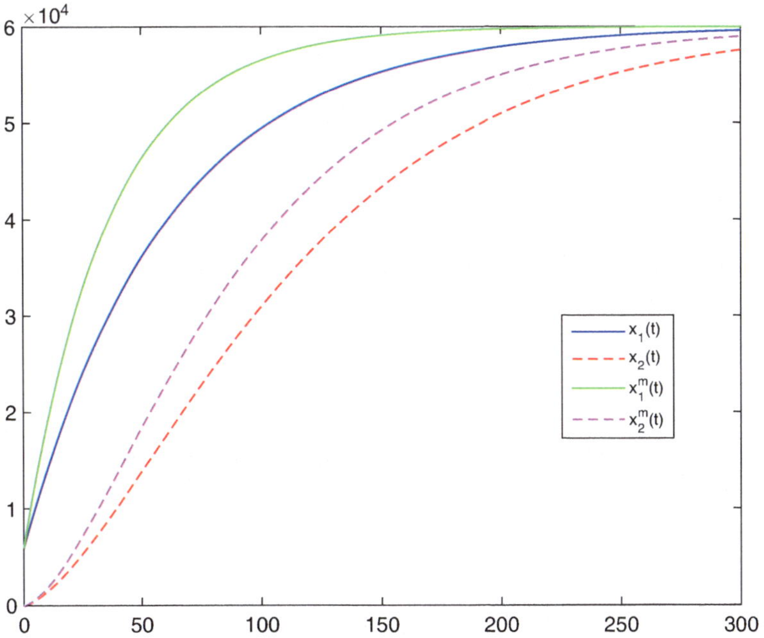


**Fig. 1** Modified Mahajam & Muller's model. Comparison cumulative adoption base and contingent products under integrated monopolistic and independent producers

With the numerical results obtained under the modified M&M model, we computed the markups and the relative profits of the base and the contingent products under the two scenarios investigated. Our results were qualitatively similar to those obtained under a finite-time horizon and did not support proposition B.

Hence, when compared to the original M&M model, the results obtained under the modified M&M model allow us to conclude that moving from a finite to an infinite planning horizon has an impact on firms strategies. Indeed, although the results on adoption levels and profit markups seem to coincide under both settings, the optimal pricing strategies of firms differ.

## 5.2 *Results with Our Model*

From a technical point of view (see Appendix for more details), our model indicates that:

- For both scenarios, and by using the same configuration of the parameters, the dynamical system of the optimality condition presents four different steady

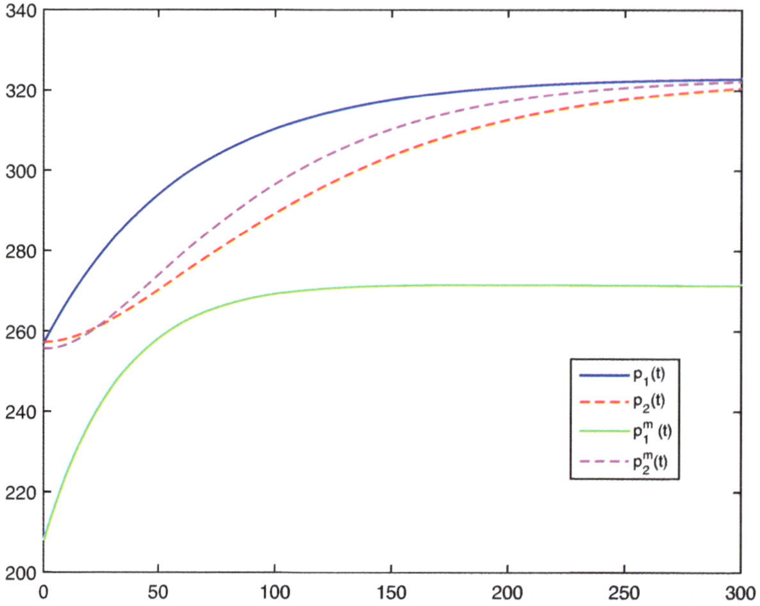


**Fig. 2** Modified Mahajam & Muller's model. Comparison retail prices of the base and contingent products under integrated monopolistic and independent producers

states. Depending on the steady state, it can be either (1) a unique optimal path converging to the steady state; (2) a one-parametric family of solutions converging to the steady state; or (3) a bi-parametric family of solutions converging to the steady state.

From the four possible steady states, we disregard three of them because one of the following reasons:

- For one of the steady states, the family of solutions converging to this steady state imposes that $x_1(t) = M, x_2(t) = M$ for all $t$ (which seems to be an uninteresting case).
- A second steady state is removed from the analysis because, along the optimal paths converging to this steady state, it imposes that $x_1(t) = M$ for all $t$.
- A third steady state is disregarded because a relationship among the initial values of the state variables, $x_{10}$ and $x_{20}$ is needed to ensure the convergence of the optimal paths to the steady state. This relationship can be viewed as a "knife-edge condition."

Therefore, we focus only on the fourth steady state for which there exists a unique optimal path converging to this steady state.

The managerial results indicate that, at this steady-state, the price of the contingent product under the integrated monopolist is identical to its price under the independent firms' scenario, while prices at the steady-state differ among the scenarios when we turn to the primary product:

$$x^m_{1ss} = x_{1ss}; \quad x^m_{2ss} = x_{2ss};\ p^m_{1ss} < p_{1ss};\ p^m_{2ss} = p_{2ss}.$$

That is, we have that in the long run (i.e., at the steady state) the comparison of the two scenarios (integrated monopolist and independent producers) is qualitatively similar with respect to the modified M&M model.[14]

As mentioned above, our numerical simulations are based on the same parameters' values as in the M&M's paper. Furthermore, we have fixed the following values of the parameters which are specific of our model's functional form:

$$\alpha_1 = \alpha_2 = 30,\ \beta_1 = \beta_2 = 0.1. \tag{8}$$

For these values of the parameters and in all the numerical simulations we have carried out we have checked that the innovation rate described in Eq. (5) is between zero and one for both products.

Figures 3 and 4 have been generated by solving our model and using the values of the parameters in (3) and (4) when applicable together with the values of parameters $\alpha$'s and $\beta$'s above.[15]

From Fig. 3 we have that under the two scenarios, the optimal cumulative adoption paths for both products increase monotonously towards the market potential level for all $t$, and $x_1(t)$ is always higher than $x_2(t)$.

Figure 4 indicates that under the two scenarios, both the optimal price paths $p_1(t)$ and $p_2(t)$ increase monotonously towards their steady-state values. In the independent producers scenario $p_2(t) < p_1(t)$ for all $t$; while in the integrated monopolist case, $p^m_1(t) < p^m_2(t)$ for all $t$. Furthermore, Fig. 4 shows for our model that $p_1(t) > p^m_1(t)$ and $p_2(t) < p^m_2(t)$ for all $t$. Therefore, the result of Proposition A in M&M holds for the price of both the contingent and primary products.

[14]However, the transitional dynamics, that is, the transition towards these steady states could be different.

[15]Qualitatively similar figures have been obtained for the following cases:

- $x_{20} = 3000, 6000$;
- $M = 40{,}000, 80{,}000$.

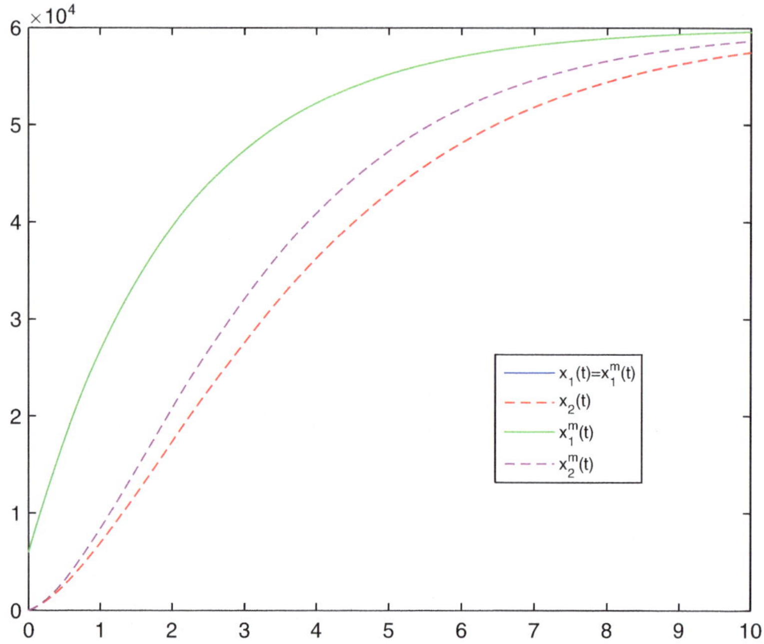


**Fig. 3** Our model. Comparison cumulative adoption base and contingent products under integrated monopolistic and independent producers

Concerning the first hypothesis in Proposition B, which was rejected in both the M&M and the modified M&M models, with our new formalization of the price effect, we found that the same result as in these models is obtained (i.e., $z_1/(z_1+z_2) > z_2/(z_1+z_2)$). Indeed, we found that by keeping fixed the parameters in (3), (4), and (8) and by changing the initial values of the cumulative adoption of the contingent product ($x_{20} \in \{0, 3000, 6000\}$), Proposition B is always rejected. Furthermore, we run some additional simulations by changing the value of the market potential $M$ and found that this value does not have an impact on this result. Independently of the value of the market potential $M$ ($M = 40{,}000$, $M = 80{,}000$) Proposition B is rejected.

Hence, these results are qualitatively similar to those obtained in the modified M&M model, which indicates that the introduction of a price-dependent innovation parameter in the diffusion model does not have a qualitative impact on the previous results obtained with diffusion models where the innovation effect is considered as a constant parameter.

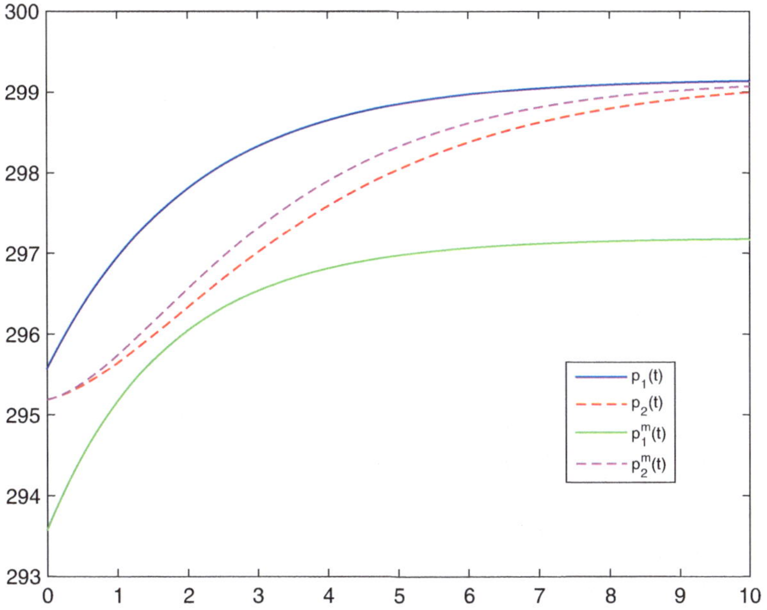


**Fig. 4** Our model. Comparison retail prices of the base and contingent products under integrated monopolistic and independent producers

# 6 Conclusion

In a paper on theoretical modeling in marketing, Moorthy (1993) provides a "logical experimentation" perspective on the building of mathematical theories in this field. This perspective is borrowed from the behavioral marketing research where empirical experiments are used to test the cause-effects relationships that could be observed between a phenomena under study, and other variables that could explain it. This process starts by building a first theoretical model to capture and explain a phenomena (i.e., the dependent variable) under a set of hypothesis. The latter could have impacts on the results and can be seen as independent variables. By modifying these hypotheses one by one, different models are tested as various treatments in a series of experiments, and by comparing the results obtained under the different models, one can understand the cause-effect relationship between the phenomena under investigation and its hypotheses and assess their impacts.

In this paper, we contribute to the modeling literature in marketing by following the logical experimentation perspective suggested by Moorthy (1993). By examining the same phenomena describing the diffusion of optional-contingent products with the M&M model, we started by extending its results from a finite to an infinite time horizon in order to isolate the effects of the planning horizon on the results. Then, we used this modified M&M model and compared its results to those obtained with a model where the innovation rate is modeled as a price-dependent variable. We were able to assess the impact of using this new model formalization on the multi-product diffusion phenomena and the firms strategies and profits.

Diffusion models have been widely studied since the introduction of the Bass model in the marketing science literature. Despite all the efforts to extend this model and to relax some of its limiting assumptions, there are still many open questions on the diffusion process. What marketing variables affect the adoption of new products, and how to model their effects on adoption are some of these open-questions. Mesak (1996) established this in a study where he classified most of the models that incorporate the effects of marketing variables in the Bass model and tested empirically the validity for some of them. He concluded his study by stating that "The issue of how marketing mix variables such as price, advertising and distribution affect the diffusion of innovations continues to be a debated issue in the literature" (Mesak 1996, p. 1011). This paper can be seen as an additional treatment in the product diffusion literature where the effect of modeling differently the price impact on diffusion and extending the planning horizon are investigated. An interesting extension of this study is to test the external validity of these multi-product models as in Mesak (1996) by performing their empirical validation with real market data on sales and prices of new products characterized by optional contingencies in their demands.

**Acknowledgements** We are grateful to two anonymous reviewers for valuable comments and suggestions on an earlier draft of this paper. This research is partially supported by Spanish MINECO under projects ECO2014-52343-P and ECO2017-82227-P (AEI) and by Junta de Castilla y León under projects VA105G18 and VA024P17 co-financed by FEDER funds (EU).

# Appendix

In the scenario where the independent producers control the pricing decisions of the primary and contingent products, the objective for the primary-product's producer is to choose the price $p_1$ in order to maximize the following functional:

$$\int_0^\infty e^{-rt}(p_1(t) - c_1)\dot{x}_1(t)\,dt$$

taking into account (6).

In order to find the first-order conditions necessary for optimality, we construct the current value Hamiltonian:

$$H^1(x_1, p_1, \lambda_1) = (p_1 - c_1)\dot{x}_1 + \lambda_1\dot{x}_1 = (p_1 - c_1 + \lambda_1)\dot{x}_1,$$

where $\lambda_1$ denotes the costate variable associated with $x_1$.

The maximization of $H^1$ with respect to $p_1$ yields $\partial H^1/\partial p_1 = 0$, and assuming $x_1$ is different from $M$, from this condition one gets:

$$p_1 = \frac{\alpha_1 + (c_1 - \lambda_1)\beta_1 + b_1 x_1}{2\beta_1}. \tag{9}$$

The Maximum Principle optimality conditions also include

$$\dot{\lambda}_1 = r\lambda_1 - \frac{\partial H^1}{\partial x_1}, \quad \lim_{t\to\infty} \lambda_1(t)x_1(t)e^{-rt} = 0,$$
$$\dot{x}_1 = (\alpha_1 - \beta_1 p_1 + b_1 x_1)(M - x_1), \quad x_1(0) = x_{10}.$$

This boundary value problem taking into account expression (9) reads:

$$\dot{\lambda}_1 = \frac{1}{4\beta_1}\Big[(3x_1 - 2M)x_1 b_1^2 + (\alpha_1 - (c_1 - \lambda_1)\beta_1)$$
$$(\alpha_1 - (c_1 - \lambda_1)\beta_1 + 2b_1(2x_1 - M)) + 4r\beta_1\lambda_1\Big],$$
$$\dot{x}_1 = \frac{1}{2}(M - x_1)(\alpha_1 - (c_1 - \lambda_1)\beta_1 + b_1 x_1).$$

Function $H^1$ is concave with respect to $p_1$.

The objective for the contingent-product's producer is to choose the price $p_2$ in order to maximize the following functional:

$$\int_0^\infty e^{-rt}(p_2(t) - c_2)\dot{x}_2(t)\, dt$$

taking into account the differential equations describing the dynamics of the cumulative adoption of the primary and contingent products (6) and (7), respectively.

The current value Hamiltonian reads:

$$H^2(x_2, p_2, \lambda_2) = (p_2 - c_2)\dot{x}_2 + \lambda_2\dot{x}_2 = (p_2 - c_2 + \lambda_2)\dot{x}_2,$$

where $\lambda_2$ denotes the costate variable associated with $x_2$.

Assuming that $x_1$ is different from $x_2$, from the optimality condition $\partial H^2/\partial p_2 = 0$, one gets:

$$p_2 = \frac{\alpha_2 + (c_2 - \lambda_2)\beta_2 + b_2 x_2}{2\beta_2}. \tag{10}$$

The Maximum Principle optimality conditions also include

$$\dot{\lambda}_2 = r\lambda_2 - \frac{\partial H^2}{\partial x_2}, \quad \lim_{t\to\infty} \lambda_2(t)x_2(t)e^{-rt} = 0,$$
$$\dot{x}_2 = (\alpha_2 - \beta_2 p_2 + b_2 x_2)(x_1 - x_2), \quad x_2(0) = x_{20}.$$

Substituting the expression of $p_2$ given by (10) into this system of differential equations we get:

$$\dot{\lambda}_2 = \frac{1}{4\beta_2}\Big[(3x_2 - 2x_1)x_2 b_2^2 + (\alpha_2 - (c_2 - \lambda_2)\beta_2)$$
$$(\alpha_2 - (c_2 - \lambda_2)\beta_2 + 2b_2(2x_2 - x_1)) + 4r\beta_2\lambda_2\Big],$$
$$\dot{x}_2 = \frac{1}{2}(x_1 - x_2)(\alpha_2 - (c_2 - \lambda_2)\beta_2 + b_2 x_2).$$

Function $H^2$ is concave with respect to $p_2$.

The characterization of the optimal time paths of the cumulative sales and prices of both products requires the solution of the differential equations for the state and costate variables $x_1, x_2, \lambda_1$ and $\lambda_2$. First of all, we focus on the characterization of the steady-state values and their asymptotically stability.

Because $\alpha_i - \beta_i p_i + b_i x_i$ for $i = 2, 2$ are strictly positive, the unique steady-state value of the cumulative sales is given by $x_{1ss} = M$ and $x_{2ss} = x_{1ss} = M$. Taking these values into account, we compute the steady-state values of the costate variables $\lambda_1$ and $\lambda_2$. It can be easily proved that the system of differential equations admits the following four different steady-state values:

$$(x_{1ss}^{(1)}, \lambda_{1ss}^{(1)}, x_{2ss}^{(1)}, \lambda_{2ss}^{(1)}) = (M, \lambda_1^{(1)}, M, \lambda_2^{(1)}),$$
$$(x_{1ss}^{(2)}, \lambda_{1ss}^{(2)}, x_{2ss}^{(2)}, \lambda_{2ss}^{(2)}) = (M, \lambda_1^{(1)}, M, \lambda_2^{(2)}),$$
$$(x_{1ss}^{(3)}, \lambda_{1ss}^{(3)}, x_{2ss}^{(3)}, \lambda_{2ss}^{(3)}) = (M, \lambda_1^{(2)}, M, \lambda_2^{(1)}),$$
$$(x_{1ss}^{(4)}, \lambda_{1ss}^{(4)}, x_{2ss}^{(4)}, \lambda_{2ss}^{(4)}) = (M, \lambda_1^{(2)}, M, \lambda_2^{(2)}),$$

where

$$\lambda_i^{(1)} = -\frac{1}{\beta_i}\left(\alpha_i - c_i\beta_i + 2r + b_i M + 2\sqrt{r(\alpha_i - c_i\beta_i + r + b_i M)}\right),$$
$$\lambda_i^{(2)} = -\frac{1}{\beta_i}\left(\alpha_i - c_i\beta_i + 2r + b_i M - 2\sqrt{r(\alpha_i - c_i\beta_i + r + b_i M)}\right), \quad i = 1, 2.$$

In order to analyze the stability of the steady states we compute the eigenvalues and associated eigenvectors of the Jacobian matrix evaluated at each of the steady states.

At the first steady state $(x_{1ss}^{(1)}, \lambda_{1ss}^{(1)}, x_{2ss}^{(1)}, \lambda_{2ss}^{(1)})$ the Jacobian matrix has two negative eigenvalues and it can be proved that there exists a bi-parametric family of solutions converging to this steady state. This family of solutions imposes that $x_1(t) = M, x_2(t) = M$ for all $t$.

At the second steady state $(x^{(2)}_{1ss}, \lambda^{(2)}_{1ss}, x^{(2)}_{2ss}, \lambda^{(2)}_{2ss})$ the Jacobian matrix has two negative eigenvalues and it can be proved that there exists a one-parametric family of solutions converging to this steady state. This family of solutions imposes that $x_1(t) = M$ for all $t$.

At the third steady state $(x^{(3)}_{1ss}, \lambda^{(3)}_{1ss}, x^{(3)}_{2ss}, \lambda^{(3)}_{2ss})$ the Jacobian matrix has two negative eigenvalues and it can be proved that a relationship among the initial values of the state variables, $x_{10}$ and $x_{20}$ is needed to ensure the convergence of the optimal paths to this steady state.

At the fourth steady state $(x^{(4)}_{1ss}, \lambda^{(4)}_{1ss}, x^{(4)}_{2ss}, \lambda^{(4)}_{2ss})$ the Jacobian matrix has two negative eigenvalues and it can be proved that there exists a unique optimal path converging to this steady state.

The numerical simulations carried out focus on this fourth scenario. In this case, the two negative eigenvalues are given by

$$\mu_i = r - \sqrt{r(\alpha_i - c_i\beta_i + r + b_i M)}, i = 1, 2.$$

Following M&M, the values of the model parameters used in the numerical simulations are assumed to be completely symmetric. Consequently, under this assumption there is a double negative eigenvalue, $\mu = \mu_i, i = 1, 2$. We have computed the associated generalized eigenvectors denoted by $\bar{v}_1 = (v_1^{(1)}, v_1^{(2)}, 0, 1)$ and $\bar{v}_2 = (v_2^{(1)}, v_2^{(2)}, 1, 0)$, with $v_i^{(j)}$ the $j$-th component of the $i$-th eigenvector (omitted for brevity). The solution of the system of differential equations read:

$$\begin{aligned}
x_1(t) &= (x_{10} - M)e^{\mu t} + M,\\
\lambda_1(t) &= w_1^{(2)}e^{\mu t} + \lambda^{(4)}_{1ss},\\
x_2(t) &= (x_{20} - M)e^{\mu t} + w_2^{(3)}te^{\mu t} + M,\\
\lambda_2(t) &= w_1^{(4)}e^{\mu t} + w_2^{(4)}te^{\mu t} + \lambda^{(4)}_{1ss},
\end{aligned}$$

where $w_i^{(k)}$ is the $k$-th component of vector $\bar{w}_i$, with

$$\bar{w}_1 = \varphi\bar{v}_1 + \eta\bar{v}_2, \quad \bar{w}_2 = (\Omega - \mu I_4)\bar{w}_1,$$

and

$$\varphi = \frac{1}{v_1^{(1)}}(x_{10} - M - (x_{20} - M)v_2^{(1)}), \quad \eta = x_{20} - M.$$

Matrices $\Omega$ and $I_4$ denote the Jacobian matrix associated with the system of differential equations evaluated at the steady state $(x^{(4)}_{1ss}, \lambda^{(4)}_{1ss}, x^{(4)}_{2ss}, \lambda^{(4)}_{2ss})$ and the fourth-order identity matrix, respectively.

The characterization of the optimal time-paths of the prices and cumulative adoption of both products in the case of the integrated monopolist follows the same steps as previously described for the scenario of two independent producers.

The objective in the case of the integrated monopolist is to choose the prices, $p_1$ and $p_2$, in order to maximize the following functional:

$$\int_0^{\infty} e^{-rt} \left[(p_1(t) - c_1)\dot{x}_1(t) + (p_2(t) - c_2)\dot{x}_2(t)\right] dt$$

taking into account the differential equations (6) and (7).

The current-value Hamiltonian reads[16]:

$$\begin{aligned} H^m(x_1, p_1, x_2, p_2, \lambda_1^m, \lambda_2^m) &= (p_1 - c_1)\dot{x}_1 + (p_2 - c_2)\dot{x}_2 + \lambda_1^m \dot{x}_1 + \lambda_2^m \dot{x}_2 \\ &= (p_1 - c_1 + \lambda_1^m)\dot{x}_1 + (p_2 - c_2 + \lambda_2^m)\dot{x}_2, \end{aligned}$$

where $\lambda_1^m$ and $\lambda_2^m$ denote the costate variables associated with $x_1$ and $x_2$, respectively.

The first-order optimality conditions for an interior solution read:

$$\frac{\partial H^m}{\partial p_i} = 0, \quad i = 1, 2,$$

$$\begin{aligned} \dot{\lambda}_i^m &= r\lambda_i^m - \frac{\partial H^m}{\partial x}, \qquad \lim_{t\to\infty} \lambda_i^m(t) x_i(t) e^{-rt} = 0 \quad i = 1, 2, \\ \dot{x}_1 &= (\alpha_1 - \beta_1 p_1 + b_1 x_1)(M - x_1), \quad x_1(0) = x_{10}, \\ \dot{x}_2 &= (\alpha_2 - \beta_2 p_2 + b_2 x_2)(x_1 - x_2), \quad x_2(0) = x_{20}. \end{aligned}$$

Assuming that $x_1$ and $x_2$ are different from $M$ and $x_1$, respectively, from the two first optimality conditions the following expressions from the prices can be derived:

$$p_i = \frac{1}{2\beta_i}(b_i x_i + \alpha_i + \beta_i(c_i - \lambda_i)), \quad i = 1, 2.$$

Substituting these expressions in the differential equations describing the time evolution of the state and costate variables, these equations read:

$$\begin{aligned} \dot{x}_1 &= \frac{1}{2}(M - x_1)(b_1 x_1 + \alpha_1 - \beta_1(c_1 - \lambda_1)), \\ \dot{x}_2 &= \frac{1}{2}(x_1 - x_2)(b_2 x_2 + \alpha_2 - \beta_2(c_2 - \lambda_2)), \end{aligned}$$

[16]The superscript stands for "monopolistic scenario."

$$\dot{\lambda}_1 = \frac{1}{4\beta_1\beta_2}\Big[\beta_2\Big(b_1^2 x_1(3x_1-2M)+(\alpha_1-\beta_1(c_1-\lambda_1))$$
$$(\alpha_1-\beta_1(c_1-\lambda_1)-2b_1(M-2x_1))+4r\beta_1\lambda_1\Big)$$
$$-\beta_1(b_2x_2+\alpha_2-\beta_2(c_2-\lambda_2))^2\Big],$$
$$\dot{\lambda}_2 = \frac{1}{4\beta_2}\Big[b_2^2 x_2(3x_2-2x_1)+(\alpha_2-\beta_2(c_2-\lambda_2))$$
$$(\alpha_2-\beta_2(c_2-\lambda_2)-2b_2(x_1-2x_2))+4r\beta_2\lambda_2\Big].$$

The characterization of the steady-states and the analysis of their stability follow the same steps as the analysis developed in the case of the independent producers. Four steady states can be characterized and the numerical simulations focus on the only steady state for which there is a unique optimal path converging to this steady state. This steady-state reads $(x_{1ss}^{(m)}, \lambda_{1ss}^{(m)}, x_{2ss}^{(m)}, \lambda_{2ss}^{(m)})$ with $x_{1ss}^{(m)} = x_{2ss}^{(m)} = M$, and

$$\lambda_{1ss}^{(m)} = -\frac{1}{\beta_1\beta_2}\left(\alpha_1 - c_1\beta_1 + 2r + b_1M - 2\sqrt{r\Gamma}\right),$$
$$\lambda_{2ss}^{(m)} = -\frac{1}{\beta_2}\left(\alpha_2 - c_2\beta_2 + 2r + b_2M - 2\sqrt{r(\alpha_2 - c_2\beta_2 + r + b_2M)}\right),$$

with $\Gamma$ given by:

$$\Gamma = \beta_2\Big[(\alpha_2 - c_2\beta_2 + 2r + b_2M - 2\sqrt{r(\alpha_2 - c_2\beta_2 + r + b_2M)})\beta_1$$
$$+(\alpha_1 - c_1\beta_1 + r + b_1M)\beta_2\Big].$$

The eigenvalues of the Jacobian matrix evaluated at this steady-state are

$$\mu_1 = r - \sqrt{r(\alpha_2 - c_2\beta_2 + r + b_2M)},$$
$$\mu_2 = r - \frac{1}{\beta_2}\sqrt{r\Gamma}.$$

The eigenvectors associated are $\bar{v}_1^m = (0, v_1^{(m2)}, v_1^{(m3)}, 1)$ and $\bar{v}_2^m = (v_2^{(m1)}, v_2^{(m2)}, v_2^{(m3)}, 1)$, with $v_i^{(mj)}$ the $j$-th component of vector $\bar{v}_i^m$ (omitted for brevity). The solution of the system of differential equations read:

$$x_1(t) = (x_{10} - M)e^{\mu_2 t} + M,$$
$$\lambda_1(t) = \xi v_1^{(m2)} e^{\mu_1 t} + \psi v_2^{(m2)} e^{\mu_2 t} + \lambda_{1ss}^{(m)},$$

$$x_2(t) = \xi v_1^{(m3)} e^{\mu_1 t} + \psi v_2^{(m3)} e^{\mu_2 t} + M,$$
$$\lambda_2(t) = \xi e^{\mu_1 t} + \psi e^{\mu_2 t} + \lambda_{2ss}^{(m)},$$

where

$$\xi = \frac{(x_{20} - M)v_2^{(m1)} - (x_{10} - M)v_2^{(m3)}}{v_1^{(m3)} v_2^{(m1)}}, \quad \psi = \frac{x_{10} - M}{v_1^{(m3)}}.$$

# References


Bass, F. M. (1969). A new product growth for model consumer durables. *Management Science, 15*(5), 215–227.

Bass, F. M., Krishnan, T. V., & Jain, D. C. (1994). Why the bass model fits without decision variables. *Marketing Science, 13*(3), 203.

Dockner, E., & Jorgensen, S. (1988). Optimal advertising policies for diffusion models of new product innovation in monopolistic situations. *Management Science, 34*(1), 119–130.

Horsky, D. (1990). A diffusion model incorporating product benefits, price, income and information. *Marketing Science*, 9(4), 279–365.

Horsky, D., & Simon, L. S. (1983). Advertising and the diffusion of new products. *Marketing Science, 2*(1), 1–17.

Jones, J. M., & Ritz, C. J. (1991). Incorporating distribution into new product diffusion models. *International Journal of Research in Marketing, 8*(2), 91–112.

Jørgensen, S., & Zaccour, G. (2004). *Differential games in marketing. International series in quantitative marketing*. Boston: Kluwer Academic Publishers.

Kalish, S. (1985). A new product adoption model with price, advertising, and uncertainty. *Management Science, 31*(12), 1569–1585.

Kort, P. M., Taboubi, S., & Zaccour, G. (2018). Pricing decisions in marketing channels in the presence of optional contingent products. *Central European Journal of Operations Research.* https://doi.org/10.1007/s10100-018-0527-x

Kotler, P. (1988). *Marketing management* (6th ed.). Englewood Cliffs: Prentice Hall.

Mahajan, V., & Muller, E. (1991). Pricing and diffusion of primary and contingent products. *Technological Forecasting and Social Change, 39*, 291–307.

Mahajan, V., Muller, E., & Bass, F. M. (1990). New product diffusion models in marketing: A review and directions for research. *Journal of Marketing, 54*(1), 1–22.

Mahajan, V., Peterson, R. A., Jain, A. K., & Malhotra, N. (1979). A new product growth model with a dynamic market potential. *Long Range Planning, 12*(4), 51–58.

Mesak, H. I. (1996). Incorporating price, advertising and distribution in diffusion models of innovation: Some theoretical and empirical results. *Computers & Operations Research, 23*(10), 1007–1023.

Moorthy, K. S. (1993). Theoretical modeling in marketing. *Journal of Marketing, 57*(2), 92–106.

Peres, R., Muller, E., & Mahajan, V. (2010). Innovation diffusion and new product growth models: A critical review and research directions. *International Journal of Research in Marketing, 27*(2), 91–106.

Peterson, R. A., & Mahajan, V. (1978). Multi-product growth models. In S. Jagdish (Ed.), *Research in marketing* (pp. 201–231). London: JAI Press.

Robinson, B., & Lakhani, C. (1975). Dynamic price models for new product planning. *Management Science, 21*(10), 1113–1122.

Teng, J., & Thompson, G. L. (1983). Oligopoly models for optimal advertising when production costs obey a learning curve. *Management Science, 29*(9), 1087–1101.

Thompson, G. L., & Teng, J. (1984). Optimal pricing and advertising policies for new product oligopoly models. *Marketing Science, 3*(2), 148–168.

# The Effects of Consumer Rebates in a Competitive Distribution Channel

**Salma Karray**

**Abstract** This research investigates the effectiveness of consumer rebates offered by competing manufacturers in a distribution channel. We consider a two-manufacturer, two-retailer channel and develop a two-period model where consumers' preferences are distributed along a Hotelling line. The rebate consists in a price discount that can be redeemed on the second period. We solve three Stackelberg games: a benchmark where no rebate is offered, a symmetric game where both manufacturers offer rebates, and an asymmetric game where only one manufacturer provides a rebate. Comparisons of equilibrium solutions show that while manufacturers should not offer rebates, they could do so due to a prisoner dilemma situation when their wholesale prices are high.

**Keywords** Rebates · Distribution channels · Competition · Pricing · Game theory

## 1 Introduction

Price promotions are widely used marketing tools aiming at stimulating demand and encouraging consumer purchase. In fact, it is estimated that about 310 billion coupons were distributed in 2014 (NCH Marketing Services) resulting in \$3.6 billion of consumer savings (Arya and Mittendorf 2013).

Firms have developed a variety of redemption policies and guidelines for these price promotions. In particular, some promotions can instantly be redeemed by customers on the same purchase occasion that they are obtained (Sigué 2008; Martín-Herrán et al. 2010). Others, commonly denoted as rebates, offer delayed price cuts that can only be redeemed on a future purchase occasion (Chen et al. 2005; Lu and Moorthy 2007). Both promotional offers are frequently used by CPG

S. Karray (✉)
Faculty of Business and IT, University of Ontario Institute of Technology, Oshawa, ON, Canada
e-mail: salma.karray@uoit.ca

 
P.-O. Pineau et al. (eds.), *Games in Management Science*,
International Series in Operations Research & Management Science 280,
https://doi.org/10.1007/978-3-030-19107-8_3

firms in a variety of categories such as grocery, apparel, hardware, and others. For example, many manufacturers offer rebates in form of on-pack promotions. These are price discounts printed on the product packaging, which can be redeemed after the initial purchase (Dhar et al. 1996).

This research focuses on rebates offered by manufacturers as discounts on future purchases (e.g., on-pack coupons). It evaluates the effectiveness of rebates for competing distribution channel members and studies their implications for pricing strategies and profits in the channel.

A large marketing literature has shown that price promotions are effective in stimulating demand by inciting consumers whose reservation prices lie below the retail price (Nijs et al. 2001; Ault et al. 2000; Arya and Mittendorf 2013). While instant promotions can immediately boost sales, the delayed redemption of rebates can lead to a different impact on demand over time. In particular, the literature points to the importance of consumer slippage arising from the non-redemption of rebates by those consumers who either forget or lose them (Gilpatric 2009; Liang et al. 2013).

Consumer slippage can make rebates attractive promotional tools to companies for two main reasons. First, rebates can increase future demand by encouraging repeat purchase. Second, rebates entice consumers to buy products for which there is promise of a future discount, which can increase product purchase at the time of rebate distribution. While rebates can be costly marketing initiatives, companies can rely on consumer slippage to lower these costs. As a proportion of distributed rebates are not redeemed on the next purchase, only a partial cost of the promotion is incurred, which results in additional savings for the firm. For example, on-pack coupons are found to generate positive effects on market share and profits that are higher than the effects generated by instant coupons (Dhar et al. 1996).

Despite the positive effects of consumer rebates, there exist some drawbacks associated with using such promotional tools. In particular, many studies point out that retailers tend to raise prices when manufacturers offer rebates (Busse et al. 2006; Khouja and Zhou 2010; Martín-Herrán and Sigué 2011; Liang et al. 2013). This is because, in a distribution channel context, rebates can impact not only the manufacturers that offer them but also their retailers. In fact, the cost associated with offering the rebate can lead manufacturers to adjust their transfer prices, which can influence the retailers' prices to consumers. Such arrangements can directly impact firms' margins and profits at each level of the channel. Further, rebates not only affect the strategies of the manufacturers issuing them and those of their retailers, but also those of firms in competing channels. For these reasons, it is important to understand the strategic implications of rebates for firms' profitability in a channel context and when competitors' actions are taken into account.

The literature that studied rebates in a distribution channel context has mainly focused on markets where manufacturers and/or retailers hold monopolistic positions. Such a setup is restrictive since competition affects the strategies of firms at both levels of the channel. It is important to address this knowledge gap since opportunistic behaviors can lead to inefficiencies in the channel and competitive interactions can significantly impact firms' decisions, which can aggravate or diminish these inefficiencies. This research addresses these issues by analyzing the strategic implications of rebates in a distribution channel with competition at both levels of the channel. We aim to address the following research questions: Are rebates profitable for manufacturers and for retailers? What are the implications of manufacturers' rebates for channel pricing over a two-period horizon?

The rest of the chapter is organized as follows. Section 2 explains the model and assumptions. In Sect. 3, we present the equilibrium solutions for three games: a benchmark where no rebates are offered, a symmetric game where both manufacturers offer rebates and an asymmetric game where only one manufacturer offers the rebate. Section 4 compares equilibrium solutions across these games. Finally, Sect. 5 concludes.

## 2 Model

Consider a market served by two competing distribution channels denoted by $a$ and $b$, each formed by one manufacturer and one retailer. In each channel, the manufacturer can offer a rebate which consists in a price discount that consumers acquire by purchasing the product and can redeem on the next purchasing occasion (Chen et al. 2005). We then consider two purchasing periods in the model.

In this market, consumers are uniformly distributed over a Hotelling line and each retailer is situated at the end of the linear market. The total market size is fixed to one in each period and is totally served by the two channels, meaning that each consumer buys one product in each period from either channel $a$ or $b$. In each period, consumers choose the product that maximizes their surplus given the prices charged by each retailer, the rebate, if any, and the disutility or travelling cost incurred by deviating from their preferred position on the Hotelling line. Each consumer buys one unit of the product as long as his/her utility surplus is positive and the utility of no consumption is normalized to zero. A summary of all notations used in the chapter is included in Table 1.

### 2.1 Consumer Utility and Demand

Following the conventional Hotelling model, in the first period, consumer utility surplus ($CS_{i1}$) for purchasing product $i$ ($i = a, b$) is affected by the retailers' prices ($p_{i1}$) and the mismatch (disutility) cost of deviating from the consumer's ideal location on the Hotelling line ($tx_{ij}$), where $x_{ij}$ is the consumer location and $t$ is the positive per unit mismatch cost (Hotelling 1929; Vickers and Armstrong 2001). When manufacturers offer rebates that can be redeemed in the next period, consumer surplus is increased by the discounted value of the rebate that will be received in period 2 ($p_{i2}r_i$). $p_{i2}$ is the price of product $i$ in period 2 and $r_i$ is the

**Table 1** Notations

| | |
|---|---|
| $i$ | Index for retailers and manufacturers, $a = 1, 2$ |
| $j$ | Index for time period, $j = 1, 2$ |
| $x_{ij}$ | Demand of retailer $i$ in period $j$, $x_{ij} \in (0, 1)$ |
| $U$ | Consumer utility, $U \geq 0$ |
| $CS_{ij}$ | Consumer surplus for product $i$ in period $j$, $CS_{ij} > 0$ |
| $v_i$ | Baseline demand of product $i$,$v_i > 0$ |
| $t$ | Hotelling transportation cost parameter, $t > 0$ |
| $g$ | Rebate redemption rate parameter, $g \in (0, 1)$ |
| $r_i$ | Rebate rate of manufacturer $i$, $r_i \in (0, 1)$ |
| $w_i$ | Wholesale price of manufacturer $i$, $w_i > 0$ |
| $p_{ic}$ | Price of retailer $i$ in period $j$, $p_{ij} > w_i$ |
| $\Pi_i$ | Total profit of manufacturer $i$, $\Pi_i > 0$ |
| $\Pi_{ij}$ | Profit of manufacturer $i$ in period $j$, $\Pi_{ij} > 0$ |
| $\pi_i$ | Total profit of retailer $i$, $\pi_i > 0$ |
| $\pi_{ij}$ | Profit of retailer $i$ in period $j$, $\pi_{ij} > 0$ |

manufacturer $i$'s rebate rate. We assume that the consumers' discount rate over the two periods of purchasing is set to one. This is a reasonable assumption given that there is usually a short period of time between purchase and receipt of the rebate (Gilpatric 2009).

Consumers are strategic in our model since they take into account their expectations of the prices in period 2 when they maximize their utility surplus in period 1. We assume that consumers have accurate expectations for the price in period 2. This means that they can form rational expectations of the firms' future prices and can then correctly predict their values at equilibrium (Singh et al. 2008; Liu and Zhang 2013). Let $v_i$ be consumer valuation of product $i$ ($v_i > 0$). Consumers' surplus for product $i$ in period 1 is then given by:

$$CS_{i1} = v_i - x_{i1}t - p_{i1} + p_{i2}r_i, \ i \in \{a, b\}.$$

The demand for each product in each period is given by its market share ($x_{ij}$ for product $i$). Since the market is fully served by both firms, the product $k$'s demand in each period is formed by the market not served by product $i$ such as $x_{kj} = 1 - x_{ij}, i, k \in \{a, b\}, i \neq k$. In period 1, product $i$'s demand is obtained by the share of consumers who get the highest surplus by buying product $i$. It is given by $x_{i1}$ for which $CS_{i1} = CS_{k1}$ such as:

$$x_{i1} = \frac{v_i - v_k + p_{k1} - p_{i1} - (p_{k2}r_k - p_{i2}r_i) - t}{2t}, \ i, k \in \{a, b\}, \ i \neq k.$$

We follow Singh et al. (2008) and consider that when rebates are offered, there is a probability $g \in (0, 1)$ that consumers who bought in period 1 would redeem the rebate in period 2. This is to represent the commonly observed consumer slippage effect due to the fact that consumers often forget or lose their coupons. The parameter $g$ is commonly denoted in the empirical literature as the redemption rate. In the model, a percentage of consumers who bought in period 1 would buy from the same firm and redeem their rebates in period 2. The rest of the market $(1 - g)$ represents the portion of switching consumers and captures the consumer slippage effect. The latter can take different values depending on consumer characteristics and on the product category. For example, Yang et al. (2010) note that slippage rates $(1 - g)$ can vary and can reach high values (e.g., 60% for electronics 90% for software).

In period 2, consumers who do not redeem their rebates will choose the product that provides the highest surplus given the prices offered by each firm in the second period. The surplus of these consumers located at point $(u_{i2})$ for product $i$ will then be $CS_{i2} = v_i - tu_{i2} - p_{i2}$. Product $i$'s share in the switching market in period 2 is then obtained at the indifferent consumer location such as $CS_{i2} = CS_{k2}$ and is given by $u_{i2} = (p_{k2} - p_{i2} - t)/2t$, and product $k$'s share in the switching market is the remaining portion $(u_{k2} = 1 - u_{i2})$. Note that we assume that the rebates are designed in such a way that consumers do not have to pay a cost for redemption. This is the case for rebates that can be redeemed at checkout in the retail store or online and do not require considerable time or effort on the part of the consumer.

Each product $i$'s demand in period 2 is given by the portion of consumers who redeem their rebates in period 2 $(gx_{i1})$ and the additional consumers who do not redeem $((1 - g)u_{i2})$. Therefore the demands in period 2 are:

$$x_{i2} = gx_{i1} + (1 - g)u_{i2}, \quad x_{k2} = 1 - x_{i2}, \quad i, k \in \{a, b\}, i \neq k.$$

## 2.2 Firms' Decisions and Profits

Each manufacturer and each retailer makes decisions over two periods. For simplicity, we assume that the manufacturers are dummy players. The retailers play Nash and choose their strategies simultaneously knowing the manufacturers' announced (given) decisions.

In period 1, the manufacturers set the rebate program, if any $(r_a, r_b \in (0, 1))$. Since rebates usually require some planning on the part of the issuing firm (e.g., printing of coupons or modification of packaging), they are usually set before the regular retail pricing decisions. We also assume that the manufacturers' wholesale prices ($w_a$ and $w_b$) are fixed, which is the case in channels with a long-term pricing agreement. Once the rebate program is set, the retailers then react to the announced rebates and decide of their retail prices in the first period $(p_{i1})$. In the second period, the retailers decide of their prices to consumers $(p_{i2})$, knowing all decisions in period 1.

**Table 2** Profit functions ($i \in (a, b)$)

| | Manufacturers | Retailers |
|---|---|---|
| Period 1 | $\Pi_{i1} = w_i x_{i1}$ | $\pi_{i1} = (p_{i1} - w_i)x_{i1}$ |
| Period 2 | $\Pi_{i2} = w_i x_{i2} - r_i p_{i2}(g x_{i1})$ | $\pi_{i2} = (p_{i2} - w_i)x_{i2}$ |
| Total profits | $\Pi_i = \Pi_{i1} + \Pi_{i2}$ | $\pi_i = \pi_{i1} + \pi_{i2}$ |

We assume that the rebate decisions do not affect the retailers' inventory levels to avoid any influence of inventory hoarding on the effectiveness of the rebate program (Ault et al. 2000). For tractability, we also assume that the marginal production costs for both manufacturers are normalized to zero.

When manufacturers' rebates are offered, the profit functions in each period and overall are given in Table 2. In period 1, each manufacturer distributes the rebates. We assume that they do so at no cost for simplicity and without loss of generality. Manufacturers only pay the rebates to consumers who redeem in period 2, which leads to a cost of rebates equal to ($r_i p_{i2}(g x_{i1})$) for each manufacturer $i$. We assume that firms have the same discount rate than consumers, which is set to one. Finally, we get the total profit of each manufacturer ($\Pi_i$) and retailer ($\pi_i$) over both periods of purchasing by summing up profits in periods 1 and 2.

## 3 Equilibrium Solutions

To assess the effects of rebates on the distribution channel firms' profits, we solve the model for three games. In each game and in each period, manufacturers are dummy players and the retailers play Nash.

- Game $NR$: This is the benchmark game where no rebates are offered ($r_a = r_b = 0$). First, the retailers play Nash and set their first-period prices simultaneously. Then, they play Nash again in the second period and decide simultaneously of their prices given their first-period prices.
- Game $SR$: This is the symmetric rebate game where both manufacturers offer rebates ($r_a, r_b \neq 0$). First, the retailers play Nash and decide simultaneously of their prices in the first period given the announced rebate rates. Then, retailers play Nash again and set their prices simultaneously in the second period given their first-period prices and the manufacturers rebate rates.
- Game $AR$: This is the asymmetric rebate game where manufacturer $a$ offers a rebate while manufacturer $b$ does not ($r_a \neq 0, r_b = 0$). First, the retailers play Nash and decide simultaneously of their prices in the first period given the announced rebate rate. Then, retailers play Nash to set their prices simultaneously in the second period given their first-period prices and manufacturer $a$'s rebate rate.

Since the game is played over two periods, the subgame perfect equilibrium is obtained by first solving the second-period problem then the first-period problem given the second-period strategies. For simplicity, we only present results in case manufacturers are symmetric, i.e., for $v_i = v$ and $w_i = w$, and for $t = 1$. The equilibrium solutions are summarized in the next three propositions.

**Proposition 1** *Assuming an interior solution, in game SR where both manufacturers offer rebates to consumers, the unique equilibrium prices, market shares, and profits for product i* $(i, k \in \{a, b\}, i \neq k)$ *in period j, for* $r_i = r$, *are:*

$$p_{i1}^{SR} = \frac{3gr + 5g - 3}{gr + 3g - 3} + w,\ p_{i2}^{SR} = \frac{1}{1 - gr - g} + w,$$

$$x_{ij}^{SR} = \frac{1}{2},\ \Pi_{i1}^{SR} = \frac{w}{2},$$

$$\Pi_{i2}^{SR} = \frac{(gr - 1)(g + gr - 1)w - gr}{2(1 - gr - g)},$$

$$\Pi_{i}^{SR} = \frac{(gr - 2)(g + gr - 1)w - gr}{2(1 - gr - g)},$$

$$\pi_{i1}^{SR} = \frac{3gr + 5g - 3}{2(gr + 3g - 3)},\ \pi_{i2}^{SR} = \frac{1}{2(1 - gr - g)},$$

$$\pi_{i}^{SR} = \frac{(g - 1)(5g - 6) + rg(8g + 3gr - 7)}{2(gr + 3g - 3)(gr + g - 1)}.$$

*Proof* To get a subgame perfect solution, we solve by backwards induction. The first-order equilibrium conditions for the retailers' problem in period 2 are $\frac{\partial \pi_{a2}}{\partial p_{a2}} = \frac{\partial \pi_{b2}}{\partial p_{b2}} = 0$. Each retailer's Hessian is semi-definite negative for $(gr_i + g - 1) < 0$, $i = a, b$. This condition is satisfied for $r_i < (1 - g)/g$ for all $g < 1$. Assuming this condition to be true, from $\frac{\partial \pi_{i2}}{\partial p_{i2}} = 0$, the following reaction functions can be derived for $i, k \in \{a, b\}, i \neq k$ :

$$p_{i2} = \frac{g[p_{i1} - p_{k1} + w(2r_i + r_k + 1)] - 3(w + 1)}{3(gr_i + g - 1)}. \tag{1}$$

Next, we replace $p_{i2}$ by their expressions in (1) in each firm's market share in period 1. The first-order conditions of the retailers' problem in period 1 are $\frac{\partial \pi_a}{\partial p_{a1}} = \frac{\partial \pi_b}{\partial p_{b1}} = 0$. Each retailer's Hessian is semi-definite negative iff:

$$-\frac{r_k g^2 (g + 3r_i) - (1 - g)\left[\left(9g + g^2 - 9\right) + 6g(r_i + r_k)\right]}{9(gr_a + g - 1)(gr_b + g - 1)} < 0.$$

Assuming this condition to be true, from $\frac{\partial \pi_i}{\partial p_{i1}} = 0$, we obtain the retailers' prices in period 1 as functions of the rebate rates $r_a$ and $r_b$. These expressions are very long, so we omit them here for simplicity. The symmetric solution in proposition 1 is obtained by setting $r_i = r$. ■

The equilibrium solution in the symmetric case shows that the retail price in the first period decreases with the rebate rate while the price in the second period increases. Note also that the feasible domain for this equilibrium solution exists only for low levels of the redemption rate ($g$) such as $g < 0.5$. This means that the equilibrium prices, profits, and margins in each period are positive and that the second-order conditions are verified when $g$ is in this range.

**Proposition 2** *Assuming an interior solution, in game $AR$ where manufacturer a offers a rebate while manufacturer b does not, the unique equilibrium prices, market shares, and profits for each product i ($i, k \in \{a, b\}, i \neq k$) in period j, for $r_a = r$, are:*

*(a) Equilibrium for channel a:*

$$p_{a1}^{AR} = \frac{-D_1 - 6r\left(3g + g^2 - 3\right)(3g + 2gr - 3) + w\,(3g + 2gr - 3)\,D_4}{\left(2g^3r + 4g^3 + 18g^2r + 23g^2 - 18gr - 54g + 27\right)(2gr + 3g - 3)},$$

$$p_{a2}^{AR} = \frac{54g + 24gr - 23g^2 - 4g^3 - 24g^2r - 4g^3r - 27 + wD_5}{\left(2g^3r + 4g^3 + 18g^2r + 23g^2 - 18gr - 54g + 27\right)(gr + g - 1)},$$

$$x_{a1}^{AR} = \frac{D_2 + wr\,(g - 1)\,(3g + 2gr - 3)^2}{2\left(2g^3r + 4g^3 + 18g^2r + 23g^2 - 18gr - 54g + 27\right)(gr + g - 1)},$$

$$x_{a2}^{AR} = \frac{(g - 1)\left[\left(27g + 4g^2 - 27\right) + wrg\,(3g + 2gr - 3)\right] + 4gr\left(6g + g^2 - 6\right)}{2\left(2g^3r + 4g^3 + 18g^2r + 23g^2 - 18gr - 54g + 27\right)},$$

$$\Pi_{a1}^{AR} = \frac{wD_2 + w^2r\,(g - 1)\,(3g + 2gr - 3)^2}{2\,(gr + g - 1)\left(2g^3r + 4g^3 + 18g^2r + 23g^2 - 18gr - 54g + 27\right)},$$

$$\pi_{ra2}^{AR} = \frac{\left[(g - 1)\left(\left(27g + 4g^2 - 27\right) + wrg\,(3g + 2gr - 3)\right) + 4gr\left(6g + g^2 - 6\right)\right]^2}{2\,(1 - gr - g)\left(2g^3r + 4g^3 + 18g^2r + 23g^2 - 18gr - 54g + 27\right)^2}.$$

*(b) Equilibrium for channel b:*

$$p_{b1}^{AR} = -\frac{D_1 + w\,(3g + 2gr - 3)\,D_3}{\left(2g^3r + 4g^3 + 18g^2r + 23g^2 - 18gr - 54g + 27\right)(2gr + 3g - 3)},$$

$$p_{b2}^{AR} = w + \frac{gr\,(2\,gr + 3\,g - 3)\,w - 4\,g^2 - 12\,gr - 27\,g + 27}{2g^3r + 4g^3 + 18g^2r + 23g^2 - 18gr - 54g + 27},$$

$$x_{b1}^{AR} = \frac{(1 - g)\left[(3g + 2gr - 3)^2\,rw + D_6\right]}{2\,(gr + g - 1)\left(2g^3r + 4g^3 + 18g^2r + 23g^2 - 18gr - 54g + 27\right)},$$

$$x_{b2}^{AR} = \frac{(1 - g)\left[wrg\,(3g + 2gr - 3) - 4g^2 - 12gr - 27g + 27\right]}{2\left(2g^3r + 4g^3 + 18g^2r + 23g^2 - 18gr - 54g + 27\right)},$$

$$\pi_{b2}^{AR} = \frac{(1-g)\left[wrg\,(3g+2gr-3) - 4g^2 - 12gr - 27g + 27\right]^2}{2\left(2g^3r + 4g^3 + 18g^2r + 23g^2 - 18gr - 54g + 27\right)^2},$$

$$\Pi_{b1}^{AR} = wx_{b1}^{AR}, \Pi_{b2}^{AR} = wx_{b2}^{AR}, \Pi_{b}^{AR} = w(x_{b1}^{AR} + x_{b2}^{AR}).$$

*where*

$$D_1 = 339g^2 - 27r - 81gr - 297g - 103g^3 - 20g^4 + 18gr^2 + 315g^2r$$
$$-189g^3r - 26g^4r + 36g^2r^2 - 72g^3r^2 - 6g^4r^2 + 81,$$
$$D_2 = 81g - 9r + 63gr - 77g^2 + 19g^3 + 4g^4 + 6gr^2 - 101g^2r + 41g^3r$$
$$+6g^4r - 24g^2r^2 + 20g^3r^2 + 2g^4r^2 - 27,$$
$$D_3 = 54g + 9r - 23g^2 - 4g^3 - 6gr^2 - 11g^2r + 6g^2r^2 + 2g^3r^2 - 27,$$
$$D_4 = 2r\left(7g^2 - 6gr - 18g + 2g^3 + 6g^2r + g^3r + 9\right) - D_3,$$
$$D_5 = D_2 + r\left(17g^2 - 6gr - 21g - 5g^3 + 8g^2r - 4g^3r + 9\right),$$
$$D_6 = 2rg\left(16g + 12gr + 3g^2 + 2g^2r - 18\right) - D_3.$$

*The remaining profit expressions are very long so we omit them here for simplicity.*

*Proof* To get a subgame perfect solution, we solve by backwards induction. The first-order equilibrium conditions for the retailers' problem in period 2 are $\frac{\partial \pi_{a2}}{\partial p_{a2}} = \frac{\partial \pi_{b2}}{\partial p_{b2}} = 0$. Retailer $a$ and $b$'s Hessians are semi-definite negative for $(gr + g - 1) < 0$ and $g - 1 < 0$. These conditions are satisfied for $r < (1-g)/g$ for all $g < 1$. Assuming this condition to be true, from $\frac{\partial \pi_{i2}}{\partial p_{i2}} = 0$, the following reaction functions can be derived for $i, k \in \{a, b\}, i \neq k$ :

$$p_{a2} = \frac{g[p_{a1} - p_{b1} + w\,(2r+3)] - 3\,(1+w)}{3\,(gr+g-1)}, \tag{2}$$

$$p_{b2} = \frac{-g[p_{a1} - p_{b1} + w\,(r+3)] - 3\,(1+w)}{3\,(g-1)}. \tag{3}$$

Next, we replace $p_{i2}$ by their expressions in Eqs. (2) and (3) in each firm's market share in period 1. The first-order conditions of the retailers' problem in period 1 are $\frac{\partial \pi_a}{\partial p_{a1}} = \frac{\partial \pi_b}{\partial p_{b1}} = 0$. Both retailers' Hessians are semi-definite negative iff $(-\frac{g^2+6gr+9g-9}{gr+g-1} < 0)$ and $(-\frac{g^3r+g^3+6g^2r+8g^2-6gr-18g+9}{(gr+g-1)(g-1)} < 0)$. Assuming these conditions to be true, from $\frac{\partial \pi_i}{\partial p_{i1}} = 0$, we obtain the retailers' prices in period 1 as functions of the rebate rate $r$. ■

The equilibrium solution in the asymmetric game is very complex. Therefore, we cannot obtain any meaningful analytical insights in this case with regard to sensitivity analyses.

**Proposition 3** *Assuming an interior solution, in game $NR$ where manufacturers do not offer rebates to consumers, the unique equilibrium prices, market shares, and profits for product $i$ ($i, k \in \{a, b\}, i \neq k$) in period $j$, are:*

$$p_{ij}^{NR} = w + 1, \;\; x_{ij}^{NR} = \frac{1}{2},$$
$$\Pi_i^{NR} = 2\Pi_{ij}^{NR} = w, \;\; \pi_i^{NR} = 2\pi_{ij}^{NR} = 1.$$

*Proof* To get a subgame perfect solution, we solve by backwards induction. The first-order equilibrium conditions for the retailers' problem in period 2 are $\frac{\partial \pi_{a2}}{\partial p_{a2}} = \frac{\partial \pi_{b2}}{\partial p_{b2}} = 0$. Each retailer's Hessian is semi-definite negative for all model's parameters. From $\frac{\partial \pi_{i2}}{\partial p_{i2}} = 0$, the following reaction functions can be derived for $i, k \in \{a, b\}, i \neq k$ :

$$p_{i2} = \frac{g\,(p_{i1} - p_{k1}) - 3}{3\,(g-1)} + w. \tag{4}$$

Next, we replace $p_{i2}$ by their expressions in (4) in each firm's market share in period 1. The first-order conditions of the retailers' problem in period 1 are $\frac{\partial \pi_a}{\partial p_{a1}} = \frac{\partial \pi_b}{\partial p_{b1}} = 0$. Each retailer's Hessian is semi-definite negative iff $g - 1 < 0$, which is satisfied for all $g < 1$. From $\frac{\partial \pi_i}{\partial p_{i1}} = 0$, we obtain the unique retailers' equilibrium prices in period 1. ■

Proposition 3 shows that when no rebates are offered, the equilibrium solution is the same in each period. This is an intuitive result because the channel members are faced with the same optimization problem in each period. Note also that when the baseline sales and wholesale prices are symmetric across manufacturers, we obtain a symmetric equilibrium solution in each period and the market share is split in this case between the two products.

## 4 The Effects of Rebates

We assess the effects of rebates on each channel member by comparing the obtained equilibrium profits and strategies in the three games: no rebates ($NR$), symmetric rebates ($SR$), and asymmetric rebate ($AR$). For tractability, we get all results for the case where the firms are symmetric ($v_a = v_b = v$ and $w_a = w_b = w$). For our comparisons to be meaningful, they must be made in a subset of parameters' values ($w$, $r$ and $g$) defined by the intersection of the three sets (feasible parameters domains) that outlines a unique interior equilibrium in the three games. For each game and in each period, we define the feasible domain for the equilibrium solution by the set of constraints on parameters that guarantee the positivity of the obtained prices, margins, demands, and profits and that satisfy the concavity conditions for the retailers' problems.

**Proposition 4** *Assuming interior equilibrium solutions in the no rebate and symmetric rebate games, comparison of equilibrium strategies in each period gives:*

$$p_{i1}^{SR} < p_{i1}^{NR}, p_{i2}^{SR} > p_{i2}^{NR},$$
$$\Pi_{i1}^{SR} = \Pi_{i1}^{NR}, \Pi_{i2}^{SR} < \Pi_{i2}^{NR}, \Pi_{i}^{SR} < \Pi_{i}^{NR},$$
$$\pi_{i1}^{SR} < \pi_{i1}^{NR}, \pi_{i2}^{SR} > \pi_{i2}^{NR}, \pi_{i}^{SR} > \pi_{i}^{NR}.$$

*Proof* Assuming interior equilibrium solutions in games $NR$ and $SR$, comparison of equilibrium prices and profits leads to:

$$p_{i1}^{SR} - p_{i1}^{NR} = \frac{2g\,(r+1)}{gr+3g-3} < 0, p_{i2}^{SR} - p_{i2}^{NR} = \frac{-g\,(r+1)}{gr+g-1} > 0,$$

$$\Pi_{i1}^{SR} - \Pi_{i1}^{NR} = 0,\ \Pi_{i2}^{SR} - \Pi_{i2}^{NR} = \frac{-gr\,(grw+gw-1-w)}{2\,(gr+g-1)} < 0,$$
$$\pi_{i1}^{SR} - \pi_{i1}^{NR} = \frac{g\,(r+1)}{gr+3g-3} < 0, \pi_{i2}^{SR} - \pi_{i2}^{NR} = \frac{-g\,(r+1)}{2\,(gr+g-1)} > 0,$$
$$\pi_{i}^{SR} - \pi_{i}^{NR} = \frac{(gr-g+1)\,(r+1)\,g}{2\,(3g+gr-3)\,(g+gr-1)} > 0.$$

■

This proposition shows that the overall profit of each manufacturer decreases with rebates. Hence, manufacturers should abstain from offering such promotional offers. The manufacturers' loss occurs mainly for two reasons. First, no matter the level of rebate offered to consumers, the symmetry setting of the game makes it impossible for the manufacturers to gain market share. Since their wholesale prices are also fixed, the rebate translates in additional cost and no additional revenue, which ultimately leads to losses. For the retailers, the rebate would be beneficial. This is due to the increase in retail price which leads to higher retail margin in the second period. This additional margin compensates for the lower retail margin in the first period and overall leads to higher total retail profit.

This result is in line with some empirical observations about rebates. For example, Dhar et al. (1996) note that package coupons, while usually redeemed by higher number of consumers than other coupons, may not have a beneficial profit impact. However, it contradicts the results in Arya and Mittendorf (2013) who considered a bilateral monopolistic channel and found that both channel members benefit from the rebate. It also contradicts results in Singh et al. (2008) who found that rebates can enhance the profits of competing firms (without channel effects). The difference in the results could be due to pricing agreements in the channel and asymmetric market structures.

Next, we compare equilibrium solutions obtained in games $NR$ and $AR$. We could not obtain analytical results in this case because the difference in profits is given by a ratio with both the numerator and the denominator being highly non-linear in parameters $g, r$ and $w$. In fact, the numerators in these comparisons are polynomials of degree 8 in these parameters. The next claim summarizes the main results.

*Claim 5* Assuming interior equilibrium solutions, comparison of equilibrium output obtained in the asymmetric rebate ($AR$) and no rebate ($NR$) scenarios gives the following results:

(a) When manufacturer $a$ offers a rebate, his profit can increase or decrease while manufacturer $b$'s profit always decreases,
(b) When manufacturer $a$ offers a rebate, retailer $a$'s profit increases while retailer $b$'s profit decreases.

Claim 5a shows that in the asymmetric game, the manufacturer who unilaterally offers the rebate might benefit from such a strategy while the competing manufacturer always ends up losing. Figure 1 shows an implicit plot of the difference in manufacturer $a$'s profits obtained in games $AR$ and $NR$ with three regions represented as follows. “UF” denotes the region of the parameter domain for which at least one of the interior equilibrium conditions in games $AR$ or $NR$ is not satisfied. “R1” denotes the region where equilibrium solutions are interior and

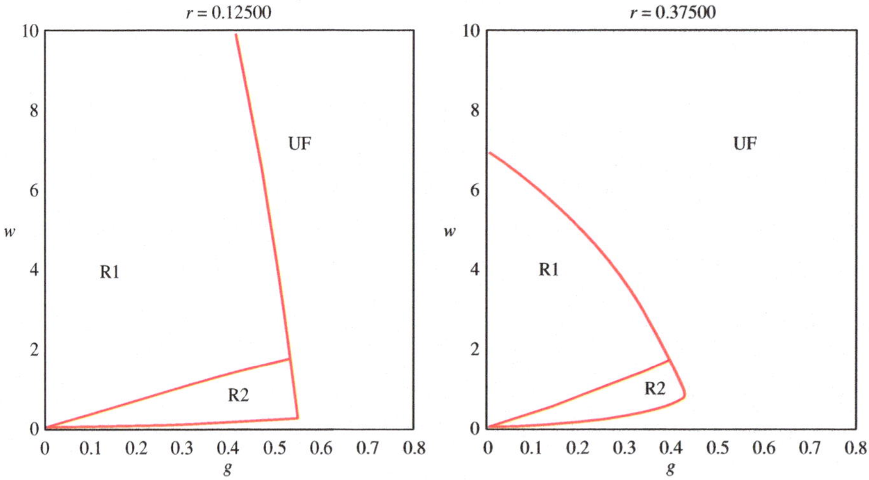


**Fig. 1** Comparison of manufacturer $a$'s profits in the $AR$ and $NR$ games ($\Pi_a^{AR} - \Pi_a^{NR}$)

$\Pi_a^{AR} < \Pi_a^{NR}$. "R2" denotes the region where equilibrium solutions are interior and $\Pi_a^{AR} > \Pi_a^{NR}$.

As we can see, manufacturer $a$ benefits from unilaterally offering a rebate (region R2) when the wholesale price is low. Also, R2 is larger for higher levels of the redemption rates ($g$) in the feasible domain and is smaller when higher rebate rates are offered. Further investigation of retail prices at equilibrium shows that when manufacturer $a$ offers the rebate, the price of its product can increase or decrease in period 1 while it increases in period 2. The retail price of manufacturer $b$'s product decreases in period 1 and increases in period 2. Looking at the effect of an asymmetric rebate on demand (market share), the rebate leads to higher demand for product $a$ in period 1 and can either increase or decrease in period 2 with high levels of $w$, leading to lower demand and vice versa. These changes in demand and prices over both periods lead ultimately to the results reported in Fig. 1. In fact, it shows that the profitability of the rebate for manufacturer $a$ is mainly driven by a demand expansion effect.

Claim 5b explains the effect of the rebate on the retailers' profits. It shows that the retailer whose manufacturer is offering the rebate always gains from such a program while the competing retailer does not. This shows that manufacturers' rebates can offer competitive advantage to retailers.

To interpret the results obtained so far in this section, we consider that the manufacturers play a strategic game where each can choose whether to offer a rebate or not. Table 3 summarizes this game in strategic form.

*Claim 6* Assuming interior equilibrium solutions in the rebate game, both manufacturers do not offer rebates in region R1 of the parameters domain in Fig. 1. In region R2, both manufacturers offer rebates and face a prisoner dilemma.

*Proof* Straightforward from Proposition 4 and Claim 5. ■

Claim 6 shows that whether manufacturers offer rebates or not depends on the values of the parameters; $w, r$ and $g$. In particular, the pricing agreement in the channel plays an important role in driving the profitability of rebates. When the manufacturers have pre-set agreements that dictate high enough wholesale prices, they should not offer rebates to consumers at equilibrium (R1 in Fig. 1). This is especially true if the consumer redemption rate is very low. However, when low wholesale pricing agreements exist in the channel, there is always an incentive for each manufacturer to unilaterally offer the rebate and gain extra profits resulting from demand cannibalization of the competitor's market share. Ultimately, this incentive leads to a prisoner dilemma as both manufacturers end up offering the rebate at equilibrium and consequently gaining lower profits.

**Table 3** Rebate game in strategic form

| | | Manufacturer $a$ | |
|---|---|---|---|
| | | No rebate | Rebate |
| Manufacturer $b$ | No rebate | $(\Pi_a^{NR}, \Pi_b^{NR})$ | $(\Pi_a^{AR}, \Pi_b^{AR})$ |
| | Rebate | $(\Pi_b^{AR}, \Pi_a^{AR})$ | $(\Pi_a^{SR}, \Pi_b^{SR})$ |

# 5 Conclusions

This research investigates the effectiveness of manufacturers' rebates that are given to consumers upon buying the product and can be redeemed on the next purchasing occasion. We develop a two-period game-theoretic model for a two-manufacturer, two-retailer channel. The market is formed by consumers distributed on a Hotelling line and wholesale prices are fixed such as the case for long-term contractual agreements in the channel. We solve three games; a first benchmark scenario where no rebate is offered, a second game where both manufacturers provide rebates to consumers, and a third game where one manufacturer offers a rebate to consumers while the other manufacturer does not.

The findings indicate that whether rebates are offered at equilibrium by competing manufacturers will depend mainly on the wholesale pricing agreement and the consumer redemption rate. In particular, manufacturers should withhold rebates when their wholesale prices are high. Alternatively, low wholesale prices combined with high enough redemption rates lead to a prisoner dilemma situation for the competing manufacturers. In these conditions, each manufacturer benefits from unilaterally offering the rebate. This incentive leads both manufacturers to offer the rebate at equilibrium and consequently lowering their overall profit. For the retailers, the manufacturers' rebate leads to losses in the first period due to shrinking margins, followed by higher margins and profits in the second period. Overall, the retailers benefit from manufacturers' rebates.

This research can be extended in many ways. For tractability, we have made the assumption that the manufacturers and retailers are symmetric. This assumption can be relaxed to include asymmetry which will result in interesting competitive reactions over market share. Further, we studied the case where the market is represented by the Hotelling line. Future research can consider other market configurations. In particular, the case where the total market size is not fixed can lead to different insights as rebates could expand the market potential (see Martín-Herrán and Sigué (2015) for an example of how to model such demand). Finally, we considered a channel where long-standing wholesale pricing agreements are implemented. Alternative pricing scenarios can be discussed in future research.

# References

Arya, A., & Mittendorf, B. (2013). Managing strategic inventories via manufacturer-to-consumer rebates. *Management Science, 59*(4), 813–818.

Ault, R. W., Beard, T. R., Laband, D. N., & Saba, R. P. (2000). Rebates, inventories, and intertemporal price discrimination. *Economic Inquiry, 38*(4), 570–578.

Busse, M., Silva-Risso, J., & Zettelmeyer, F. (2006). $1,000 Cash back: The pass-through of auto manufacturer promotions. *The American Economic Review, 96*, 1253–1270.

Chen, Y., Moorthy, S., & Zhang, Z. J. (2005). Research note—Price discrimination after the purchase: Rebates as state-dependent discounts. *Management Science, 51*(7), 1131–1140.

Dhar, S. K., Morrison, D. G., & Raju, J. S. (1996). The effect of package coupons on brand choice: An epilogue on profits. *Marketing Science, 15*(2), 192–203.

Gilpatric, S. M. (2009). Slippage in rebate programs and present-biased preferences. *Marketing Science, 28*(2), 229–238.

Hotelling, H. (1929). Stability in competition. *The Economic Journal, 39*(153), 41–57.

Khouja, M., & Zhou, J. (2010). The effect of delayed incentives on supply chain profits and consumer surplus. *Production and Operations Management, 19*, 172–197.

Liang, D., Li, G., Sun, L., & Chen, Y. (2013). The role of rebates in the hybrid competition between a national brand and a private label with present-biased preferences. *International Journal of Production Economics, 145*(1), 208–219.

Liu, Q., & Zhang, D. (2013). Dynamic pricing competition with strategic customers under vertical product differentiation. *Management Science, 59*(1), 84–101.

Lu, Q., & Moorthy, S. (2007). Coupons versus rebates. *Marketing Science, 26*(1), 67–82.

Martín-Herrán, G., & Sigué, S. P. (2011). Prices, promotions, and channel profitability: Was the conventional wisdom mistaken? *European Journal of Operational Research, 211*(2), 415–425.

Martín-Herrán, G., & Sigué, S. P. (2015). Trade deals and/or on-package coupons. *European Journal of Operational Research, 241*(2), 541–554.

Martín-Herrán, G., Sigué, S. P., & Zaccour, G. (2010). The dilemma of pull and push-price promotions. *Journal of Retailing, 86*(1), 51–68.

Nijs, V. R., Dekimpe, M. G., Steenkamps, J. E. M., & Hanssens, D. M. (2001). The category-demand effects of price promotions. *Marketing Science, 20*(1), 1–22.

Sigué, S. P. (2008). Consumer and retailer promotions: Who is better off? *Journal of Retailing, 84*(4), 449–460.

Singh, S. S., Jain, D. C., & Krishnan, T. V. (2008). Research note-Customer loyalty programs: Are they profitable? *Management Science, 54*(6), 1205–1211.

Vickers, J., & Armstrong, M. (2001). Competitive price and discrimination. *The Rand Journal of Economics, 32*(4), 579–593.

Yang, S., Munson, C. L., & Chen, B. (2010). Using MSRP to enhance the ability of rebates to control distribution channels. *European Journal of Operational Research, 205*(1), 127–135.

# On the Coordination of Static and Dynamic Marketing Channels in a Duopoly with Advertising

**Luca Lambertini**

**Abstract** A *leitmotiv* of the analysis of marketing channels' behaviour is the possibility of designing contractual relations so as to replicate the performance of vertically integrated firms, whenever this is efficient for firms. This is particularly relevant when the vertical externality provokes distortions in the firms' incentives to invest in R&D or advertising. The present model illustrates the possibility of using two-part tariffs endogenously defined as linear functions of firms' efforts to sterilize the vertical externality altogether in a duopoly where firms' invest in advertising to increase brand equity. This is done first in a static model and then replicated in the differential game based upon the same building blocks.

**Keywords** Supply chain · Vertical relations · Vertical integration · Advertising

## 1 Introduction

The structure of a firm or a supply chain has been traditionally characterized as a nexus of contracts, in the literature belonging to both industrial economics (Williamson 1971; Grossman and Hart 1986) and business and management (Zusman and Etgar 1981). This approach has produced its major efforts to understand how to correct the distortion created by the hold-up problem affecting investment incentives under vertical separation, either in R&D (Grout 1984; Rogerson 1992; MacLeod and Malcomson 1993) or in advertising (Jeuland and Shugan 1983, 1988; Moorthy 1987; Ingene and Parry 1995). Recent extensions rely on the adoption of

L. Lambertini (✉)
Department of Economics, University of Bologna, Bologna, Italy
e-mail: luca.lambertini@unibo.it



P.-O. Pineau et al. (eds.), *Games in Management Science*,
International Series in Operations Research & Management Science 280,
https://doi.org/10.1007/978-3-030-19107-8_4

revenue- or cost-sharing contracts (Giannoccaro and Pontrandolfo 2004; Cachon and Lariviere 2005; Leng and Parlar 2010).[1]

The backbone of the analysis of supply chains' strategic interplay in oligopolistic models is in McGuire and Staelin (1983), Coughlan (1985) and Bonanno and Vickers (1988). An essential feature of the model investigated in their contributions is that industry-wide vertical integration (1) may not be the unique equilibrium, and (2) may not be Pareto-efficient from the firms' viewpoint. In particular, even if unique, the pure vertically integrated industry structure can be the outcome of a prisoner's dilemma when goods are not sufficiently differentiated, in which case vertical separation helps sustaining prices compensating the low degree of product differentiation with double marginalization. If vertical integration is efficient for firms (or, at least, for suppliers), its performance can be replicated via two-part tariffs (TPTs).

While the early analysis of supply chains (and their coordination) has focussed on setups involving the choice of wholesale and market prices, the ensuing literature has frequently discussed the bearings of other strategic variables, such as, most frequently, advertising activities. This has been done with elastic market demands (from Lee and Staelin (1997) and Kim and Staelin (1999), onwards) and inelastic market demands typically associated with spatial competition models (see Wang et al. (2011) and Karray (2015), *inter alia*).

The present contribution is related with the optimal design of two-part tariffs (TPTs) to coordinate supply chains operating in oligopolistic markets in which elastic demand functions can be shifted outwards by firms via advertising efforts, as in the model by Jeuland and Shugan (1983) and its follow-ups, including, in particular, those of Zaccour (2008) and Lambertini (2014, 2018), where a single supply chain operates in the market, with firms or divisions investing in advertising to increase brand equity, through an outward shift of demand. The aim of the ensuing analysis is to extend the results illustrated in these two papers to a scenario in which competition is present both upstream and downstream. Indeed, it is shown that, provided the vertically integrated outcome is efficient for firms, the adoption of a TPT involving a wholesale price equal to marginal cost and a fixed fee defined as a linear function of firms' advertising effort allows a vertically separated supply chain to replicate the performance of a vertically integrated firm. This holds in the static and the differential game as well. Additionally, although the model is restricted to duopolistic competition, this finding intuitively extends to the oligopoly case, with a generic number of marketing channels supplying the market.

The remainder of the paper is structured as follows. The building blocks of the model and the static game are illustrated in Sect. 2. Section 3 contains a compact analysis of the differential game. Concluding remarks are in Sect. 3.

[1]Exhaustive accounts of the large debate on this matter can be found in Cachon (2003), Ingene and Parry (2004), Nagarajan and Sošić (2008) and Ingene et al. (2012).

## 2 The Static Game

The model describes a market served by two supply chains, in such a way that there exist (at most) four firms, two upstream suppliers (manufacturers) and two downstream sellers (retailers). Each supply chain identified by $i = 1, 2$ offers a differentiated variety of the same good, whose inverse demand system is defined by

$$p_i = a_i - q_i - sq_j \tag{1}$$

in which $a_i$ is the representative consumer's reservation (or choke) price, and $s \in (0, 1]$ measures the degree of product substitutability, as in Singh and Vives (1984). The wholesale price along supply chain $i$ is $w_i$, and the marginal and average production cost is $c \in (0, a_i)$, the same in both supply chains. The demand for variety $i$ can be shifted up by advertising investments carried out by the firms (under vertical separation) or divisions of the same firms (under vertical integration), in such a way that

$$a_i = a_0 + k_{iU} + k_{iD} \tag{2}$$

where $k_{iJ}$, $J = D, U$ is the effort exerted by each division or firm along supply chain $i$. Hence, as in Zaccour (2008) and Lambertini (2014), (2) describes a scenario in which advertising (if any) enhances brand equity by increasing the representative consumer's willingness to pay, as measured by the choke price; or, equivalently, by shifting the demand function outwards, leaving the slope and the degree of product differentiation unchanged. As in Zaccour (2008), Lambertini (2014) and Lambertini and Zaccour (2015), the associated cost is $C(k_{iJ}) = bk_{iJ}^2$, with $b > 0$, to guarantee concavity and account for the presence of decreasing returns in the advertising activity.[2] Leaving aside for the moment the role of TPTs, the profit functions of the firms/divisions are defined as follows:

$$\begin{aligned} \pi_{iU} &= (w_i - c)\, q_i - bk_{iU}^2 \\ \pi_{iD} &= (p_i - w_i)\, q_i - bk_{iD}^2 \end{aligned} \tag{3}$$

so that, under vertical integration, the profits of firm $i$ are

$$\Pi_i = (p_i - c)\, q_i - b\left(k_{iU}^2 + k_{iD}^2\right) \tag{4}$$

[2]The same quadratic functional form is also adopted in several other models discussing advertising campaigns (see Chu and Desai 1995; Jørgensen et al. 2001; Karray 2015, among many others). Of course, a linear advertising cost coupled with a concave impact (usually, the square root of the effort) also ensures concavity while at the same time accounting for decreasing returns (see Kim and Staelin 1999; Karray and Zaccour 2006).

Strategic variables are advertising efforts, the wholesale price (under vertical separation) and the output level.[3] Upstream (resp., downstream) firms or divisions choose $w_i$ and $k_{iU}$ (resp., $q_i$ and $k_{iD}$). There exist three possible scenarios: (1) both channels are vertically integrated; (2) both channels are vertically separated; (3) one channel is vertically integrated while the other is vertically separated. In the first case, the game has two-stage structure: firms choose advertising efforts in the first stage, and then compete in output levels in the second. In the second case, the game consists of four stages: upstream firms set their own advertising efforts in the first stage and wholesale prices in the second; then, downstream firms choose their advertising efforts in the third stage and compete in outputs in the fourth. The third case also has four stages, except that in the second there is a single player (the upstream firm along the vertically separated channel) setting its wholesale price in isolation. Independently of the vertical structure of both channels, the solution concept is, as usual, subgame perfection attained through backward induction.

## 2.1 Vertical Integration

This case has the structure of a two-stage game. In the first, the two firms noncooperatively and simultaneously choose their respective advertising efforts,[4] in the second they compete on the market place *à la* Cournot-Nash. The first order condition (FOC) for the maximization of (4) w.r.t. $q_i$ yields the coordinates of the downstream equilibrium for any vector of efforts $\{k_{iU}, k_{iD}\}$:

$$q_i^{II} = \frac{(a_0 - c)(2 - s) + 2(k_{iU} + k_{iD}) - s\left(k_{jU} + k_{jD}\right)}{(2 + s)(2 - s)} \tag{5}$$

where superscript $II$ indicates that firms are vertically integrated. The above pair of outputs must be plugged into the profit functions which are relevant at the first stage, with each of the four divisions choosing the following symmetric advertising effort:

$$k^{II} = \frac{2(a_0 - c)}{b(2 - s)(2 + s)^2 - 4} > 0 \,\forall\, b > \frac{4}{(2 - s)(2 + s)^2} \tag{6}$$

with the same condition on $b$ also ensuring $q_i^{II} > 0$ and concavity. At the subgame perfect equilibrium,

[3] I have intentionally assumed quantity to be the relevant market variable, for two unrelated reasons. The first is that the resulting expressions are slightly more compact, while the second is that the formal structure and properties of the model, in particular, the emergence of a prisoner's dilemma, do not depend on whether retailers (or downstream divisions) are price or quantity setters.

[4] Profit functions (3) are additively separable in the vector of advertising efforts. Therefore, the simultaneous and sequential choice of $k_{iU}$ and $k_{iD}$ yield the same equilibrium outcome.

$$a^{II} = a_0 + \frac{4(a_0 - c)}{b(2-s)(2+s)^2 - 4} \tag{7}$$

and

$$\Pi^{II} = \frac{b(a_0 - c)^2\left[b\left(4 - s^2\right)^2 - 8\right]}{\left[b(2-s)(2+s)^2 - 4\right]^2} > 0 \, \forall b > \frac{8}{\left(4 - s^2\right)^2} \tag{8}$$

Since

$$\frac{8}{\left(4 - s^2\right)^2} > \frac{4}{(2-s)(2+s)^2} \tag{9}$$

for all $s \in (0, 1]$, the solution of the vertically integrated case is summarized in the following

**Proposition 1** *Provided* $b > 8/\left(4 - s^2\right)^2$, *under industry-wide vertical integration the vector of profit-maximizing advertising efforts is identified by* $k^{II} = 2(a_0 - c)/\left[b(2-s)(2+s)^2 - 4\right]$.

Provided this outcome is the most efficient one for both marketing channels (which, as we shall see below, may not be true), the task the ensuing analysis has to perform is to outline the structure of at least one contract allowing the vertically separated channel to perform as a vertically integrated entity.[5]

## 2.2 *Vertical Separation*

Here we examine the scenario in which both supply chains are vertically separated, bypassing the detailed exposition of double marginalization for the sake of brevity. If both supply chains aim at eliminating the vertical externality, they have to design TPTs so as to replicate the vertically integrated outcome in full. However, unlike what happens when a single marketing channel supplies the market, if the latter is oligopolistic the presence of double marginalization might mean good news from the firms' standpoint. This fact is known since McGuire and Staelin (1983), Coughlan (1985) and Bonanno and Vickers (1988), and can be quickly sketched here on the basis of wholesale prices and outputs only, leaving advertising efforts out of the picture for a moment.

[5] Additionally, a few words suffice to stress an aspect which is usually left aside: the presence of decreasing returns to advertising, embodied in the convex cost functions, implies that it is surely efficient to smooth investments onto divisions (or separated firms) by Jensen's inequality.

In this case, $a_i = a_0$ and the vertically integrated case coincides with the duopoly equilibrium in (Singh and Vives 1984):

$$q^{II} = \frac{a_0 - c}{2 + s}\,;\ \pi^*_{VI} = \frac{(a_0 - c)^2}{(2 + s)^2} \tag{10}$$

When channels are vertically separated, the equilibrium outcome is

$$\begin{aligned} q^{SS} &= \frac{2\,(a_0 - c)}{(4 - s)\,(2 + s)}\,;\ w^{SS} = \frac{a_0\,(2 - s) + 2c}{4 - s} > c \\ \pi^{SS}_{iU} + \pi^{SS}_{iD} \equiv \Pi^{SS} &= \frac{2\,(a_0 - c)^2\,(6 - s^2)}{(4 - s)^2\,(2 + s)^2} \end{aligned} \tag{11}$$

where superscript $SS$ indicates that both channels are vertically separated. A little algebra suffices to verify that $q^{II} > q^{SS}$ for all $s \in (0, 1]$ while $\Pi^{II} \geq \Pi^{SS}$ for all $s \in (0, 2/3]$ and conversely outside this range. The first fact is a direct consequence of double marginalization, which implies that downstream firms sell less than vertically integrated entities, at a higher price. Yet, this output restriction goes along with an increase in price, with $p^{SS} > p^{II}$ for all $s \in (0, 1]$; as soon as substitutability is large enough, the effect of the price increase outweighs the corresponding impact of output restriction, thereby generating a reversal in the profit ranking, in favour of vertical separation. In other words, when differentiation is low, double marginalization is a remedy to the demand externality (or, equivalently, to the intensity of market competition) and makes separation more profitable than integration. This, except for the use of outputs instead of profits in the market stage, replicates what we know from McGuire and Staelin (1983, p. 124) and Bonanno and Vickers (1988, pp. 262–263): the adoption of wholesale prices above marginal cost under vertical separation may represent a quasi-collusive device, which makes the replication of vertical integration no longer desirable.

Resorting to numerical simulations, it can be shown that the same conclusion extends to the case in which firms invest in advertising to increase the choke price, although in such a case the threshold value of $s$ endogenously depends on the steepness of the advertising cost, i.e., on parameter $b$. For the sake of brevity, we may confine ourselves to the equilibrium expressions of outputs and advertising efforts:

$$q^{SS} = \frac{a_0 + k^{SS}_{iU} + k^{SS}_{iD} - w^{SS}}{2 + s} \tag{12}$$

$$k^{SS}_U = \frac{2\,(a_0 - c)\,(4 - s^2)\left[b\,(4 - s^2)\,\Theta - 8\right]}{b^4\,(4 - s^2)\,(4 + s)\,(2 - s)^4\,(2 + s)^5 + 16\,(12 - s^2) + \Omega} \tag{13}$$

$$k^{SS}_D = \frac{4\,(a_0 - c)\left[b\,(4 - s^2)\,\Upsilon - 16\right]}{b^4\,(4 - s^2)\,(4 + s)\,(2 - s)^4\,(2 + s)^5 + 16\,(12 - s^2) + \Omega}$$

where

$$\Theta \equiv 24 + b\left(4 - s^2\right)\left[b\left(4\left(8 - 3s^2\right) + s^4\right) - 24 + s^2\right] \tag{14}$$

$$\Upsilon \equiv 48 + b\left(4 - s^2\right)\left[b\left(4\left(16 - 5s^2\right) + s^4\right) - 48 + s^2\right]$$

$$\Omega \equiv 16b\left(4 - s^2\right)\left[s\left(3s - 2\right) - 44\right] + 2b^2\left(4 - s^2\right)^2\left[480 + s\left(8\left(6 - 5s\right) + s^3\right)\right]$$

$$+2b^3\left(4 - s^2\right)^3\left[288 + s\left(48 - s\left(36 + s\left(1 - s\right)\right)\right)\right] \tag{15}$$

The conditions for concavity and non-negativity of all the relevant magnitudes are omitted for the sake of brevity. However, numerical calculations show that they are systematically met if $b > 8/\left(4 - s^2\right)^2$. In the same parameter range, it turns also out that $k_U^{SS} < k_D^{SS} < k^{II}$, i.e., the presence of double marginalization reduces the channel's incentives to invest, in particular those of the upstream firm.

## 2.3 *The Mixed Case*

The asymmetric case in which channel $i$ is a single integrated firm with two divisions while channel $j$ is vertically separated is, intuitively, more involved, having a four-stage structure with a single agent setting the wholesale price $w_j$ at the second stage—with the unpleasant consequence that the expressions of all equilibrium magnitudes are asymmetric and overlong. This subgame is entirely omitted for brevity, although calculations are available upon request. For later reference, we may label the resulting equilibrium profits as $\Pi^{IS}$ (for the integrated firm) and $\pi_D^{SI} + \pi_U^{SI} \equiv \Pi^{SI}$ (for the separated channel).

## 2.4 *The Upstream Stage*

As in McGuire and Staelin (1983), one may complete the analysis by looking at a pre-play stage taking place in discrete strategies, *integration* ($I$) and *separation* ($S$). This task can be performed from two alternative standpoints, that of supply chains or that of upstream firms (or, manufacturers). The first case is portrayed by Matrix 1.

| | | 2: $I$ | 2: $S$ |
|---|---|---|---|
| 1 | $I$ | $\Pi^{II}, \Pi^{II}$ | $\Pi^{IS}, \Pi^{SI}$ |
| | $S$ | $\Pi^{SI}, \Pi^{IS}$ | $\Pi^{SS}, \Pi^{SS}$ |

**Matrix 1**

Leaving aside the measure of initial market size $(a_0 - c)$, which only exerts a pure scale effect on all profits alike, numerical simulations on parameters $b$ and $s$ accounting for concavity and non-negativity conditions show that $\Pi^{II} > \Pi^{SI}$ and $\Pi^{IS} > \Pi^{SS}$ everywhere, while $\Pi^{II} > \Pi^{SS}$ for all $s \in (0, \widetilde{s})$, while $\Pi^{II} \leq \Pi^{SS}$ for all $s \in [\widetilde{s}, 1]$, as it appears from the relationship between channel profits and product substitutability drawn in Fig. 1.

The critical threshold of substitutability increases in the steepness of the advertising cost function, with $\widetilde{s} \in [0.528, 2/3)$ for $b \in [2, \infty)$.[6] These findings imply the following:

**Proposition 2** *From the supply chains' standpoint, $(I, I)$ is the unique pure-strategy Nash equilibrium of the upstream stage, attained at the intersections of dominant strategies. If product differentiation is high enough, it is also privately efficient. Otherwise, it is the outcome of a prisoner's dilemma.*

The analysis of this case reveals that switching from price to quantity competition makes a difference, generated by the softness of Cournot behaviour as compared to Bertrand's: in McGuire and Staelin (1983, p. 124) the fully integrated outcome is an equilibrium for all degrees of differentiation, while vertical separation emerges as an equilibrium when differentiation is low, the reason being that double marginalization helps suppliers sustain higher prices when differentiation is insufficient.

**Fig. 1** Channel profits and product substitutability

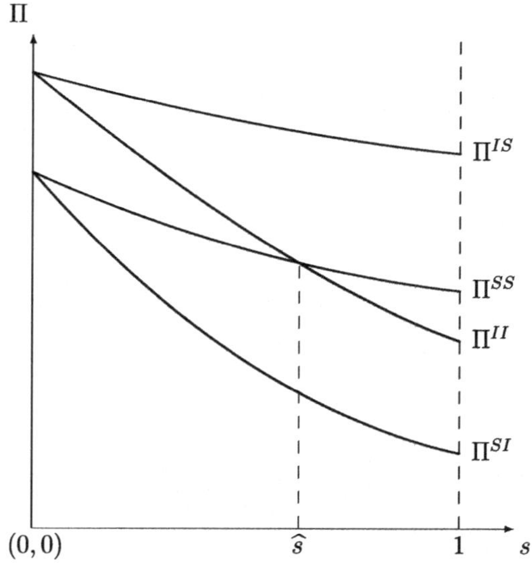


[6] The lower bound to $b$ has been chosen so as to satisfy all concavity and non-negativity requirements. It is also worth noting that the asymptotic value of $\widetilde{s}$ as $b$ tends to infinity coincides with that identified in Sect. 2.2 in absence of advertising. This is due to the fact that all equilibrium advertising efforts are monotonically decreasing in $b$, with $\lim_{b \to \infty} k^{II}, k_U^{SS}, k_D^{SS} = 0$.

The second perspective is portrayed in Matrix 2, where the relevant players are the two upstream firms and therefore profits are either those generated by the entire supply chain when at least one firm is vertically integrated or those of the sole supplier.

| | | $U_2$ | |
|---|---|---|---|
| | | $I$ | $S$ |
| $U_1$ | $I$ | $\Pi^{II}, \Pi^{II}$ | $\Pi^{IS}, \pi_U^{SI}$ |
| | $S$ | $\pi_U^{SI}, \Pi^{IS}$ | $\pi_U^{SS}, \pi_U^{SS}$ |

**Matrix 2**

Here $\Pi^{II} > \pi_U^{SI}$, $\Pi^{IS} > \pi_U^{SS}$ and $\Pi^{II} > \pi_U^{SS}$ for all $b$ and $s$, implying

**Proposition 3** *From the suppliers' standpoint,* $(I, I)$ *is the unique pure-strategy Nash equilibrium of the upstream stage, attained at the intersections of dominant strategies. It is privately efficient for all* $s \in (0, 1]$.

Also in this case, Cournot competition makes a difference (for the aforementioned reason) with respect to the findings illustrated in McGuire and Staelin (1983, p. 123), whereby under Bertrand competition industry-wide vertical integration is indeed the unique equilibrium but the upstream stage equivalent to Matrix 2 is a prisoner's dilemma when product substitutability is high.

## 2.5 *The Two-Part Tariff*

We are now in a position to deal with the case in which industry-wide vertical integration is efficient for firms, in particular, when it is so from the supply chains' standpoint. This happens for $s \in (0, \widetilde{s})$, in which case vertically separated firms would like to replicate the performance of vertically integrated firms through an appropriate contractual design coordinating vertically separated channels. As shown by Zaccour (2008), the classical TPT with $w_i = c$ accompanied by an exogenous fixed fee $F$ cannot do the job. Although this result was obtained in a model envisaging the presence of a single marketing channel, it clearly holds true here as well. Consequently, one may follow Lambertini (2014, 2018) and propose an *endogenous* TPT defined by $w_i = c$ and $F_i = x_i + y_i k_{iU}$, in such a way that the profits of the firms involved in supply chain $i$ write as follows:

$$\begin{aligned} \pi_{iU} &= -bk_{iU}^2 + x_i + y_i k_{iU} \\ \pi_{iD} &= (p_i - c)\, q_i - bk_{iD}^2 - x_i - y_i k_{iU} \end{aligned} \tag{16}$$

where auxiliary variables $x_i$ and $y_i$ are to be determined so as to (1) replicate the performance of a vertically integrated supply chain and (2) assigning profits to firms along the supply chain (thus reflecting the spirit of the original approach to TPTs). Moreover, looking at $\pi_{iU}$, it appears that the reason for adopting such a TPT is that of restoring the concavity of the upstream firm's profit function and therefore also the same firm's incentive to invest in advertising, which would disappear altogether should the traditional TPT be adopted. Last but not least, the effect of this unconventional TPT has some elements in common with the concept of potential function and its arising in potential games (Monderer and Shapley 1996), as here the TPT has the task of reproducing the same FOCs and therefore also the same equilibrium as under vertical integration. We may look at the optimal advertising effort of upstream firms,

$$k_{iU}^{TPT} = \frac{1}{2b}\left[y_i + \frac{4z_i\left(1 - b\left(4 - s^2\right)\right)}{b\left(4 - s^2\right)\left(8 - b\left(4 - s^2\right)^2\right) - 4}\right] \tag{17}$$

to find out that $k_{iU}^{TPT} = k^{II}$ in correspondence of

$$y^* = \frac{4b\left(a_0 - c\right)}{b\left(2 - s\right)\left(2 + s\right)^2 - 4} \tag{18}$$

which also ensures $k_{iD}^{TPT} = k^{II}$. Given that, by construction, equilibrium outputs are defined as in (5), adopting $y^*$ allows firms to replicate also the sales volumes and the total profits of vertically integrated firms with the same cost structure. As anticipated above, the only remaining magnitude, $x_i$, has to be negotiated upon to distribute profits along the channel, as firms' profits at the symmetric subgame perfect equilibrium are

$$\begin{aligned} \pi_U^{TPT} &= \frac{4b\left(a_0 - c\right)^2}{\left[b\left(2 - s\right)\left(2 + s\right)^2 - 4\right]^2} + x \\ \pi_U^{TPT} &= \frac{b\left(a_0 - c\right)^2\left[b\left(4 - s^2\right)^2 - 12\right]}{\left[b\left(2 - s\right)\left(2 + s\right)^2 - 4\right]^2} - x \end{aligned} \tag{19}$$

The analysis carried out in this subsection boils down to

**Proposition 4** *For all $s \in (0, \widetilde{s})$, the vertically separated marketing channel may replicate the performance of the vertically integrated firm through the adoption of a TPT defined as an appropriate function of the upstream firm's advertising effort.*

Of course the same result can be reached along other routes, as already shown in Lambertini (2014, 2018). An obvious one is to make the fixed fee a function of the advertising effort of both firms along the same channel, whereby $F_i = x_i + y_i k_{iU} + z_i k_{iD}$, but the foregoing analysis implies that the presence of the downstream firm's investment is redundant.

## 3 The Differential Game

If the game takes a properly dynamic structure, the building blocks of the model remain the same as in the previous section, except that all the relevant functions evolve over time, choke prices become state variables and the advertising technology (2) must be replaced by state equations. Assume the game unravels over continuous time $t \in [0, \infty)$, and the dynamics of state $i$ is

$$\dot{a}_i(t) = k_{iU}(t) + k_{iD}(t) - \delta a_i(t) \tag{20}$$

where $\delta > 0$ is a time-invariant decay rate common to both states. The above dynamics states that both firms or divisions located along the marketing channel may contribute to the increase in demand (or brand equity) by shifting up the reservation price, and do so with the same effectiveness (since both efforts enter (20) with the same coefficient, which is set equal to one by an appropriate choice of units, and without further loss of generality).

Except for the presence of the time argument, instantaneous demand and profit functions are defined as in (1) and (3)–(4), respectively. Firms (or divisions) noncooperatively maximize their discounted profit flows using the same time-invariant discount rate $\rho$, under the constraint posed by (20).[7]

As in the static setup, we may first take a look at the main features of the vertically integrated case.

### 3.1 Vertical Integration

The Hamilton-Jacobi-Bellman (HJB) equation of firm $i$ is

$$\rho V_i\left(a_i(t), a_j(t)\right) = \max_{q_i(t), k_{iU}(t), k_{iD}(t)} \left\{[p_i(t) - c]\, q_i(t) - b\left[k_{iU}^2(t) + k_{iD}^2(t)\right] + \right.$$

$$\left. V'_{ii}\left(a_i(t), a_j(t)\right)\dot{a}_i(t) + V'_{ij}\left(a_i(t), a_j(t)\right)\dot{a}_j(t)\right\} \tag{21}$$

where $V'_{ij} \equiv \partial V_i\left(a_i(t), a_j(t)\right)/\partial a_j(t)$, $i, j = 1, 2$. From (21), one obtains the following FOCs:

$$a_i(t) - c - 2q_i(t) - s q_j(t) = 0 \tag{22}$$

$$V'_{ii}\left(a_i(t), a_j(t)\right) - 2b k_{iU}(t) = 0 \tag{23}$$

[7]It is worth stressing that, should divisions along the same supply chain cooperate, i.e., maximize joint profits, this would obviously replicate the outcome of the vertically integrated case.

$$V'_{ii}\left(a_i(t), a_j(t)\right) - 2bk_{iD}(t) = 0 \tag{24}$$

which of course imply that the resulting Cournot-Nash equilibrium has a quasi-static nature:

$$q_i^*(t) = \frac{2a_i(t) - sa_j(t) - c(2-s)}{4-s^2} \tag{25}$$

and guessing a linear-quadratic value function (since the game has a linear-quadratic form)

$$V_i\left(a_i(t), a_j(t)\right) = \varepsilon_1 a_i^2(t) + \varepsilon_2 a_j^2(t) + \varepsilon_3 a_i(t) a_j(t) + \varepsilon_4 a_i(t) + \varepsilon_5 a_j(t) + \varepsilon_6 \tag{26}$$

one has that $V'_{ii}\left(a_i(t), a_j(t)\right) = 2\varepsilon_1 a_i(t) + \varepsilon_3 a_j(t) + \varepsilon_4$ and $V'_{ij}\left(a_i(t), a_j(t)\right) = 2\varepsilon_2 a_j(t) + \varepsilon_3 a_i(t) + \varepsilon_5$ and, therefore,

$$k_{iJ}^{VI}(t) = \frac{2\varepsilon_1 a_i(t) + \varepsilon_3 a_j(t) + \varepsilon_4}{2b}; \; J = D, U \tag{27}$$

Then, plugging quantities (25) together with the optimal advertising efforts (27) into the HJB equation (21) and simplifying the latter, one can write the system of six algebraic Riccati equations whose solution identifies the vector of coefficients $\{\varepsilon_1, \varepsilon_2, \varepsilon_3, \varepsilon_4, \varepsilon_5, \varepsilon_6\}$. As it commonly happens (except in special cases, as in Wirl (2010)) in LQ games with more than one state variable, solving the system of Riccati equations may require the use of trigonometric relations between pairs of undetermined coefficients and this, in turn, heavily limits both the interpretation and the application of the solution itself (see, e.g., Reynolds 1987). The source of this problem can be briefly illustrated, as the system of Riccati equations is solvable up to the coefficient $\varepsilon_3$ pertaining to the product of the two states. The sequence of solutions is

$$\varepsilon_6 = \frac{\varepsilon_4(\varepsilon_4 + 2\varepsilon_5) + 2bc^2}{2b\rho(2+s)^2} \tag{28}$$

$$\varepsilon_5 = \frac{2bcs + \varepsilon_4(2\varepsilon_2 + \varepsilon_3)(2-s)(2+s)^2}{(2-s)(2+s)^2[b(\delta+\rho) - 2\varepsilon_1]} \tag{29}$$

$$\varepsilon_4 = \frac{2bc[2b(\delta+\rho) - 4\varepsilon_1 - s\varepsilon_3]}{2\varepsilon_1[2\varepsilon_1 + \varepsilon_3 - 2b(\delta+\rho)] - \varepsilon_3(2\varepsilon_2 + \varepsilon_3) + b(\delta+\rho)[b(\delta+\rho) - \varepsilon_3]} \tag{30}$$

$$\varepsilon_3 = \frac{4bs}{\left(4-s^2\right)^2[2(2\varepsilon_1 + \varepsilon_2) - b(2\delta+\rho)]} \tag{31}$$

Then, the two remaining equations simplify as follows:

$$2a_i^2\left[b\left(4+\varepsilon_1\left(4-s^2\right)^2(2\delta+\rho)\right)+2\left(4-s^2\right)^2\left(\varepsilon_1^2+\frac{bs^2\Lambda}{\left(4-s^2\right)^2}\right)\right] \tag{32}$$

$$a_j^2\left[-8\varepsilon_1\varepsilon_2\left(4-s^2\right)^2+2b\left(\varepsilon_2\left(4-s^2\right)^2(2\delta+\rho)-s^2(1+\Lambda)\right)\right] \tag{33}$$

where

$$\Lambda\equiv\frac{8b}{\left(4-s^2\right)^2\left[2\left(2\varepsilon_1+\varepsilon_2\right)-b\left(2\delta+\rho\right)\right]^2} \tag{34}$$

Now, noting that $\varepsilon_1^4$ and $\varepsilon_2^3$ appear, respectively, in (32) and (33), the analytical solution can be drawn resorting to trigonometric functions as in Reynolds (1987); otherwise, one may revert to numerical methods to solve (21). Either way, the resulting solutions are not easy to interpret or to use.

However, the aim of the present analysis is not that of delivering the full analytical characterization of feedback solutions for the vertically integrated and separated industries, but rather that of showing how to design the TPT in order to ensure that the two solutions coincide. As in the static setup, this shares some essential features with the concept of potential game (see Dragone et al. 2015). With this in mind, we may turn to the vertically separated case.

## 3.2 *Vertical Separation*

If the two marketing channels are vertically separated, the two firms operating along channel $i=1,2$ must solve, respectively:

$$\rho V_{iU}\left(a_i(t),a_j(t)\right)=\max_{w_{iU}(t),k_{iU}(t)}\left\{\left[w_i(t)-c\right]q_i(t)-bk_{iU}^2(t)+\right.$$

$$\left.V'_{iUi}\left(a_i(t),a_j(t)\right)\dot{a}_i(t)+V'_{iUj}\left(a_i(t),a_j(t)\right)\dot{a}_j(t)\right\} \tag{35}$$

$$\rho V_{iD}\left(a_i(t),a_j(t)\right)=\max_{q_{iD}(t),k_{iD}(t)}\left\{\left[p_i(t)-c\right]q_i(t)-bk_{iD}^2(t)+\right.$$

$$\left.V'_{iDi}\left(a_i(t),a_j(t)\right)\dot{a}_i(t)+V'_{iDj}\left(a_i(t),a_j(t)\right)\dot{a}_j(t)\right\} \tag{36}$$

when playing noncooperatively. In (35)–(36), $V'_{iJj}\equiv\partial V_{iJ}\left(a_i(t),a_j(t)\right)/\partial a_j(t)$, with $i,j=1,2$ and $J=D,U$.

If instead they want to replicate the performance of the vertically integrated firms—because it is privately efficient to do so—they may design a contract specifying the two components of the TPT as $w_i(t) = c$ and $F_i(t) = x_i + y_i(t)\, k_{iU}(t) + z_i(t)\, k_{iD}(t)$ and then solve the feedback game by backward induction, starting with the characterization of the solution at the downstream stage, whose FOCs yield the same output levels as in (25) and[8]

$$k_{iD}^{TPT} = \frac{V'_{iDi}\left(a_i, a_j\right) - z_i}{2b} \tag{37}$$

Since the wholesale price is set at marginal cost, the solution of the upstream stage confines to finding the optimal advertising efforts, which correspond to

$$k_{iU}^{TPT} = \frac{V'_{iUi}\left(a_i, a_j\right) - y_i}{2b} \tag{38}$$

Given that optimal outputs (25) are defined in the same way in the two scenarios, the replication of the vertically integrated equilibrium requires $k_{iD}^{TPT} = k_{iU}^{TPT} = k_{iJ}^{VI}$, which in turn implies $V'_{iDi}\left(a_i, a_j\right) - z_i = V'_{iUi}\left(a_i, a_j\right) - y_i = V'_{ii}\left(a_i, a_j\right)$. More explicitly, suppose

$$V_{iD}\left(a_i, a_j\right) = \gamma_1 a_i^2 + \gamma_2 a_j^2 + \gamma_3 a_i a_j + \gamma_4 a_i + \gamma_5 a_j + \gamma_6 \tag{39}$$

$$V_{iU}\left(a_i, a_j\right) = \eta_1 a_i^2 + \eta_2 a_j^2 + \eta_3 a_i a_j + \eta_4 a_i + \eta_5 a_j + \eta_6 \tag{40}$$

so that $V'_{iDi}\left(a_i, a_j\right) = 2\gamma_1 a_i + \gamma_3 a_i + \gamma_4$ and $V'_{iUi}\left(a_i, a_j\right) = 2\eta_1 a_i + \eta_3 a_i + \eta_4$ and therefore (37)–(38) can be rewritten as follows:

$$\begin{aligned} k_{iD}^{TPT} &= \frac{2\gamma_1 a_i + \gamma_3 a_i + \gamma_4 - z_i}{2b} \\ k_{iU}^{TPT} &= \frac{2\eta_1 a_i + \eta_3 a_i + \eta_4 + y_i}{2b} \end{aligned} \tag{41}$$

Now it is immediate to check that

$$k_{iJ}^{VI} - k_{iD}^{TPT} = \frac{2\left(\varepsilon_1 - \gamma_1\right) a_i + \left(\varepsilon_3 - \gamma_3\right) a_j + \varepsilon_4 - \gamma_4 + z_i}{2b} = 0 \tag{42}$$

in $z_i = \gamma_4 - 2\left(\varepsilon_1 - \gamma_1\right) a_i - \left(\varepsilon_3 - \gamma_3\right) a_j - \varepsilon_4$ and

$$k_{iJ}^{VI} - k_{iU}^{TPT} = \frac{2\left(\varepsilon_1 - \eta_1\right) a_i + \left(\varepsilon_3 - \eta_3\right) a_j + \varepsilon_4 - \eta_4 - y_i}{2b} = 0 \tag{43}$$

[8] Henceforth, the time argument will be omitted for the sake of brevity.

in $y_i = 2(\varepsilon_1 - \eta_1) a_i + (\varepsilon_3 - \eta_3) a_j + \varepsilon_4 - \eta_4$.

The analysis carried out in this section can be summarized by formulating the following.

**Proposition 5** *In a dynamic duopoly with vertically separated firms investing in advertising to increase choke prices, the vertically integrated outcome—if privately efficient—can be reproduced through TPTs whose fixed fee is linear in advertising controls.*

Of course, once again the corresponding systems of algebraic Riccati equations generated by the HJB equations of firms along either marketing channel might not be solvable (as is the case here) unless one finds a suitable set of trigonometric functions establishing pairwise relations between undetermined coefficients. However, the foregoing procedure illustrates the existence of a solution (possibly not the only one, as is the case in the static game) to the problem generated by the vertical externality created by double marginalization, irrespective of whether the fully analytical solution of the HJB equations remains out of reach. Moreover, the equivalent of system (42)–(43) can be replicated for a generic oligopoly, i.e., for any number of marketing channels or vertically integrated firms.

## 4 Concluding Remarks

The models illustrated above show that the adoption of TPTs contemplating endogenous fixed fees defined as a function of firms' controls (as here) or states (as in Lambertini 2014) may drive the replication of the vertically integrated outcome also when oligopolistic competition takes place in an industry, although in such settings double marginalization may be welcome on the part of firms when product differentiation is low enough.

Leaving aside the existence of a contract (or a set of alternative contracts) reproducing the vertically integrated equilibrium, two last considerations are in order. On the one hand, the intrinsic limit of the analysis carried out in games like the above one seems not to be the applicability of the remedy to double marginalization and the resulting hold-up problem but, rather, the analytical treatment of the Bellman equations of the feedback game, even in settings defined in a linear-quadratic form, as the one used in this paper. On the other hand, however, the foregoing analysis illustrates a method for designing TPTs able to eliminate (if necessary) the vertical externality which is altogether independent of specific elements, for instance, a given shape of the demand system, which may not be linear. That is, the same procedure intuitively holds as well in case of parabolic (and therefore isoelastic) demand functions, as in Lambertini (2010). Likewise, the assumption of full symmetry could be abandoned to generalize the analysis to firms endowed with asymmetric brand equity stocks or asymmetric advertising technologies (or state equations). Except for the unavoidable additional complication in calculations, the procedure illustrated in Sects. 2 and 3 could be replicated step by step.

**Acknowledgements** The author would like to thank Davide Dragone, Arsen Palestini and two anonymous referees for helpful comments and suggestions. The usual disclaimer applies.

## References

Bonanno, G., & Vickers, J. (1988). Vertical separation. *Journal of Industrial Economics, 36*, 257–265.

Cachon, G. P. (2003). Supply chain coordination with contracts. In S. Graves & T. de Kok (Eds.), *Handbook in operation research and management science: Supply chain management*. Amsterdam: North-Holland.

Cachon, G. P., & Lariviere, M. A. (2005). Supply chain coordination with revenue-sharing contracts: Strengths and limitations. *Management Science, 51*, 30–44.

Chu, W., & Desai, P. S. (1995). Channel coordination mechanisms for customer satisfaction. *Marketing Science, 14*, 343–359.

Coughlan, A. (1985). Competition and cooperation in marketing channel choice: Theory and application. *Marketing Science, 4*, 110–129.

Dragone, D., Lambertini, L., Leitmann, G., & Palestini, A. (2015). Hamiltonian potential functions for differential games. *Automatica, 62*, 134–138.

Giannoccaro, I., & Pontrandolfo, P. (2004). Supply chain coordination by revenue sharing contracts. *International Journal of Production Economics, 89*, 131–139.

Grossman, S., & Hart, O. (1986). The costs and benefits of ownership: A theory of vertical and lateral integration. *Journal of Political Economy, 94*, 691–719.

Grout, P. (1984). Investment and wages in the absence of binding contracts: A Nash bargaining approach. *Econometrica, 52*, 449–460.

Ingene, C. A., & Parry, M. E. (1995). Channel coordination when retailers compete. *Marketing Science, 14*, 360–377.

Ingene, C. A., & Parry, M. E. (2004). *Mathematical models of distribution channels*. Boston, MA: Kluwer.

Ingene, C. A., Taboubi, S., & Zaccour, G. (2012). Game-theoretic coordination mechanisms in distribution channels: Integration and extensions for models without competition. *Journal of Retailing, 88*, 476–496.

Jeuland, A. P., & Shugan, S. M. (1983). Managing channel profits. *Marketing Science, 2*, 239–272.

Jeuland, A. P., & Shugan, S. M. (1988). Reply to: Managing channel profits: Comment. *Marketing Science, 7*, 103–106.

Jørgensen, S., Taboubi, S., & Zaccour, G. (2001). Cooperative advertising in a marketing channel. *Journal of Optimization Theory and Applications, 110*, 145–158.

Karray, S. (2015). Modeling brand advertising with heterogeneous consumer response: Channel implications. *Annals of Operations Research, 233*, 181–199.

Karray, S., & Zaccour, G. (2006). Could co-op advertising be a manufacturer's counterstrategy to store brands? *Journal of Business Research, 59*, 1008–1015.

Kim, S. Y., & Staelin, R. (1999). Manufacturer allowances and retailer pass-through rates in a competitive environment. *Marketing Science, 18*, 59–76.

Lambertini, L. (2010). Oligopoly with hyperbolic demand: A differential game approach. *Journal of Optimization Theory and Applications, 145*, 108–119.

Lambertini, L. (2014). Coordinating static and dynamic supply chains with advertising through two-part tariffs. *Automatica, 50*, 565–569.

Lambertini, L. (2018). Coordinating research and development efforts for quality improvement along a supply chain. *European Journal of Operational Research, 270*, 599–605.
Lambertini, L., & Zaccour, G. (2015). Inverted-U aggregate investment curves in a dynamic game of advertising. *Economics Letters, 132*, 34–38.
Lee, E., & Staelin, R. (1997). Toward the general theory of channel management strategy. *Marketing Science, 1b*, 185–207.
Leng, M., & Parlar, M. (2010). Game-theoretic analyses of decentralized assembly supply chains: Non-cooperative equilibria vs coordination with cost-sharing contracts. *European Journal of Operational Research, 204*, 96–104.
MacLeod, W. B., & Malcomson, J. M. (1993). Investments, holdup, and the form of market contracts. *American Economic Review, 83*, 811–837.
McGuire, T., & Staelin, R. (1983). An industry equilibrium analysis of downstream vertical integration. *Marketing Science, 2*, 161–191.
Monderer, D., & Shapley, L. S. (1996). Potential games. *Games and Economic Behavior, 14*, 124–143.
Moorthy, K. S. (1987). Managing channel profits: Comment. *Marketing Science, 6*, 375–379.
Nagarajan, M., & Sošić, G. (2008). Game-theoretic analysis of cooperation among supply chain agents: Review and extensions. *European Journal of Operational Research, 187*, 719–745.
Reynolds, S. (1987). Preemption and commitment in an infinite horizon model. *International Economic Review, 28*, 69–88.
Rogerson, W. P. (1992). Contractual solutions to the hold-up problem. *Review of Economic Studies, 59*, 777–793.
Singh, N., & Vives, X. (1984). Price and quantity competition in a differentiated duopoly. *RAND Journal of Economics, 15*, 546–554.
Wang, C.-J., Chen, Y.-J., & Wu, C.-C. (2011). Advertising competition and industry channel structure. *Marketing Letters, 22*, 79–99.
Williamson, O. E. (1971). The vertical integration of production: Market failure considerations. *American Economic Review, 61*, 112–123.
Wirl, F. (2010). Dynamic demand and noncompetitive intertemporal output adjustments. *International Journal of Industrial Organization, 28*, 220–229.
Zaccour, G. (2008). On the coordination of dynamic marketing channels and two-part tariffs. *Automatica, 44*, 1233–1239.
Zusman, P., & Etgar, M. (1981). The marketing channel as an equilibrium set of contracts. *Management Science, 27*, 284–302.

# Product Recalls and Channel Pricing

**Olivier Rubel**

**Abstract** We propose a stochastic differential game between a manufacturer and a retailer to investigate how the risk of facing a product recall impacts pricing strategies in marketing channels. By doing so, we analyze whether vendor agreements between manufacturers and retailers, which are signed before any unit is sold, could distort channel profits by aggravating double marginalization. We characterize the equilibrium pricing strategies in closed form for both linear and quadratic costs of recall. We find that the manufacturer and the retailer respond differently to certain clauses of the vendor agreement, but that in equilibrium, such agreements do not distort channel profit, even when costs of recall are quadratic.

**Keywords** Marketing channel · Pricing · Product recall · Stochastic differential games

## 1 Introduction

Product recalls hurt not only manufacturers, but also the retailers that carry the recalled products. To protect themselves against recall costs, retailers often require sellers to sign vendor agreements that define ex ante how manufacturers will reimburse retailers' expenses related to recalls. In the USA for instance, Safeway specifies that "Seller shall be responsible for all reasonable costs and expenses associated with the recall."[1] Other retailers are even more specific and provide a breakdown of recalls costs, which are usually divided into two categories, i.e., cost of goods sold and other administrative and logistical costs. Albertsons companies,

[1] https://suppliers.safeway.com.

O. Rubel (✉)
Graduate School of Management, University of California, Davis, CA, USA
e-mail: orubel@ucdavis.edu



P.-O. Pineau et al. (eds.), *Games in Management Science*,
International Series in Operations Research & Management Science 280,
https://doi.org/10.1007/978-3-030-19107-8_5

for example, define the cost of goods as "number of units × cost," whereas Kroger details that "Cost of goods will be billed using the most recent FCB cost times the number of retail units submitted through the Quick Recall application."[2] Meanwhile, Albertsons companies specify that recall/withdrawal logistics costs include retailer labor fee ($\$40/store$), reclamation fee ($\$0.84/unit$), hazardous material fee ($\$0.84/unit$), and warehouse recovery fee ($\$40/case$).

Even though such contractual agreements are common in practice, there is no study, to the best of our knowledge, that investigates how they would impact channel pricing strategies and most importantly profits. Despite being rich, the literature on product recalls (e.g., Dawar and Pillutla 2000; Roehm and Tybout 2006; Van Heerde et al. 2007; Rubel et al. 2011; Liu and Shankar 2015; Borah and Tellis 2016; Eilert et al. 2017; Rubel 2017) does not provide any insight on how manufacturers and retailers should strategically set prices in marketing channels when product recalls can occur. In particular, it is unclear how contractual agreements demanded by retailers from manufacturers improve or degrade profits.

To investigate such issues, we propose a stochastic differential game between a manufacturer and a retailer. The game is stochastic because the time at which the recall will take place is unknown. In such contexts, the manufacturer sets the wholesale price and the retailer sets the retail price paid by consumers. The proposed model considers two state variables. The first state variable is the number of products that must be recalled when the crisis occurs, as in Rubel (2017), which will determine the logistics costs faced by the manufacturer and the retailer. The second state variable keeps tracks of the financial transfer that will take place between the channel partners with respect to the "cost of goods" clause.

We derive the optimal pricing strategies and value functions in closed forms for both linear and quadratic costs, which allows to obtain several insights. We first find that the manufacturer and the retailer respond differently to clauses in the vendor agreement. For instance, we find that the equilibrium retail price does not vary whether the "cost of goods" clause is included or not in the vendor agreement, whereas the wholesale price does. Furthermore, we find that when the costs of recalls are linear, the optimal pricing strategies are constant over time and do not depend on accumulation of units that would have to be recalled when a crisis happens. Conversely, when the costs of recalls are quadratic, firms should implement dynamic pricing strategies that vary with the number of units that would have to be recalled when a crisis happens. We find, however, that in this case the retailer and the manufacturer could respond differently. Specifically, while the retailer would always increase the retail price as the number of units to be recalled increases, we find that the manufacturer can increase or decrease the wholesale price. Finally, we find that vendor agreements do not alter total channel profits, but serve as a way to allocate costs between the two channel partners. Our results thus add to the literature on product recalls by considering the decentralized channel structure and its impact on pricing strategies when envisioning a product harm crisis.

[2] https://www.thekrogerco.com/.

We also contribute to the differential games literature in marketing (e.g., Jørgensen and Zaccour 2004; Breton et al. 2006; Martin-Herran et al. 2008; Rubel 2013) and in particular on marketing channels (e.g., Jørgensen et al. 2000; Rubel and Zaccour 2007; Zaccour 2008) by investigating the interplay between pricing strategies and uncertain product recalls.

In the next section, we present the model. In Sect. 3, we present our results when costs of recall are linear and in Sect. 4, we provide new insights when costs of recall are quadratic. Finally, we conclude in Sect. 5.

## 2 Model

We consider a dynamic marketing channel comprised of a manufacturer and a retailer. Time $(t)$ is continuous with $t \in [0, T]$, where $T$ is the uncertain time at which the manufacturer's products have to be recalled because of the potential harm products could impose on consumers. Thus $T$ is a random variable that is characterized by the stochastic process $\{\Gamma(t) : t \geq 0\}$ defined as follows:

$$\lim_{dt \to 0} \frac{P[\Gamma(t+dt) = 1 | \Gamma(t) = 0]}{dt} = \chi, \quad \lim_{dt \to 0} \frac{P[\Gamma(t+dt) = 0 | \Gamma(t) = 1]}{dt} = 0, \tag{1}$$

where $\chi$ is the intensity of the jump process governing $T$ (see, e.g., Haurie and Moresino 2006).

Demand for the manufacturer's product $q(p)$ is assumed to be linear such that

$$q(p) = v - \alpha \times p, \tag{2}$$

with $v > 0$, $\alpha > 0$ and where $p$ is the retail price chosen by the retailer. Our first state variable is the volume of units that must be recalled at the time of recall, i.e. $x(T)$, such that its dynamics is

$$\frac{dx}{dt} = q(p) - \delta x(t), \tag{3}$$

with $x(0) = 0$. Similar to Rubel (2017), we note that the total number of units to be recalled is not necessarily equal to cumulative sales due to possible attrition, which we capture through $\delta > 0$. The second state variable that we consider is the financial transfer that the manufacturer will pay to the retailer to cover the costs of goods sold, i.e., $B(T)$, such that

$$\frac{dB}{dt} = q(p) \times w \times \mathbb{E}(\theta(t)), \tag{4}$$

with $B(0) = 0$ and where $w$ is the manufacturer's wholesale price. Furthermore, $\mathbb{E}(\theta(t))$ captures the expected proportion of goods sold at time $t$, i.e., $q(.)$, that will be recalled. Specifically, due to the parameter $\delta$ in (3), not all products sold will be recalled; the volume of goods recalled depends on $\delta$ and the *uncertain* time at which the recall is issued such that $\theta(t) = q(p(t)) \times e^{-\delta(T-t)}$, where $T$ and consequently $T - t$ is a random variable. As a result, $\mathbb{E}(\theta(t)) = \frac{\chi}{\delta+\chi}$, which is equal to 1 when $\delta = 0$, as it should be.

The manufacturer's and retailer's objective functions are defined as

$$\Pi_M(t, x, B) = \mathbb{E}_\Gamma \left\{ \int_0^T e^{-rt} q(p) \times (w - c)dt - (B(T) + \kappa_M x(T))\, e^{-rT} \right\}, \tag{5}$$

and

$$\Pi_R(t, x, B) = \mathbb{E}_\Gamma \left\{ \int_0^T e^{-rt} q(p) \times (p - w)dt - (-B(T) + \kappa_R x(T))\, e^{-rT} \right\}, \tag{6}$$

respectively, where $r$ is the discount rate, $c$ is the manufacturer's marginal cost of production, and $\kappa_i$ is the cost parameter of player $i = \{M, R\}$ related to the recall. Recall that an important clause of vendor agreements pertains to the logistical costs engendered by the recall of $x$ units. Specifically, such clauses specify how much of the retailer's recall costs will be reimbursed by the manufacturer. Formally, such a clause is equivalent to writing $\kappa_M = k_M + \phi k_R$ and $\kappa_R = (1 - \phi)k_R$, where $k_M$ and $k_R$ are the actual costs incurred by both channel partners and where $0 < \phi < 1$ is the share of the retailer's costs that is reimbursed by the manufacturer. For instance, setting $\phi$ to one, i.e., the manufacturer covers all the costs of the retailers, implies that $\kappa_R = 0$.

To solve the stochastic problems faced by the manufacturer and the retailer, we rewrite their respective objective functions as infinite horizon deterministic problems via integration by parts (see, e.g., Haurie and Moresino 2006; Rubel 2013), such that (5) and (6) become

$$\Pi_M(x, B) = \int_0^\infty e^{-(r+\chi)t} \{q(p) \times (w - c) - \chi\, (B + \kappa_M x)\}\, dt, \tag{7}$$

and

$$\Pi_M(x, B) = \int_0^\infty e^{-(r+\chi)t} \{q(p) \times (p - w) - \chi\, (-B + \kappa_R x)\}\, dt, \tag{8}$$

respectively. Assuming that the manufacturer and the retailer play Stackelberg, with the retailer as the follower, we define the retailer's value function as $J_R(x) = \underset{p}{Max}\, \Pi_R$ for any wholesale price $w$ announced by the manufacturer. Consequently, the retailer's Hamilton-Jacobi-Bellman (HJB) equation is

$$(r+\chi)J_R(x,B) = \underset{p}{Max}\left\{q(p)\times(p-w)-\chi\left(-B+\kappa_R x\right)\right.$$
$$\left.+\frac{\partial J_R}{\partial B}\left(q(p)w\frac{\chi}{\delta+\chi}\right)+\frac{\partial J_R}{\partial x}\left(q(p)-\delta x\right)\right\}. \quad (9)$$

Differentiating (9) with respect to $p$ and equating the resulting equation to zero yields that the retailer's optimal pricing strategy is

$$p^* = \frac{v+w\times\alpha}{2\alpha}-\frac{1}{2}\times\left(w\times\frac{\chi}{\delta+\chi}\times\frac{\partial J_R}{\partial B}+\frac{\partial J_R}{\partial x}\right), \quad (10)$$

where $\frac{v+w\times\alpha}{2\alpha}$ is the static retail price and $\frac{1}{2}\times\left(w\times\frac{\chi}{\delta+\chi}\times\frac{\partial J_R}{\partial B}+\frac{\partial J_R}{\partial x}\right)$ is the additional term that adjusts the retail price in anticipation of the product recall. Next, we replace the retailer's pricing strategy (10) in the manufacturer's value function $J_M(x) = \underset{w}{Max}\,\Pi_M$, such that the manufacturer's HJB equation is

$$(r+\chi)J_M(x,B) = \underset{w}{Max}\left\{q(p^*)\times(w-c)-\chi\left(B+\kappa_M x\right)\right.$$
$$\left.+\frac{\partial J_M}{\partial B}\left((v-\alpha p^*)\frac{w\chi}{\delta+\chi}\right)+\frac{\partial J_M}{\partial x}(v-\alpha p^*)\right\}. \quad (11)$$

Differentiating (11) with respect to $w$ and equating the resulting expression to zero yields that

$$w^* = \frac{v+\alpha\left(c+\frac{\partial J_R}{\partial x}-\frac{\partial J_M}{\partial x}\right)+\frac{\chi}{\delta+\chi}\times\left(\frac{\partial J_M}{\partial B}\left(v+\alpha\frac{\partial J_R}{\partial x}\right)+\alpha\frac{\partial J_R}{\partial B}\left(\frac{\partial J_M}{\partial x}-c\right)\right)}{2\alpha\times\left(\frac{\chi}{\delta+\chi}\times\frac{\partial J_R}{\partial B}-1\right)\times\left(\frac{\chi}{\delta+\chi}\frac{\partial J_M}{\partial B}+1\right)}. \quad (12)$$

## 3 Results

To characterize the retailer's value function, we first replace (10) in (8), then replace $w$ by $w^*$ in the resulting expression. Similarly, to characterize the manufacturer's value function, we first replace (10) in (11), then replace $w$ by $w^*$ in the resulting expression. Next, owing to its structure, we note that the game played by the retailer and the manufacturer is of the linear-state variety and thus conjecture that

$$J_M(x,B) = M_1+M_2B+M_3x \quad \text{and} \quad J_R(x,B) = R_1+R_2B+R_3x, \quad (13)$$

where $M_1$, $M_2$, $M_3$, $R_1$, $R_2$, and $R_3$ solve the following system of equations

$$(r+\chi)M_1 =$$

$$-\frac{(\alpha(c(\frac{\chi}{\delta+\chi}R_2-1)+\frac{\chi}{\delta+\chi}M_2R_3+\frac{\chi}{\delta+\chi}M_3(-R_2)+M_3+R_3)+\frac{\chi}{\delta+\chi}M_2v+v)^2}{8\alpha(\frac{\chi}{\delta+\chi}M_2+1)(\frac{\chi}{\delta+\chi}R_2-1)}$$

$$(r+\chi)M_2 = -\chi$$

$$(r+\chi)M_3 = -M_3\delta - \kappa_M\chi$$

$$(r+\chi)R_1 =$$

$$\frac{(\alpha(c(\frac{\chi}{\delta+\chi}R_2-1)+\frac{\chi}{\delta+\chi}M_2R_3+\frac{\chi}{\delta+\chi}M_3(-R_2)+M_3+R_3)+\frac{\chi}{\delta+\chi}M_2v+v)^2}{16\alpha(\frac{\chi}{\delta+\chi}M_2+1)^2}$$

$$(r+\chi)R_2 = \chi$$

$$(r+\chi)R_3 = -R_3\delta - \kappa_R\chi.$$

As a result, we obtain the following proposition.

**Proposition 1** *The manufacturer's and retailer's value functions are*

$$J_M(x,B) = M_1 + M_2B + M_3x \quad \text{and} \quad J_R(x,B) = R_1 + R_2B + R_3x, \tag{14}$$

*with* $M_1 = \frac{(\alpha c(\delta+\rho+\chi)+\alpha\chi(\kappa_M+\kappa_R)-v(\delta+\rho+\chi))^2}{8\alpha(\rho+\chi)(\delta+\rho+\chi)^2}$, $M_2 = -\frac{\chi}{\delta+\chi}$, $M_3 = -\frac{\kappa_M\chi}{\delta+\chi+\rho}$, $R_1 = \frac{M_1}{2}$, $R_2 = \frac{\chi}{\delta+\chi}$ *and* $R_3 = -\frac{\kappa_R\chi}{\delta+\chi+\rho}$.

*As a result, the optimal wholesale and retail prices are*

$$w^* = \frac{(\delta+\chi)(\rho+\chi)((v+\alpha c)(\delta+\rho+\chi)+\alpha\chi(\kappa_M-\kappa_R))}{2\alpha(\delta+\rho+\chi)(\delta(\rho+\chi)+\rho\chi)} \tag{15}$$

*and*

$$p^* = \frac{\alpha c(\delta+\rho+\chi)+\alpha\chi(\kappa_M+\kappa_R)+3v(\delta+\rho+\chi)}{4\alpha(\delta+\rho+\chi)}, \tag{16}$$

*respectively.*

**Corollary 1** *If the vendor agreement were not to include a clause on the cost of goods sold, the manufacturer's and retailer's value functions would be such that* $M_2 = R_2 = 0$*, and as a result, the optimal pricing strategies become*

$$\tilde{w}^* = \frac{(v + \alpha c)(\delta + \rho + \chi) + \alpha\chi(\kappa_M - \kappa_R)}{2\alpha(\delta + \rho + \chi)} \tag{17}$$

*and*

$$\tilde{p}^* = \frac{\alpha c(\delta + \rho + \chi) + \alpha\chi(\kappa_M + \kappa_R) + 3v(\delta + \rho + \chi)}{4\alpha(\delta + \rho + \chi)}. \tag{18}$$

Proposition 1 and its corollary allow us to derive several insights. We first learn that the manufacturer and the retailer respond differently to the clauses of the vendor agreement since the optimal pricing strategies are such that the *optimal* retail price should not change due to the inclusion of the cost of goods clause, whereas the optimal wholesale price should; formally,

$$\tilde{p}^* = p^* \quad \text{whereas} \quad \tilde{w}^* < w^*.$$

From a managerial perspective, these two relationships mean that even though the manufacturer increases the wholesale price in response to the inclusion of the clause on the cost the goods sold, the retailer does not. This result is surprising as economic intuition would have suggested that the retailer should have instead increased the retail price in response to a higher wholesale price.

The second new insight that Proposition 1 reveals is that the vendor agreement does not distort the total channel profit $J_D = J_M + J_R$. Instead, it only reallocates profit (and costs) between the two channel partners in case of a recall, as it should be. To see this point, we first consider the impact of clause related to the cost of goods sold on the total channel profit. Specifically,

$$J_D = \frac{3}{2}M_1 - (k_M + k_R)\frac{\chi}{\delta + r + \chi}x,$$

which implies that $\frac{\partial J_D}{\partial B} = 0$. Furthermore, we recall that the second clause that is usually included in vendor agreements pertains to the logistical costs engendered by the recall of $x$ units and how much of the retailer's costs are supported by the manufacturer, i.e., $0 < \phi < 1$. In equilibrium, it is interesting to note that

$$\frac{\partial J_D}{\partial \phi} = \frac{dJ_D}{d\phi} = 0,$$

not only because $\frac{\partial M_1}{\partial \phi} = 0$, but also because $\frac{\partial x}{\partial \phi} = 0$ since the equilibrium price $p^*$ does not vary with $\phi$. Therefore, one important finding from Proposition 1 and its corollary is that vendor agreements *do not distort* channel profit by aggravating double marginalization, a finding that is absent from the extant literature on channel.

## 4 Equilibrium Under Quadratic Recall Costs

We now analyze the case where costs of recall are quadratic in $x$, i.e., $\kappa_M x^2$ and $\kappa_R x^2$. As a result, the retailer's HJB equation, for any $w$, is

$$(r+\chi)J_R(x,B) = \underset{p}{Max}\left\{q(p)\times(p-w)-\chi\left(-B+\kappa_R x^2\right)+\frac{\partial J_R}{\partial B}\frac{dB}{dt}+\frac{\partial J_R}{\partial x}\frac{dx}{dt}\right\}. \quad (19)$$

The first order condition for (19) yields that

$$p^* = \frac{v+w\times\alpha}{2\alpha}-\frac{1}{2}\times\left(w\times\frac{\chi}{\delta+\chi}\times\frac{\partial J_R}{\partial B}+\frac{\partial J_R}{\partial x}\right). \quad (20)$$

We then write the manufacturer's HJB as

$$(r+\chi)J_M(x,B) = \underset{w}{Max}\left\{q(p^*)\times(w-c)-\chi\left(B+\kappa_M x^2\right)+\frac{\partial J_M}{\partial B}\left((v-\alpha p^*)\frac{w\chi}{\delta+\chi}\right)+\frac{\partial J_M}{\partial x}(v-\alpha p^*)\right\}. \quad (21)$$

The first order condition for the manufacturer's optimal wholesale price yields that $w^*$ is identical to (12), as it should be. We then replace the optimal strategies in the HJB equations and note that owing to the linear-quadratic structure of the Stackelberg game, we posit that

$$J_M(x,B) = N_0+N_1x+\frac{N_2}{2}x^2+N_3B \quad \text{and} \quad J_R(x,B) = S_0+S_1x+\frac{S_2}{2}x^2+S_3B, \quad (22)$$

where $(\rho+\chi)S_3=\chi \implies S_3=\frac{\chi}{\chi+\rho}$ and $(\rho+\chi)N_3=-\chi \implies N_3=-\frac{\chi}{\chi+\rho}$, and where the other coefficients solve the system of equations

$$\frac{\chi+\rho}{2}N_2 = \frac{\alpha\left(N_2+S_2\right)^2}{8}-\delta N_2-\kappa_M\chi \quad (23)$$

$$\frac{\chi+\rho}{2}S_2 = \frac{\alpha\left(N_2+S_2\right)^2}{16}-\delta S_2-\kappa_R\chi \quad (24)$$

$$(\chi+\rho)N_1 = \frac{\left(N_2+S_2\right)\left(v+\alpha(N_1+S_1-c)\right)}{4}-N_1\delta \quad (25)$$

$$(\chi+\rho)S_1 = \frac{\left(N_2+S_2\right)\left(v+\alpha(N_1+S_1-c)\right)}{8}-S_1\delta \quad (26)$$

$$(\chi + \rho)N_0 = \frac{(v + \alpha (N_1 + S_1 - c))^2}{8\alpha} \tag{27}$$

$$(\chi + \rho)S_0 == \frac{(v + \alpha (N_1 + S_1 - c))^2}{16\alpha}. \tag{28}$$

**Proposition 2** *Under quadratic costs, the value functions of the manufacturer and the retailers are*

$$J_M(x, B) = N_0 + N_1 x + \frac{N_2}{2}x^2 + N_3 B \quad \text{and} \quad J_R(x, B) = S_0 + S_1 x + \frac{S_2}{2}x^2 + S_3 B, \tag{29}$$

*respectively, with*

$$N_3 = -\frac{\chi}{r + \chi}, \quad S_3 = -N_3, \quad N_2 = \frac{\alpha Y^2 - 8\kappa_M \chi}{4\,(2\delta + r + \chi)}, \quad S_2 = \frac{\alpha Y^2 - 16\kappa_R \chi}{8\,(2\delta + r + \chi)}, \tag{30}$$

$$N_1 = -\frac{2(v - \alpha c)\left(3\alpha Y^2 - 16\chi(\kappa_M + \kappa_R)\right)}{9\alpha^2 Y^2 - 16(4(\delta + r + \chi)(2\delta + r + \chi) + 3\alpha\chi(\kappa_1 + \kappa_2))} \tag{31}$$

$$S_1 = \frac{N1}{2}, \quad N_0 = \frac{\left(v - c\alpha + \alpha\frac{3N_1}{2}\right)^2}{8\alpha(r + \chi)}, \quad S_0 = \frac{N_0}{2}, \tag{32}$$

$$\textit{with} \quad Y = \frac{4\left(2\delta + r + \chi - \sqrt{3(\kappa_M + \kappa_R)\alpha\chi + (2\delta + r + \chi)^2}\right)}{3\alpha}. \tag{33}$$

*As a result, the optimal pricing strategies are*

$$w^* = \frac{(2(v + c\alpha) - N_1\alpha)(r + \chi)}{4\alpha r} + \frac{(S_2 - N_2)(r + \chi)}{2r} \times x \tag{34}$$

$$p^* = \frac{6v + 2c\alpha - 3N_1\alpha}{8\alpha} - \frac{N_2 + S_2}{4} \times x \tag{35}$$

*Proof* Applying the method of undetermined coefficients to identify the values of the coefficients revealed that $Y$ could take two values, i.e.,

$$\frac{4\left(2\delta + r + \chi + \sqrt{3(\kappa_M + \kappa_R)\alpha\chi + (2\delta + r + \chi)^2}\right)}{3\alpha}$$

and

$$\frac{4\left(2\delta + r + \chi - \sqrt{3(\kappa_M + \kappa_R)\alpha\chi + (2\delta + r + \chi)^2}\right)}{3\alpha},$$

We keep the second root to ensure convergence to a stable steady state. ■

Proposition 2 displays how the manufacturer and the retailer should dynamically vary their prices based on the volume of units that should be recalled in the event of a crisis. Such pricing strategies depart from those obtained in Proposition 1 where optimal pricing strategies were constant. Given the feedback structure of the pricing strategies in Proposition 2, we learn that the manufacturer and the retailer could respond differently to changes in $x$. Specifically, it is apparent from Proposition 2 that the optimal retail price always increases as $x$ increases since $N_2 + S_2 < 0$, i.e., $\frac{\partial p^*}{\partial x} > 0$. Conversely, however, the optimal wholesale price can either decrease or increase as $x$ increases, since

$$S_2 - N_2 = \frac{4\left(3\alpha\chi(\kappa_M - 2\kappa_R) - (2\delta + r + \chi)^2\right)}{9\alpha(2\delta + r + \chi)} + \frac{4\sqrt{(2\delta + r + \chi)^2 + 3\alpha\chi(\kappa_M + \kappa_R)}}{9\alpha},$$

which could be positive or negative depending on parameters' values.[3]

Finally, Proposition 2 allows us to confirm that even when recall costs are quadratic in the number of units, the vendor agreement does not distort the total channel profit $J_D = J_M + J_R$, since $\frac{\partial J_D}{\partial B} = \frac{\partial J_D}{\partial \phi} = 0$.

## 5 Concluding Remarks

The proposed model informs how a manufacturer and a retailer should strategically price a product that is likely to be recalled at a future, yet uncertain, date. The proposed model allows us to analyze the role of vendor agreements on pricing strategies and most importantly profits. We find that the manufacturer and the retailer respond differently to the clauses found in such agreements. For instance, when recall costs are linear in the number of units that have to be recalled, we find that the retailer does not vary the retail price depending on whether the clause on cost of goods sold is included or not in the agreement, while the manufacturer does. When recall costs are quadratic, we find that the retailer always increases the retail price, while the manufacturer can either decrease or increase the wholesale price depending on parameters' values. Meanwhile, we are also able to conclude, based on the proposed model, that vendor agreements do not distort channel profits whether costs are linear or quadratic. This result adds to the marketing channel literature (see, e.g., Ingene et al. 2012) by considering how the risk of a major recall might affect optimal pricing strategies by a manufacturer and a retailer.

[3] For instance, when $\chi = 0.001$, $\rho = 0.05$, $\alpha = 1$, $\delta = 0.01$, $\kappa_M = 0.1$ and $\kappa_R = 0.1$, then $S_2 - N_2 < 0$; however, changing $\kappa_M$ from 0.1 to 0.15 yields $S_2 - N_2 > 0$.

## References

Borah, A., & Tellis, G. J. (2016). Halo (Spillover) effects in social media: Do I? *Journal of Marketing Research, 53*, 143–160.

Breton, M., Jarrar, R., & Zaccour, G. (2006). A note on feedback sequential equilibria in a Lanchester model with empirical application. *Management Science, 52*, 804–811.

Dawar, N., & Pillutla, M. M. (2000). Impact of product-harm crises on brand equity: The moderating role of consumer expectations. *Journal of Marketing Research, 37*, 215–226.

Eilert, M., Jayachandran, S., Kalaignanam, K., & Swartz, T. A. (2017). Does it pay to recall your product early? An empirical investigation in the automobile industry. *Journal of Marketing, 81*(3), 111–129.

Haurie, A., & Moresino, F. (2006). A stochastic control model of economic growth with environmental disaster prevention. *Automatica, 42*, 1417–1428.

Ingene, C. A., Taboubi, S., & Zaccour, G. (2012). Game-theoretic coordination mechanisms in distribution channels: Integration and extensions for models without competition. *Journal of Retailing, 88*, 476–496.

Jørgensen, S., Sigue, S. P., & Zaccour, G. (2000). Dynamic cooperative advertising in a channel. *Journal of Retailing, 76*, 71–92.

Jørgensen, S., & Zaccour, G. (2004). *Differential games in marketing*. Dordrecht: Kluwer Academic Publishers.

Liu, Y., & Shankar, V. (2015). The dynamic impact of product-harm crises on brand preference and advertising effectiveness: An empirical analysis of the automobile industry. *Management Science, 61*, 2514–2535.

Martin-Herran, G., Rubel, O., & Zaccour, G. (2008). Competing for consumer's attention. *Automatica, 44*, 361–370.

Roehm, M. L., & Tybout, A. M. (2006). When will a brand scandal spill over, and how should competitors respond? *Journal of Marketing Research, 43*, 366–373.

Rubel, O. (2013). Stochastic competitive entries and dynamic pricing. *European Journal of Operational Research, 231*, 381–392.

Rubel, O. (2017). Profiting from product-harm crises in competitive markets. *European Journal of Operational Research, 265*, 219–227.

Rubel, O., Naik, P. A., & Srinivasan, S. (2011). Optimal advertising when envisioning a product-harm crisis. *Marketing Science, 30*, 1048–1065.

Rubel, O., & Zaccour, G. (2007). *A differential game of a dual distribution channel* (pp. 547–568). Boston: Birkhäuser.

Van Heerde, H., Helsen, K., & Dekimpe, M. G. (2007). The impact of a product-harm crisis on marketing effectiveness. *Marketing Science, 26*, 230–245.

Zaccour, G. (2008). On the coordination of dynamic marketing channels and two-part tariffs. *Automatica, 44*, 1233–1239.

# Coordination in Closed-Loop Supply Chain with Price-Dependent Returns

**Pietro De Giovanni and Talat S. Genc**

**Abstract** This paper proposes two Closed-loop Supply Chain (CLSC) games in which a manufacturer sets some green activity programs efforts and a retailer sets the selling price. Both strategies influence the return rate, which is a state variable. The pricing strategy plays a key role in the identification of the best contract to achieve coordination as well as in achieving environmental objectives. The pricing strategy influences the return rate negatively, as consumers delay the return of their goods when the purchasing (and repurchasing) price is high. We then compare a wholesale price contract (WPC) and a revenue sharing contract (RSC) mechanism as both have interesting pricing policy implications. Our result shows that firms coordinate the CLSC through a (WPC) when the sharing parameter is too low while the negative effect of pricing on returns is too severe. In that case, the low sharing parameter deters the manufacturer to accept any sharing agreements. Further, firms coordinate the CLSC when the sharing parameter is medium independent of the negative impact of pricing on returns. When the sharing parameter is too high the retailer never opts for an RSC. We find that the magnitude of pricing effect on returns determines the contract to be adopted: For certain sharing parameter, firms prefer an RSC when the price effect on return is low and a WPC when this effect is high. In all other cases, firms do not have a consensus on the contract to be adopted and coordination is then not achieved.

**Keywords** Closed-loop supply chain · Dynamic return rate · Coordination · Wholesale price contract · Revenue sharing contract

---

P. De Giovanni (✉)
Department of Operations Management, ESSEC Business School, Paris, France
e-mail: pietro.degiovanni@essec.edu

T. S. Genc
Department of Economics, University of Guelph, Guelph, ON, Canada
e-mail: tgenc@uoguelph.ca



P.-O. Pineau et al. (eds.), *Games in Management Science*,
International Series in Operations Research & Management Science 280,
https://doi.org/10.1007/978-3-030-19107-8_6

# 1 Introduction

A closed loop supply chain (CLSC) integrates forward and backward flows into a unique system, including product acquisition, reverse logistics, points of use and disposal, testing, sorting, refurbishing, recovery, recycling, re-marketing, and re-selling. These activities have to be integrated in a classical forward system (Guide and Van Wassenhove 2009; Fleischmann et al. 2001). The recent trend of closed-loop supply chain (CLSC) has highlighted three main aspects that have the merit to be investigated in such a framework: (1) the return rate is a dynamic phenomenon and should be evaluated accordingly (2) consumers return end-of-life/end-of-use products according to the purchasing price; (3) firms partnering in a CLSC always look to the best contracts to be implemented so as to achieve higher economic and environmental performance (see De Giovanni and Zaccour (2019) for a recent survey).

Firms look for the implementation of a CLSC since the returns have some residual value that contributes to the margins. For instance, producing by means of virgin material is always more expensive than producing by means of returned products. Savings vary according to the industry. In the car engines industry, Volkswagen can save up to 70% of costs (Volkswagen 2011). Kodak saves 40–60% of production costs because it manufacturers by means of returned cameras rather than using raw material (Savaskan et al. 2004). Fleischmann et al. (2003) reported that remanufacturing costs at IBM are much lower than those associated with buying new parts, sometimes as much as 80% lower. Duracell saves 40% of the production costs when producing by means of returned batteries (De Giovanni 2017). Xerox saves 40–65% of its manufacturing costs by reusing parts, components, and materials from returned products (Savaskan et al. 2004). Remanufactured cartridges cost 30–60% less on a per- copy basis than non-remanufactured cartridges. TriNet has been purchasing remanufactured toner cartridges, saving 25–60% in costs over the price of new cartridges within 5 years (www.stopwaste.com). Interface, Inc., is the world's largest provider of commercial carpet tile. To create efficiency in the CLSC, the company has decided to lease carpets instead of selling them; the ownership of off-lease products provides Interface motivations to close the loop and recover the residual value of these products (Agrawal and Tokay 2010). Dell saves 30% of the production cost when recycling their returns (De Giovanni and Zaccour 2018). Manufacturers have high economic interests for performing the backward logistics activities and closing the loop, because the residual value that returns carried out positively contributes to their profits.

Our first contribution is in line with the investigation of the green activity program (GAP) strategy as well as the selling price on returns. The latter represents a novelty with respect to the earlier mentioned literature. The rationale behind this approach is that when the selling price is high, consumers should make considerable sacrifices to purchase it (De Giovanni 2018). Therefore, they delay the product return to exploit the good as much as they can. This intuition fulfills a research gap in the literature of dynamic games, in which the relationship between returns and pricing has been mainly disregarded. Rather, the static literature reports the negative relationships between pricing and returns. For example, De Giovanni and Zaccour (2018) show that a decision maker can be confronted with sophisticated consumers. When introducing a new product in the market, consumers decide whether to return their product and purchase the updated one according to the price: The higher the price of a product, the lower the return rate, because the lower is the consumers' willingness to return their product and purchase a new one. This model is based on two cases, one from the automotive industry and one from the high-tech industry. The findings obtained apply in both cases; therefore, the results in De Giovanni and Zaccour (2018) corroborate our assumptions on the relationship between pricing and returns. Finally, other research like Zhou et al. (2017), Ramani and De Giovanni (2017), and Miao et al. (2017) model a trade-off between pricing and returns.

Following the early intuitions, we consider a benchmark CLSC setup where the manufacturer optimally sets the green efforts and the retailer sets the selling price in a dynamic framework. Notice that the retailer participates in determining the return rate through the pricing strategy, but she does not really exploit its potentiality. Then, we contrast the results of the benchmark (WPC) game with an RSC game in which the retailer transfers a share of her revenues to the manufacturer while paying no wholesale price. The manufacturer can find an RSC very interesting to increase the return rate and enjoy the returns' residual value even more. Further, the manufacturer seeks to exploit the property of the RSC that leads to a price reduction (Cachon and Lariviere 2005).

To recapitulate, we wish to answer the following research questions:

- how do firms in a CLSC set their pricing and GAP strategies when the return rate depends on both pricing and green efforts?
- how do firms' strategies and profits change when moving from a WPC to an RSC setting?
- how do returns change when moving from one setting to another?

Our findings demonstrate that the manufacturer prefers an RSC when more investments in green efforts under an RSC are needed. At the same time, the CLSC performs higher returns as the green efforts have a dominant effect on pricing. When fixing the sharing parameter at a high level, the manufacturer always prefers the RSC. When the sharing parameter is low, its preferences are fully dependent on the pricing effect on returns. When the latter is low, the manufacturer opts for an RSC; otherwise, he will opt for a WPC. The retailer never prefers an RSC when the sharing parameter is too high, because she transfers too much revenues to the manufacturer. In contrast, when the sharing parameter is too low, the retailer's preferences mainly depend on the pricing effect on returns. When this is too severe, she will opt for a WPC because the low sharing parameter is not sufficient to properly involve the manufacturer to accept an RSC. Finally, we identify two Pareto-improving regions in which firms reach coordination in CLSC. The first is represented by a low sharing parameter and a high negative effect of pricing on returns, according to which coordination is reached through a WPC. The second region is represented by the case in which the sharing parameter is medium, which calls for the adoption of an RSC, independent of the negative effect of pricing on returns. For certain sharing parameter values, the negative effect of pricing on returns determines the contract to be adopted: An RSC, when this effect is low, and a WPC when this effect is high. In all other cases, the firms have divergent contractual preferences and therefore coordination is never reached.

The remaining of the paper is structured as follows. Section 2 proposes the dynamic games to be analyzed. Section 3 proposes the solutions for all models, while Sect. 4 compares the games' outcomes. Section 5 briefly concludes.

## 2 Dynamic Games

All notations that we use in this paper are listed in Table 1. A Closed-loop Supply Chain (CLSC) is composed of one Manufacturer, firm $M$, and one Retailer, firm $R$. Both firms are involved in the management of both the forward and the reverse flows. $M$ decides the green activity programs to be carried out, namely $A(t)$, which includes, among others, investments in green technologies, green process innovation, green advertising, green marketing, and reverse logistics. $R$ is a pure seller that purchases goods from $M$ at a wholesale price, $\omega(t)$, and sets the optimal selling price, $p(t)$, according to the WPC. $R$'s marginal profits are given by $\pi_R = p(t) - \omega(t)$, while $M$'s marginal profits are given by $\pi_M = \omega(t) + \Delta r(t)$. $\Delta r(t)$ represents the marginal benefit that $M$ receives when collecting used-product from the market. We assume that $\Delta > 0$ otherwise $M$ would not have any interest in collecting used products. Thus, $\Delta$ is the per-return residual value while $r(t)$ is the fraction of sold goods that will be returned. $r(t)$ is a dynamic stock that takes the form:

**Table 1** Notations

| Notation | Description |
|---|---|
| $M, R$ | Manufacturer, retailer |
| $t$ | Time |
| $\omega(t)$ | Wholesale price at time $t$ |
| $p(t)$ | Price at time $t$ |
| $A(t)$ | Green activity programs at time $t$ |
| $\phi$ | Revenue sharing parameter |
| $\pi_M, \pi_R$ | Marginal revenues for $M$ and $R$ |
| $\Delta$ | Returns' residual value |
| $r(t)$ | Return rate |
| $k$ | Green activity programs sensitivity to $r$ |
| $\lambda$ | Price programs sensitivity to $r$ |
| $\alpha$ | Market potential |
| $\beta$ | Consumers sensitivity to price |
| $D$ | Demand |
| $\rho$ | Discount factor |

$$\dot{r}(t) = kA(t) - \lambda p(t)\sqrt{r(t)} - \delta r(t) \tag{1}$$

such that $r(t) \in (0, 1)$. When $r(t) = 0$, $M$ does not enjoy any return, while when $r(t) = 1$ all consumers return their good when it reaches the end-of-use/life stage. De Giovanni and Zaccour (2013) and De Giovanni et al. (2016) have investigated the return rate as a dynamic equation that mainly evolves according to the GAP investments. Hereby, we introduce also the price effect in the state dynamics following the idea that when consumers pay a high price for their goods, they attempt to substitute it very late over time. First, they try to enjoy the product for as long as they can after the economic sacrifice requested to purchase it. Second, consumers will eventually need to purchase a new product to continue satisfying their needs; therefore, they will again face an important sacrifice when purchasing the product again. According to Eq. (1), firms can make consumers aware of their green investments and sensibilize consumers in returning their products according to the parameter $k > 0$, which exemplifies the return rate changes according to the GAP investments that $M$ affords. Further, consumers can delay their return because of the pricing strategy according to the parameter $\lambda > 0$, which describes the return rate changes according to $R$'s pricing strategy. Finally, there is a natural decay rate $\delta$ affecting the return rate state, as consumers forget to send back their end-of-use/life products if they are not exposed to the GAP efforts.

Both the returns and the pricing strategies have an effect on the demand, which can be described as follows:

$$D(r(t), p(t)) = \alpha\sqrt{r(t)} - \beta p(t) \tag{2}$$

where $\alpha$ is the market potential and explains the amount of consumers exposed to the product and $\beta > 0$ is the consumers' sensitivity to price. Interestingly, $M$'s gives a positive contribution to both the state and the demand, while the opposite applies for $R$'s strategy. We model the GAP efforts strategy using a classical quadratic cost function, e.g., $C(A(t)) = \frac{[A(t)]^2}{2}$.

Following the same assumptions as in De Giovanni (2017), we start our analysis from a Wholesale Price Contract (WPC) scenario, in which $M$ sells some goods to $R$ at $\omega$ while $R$ sets $p(t) > \omega$ causing an issue of double marginalization. We will use the superscript $\mathcal{W}$ to refer to the WPC game. Accordingly, the firms' profit functions take the following forms:

$$J_M^{\mathcal{W}} = \max_{\omega^{\mathcal{W}}(t), A^{\mathcal{W}}(t)} \int_0^{+\infty} e^{-\rho t}\left[\left(\alpha\sqrt{r^{\mathcal{W}}(t)} - \beta p^{\mathcal{W}}(t)\right)\left(\omega^{\mathcal{W}}(t) + \Delta\sqrt{r^{\mathcal{W}}(t)}\right) - C^{\mathcal{W}}\left(A^{\mathcal{W}}(t)\right)\right] dt \tag{3}$$

$$J_R^{\mathcal{W}} = \max_{p^{\mathcal{W}}(t)} \int_0^{+\infty} e^{-\rho t}\left[\left(\alpha\sqrt{r^{\mathcal{W}}(t)} - \beta p^{\mathcal{W}}(t)\right)\left(p^{\mathcal{W}}(t) - \omega^{\mathcal{W}}(t)\right)\right] dt \tag{4}$$

where $\rho$ is the discount factor that we assume to be the same for the two firms.

One of the main novelties presented in Eq. (1) is the relationship between the return rate, $r(t)$, and the pricing strategy, $p(t)$. Therefore, we evaluate the effectiveness of an RSC within this framework. $M$'s wholesale price takes null values, i.e., $\omega = 0$, while $R$ still sets the selling price. We use the superscript $\mathcal{R}$ to refer to the RSC game. Accordingly, the firms marginal profits are given by $\pi_M^{\mathcal{R}} = p^{\mathcal{R}}(t)\phi + \Delta\sqrt{r^{\mathcal{R}}(t)}$ and $\pi_R^{\mathcal{R}} = p^{\mathcal{R}}(t)(1-\phi)$, where $\phi$ is the sharing parameter and informs on the firms' negotiation on how revenues are shared. Finally, the firms' profits in the $\mathcal{R}$-game are given by:

$$J_M^{\mathcal{R}} = \max_{A^{\mathcal{R}}(t)} \int_0^{+\infty} e^{-\rho t}\left[\left(\alpha\sqrt{r^{\mathcal{R}}(t)} - \beta p^{\mathcal{R}}(t)\right)\left(p^{\mathcal{R}}(t)\phi + \Delta\sqrt{r^{\mathcal{R}}(t)}\right) - C^{\mathcal{R}}\left(A^{\mathcal{R}}(t)\right)\right]dt \tag{5}$$

$$J_R^{\mathcal{R}} = \max_{p^{\mathcal{R}}(t)} \int_0^{+\infty} e^{-\rho t}\left[\left(\alpha\sqrt{r^{\mathcal{R}}(t)} - \beta p^{\mathcal{R}}(t)\right) p^{\mathcal{R}}(t)(1-\phi)\right]dt \tag{6}$$

Both the $\mathcal{W}$*-game* and the $\mathcal{R}$*-game* are played á la Stackelberg. We resolve the game by assuming that the players use a stationary feedback strategy, which is standard in differential games over the infinite time horizon (Dockner et al. 2000). When modeling CLSCs, one would resolve the game using an open-loop strategy due to the complex relationships between controls and states (Genc and De Giovanni 2017). Although we cannot obtain an analytical solution by using feedback strategies, we can derive a time-consistent equilibrium. Moreover, the value of the information obtained in the feedback strategy is much more appropriate from a managerial perspective in both channel and supply chain management studies.

## 3 Equilibria

In this section, we present the solutions for the dynamic games earlier introduced. In both cases, we solve the games á la Stackelberg, where $M$ is the leader. As for conventional solutions in dynamic games with infinite time horizon, all strategies and value functions are written as a function of the state.

### *3.1 A Dynamic CLSC Using a Wholesale Price Contract: $\mathcal{W}$-Scenario*

In this section, we present the solution of the $W$-game. Hereby, $M$ decides the green efforts, $A(t)$, along with the wholesale price, $\omega(t)$, when selling the products to $R$. The latter charges a selling price, $p(t)$, such that $p(t) > \omega(t)$. $M$ announces that the chain uses a WPC to regulate the financial flow; $R$ considers this announcement and decides the optimal selling price, $p(t)$; $M$ takes $p(t)$ into consideration and optimally sets its controls. The firms' strategies and profits are summarized in the following proposition.

**Proposition 1** *The equilibrium strategies in the $\mathcal{W}$-Scenario are given by:*

$$\omega^{\mathcal{W}*} = \frac{\left(\lambda L_1^{\mathcal{W}} - \left(\Delta\beta + \lambda B_1^{\mathcal{W}}\right)\right)}{\beta}\sqrt{r^{\mathcal{W}}} \tag{7}$$

$$p^{\mathcal{W}*} = \frac{\left(\alpha - \Delta\beta - \lambda B_1^{\mathcal{W}}\right)}{2\beta}\sqrt{r^{\mathcal{W}}} \tag{8}$$

$$A^{\mathcal{W}*} = \frac{kB_1^{\mathcal{W}}}{\mu} \tag{9}$$

*where the pairs $\left(B_i^{\mathcal{W}}, L_i^{\mathcal{W}}\right)$ with $i = 1, 2$ are the coefficients of the value functions $V_M^{\mathcal{W}}\left(r^{\mathcal{W}}\right)$ and $V_R^{\mathcal{W}}\left(r^{\mathcal{W}}\right)$ ,which are given by:*

$$V_M^{\mathcal{W}*} = B_1^{\mathcal{W}} r^{\mathcal{W}} + B_2^{\mathcal{W}} \tag{10}$$

$$V_R^{\mathcal{W}*} = L_1^{\mathcal{W}} r^{\mathcal{W}} + L_2^{\mathcal{W}} \tag{11}$$

*These value functions describe the optimal profits along the optimal return rate trajectory, $r^{\mathcal{W}}(t)$ . The optimal time-path of the return rate reads as follows:*

$$r^{\mathcal{W}}(t) = \left(r_0 - r_{SS}^{\mathcal{W}}\right) e^{-t\left[\mu\left(\left(\alpha - \Delta\beta - \lambda B_1^{\mathcal{W}}\right)\lambda + 2\beta\delta\right)\right]} + r_{SS}^{\mathcal{W}} \tag{12}$$

*where $r_{SS}^{\mathcal{W}}$ is the steady-state return rate and is given by:*

$$r_{SS}^{\mathcal{W}} = \frac{2\beta B_1^{\mathcal{W}} k^2}{\mu\left(\left(\alpha - \Delta\beta - \lambda B_1^{\mathcal{W}}\right)\lambda + 2\beta\delta\right)} \tag{13}$$

*Proof* See the Appendix. □

From the equilibria, one can see that value functions are linear in the state variable $r^{\mathcal{W}}$, although the state equation contains the square root of $r$. The reason is that the present games is a special case of the framework dating back to Sethi (1983) and also discussed in Sethi and Thompson (2000) and Dockner et al. (2000).

From the Appendix, we can see that $B_1^{\mathcal{W}}$ has two roots, specifically:

$$B_1^{\mathcal{W}} = \frac{\left((\alpha - \Delta\beta)\lambda + 2\beta(\delta + \rho) \pm 2\sqrt{\beta(\delta + \rho)\left((\alpha - \Delta\beta)\lambda + \beta(\delta + \rho)\right) - \Delta\alpha\beta\lambda^2}\right)}{\lambda^2} \tag{14}$$

Both of them are positive as the price, $p$, is always larger than the returns' value, $\Delta$. Thus, $\alpha - \Delta\beta > 0$. We can take the negative root out. This guarantees that $A_{SS}^{\mathcal{W}*} > 0$. Notice that the green efforts strategies are state independent, meaning that $M$ sets the green efforts while disregarding the value of the stock. Intuitively, the impact of green efforts efficiency, exemplified by $k$ in the state and $\mu$ in the cost function, suggests that when $M$ should increase or decrease these efforts. From the Appendix, one can see that $L_1^{\mathcal{W}}$ is given as follows:

$$L_1^{\mathcal{W}} = \frac{\mu\left(\alpha + \Delta\beta + \lambda B_1^{\mathcal{W}}\right)^2}{4\mu\left(\alpha\lambda - \beta\left(\delta + \rho\right)\right)} \tag{15}$$

This guarantees that $p^{\mathcal{W}*} > \omega^{\mathcal{W}*}$ always holds at the steady-state, as $p^{\mathcal{W}*} - \omega^{\mathcal{W}*} = \frac{\left(\alpha+\Delta\beta+\lambda B_1^{\mathcal{W}}-2\lambda L_1^{\mathcal{W}}\right)}{2\beta} > 0$. Both the price and wholesale price strategies depend on the return rate. Therefore, when the CLSC performs the return rate, firms know that the portfolio of consumers increases, generating more market potential. In this case, increasing the prices will not be detrimental for sales. The latter takes the following form:

$$D^{\mathcal{W}} = \frac{\left(\alpha + \left(2\Delta\beta + \lambda B_1^{\mathcal{W}}\right)\right)}{2}\sqrt{r^{\mathcal{W}}} \tag{16}$$

Interestingly, we can see that there is a trade-off between environmental performance, given by $r^{\mathcal{W}}$, and the double marginalization effect, $p^{\mathcal{W}*} - \omega^{\mathcal{W}*}$, as increasing the return rate is detrimental for consumers, who are subject to higher prices. Since $B_1^{\mathcal{W}} > 0$ and $L_1^{\mathcal{W}} > 0$, both firms have a certain convenience in increasing the returns. This result has a direct and positive effect on their profits, given that $\frac{\partial V_M^{\mathcal{W}*}}{\partial r^{\mathcal{W}}} = B_1^{\mathcal{W}} > 0$ and $\frac{\partial V_R^{\mathcal{W}*}}{\partial r^{\mathcal{W}}} = L_1^{\mathcal{W}} > 0$. From the Appendix, we can also see that

$B_2^{\mathcal{W}}$ and $L_2^{\mathcal{W}}$ are always positive, given that:

$$B_2^{\mathcal{W}} = \frac{B_1^2 k^2}{2\mu\rho} \text{ and } L_2^{\mathcal{W}} = \frac{L_1^{\mathcal{R}} B_1^{\mathcal{W}} k^2}{\mu\rho} \tag{17}$$

Consequently, the CLSC business is economically interesting for both firms even if the return rate is negligible. Finally, $r_{SS}^{\mathcal{W}}$ is globally asymptotically stable because $\left(\alpha - \Delta\beta - \lambda B_1^{\mathcal{R}}\right)\lambda + 2\beta\delta > 0$. Accordingly, the return rate is positive at the steady-state. Note that $\lim_{\lambda \to 0} r_{SS}^{\mathcal{W}} = \frac{B_1^{\mathcal{W}} k^2}{\mu\delta}$, thus highlighting the considerable impact of pricing on the returns. These circumstances create the basis to evaluate an alternative price-based mechanism to mitigate this negative effect.

## 3.2 A Dynamic CLSC Using a Revenue Sharing Contract: $\mathcal{R}$-Scenario

Next we solve the $\mathcal{R}$-game, in which firms use the revenue sharing contract to coordinate their financial flows. The WPC earlier described leaves the issue of double marginalization effect, as $p(t) > \omega(t)$ always holds. This can be very detrimental for sales, as the double marginalization leads to lower sales. In addition, since pricing is also influencing negatively the returns (e.g., Eq. (1)), there is a further interest to look into a mechanism to mitigate the negative effects on the return rate. In this setting, $M$ only decides the green efforts, $A(t)$, while $R$ charges a selling price, $p(t)$ without being subject to the constraint $p(t) > \omega(t)$. In addition, firms share the revenues generated by the business through the sharing parameter $\phi \in (0, 1)$. $M$ announces that the chain uses an RSC to regulate the financial flows; $R$ considers this announcement and decides the optimal selling price, $p(t)$; $M$ takes $p(t)$ into consideration and optimally sets $A(t)$. The firms' strategies and profits are summarized in the following proposition.

**Proposition 2** *The equilibrium strategies in the $\mathcal{R}$-Scenario are given by:*

$$p^{\mathcal{R}*} = \frac{\alpha(1-\phi) - \lambda L_1^{\mathcal{R}}}{2\beta(1-\phi)}\sqrt{r} \tag{18}$$

$$A^{\mathcal{R}*} = \frac{kB_1^{\mathcal{R}}}{\mu} \tag{19}$$

*where the pairs $\left(B_i^{\mathcal{R}}, L_i^{\mathcal{R}}\right)$ are the coefficients of the value functions $V_M^{\mathcal{R}}\left(r^{\mathcal{R}}\right)$ and $V_R^{\mathcal{R}}\left(r^{\mathcal{R}}\right)$, which are given by:*

$$V_M^{\mathcal{R}} = B_1^{\mathcal{R}} r^{\mathcal{R}} + B_2^{\mathcal{R}} \tag{20}$$

$$V_R^{\mathcal{R}} = L_1^{\mathcal{R}} r^{\mathcal{R}} + L_2^{\mathcal{R}} \tag{21}$$

*These value functions describe the optimal profits along the optimal return rate trajectory, $r^{\mathcal{R}}(t)$. The optimal time-path of the return rate reads as follows:*

$$r^{\mathcal{R}}(t) = \left(r_0 - r_{SS}^{\mathcal{R}}\right)e^{-t[\mu((\alpha\lambda+2\beta\delta)(1-\phi)-\lambda^2 L_1)]} + r_{SS}^{\mathcal{R}} \tag{22}$$

*where $r_{SS}^{\mathcal{R}}$ is the steady-state return rate and is given by:*

$$r_{SS}^{\mathcal{R}} = \frac{2(1-\phi)\beta B_1 k^2}{\mu\left((\alpha\lambda + 2\beta\delta)(1-\phi) - \lambda^2 L_1^{\mathcal{R}}\right)}. \tag{23}$$

*Proof* See the Appendix. □

From the Appendix, we can see that $B_1^{\mathcal{R}}$ has one unique solution, specifically:

$$B_1^{\mathcal{R}} = \frac{\mu\left(\lambda L_1^{\mathcal{R}} + \alpha(1-\phi)\right)\left((2\Delta\beta + \alpha\phi)(1-\phi) - \lambda\phi L_1^{\mathcal{R}}\right)}{2\mu(1-\phi)\left((\lambda\alpha + 2\beta(\delta+\rho))(1-\phi) - \lambda^2 L_1^{\mathcal{R}}\right)}. \tag{24}$$

This is a positive expression, which guarantees that $A_{SS}^{\mathcal{R}*} > 0$. Note that, even in the revenue sharing setting, the green efforts strategies are state independent, meaning that $M$ always disregards the value of the stock when setting the green efforts, independent of the contract the CLSC uses. Further, the green efforts structure is similar to the one derived in the WPC contract, thus we can refer to the previous discussion. From the Appendix, one can see that $L_1^{\mathcal{W}}$ has two roots, one of which is positive and one of which is negative. To have positive returns, we take the negative root out, which is given by:

$$L_1^{\mathcal{R}} = \frac{(\alpha\lambda + 2\beta(\delta+\rho))(1-\phi) - 2\sqrt{\beta(1-\phi)^2(\delta+\rho)(\alpha\lambda + \beta(\delta+\rho))}}{\lambda^2}. \tag{25}$$

Further, this guarantees that $r^{\mathcal{R}} \in (0, 1)$ and $p^{\mathcal{R}} > 0$. The price positively depends on the return rate. Therefore, when the CLSC performs the return rate, firms can charge a higher price. Interestingly, when firms enjoy high returns, they also enjoy a higher market potential, which allows them to increase the price to extract more economic value from the market. Also, this result informs researchers on the importance of dynamic elements: the contribution that the state, $r^{\mathcal{R}}(t)$, gives to the firms' profit function is more important than a single strategy, e.g., pricing. Therefore, CLSC is a dynamic phenomenon and should be studied as such. Consequently, the sales under a revenue sharing contract are given by:

$$D^{\mathcal{R}} = \frac{\alpha(1-\phi) + \lambda L_1^{\mathcal{R}}}{2(1-\phi)}\sqrt{r^{\mathcal{R}}} \tag{26}$$

Since $B_1^{\mathcal{R}} > 0$ and $L_1^{\mathcal{R}} > 0$, both firms have a certain convenience in contributing for increasing the returns. This result has a direct and positive effect on their profits, given that $\frac{\partial V_M^{\mathcal{R}*}}{\partial r^{\mathcal{R}}} = B_1^{\mathcal{R}} > 0$ and $\frac{\partial V_R^{\mathcal{R}*}}{\partial r^{\mathcal{R}}} = L_1^{\mathcal{R}} > 0$. From the Appendix, we can also see that $B_2^{\mathcal{R}}$ and $L_2^{\mathcal{R}}$ are always positive, given that:

$$B_2^{\mathcal{R}} = \frac{B_1^{\mathcal{R}^2} k^2}{2\mu\rho} \text{ and } L_2^{\mathcal{R}} = \frac{L_1^{\mathcal{R}} B_1^{\mathcal{R}} k^2}{\mu\rho} \tag{27}$$

As in the WPC, the CLSC business is economically interesting for both firms even if the return rate is negligible. Finally, $r_{SS}^{\mathcal{R}}$ is globally asymptotically stable when $\mu\left((\alpha\lambda + 2\beta\delta)(1-\phi) - \lambda^2 L_1^{\mathcal{R}}\right) > 0$. Accordingly, the return rate is positive at the steady-state. Note that $\lim\limits_{\lambda->0} r_{SS}^{\mathcal{R}} = \frac{B_1^{\mathcal{R}} k^2}{\delta\mu}$, thus highlighting the considerable impact of pricing on the returns. Interestingly, when the returns are not affected by pricing, the return rate takes the same structure as in the $\mathcal{W}$-game. We seek now to evaluate the impact of the sharing parameter, $\phi$, on the firms' strategies and profits as well as on the sales.

Numerically, we can see that $\frac{\partial B_1^{\mathcal{R}}}{\partial \phi} > 0$ and $\frac{\partial L_1^{\mathcal{R}}}{\partial \phi} < 0$. Figure 1 displays the relationship between the coefficients and the sharing parameter, $\phi$.[1]

Consequently, we can formulate the following corollary:

**Corollary 1** *The behavior of strategies, demand and profits with respect to the sharing parameter, $\phi$, is as follows:* $\frac{\partial A_{SS}^{\mathcal{R}}}{\partial \phi} > 0$, $\frac{\partial r_{SS}^{\mathcal{R}}}{\partial \phi} > 0$, $\frac{\partial p_{SS}^{\mathcal{R}}}{\partial \phi} > 0$, $\frac{\partial D_{SS}^{\mathcal{R}}}{\partial \phi} > 0$; $\frac{\partial V_{M_{SS}}^{\mathcal{R}}}{\partial \phi} > 0$ *and* $\frac{\partial V_{R_{SS}}^{\mathcal{R}}}{\partial \phi} \gtrless 0$, $\forall \phi \in (0, \overline{\phi}]$.

*Proof* See the Appendix. □

The results of Corollary 1 are clearly displayed in Figs. 2 and 3.

Accordingly, we can leave the following remarks. The RSC does not provide the usual benefits of decreased price as claimed by Cachon (2003). This is due to the fact that the pricing influences the state variable, which decreases due to the higher price. This implication calls $M$ to invest more in green efforts, $A^{\mathcal{R}}$, given the fact that he is receiving a share: higher share implies higher economic availability, and thus a larger chance to increase the green efforts. At the same time, $R$ needs more

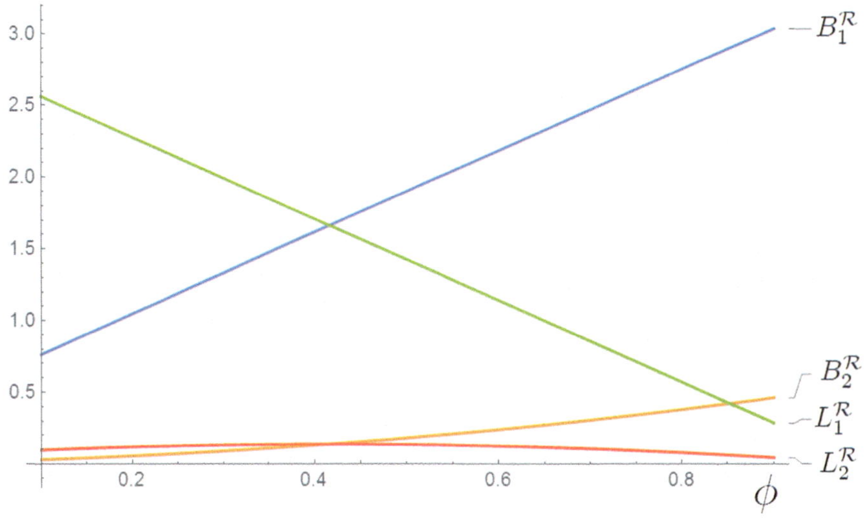


**Fig. 1** The relationship between the coefficients $B_i$, $L_i$ with respect to $\phi$

[1]We carry out the numerical analysis by setting the parameters at the following values: $\alpha = 2$, $\beta = 0.6$, $\delta = 0.2$, $\rho = 0.1$, $\lambda = 0.2$, $\mu = 1$, $\Delta = 0.2$, $k = 0.5$. Instead, we leave $\phi$ as free.

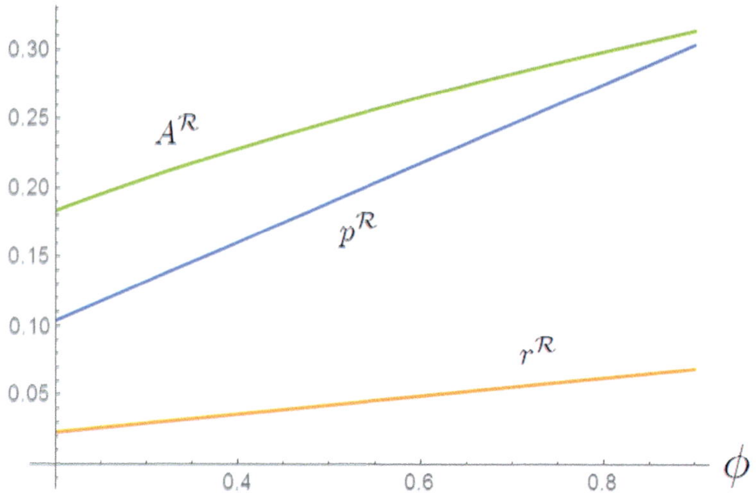


**Fig. 2** Price, green efforts, and returns with respect to $\phi$

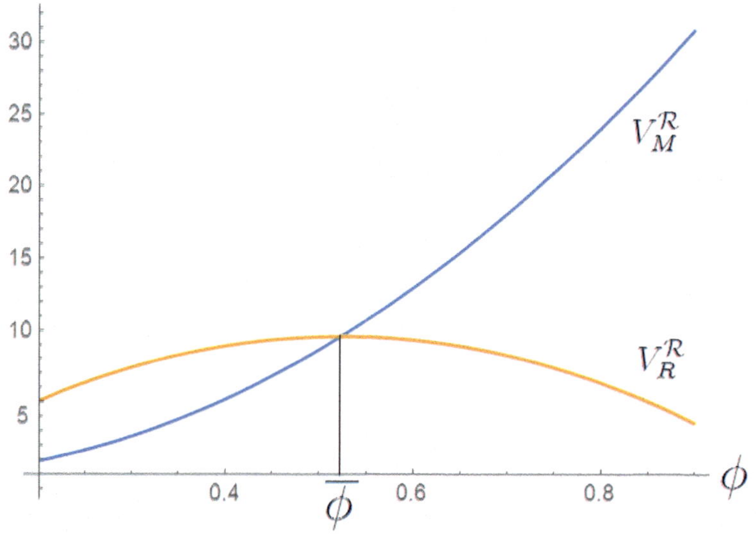


**Fig. 3** The relationship between the $V_M$ and $V_R$ with respect to $\phi$

economic resources now, since a part of her revenues is transferred to $M$. The joint effect of pricing and green efforts translates into higher returns and sales, thus we assist to a positive reaction from the market that returns used goods and purchases new ones. Finally, there is a tough negotiation to be carried out before starting the game on the sharing parameter. While $M$ is always happy to receive a share, $R$ benefits from it till a certain level. In fact, when the sharing parameter is too high, e.g., $\phi > \overline{\phi}$, $R$ transfers too much revenues to $M$ with the results that her profits decrease. Interestingly, when $\phi = \overline{\phi}$, the firms gain the same profits.

## 4 Comparison Between Games

In this section we compare the outcomes of both the WPC and the RSC games to answer our initial research questions. We use the same benchmark parameters that have been previously set while we focus in the $(\phi, \lambda) - space$ analysis. The motivations for doing so are twofold. First, the analysis of the sharing parameter will inform on the efficiency of an RSC with respect to a WPC in the exchange of financial flows. Second, the sharing parameter has a direct effect on pricing. Thus, the impact of pricing on the return rate adds new insights in this literature frame.

From Fig. 4, it is interesting to see that $M$ adjusts the green efforts according to both the sharing parameter value and the impact of pricing on returns. When these two effects are low, $M$ is not very much interested in the business; instead, he does care about $R$'s power. Even when he gets a minimal revenue and/or there is

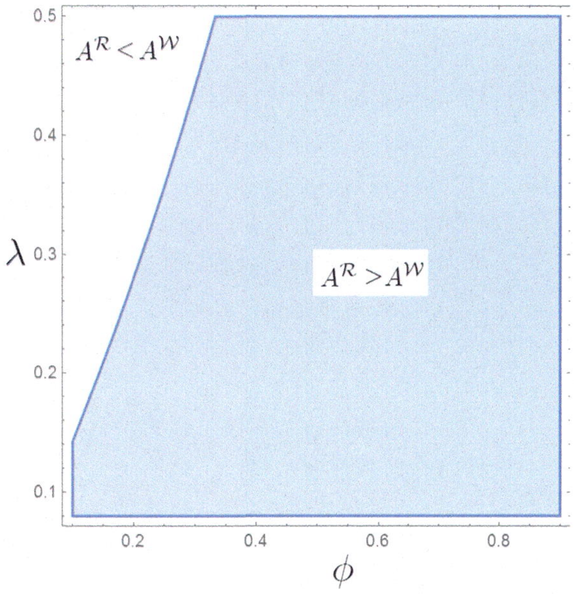


**Fig. 4** Comparison between green efforts, $A$

no impact of $R$ on the returns. When the sharing parameter is high, $M$ has a large incentive for investing in green efforts. When the share increases, he invests more than in the WPC. We can see that when the pricing strategy becomes very severe (for example, when the pricing strategy in revenue sharing is much larger than the pricing strategy in the wholesale price contract) the manufacturer wishes to invest more in green efforts in order to overcome the negative effect of the pricing strategy. Nevertheless, this only happens when the business is interesting for $M$, that is, when the sharing parameter is high. Otherwise, when the sharing parameter is low, $M$ does not invest more in green efforts when a revenue sharing contract is implemented.

Figure 5 displays the areas in which $p^{\mathcal{R}} > p^{\mathcal{W}}$. Intuitively, when the sharing parameter is too large, $R$ transfers a considerable amount to $M$; therefore, she needs to set a higher price to make the business profitable. In this case, the CLSC obtains the reverse effect of what we expect from an RSC, that is, the price reduction compared to the WPC setting. We can see that this only happens when the sharing parameter, $\phi$, is sufficiently small. Further, $R$ sets higher prices in RSC when the negative impact of pricing on returns increases. This is a counter-intuitive result as we expect $R$ to lower the price for high values of $\lambda$. In fact, this happens only when $R$ retains a sufficiently large amount of revenues. For example, when the sharing parameter is low, $R$ retains more revenues than $M$, thus she sets an RSC price lower than a WPC price. Nevertheless, when her fraction is low, she needs more economic resources. Then, she opts for charging larger prices while expecting $M$ to invest more in green efforts.

Figure 6 displays the comparison between the return rate at the steady-state in the two proposed games. Here, the returns most likely reflect the green efforts and

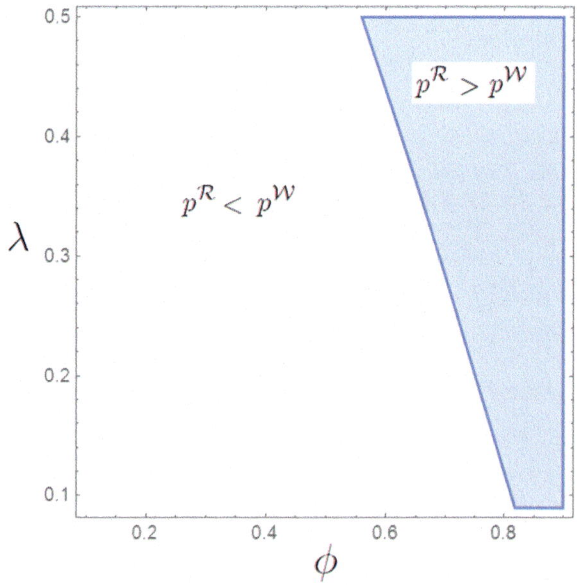


**Fig. 5** Comparison between pricing strategies

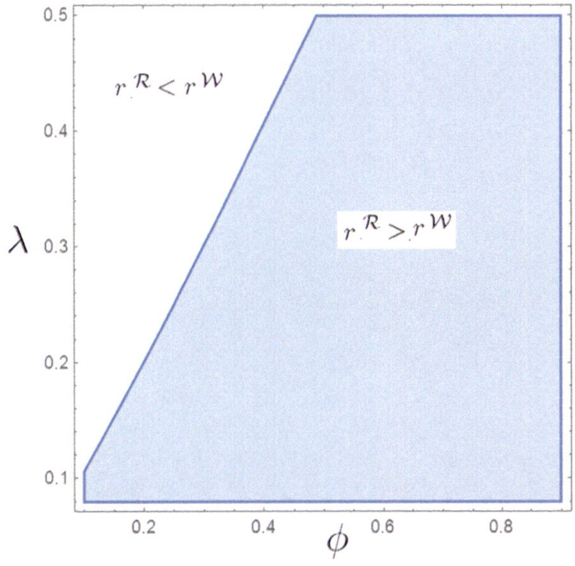


**Fig. 6** Comparison between return rates, *r*

the pricing strategies. We can see that when the latter becomes more important, the returns decrease more under an RSC. This also happens when the sharing rate is low. Therefore, *M* does not invest too much in green activity efforts because the business is not appealing, while the opportunities for building an effective reverse flow are minimal. We can see that, in order to enjoy a very efficient CLSC, the sharing parameters should be high while the impact of pricing on returns should be minimal. Supply chains will be very much efficient in managing their own returns if and only if the customers perceive that the pricing doesn't affect the returns. In this sense the CLSC can give an interesting rebate to customers to increase their willingness to return the products and, therefore, have a lower pricing effect on returns (Genc and De Giovanni 2017).

Figure 7 compares the sales in the two games that we investigate. Accordingly, the adoption of an RSC allows the CLSC to sell more products in the market in most of the cases. In this regard, an RSC is more socially sustainable as more people access to the product. It is interesting to see that this happens also when $p^{\mathcal{R}} > p^{\mathcal{W}}$, thus firms can optimally adjust the green efforts to create a compensation effect when the price is detrimental for both the sales and the returns. Finally, there is only one case in which the sales under an RSC are lower than the sales under a WPC, that is, when $\lambda$ is very high. In this case, the negative effects generated by a pricing strategy are too severe and the green efforts are not sufficiently high. Figure 8 displays the *M*'s preferences in the selection of the coordination mechanism. This shape is very similar to the one in Fig. 4 and highlights a clear message. When the sharing parameter is sufficiently large, *M* always supports the adoption of an RSC. This is very much intuitive as the larger the sharing parameter, the higher the fraction of revenues that goes in his pocket. Interestingly, this finding holds true

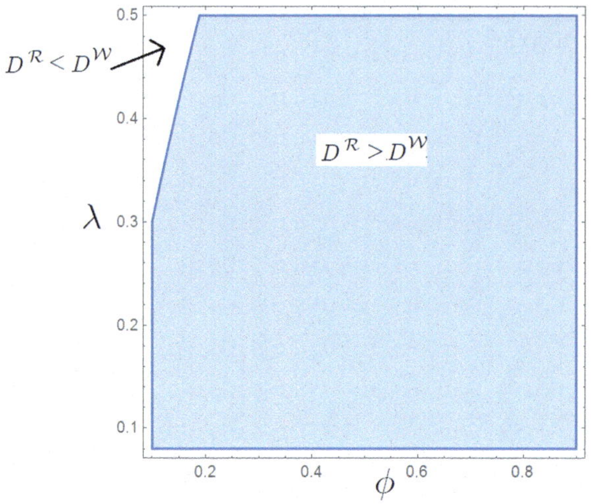


**Fig. 7** Comparison between sales, $D$

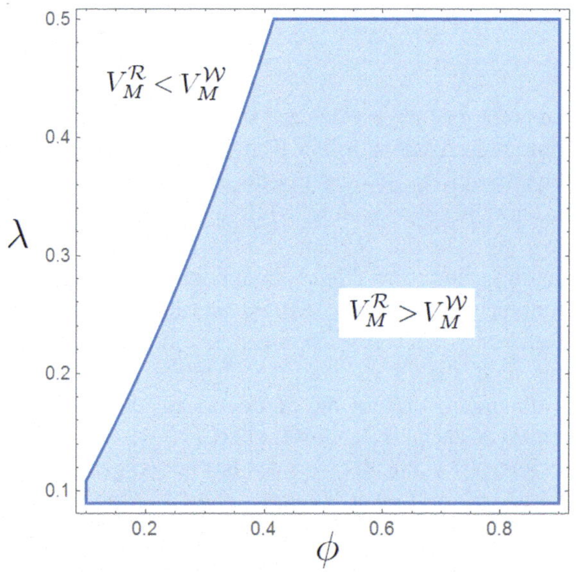


**Fig. 8** Comparison between $M$'s profits, $V_M$

independent of the negative effect that pricing has on returns. Therefore, under these circumstances, $M$ is always willing to invest more in green efforts to overcome the negative effects generated by pricing, as he gains sufficiently large revenues to be reinvested in green initiatives. Instead, when the sharing parameter is medium vs. low, the convenience of adopting an RSC highly depends on $\lambda$: when $\lambda$ is low, $M$ always opts for an RSC; when $\lambda$ is high, $M$'s preferences for an RSC decrease according to increasing values for $\lambda$. In the latter case, the negative impact of pricing

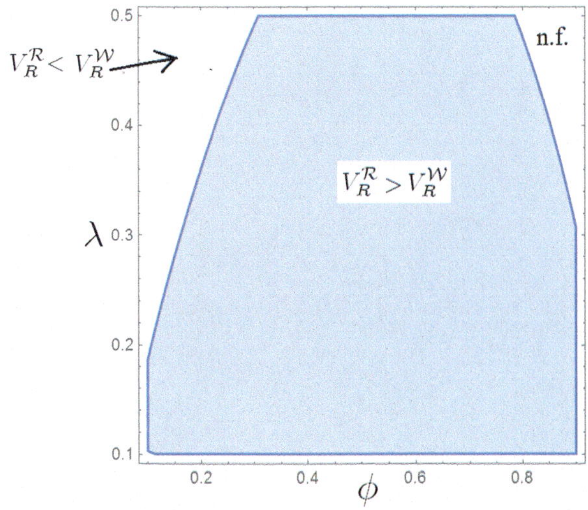


**Fig. 9** Comparison between $R$'s profits, $V_R$

on returns is too severe, and investing more in green efforts translates in a marginally convenient option.

Figure 9 displays $R$'s preferences with respect to the coordination mechanism to be adopted. We can mainly identify three regions:

1. The sharing parameter, $\phi$, and the impact of pricing on returns, $\lambda$, are too high and the RSC is a non-feasible (n.f.) option for $R$. Hereby, the revenues transferred to $M$ are too high and her profits become negative. So, when the sharing parameter is too high, $R$ will never be interested in the business if $\lambda$ is also high. In contrast, when $\lambda$ is low, $R$ profits are positive and she opts for the adoption of an RSC. Therefore, the impact of pricing on returns plays a key role in determining whether $R$ finds an RSC convenient for coordinating the CLSC.
2. The sharing parameter, $\phi$, is very low and the impact of pricing on returns, $\lambda$, is high. In this region, $R$ should be very much happy to coordinate the CLSC through an RSC because her fraction of revenues is the largest. Nevertheless, the negative impact of pricing on returns has a very detrimental effect on this preference; thus, $R$ opts for a WPC when the effect of pricing on returns is too severe. Again, it is interesting to see that the negative effect that pricing exerts on returns by delaying the moment in which consumers return the product has a considerable weight on the $R$'s preferences.
3. In all cases that are not contemplated in 1 and 2, $R$ always opt for an RSC, independent of $\lambda$.

Finally, Fig. 10 informs on the regions in which a Pareto-improving situation is realized. Four regions are identified.

1. The blue region highlights that both firms have a preference for an RSC. As mentioned, the role of $\lambda$ is vital to determine whether the adoption of an RSC leads to a Pareto-improving situation.

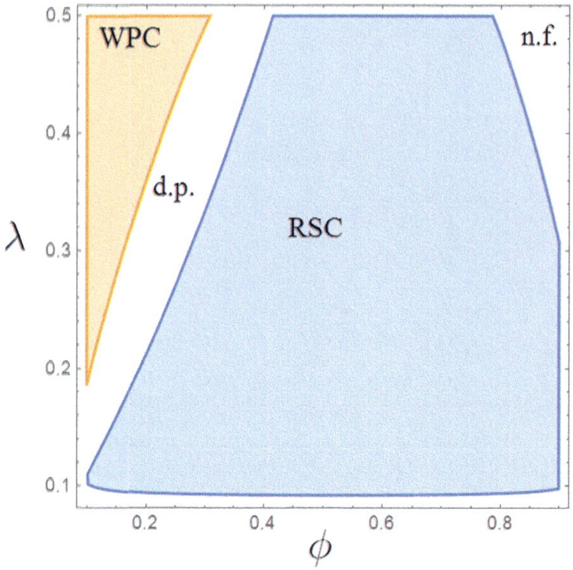


**Fig. 10** Pareto-improving region

2. The orange region indicates that both firms have a preference for a WPC. Hereby, the sharing parameter is too low to engage $M$ in investing in green efforts such that the whole CLSC gets benefits. Also, the impact of pricing on returns is too detrimental, thus backward activities simply become less important.
3. Firms have divergent preferences (d.f.), specifically: $R$ opts for an RSC while $M$ would prefer a WPC. The sharing parameter seems to be too low for $M$ to convince him in implementing an RSC.
4. There exists a non-feasible (n.f.) region in which $R$ would opt for a WPC while $M$ would select an RSC. The latter is not feasible for $R$, thus firms have divergent preferences.

We would highlight the role that the parameter $\lambda$ can have in moving firms' preferences for a contract to another. Let's take, for example, the case of $\phi = 0.2$. When $\lambda$ is low, firms reach coordination through an RSC: although $M$ receives a low share, the marginal impact of pricing does not hurt the return rate. Thus, he does not need to invest too much in green efforts to perform the return rate. When $\lambda$ is medium, firms have divergent preferences and coordination is never reached. In this case, $M$ gains through the $R$'s share of revenues is too low to induce higher investments in green efforts. When $\lambda$ is very high, both firms prefer the adoption of a WPC; $M$ will invest the minimum efforts in the return rate and, consequently, backward activities and closing the loop become negligible targets.

# 5 Conclusions

This paper studies a dynamic CLSC that is involved in managing both the backward and the forward flows of goods. The firms being part of the CLSC, namely, manufacturer and retailer, coordinate their financial flows by choosing either a wholesale price contract (WPC) or a revenue sharing contract (RSC). The manufacturer is the chain leader and fully benefits from the returns' residual value. Nevertheless, the return rate, which consists of the fraction of past sold products that comes back to the manufacturer's plants to be either reused or recycled, is negatively influenced by the pricing strategy. The latter is set by the retailer. This is the main contribution to the literature of dynamic games in CLSC, which focuses on increasing the return rate by some green efforts, generally set by a manufacturer. In addition to that, we also model a return rate that is negatively influenced by the retailer's pricing. The motivations for this assumption lie on the relationship between consumers' willingness to return a product and the product value. When consumers pay high prices for purchasing goods, they will be more parsimonious in returning them. In particular, they will postpone the return with the purpose of exploiting the returns residual value as much as possible. According to this assumption, we first model a WPC game, in which the manufacturer sets the green efforts along with the wholesale price, while the retailer sets the pricing strategy. Later we model an RSC game, in which the manufacturer does not set the wholesale price anymore, while the firms share the revenues generated within the CLSC according to an exogenous sharing parameter. The motivations for contrasting a wholesale price to a revenue sharing contract lies in the fact that the implementation of an RSC generally leads to a decrease of the selling price, with the purpose of increasing the demand, thus generating more revenues. The reduction of the selling price in the CLSC can be a driver to limit the negative effect of pricing on the return rate.

Our results show that the manufacturer invests more in a CLSC using an RSC when the sharing parameter is sufficiently high. In fact, this investment fully depends on the amount of money that the manufacturer receives. In such cases, the negative effect of pricing on returns is not a problem. Nevertheless, when the latter effect is severe, the manufacturer does not invest more in an RSC when the sharing parameter is low. In that case, his revenues will be too low to allow the green efforts to compensate for the negative effect of pricing. The CLSC enjoys the positive effect of price reduction when the sharing parameter is not too high. Therefore, there is a need to negotiate the sharing parameter before starting the game in order to get the desired effect. In fact, when the sharing parameter is too high, the retailer needs to charge a higher price than the WPC price because she needs more economic resources to consider the business as interesting. The two strategies (green efforts and pricing) contribute to the return rate with opposite sign. Nevertheless, we find that the return rate shape follows the green efforts shape, which seems to have a dominant effect with respect to the pricing strategy. Thus, any time the manufacturer invests more due to the implementation of an RSC, the returns turn out to be higher than the WPC case. Instead, the consumers enjoy an efficient combination of pricing and green efforts when an RSC is adopted. Thus, the RSC is socially sustainable in most of the cases. The manufacturer prefers the adoption of an RSC any time he invests more in green efforts under an RSC while the CLSC performs higher returns. The sharing parameter plays a key role. It is sufficient to have a high sharing parameter to make the manufacturer always prefers the RSC. When the sharing parameter is low, his preferences will be fully dictated by the influence of pricing on returns. When the latter effect is low, the manufacturer prefers an RSC; otherwise, he will opt for a WPC. On her side, the retailer never opts for an RSC when the sharing parameter is too high, because she transfers a fraction that is not economically viable for her. In contrast, when the sharing parameter is too low, the retailer's preferences are mainly driven by the effect of pricing on returns. When this is too severe, she will opt for a WPC because the sharing parameter will be too low to convince the manufacturer to accept the deal. Finally, we identify two Pareto-improving regions, which represent the areas in which firms reach coordination in CLSC. The first is represented by a low sharing parameter and a high negative effect of pricing on returns. In that case, both firms prefer a WPC. The second area is represented by the case in which the sharing parameter is medium. In that case, both firms will prefer an RSC, independent of the negative effect of pricing on returns. For certain sharing parameter values, the target of coordination fully depends on the negative effect of pricing on returns: when this is low, an RSC allows firms to reach coordination; when this effect is high, a WPC allows firms to achieve coordination. In all other cases, the firms have divergent contractual preferences and coordination is never reached.

This research is not free of limitations, which are listed here to inspire future research in this direction. We assume that the CLSC does not experience any competitive effect within each tier. Introducing competition in the manufacturer and/or at the retailer levels will probably change some of our results. We model coordination while considering that the retailer never benefits of returns, as consumers directly send back products to the manufacturer. Having the retailer involved in the collection process will need a further reflection on the coordination mechanisms to be adopted. We assume that the returns' residual value is exogenous and fixed, while it most likely depends on how consumers used the product during the residence time. Therefore, the consumers' behavior also matters in the determination of the goods' residual value. We have modeled a negative impact of pricing on returns. Nevertheless, additional research can be carried out to show the positive effect of pricing on a state variable like green goodwill and, consequently, modeling the return rate as a function of the stock. This would definitely lead to completely different results as pricing would positively influence the returns. This is an ongoing research direction that the authors are exploring.

# Appendix

*Proof of Proposition 1* We search for a pair of bounded and continuously differentiable value functions $V_M^{\mathcal{W}}(r^{\mathcal{W}})$ and $V_R^{\mathcal{W}}(r^{\mathcal{W}})$ for which a unique solution for $r^{\mathcal{W}}(t)$ does exist, and the HJB equations:

$$\rho V_M^{\mathcal{W}} = \left(\alpha\sqrt{r^{\mathcal{W}}} - \beta p^{\mathcal{W}}\right)\left(\omega^{\mathcal{W}} + \Delta\sqrt{r^{\mathcal{W}}}\right) - \frac{\mu\left(A^{\mathcal{W}}\right)^2}{2} + V_M^{\mathcal{W}\prime}\left(kA^{\mathcal{W}} - \lambda p^{\mathcal{W}}\sqrt{r^{\mathcal{W}}} - \delta r^{\mathcal{W}}\right) \tag{28}$$

$$\rho V_R^{\mathcal{W}} = \left(\alpha\sqrt{r^{\mathcal{W}}} - \beta p^{\mathcal{W}}\right)\left(p^{\mathcal{W}} - \omega^{\mathcal{W}}\right) + V_R^{\mathcal{W}\prime}\left(kA^{\mathcal{W}} - \lambda p^{\mathcal{W}}\sqrt{r^{\mathcal{W}}} - \delta r^{\mathcal{W}}\right) \tag{29}$$

are satisfied for any value of $r^{\mathcal{W}}(t) \in (0, 1]$. We solve the game á la Stakelberg, where M is the leader. Therefore, we start by solving the $R$'s optimization problem. The optimization of $R$'s HJB with respect to the pricing strategy leads to:

$$p^{\mathcal{W}} = \frac{\omega^{\mathcal{W}}\beta - \sqrt{r^{\mathcal{W}}}\left(\lambda V_R^{\mathcal{W}\prime} - \alpha\right)}{2\beta} \tag{30}$$

Substituting Eq. (30) in the $M$'s HJB gives:

$$\rho V_M^{\mathcal{W}} = \left(\alpha\sqrt{r^{\mathcal{W}}} - \beta\frac{\omega^{\mathcal{W}}\beta - \sqrt{r^{\mathcal{W}}}\left(\lambda V_R^{\mathcal{W}\prime} - \alpha\right)}{2\beta}\right)\left(\frac{\omega^{\mathcal{W}}\beta - \sqrt{r^{\mathcal{W}}}\left(\lambda V_R^{\mathcal{W}\prime} - \alpha\right)}{2\beta} + \Delta\sqrt{r^{\mathcal{W}}}\right) - \frac{\mu\left(A^{\mathcal{W}}\right)^2}{2} + V_M^{\mathcal{W}\prime}\left(kA^{\mathcal{W}} - \lambda\frac{\omega^{\mathcal{W}}\beta - \sqrt{r^{\mathcal{W}}}\left(\lambda V_R^{\mathcal{W}\prime} - \alpha\right)}{2\beta}\sqrt{r^{\mathcal{W}}} - \delta r^{\mathcal{W}}\right) \tag{31}$$

Maximizing with respect to green efforts, $A^{\mathcal{W}}$, and wholesale price, $\omega^{\mathcal{W}}$, gives:

$$A^{\mathcal{W}} = \frac{kV_M^{\mathcal{W}\prime}}{\mu} \tag{32}$$

$$\omega^{\mathcal{W}} = \frac{\left(\lambda V_R^{\mathcal{W}\prime} - \left(\Delta\beta + \lambda V_M^{\mathcal{W}\prime}\right)\right)}{\beta}\sqrt{r^{\mathcal{W}}} \tag{33}$$

Plugging Eq. (33) in Eq. (30), we obtain the optimal price:

$$p^{\mathcal{W}} = \frac{\left(\alpha - \Delta\beta - \lambda V_M^{\mathcal{W}\prime}\right)}{2\beta}\sqrt{r^{\mathcal{W}}} \tag{34}$$

Substituting the optimal strategies inside Eqs. (29) and (31) and simplifying we obtain:

$$\begin{aligned} 4\beta\mu\rho V_M^{\mathcal{W}} &= \mu\left(\alpha + \Delta\beta - \lambda V_M^{\mathcal{W}\prime}\right)\left(\alpha + \Delta\beta + \lambda V_M^{\mathcal{W}\prime}\right)r^{\mathcal{W}} + 2\beta V_M^{\mathcal{W}\prime 2}k^2 \\ &\quad + 2\mu V_M^{\mathcal{W}\prime}\left(\lambda\left(\Delta\beta - \alpha + \lambda V_M^{\mathcal{W}\prime}\right) - 2\beta\delta\right)r^{\mathcal{W}} \end{aligned} \tag{35}$$

$$4\beta\mu\rho V_R^{\mathcal{W}} = \mu\left(\left(\alpha + \Delta\beta + \lambda V_M^{\mathcal{W}\prime}\right)^2 - 4\alpha\lambda V_R^{\mathcal{W}\prime}\right)r + 4\beta V_R^{\mathcal{W}\prime}\left(k^2 V_M^{\mathcal{W}\prime} - \delta\mu r^{\mathcal{W}}\right) \tag{36}$$

To solve the previous pair of equations, we can conjecture linear value functions $V_M^{\mathcal{W}} = B_1^{\mathcal{W}} r^{\mathcal{W}} + B_2^{\mathcal{W}}$ and $V_R^{\mathcal{W}} = L_1^{\mathcal{W}} r^{\mathcal{W}} + L_2^{\mathcal{W}}$. Substituting these conjectures and their derivatives inside Eqs. (35) and (36) gives:

$$\begin{aligned} 4\beta\mu\rho\left(B_1^{\mathcal{W}} r^{\mathcal{W}} + B_2^{\mathcal{W}}\right) &= \mu\left(\alpha + \Delta\beta - \lambda B_1^{\mathcal{W}}\right)\left(\alpha + \Delta\beta + \lambda B_1^{\mathcal{W}}\right)r^{\mathcal{W}} \\ &\quad + 2\beta B_1^{\mathcal{W}2}k^2 + 2\mu B_1^{\mathcal{W}}\left(\lambda\left(\Delta\beta - \alpha + \lambda B_1^{\mathcal{W}}\right) - 2\beta\delta\right)^{\mathcal{W}} r \end{aligned} \tag{37}$$

$$\begin{aligned} 4\beta\mu\rho\left(L_1^{\mathcal{W}} r^{\mathcal{W}} + L_2^{\mathcal{W}}\right) &= \mu\left(\left(\alpha + \Delta\beta + \lambda B_1^{\mathcal{W}}\right)^2 - 4\alpha\lambda L_1^{\mathcal{W}}\right)r^{\mathcal{W}} \\ &\quad + 4\beta L_1^{\mathcal{W}}\left(k^2 B_1^{\mathcal{W}} - \delta\mu r^{\mathcal{W}}\right) \end{aligned} \tag{38}$$

By identification, we obtain the following system of equations:

$$\begin{aligned} &-4\beta\mu\rho B_1^{\mathcal{W}} + \mu\left(\alpha + \Delta\beta - \lambda B_1^{\mathcal{W}}\right)\left(\alpha + \Delta\beta + \lambda B_1^{\mathcal{W}}\right) \\ &+ 2\mu B_1^{\mathcal{W}}\left(\lambda\left(\Delta\beta - \alpha + \lambda B_1^{\mathcal{W}}\right) - 2\beta\delta\right) = 0 \end{aligned} \tag{39}$$

$$-4\beta\mu\rho B_2^{\mathcal{W}} + 2\beta\left(B_1^{\mathcal{W}}\right)^2 k^2 = 0 \tag{40}$$

$$-4\beta\mu\rho L_1^{\mathcal{W}} + \mu\left(\left(\alpha + \Delta\beta + \lambda B_1^{\mathcal{W}}\right)^2 - 4\alpha\lambda L_1^{\mathcal{W}}\right) + 4\beta L_1^{\mathcal{W}}\delta\mu = 0 \tag{41}$$

$$-4\beta\mu\rho L_2^{\mathcal{W}} + 4\beta L_1^{\mathcal{W}} k^2 B_1^{\mathcal{W}} = 0 \tag{42}$$

We can select the negative root of $B_1^{\mathcal{W}}$, which is given by

$$B_1^{\mathcal{W}} = \frac{\left((\alpha - \Delta\beta)\lambda + 2\beta(\delta+\rho) - 2\sqrt{\beta(\delta+\rho)((\alpha-\Delta\beta)\lambda + \beta(\delta+\rho)) - \Delta\alpha\beta\lambda^2}\right)}{\lambda^2} \tag{43}$$

Then, the remaining parameters are given by:

$$B_2^{\mathcal{W}} = \frac{\left(B_1^{\mathcal{W}}\right)^2 k^2}{2\mu\rho} \tag{44}$$

$$L_1^{\mathcal{W}} = \frac{\mu\left(\alpha + \Delta\beta + \lambda B_1^{\mathcal{W}}\right)^2}{4\mu(\alpha\lambda - \beta(\delta+\rho))} \tag{45}$$

$$L_2^{\mathcal{W}} = \frac{L_1^{\mathcal{W}} B_1^{\mathcal{W}} k^2}{\mu\rho} \tag{46}$$

□

*Proof of Corollary 1* The results in Corollary 1 follow the following derivatives:

- $\frac{\partial A_{SS}^{\mathcal{R}}}{\partial\phi} = \frac{k}{\mu}\frac{\partial B_1^{\mathcal{R}}}{\partial\phi} > 0$
- $\frac{\partial r_{SS}^{\mathcal{R}}}{\partial\phi} = \left\{2\beta k^2\left(-B_1^{\mathcal{R}} + \frac{\partial B_1^{\mathcal{R}}}{\partial\phi}(1-\phi)\right)DEN[r_{SS}^{\mathcal{R}}] + \mu NUM[r_{SS}^{\mathcal{R}}]\right.$ $\left.\left((\alpha\lambda + 2\beta\delta) + \lambda^2\frac{\partial L_1^{\mathcal{R}}}{\partial\phi}\right)\right\} / DEN[r_{SS}^{\mathcal{R}}]^2 > 0$
- $\frac{\partial p_{SS}^{\mathcal{R}}}{\partial\phi} = \frac{\left[\left(-\alpha - \frac{\partial L_1^{\mathcal{R}}}{\partial\phi}\lambda\right)\sqrt{r} + \left(\alpha(1-\phi) - \lambda L_1^{\mathcal{R}}\right)\sqrt{\frac{\partial r_{SS}^{\mathcal{R}}}{\partial\phi}}\right] DEN[p_{SS}^{\mathcal{R}}] + 2\beta NUM[p_{SS}^{\mathcal{R}}]}{[2\beta(1-\phi)]^2} > 0;$
- $\frac{\partial D_{SS}^{\mathcal{R}}}{\partial\phi} = \alpha\sqrt{\frac{\partial r_{SS}^{\mathcal{R}}}{\partial\phi}} - \beta\frac{\partial p_{SS}^{\mathcal{R}}}{\partial\phi} > 0;$
- $\frac{\partial V_{M_{SS}}^{\mathcal{R}}}{\partial\phi} = \frac{\partial B_1^{\mathcal{R}}}{\partial\phi} r_{SS}^{\mathcal{R}} + \frac{\partial r_{SS}^{\mathcal{R}}}{\partial\phi} B_1^{\mathcal{R}} + \frac{\partial B_2^{\mathcal{R}}}{\partial\phi} > 0$
- $\frac{\partial V_{R_{SS}}^{\mathcal{R}}}{\partial\phi} = \frac{\partial L_1^{\mathcal{R}}}{\partial\phi} r_{SS}^{\mathcal{R}} + \frac{\partial r_{SS}^{\mathcal{R}}}{\partial\phi} L_1^{\mathcal{R}} + \frac{\partial L_2^{\mathcal{R}}}{\partial\phi}\{\begin{matrix} \geqslant 0, \forall\phi \in (0, \overline{\phi}] \\ < 0, \text{otherwise} \end{matrix}.$

□

*Proof of Proposition 2* We search for a pair of bounded and continuously differentiable value functions $V_M^{\mathcal{R}}(r^{\mathcal{R}})$ and $V_R^{\mathcal{R}}(r^{\mathcal{R}})$ for which a unique solution for $r^{\mathcal{R}}(t)$ does exist, and the HJB equations:

$$\begin{aligned} \rho V_M^{\mathcal{R}} &= \left(\alpha\sqrt{r^{\mathcal{R}}} - \beta p^{\mathcal{R}}\right)\left(p^{\mathcal{R}}\phi + \Delta\sqrt{r^{\mathcal{R}}}\right) \\ &\quad -\frac{\mu\left(A^{\mathcal{R}}\right)^2}{2} + V_M^{\mathcal{R}\prime}\left(kA^{\mathcal{R}} - \lambda p^{\mathcal{R}}\sqrt{r^{\mathcal{R}}} - \delta r^{\mathcal{R}}\right) \end{aligned} \tag{47}$$

$$\rho V_R^{\mathcal{R}} = \left(\alpha\sqrt{r^{\mathcal{R}}} - \beta p^{\mathcal{R}}\right) p^{\mathcal{R}} (1-\phi) + V_R^{\mathcal{R}\prime}\left(kA^{\mathcal{R}} - \lambda p^{\mathcal{R}}\sqrt{r^{\mathcal{R}}} - \delta r^{\mathcal{R}}\right) \quad (48)$$

are always satisfied for any value of $r^{\mathcal{R}}(t) \in (0,1]$. We solve the game á la Stakelberg, where $M$ is the leader. Nevertheless, the pricing and green efforts strategies are independent; therefore, solving the Stakelberg game corresponds to solving the Nash game. In fact, the firms reaction functions are given by

$$\frac{\partial V_M^{\mathcal{R}}}{\partial A^{\mathcal{R}}} = kV_M^{\mathcal{R}\prime} - A^{\mathcal{R}}\mu \quad (49)$$

$$\frac{\partial V_R^{\mathcal{R}}}{\partial p^{\mathcal{R}}} = p^{\mathcal{R}}\beta\phi - p^{\mathcal{R}}\beta - \sqrt{r^{\mathcal{R}}}\lambda V_R^{\mathcal{R}\prime} + (1-\phi)\left(\sqrt{r^{\mathcal{R}}}\alpha - p^{\mathcal{R}}\beta\right) \quad (50)$$

Therefore, the optimal strategies result as follows:

$$A^{\mathcal{R}} = \frac{kV_M^{\mathcal{R}\prime}}{\mu} \quad (51)$$

$$p^{\mathcal{R}} = \frac{\alpha(1-\phi) - \lambda V_R^{\mathcal{R}\prime}}{2\beta(1-\phi)}\sqrt{r^{\mathcal{R}}} \quad (52)$$

Substituting the optimal strategies inside the firms' HJBs gives:

$$\begin{aligned}4\mu\beta(1-\phi)^2\rho V_M^{\mathcal{R}} = {} & \mu\left((1-\phi)(2\Delta\beta+\alpha\phi) - \lambda\phi V_R^{\mathcal{R}\prime}\right)\left(\alpha(1-\phi)+\lambda V_R^{\mathcal{R}\prime}\right)r^{\mathcal{R}} \\ & +2\mu(1-\phi)V_M^{\mathcal{R}\prime}\left(-\lambda\left(\alpha(1-\phi)-\lambda V_R^{\mathcal{R}\prime}\right) - 2\beta\delta(1-\phi)\right)r^{\mathcal{R}} \\ & +2\beta(1-\phi)^2k^2V_M^{\mathcal{R}\prime 2} \quad (53)\end{aligned}$$

$$4(1-\phi)\beta\mu\rho V_R^{\mathcal{R}} = \mu\left(\alpha\phi - \alpha + \lambda V_R^{\mathcal{R}\prime}\right)^2 r^{\mathcal{R}} + 4(1-\phi)\beta V_R^{\mathcal{R}\prime}\left(k^2V_M^{\mathcal{R}\prime} - \delta\mu r^{\mathcal{R}}\right) \quad (54)$$

To solve the previous pair of equations, we can conjecture linear value functions $V_M^{\mathcal{R}} = B_1^{\mathcal{R}}r^{\mathcal{R}} + B_2^{\mathcal{R}}$ and $V_R^{\mathcal{R}} = L_1^{\mathcal{R}}r^{\mathcal{R}} + L_2^{\mathcal{R}}$. Substituting these conjectures and their derivatives inside Eqs. (53) and (54) gives:

$$\begin{aligned}4\mu\beta(1-\phi)^2\rho\left(B_1^{\mathcal{R}}r^{\mathcal{R}}+B_2^{\mathcal{R}}\right) = {} & \mu\left((1-\phi)(2\Delta\beta+\alpha\phi) - \lambda\phi L_1^{\mathcal{R}}\right) \\ & \left(\alpha(1-\phi)+\lambda L_1^{\mathcal{R}}\right)r^{\mathcal{R}} + 2\mu(1-\phi)B_1^{\mathcal{R}} \quad (55) \\ & \left(-\lambda\left(\alpha(1-\phi)-\lambda L_1^{\mathcal{R}}\right) - 2\beta\delta(1-\phi)\right)r \\ & +2\beta(1-\phi)^2k^2\left(B_1^{\mathcal{R}}\right)^2 \quad (56)\end{aligned}$$

$$4(1-\phi)\beta\mu\rho\left(L_1^{\mathcal{R}}r^{\mathcal{R}}+L_2^{\mathcal{R}}\right)=\mu\left(\alpha\phi-\alpha+\lambda L_1^{\mathcal{R}}\right)^2 r^{\mathcal{R}}$$
$$+4(1-\phi)\beta L_1^{\mathcal{R}}\left(k^2B_1^{\mathcal{R}}-\delta\mu r^{\mathcal{R}}\right) \tag{57}$$

By identification, the model parameters are:

$$-4\mu\beta(1-\phi)^2\rho B_1^{\mathcal{R}}+\mu\left((1-\phi)(2\Delta\beta+\alpha\phi)-\lambda\phi L_1^{\mathcal{R}}\right)\left(\alpha(1-\phi)+\lambda L_1^{\mathcal{R}}\right)$$
$$+2\mu(1-\phi)B_1^{\mathcal{R}}\left(-\lambda\left(\alpha(1-\phi)-\lambda B_1^{\mathcal{R}}\right)-2\beta\delta(1-\phi)\right)=0 \tag{58}$$

$$-4\mu\beta(1-\phi)^2\rho B_2^{\mathcal{R}}+2\beta(1-\phi)^2k^2\left(B_1^{\mathcal{R}}\right)^2=0 \tag{59}$$

$$-4(1-\phi)\beta\mu\rho L_1^{\mathcal{R}}+\mu\left(\alpha\phi-\alpha+\lambda L_1^{\mathcal{R}}\right)^2-4\delta\mu(1-\phi)\beta L_1^{\mathcal{R}}=0 \tag{60}$$

$$-4(1-\phi)\beta\mu\rho B_2^{\mathcal{R}}+4(1-\phi)\beta L_1^{\mathcal{R}}k^2B_1^{\mathcal{R}}=0 \tag{61}$$

We can see that there exists one solution only for $B_1^{\mathcal{R}}$ while we take the negative root for $R_1^{\mathcal{R}}$. The solution is given as follows:

$$B_1^{\mathcal{R}}=\frac{\mu\left(\lambda L_1^{\mathcal{R}}+\alpha(1-\phi)\right)\left((2\Delta\beta+\alpha\phi)(1-\phi)-\lambda\phi L_1^{\mathcal{R}}\right)}{2\mu(1-\phi)\left((\lambda\alpha+2\beta(\delta+\rho))(1-\phi)-\lambda^2L_1^{\mathcal{R}}\right)} \tag{62}$$

$$B_2^{\mathcal{R}}=\frac{k^2\left(B_1^{\mathcal{R}}\right)^2}{2\mu\rho} \tag{63}$$

$$L_1^{\mathcal{R}}=\frac{(\alpha\lambda+2\beta(\delta+\rho))(1-\phi)-2\sqrt{\beta(1-\phi)^2(\delta+\rho)(\alpha\lambda+\beta(\delta+\rho))}}{\lambda^2} \tag{64}$$

$$L_2^{\mathcal{R}}=\frac{k^2B_1^{\mathcal{R}}L_1^{\mathcal{R}}}{\mu\rho} \tag{65}$$

□

## *Pareto Analysis on Different Sets*

Hereby, we carry out the Pareto analysis on two different parameter sets to demonstrate the robustness of our findings. Specifically, we use the following two parameter sets:

- high parameter values, by fixing the parameters as follows: $\alpha=3$, $\beta=0.8$, $k=0.15$, $\Delta=0.3$, $\delta=0.3$, $\rho=0.15$ and $\mu=1.5$;

- low parameter values, by fixing the parameters as follows: $\alpha = 1$, $\beta = 0.4$, $k = 0.05$, $\Delta = 0.1$, $\delta = 0.1$, $\rho = 0.05$ and $\mu = 0.5$.

As we display in Figs. 11 and 12, the findings that we obtain in Fig. 10 are confirmed when taking different parameter sets.

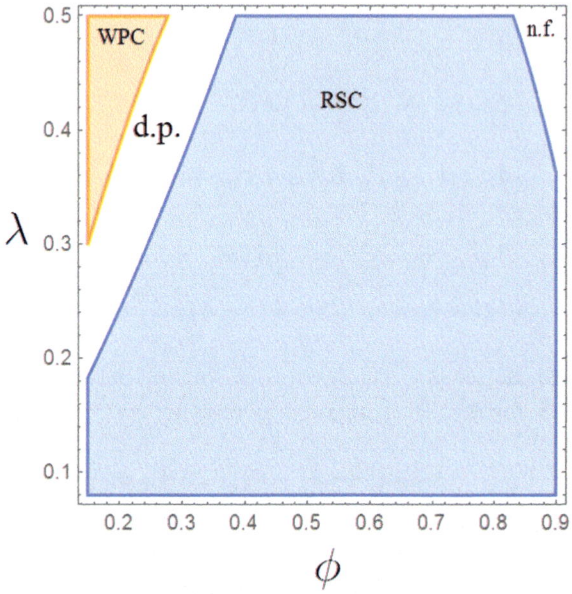


**Fig. 11** Pareto improving region with high parameter values

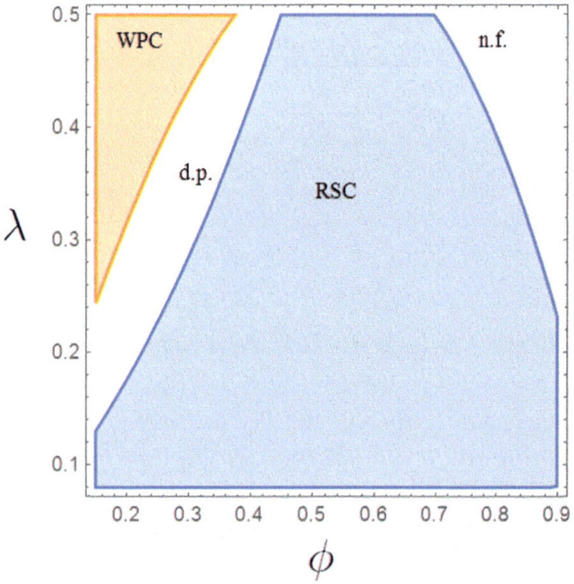


**Fig. 12** Pareto improving region with low parameter values

## References


Agrawal, V., & Tokay, L. B. (2010). Interdisciplinarity in closed-loop supply chain management research. In F. E. Ferguson & G. C. Souza (Eds.), *Closed-loop supply chains: New developments to improve the sustainability of business practices*. Boca Raton: CRC Press.

Cachon, G. P. (2003). Supply chain coordination with contracts. In S. Graves & T. de Kok (Eds.), *Handbooks in operations research and management science: Supply chain management*. Amsterdam: North- Holland.

Cachon, G. P., & Lariviere, M. A. (2005). Supply chain coordination with revenue-sharing contracts: Strengths and limitations. *Management Science, 51*(1), 30–44.

De Giovanni, P. (2017). Closed-loop supply chain coordination through incentives with asymmetric information. *Annals of Operations Research, 253*(1), 133–167.

De Giovanni, P. (2018). A joint maximization incentive in closed-loop supply chains with competing retailers: The case of spent-battery recycling. *European Journal of Operational Research, 268*(1), 128–147.

De Giovanni, P., & Ramani, V. (2018). Product cannibalization and the effect of a service strategy. *Journal of the Operational Research Society, 69*(3), 340–357.

De Giovanni, P., Reddy, P. V., & Zaccour, G. (2016). Incentive strategies for an optimal recovery program in a closed-loop supply chain. *European Journal of Operational Research, 249*(2), 605–617.

De Giovanni, P., & Zaccour, G. (2013). Cost–revenue sharing in a closed-loop supply chain. Advances in dynamic games. *Annals of the International Society of Dynamic Games, 12*, 395–421.

De Giovanni, P., & Zaccour, G. (2018). Optimal quality improvements and pricing strategies with active and passive product returns. Omega (in press).

De Giovanni, P., & Zaccour, G. (2019). A selective survey of game-theoretic models of closed-loop supply chains. *4OR*, 1–44.

Dockner, E. J., Jorgensen, S., Van Long, N., & Sorger, G. (2000). *Differential games in economics and management science*. Cambridge: Cambridge University Press.

Fleischmann, M., Beullens, P., Bloemhof-Ruwaard, J. M., & Van Wassenhove, L. N. (2001). The impact of product recovery on logistics network design. *Production and Operations Management, 10*(2), 156–173.

Fleischmann, M., Van Nunen, J. A., & Gräve, B. (2003). Integrating closed-loop supply chains and spare-parts management at IBM. *Interfaces, 33*(6), 44–56.

Genc, T. S., & De Giovanni, P. (2017). Trade-in and save: A two-period closed-loop supply chain game with price and technology dependent returns. *International Journal of Production Economics, 183*, 514–527.

Guide, V. D. R. Jr., & Van Wassenhove, L. N. (2009). OR FORUM—The evolution of closed-loop supply chain research. *Operations Research, 57*(1), 10–18.

Miao, Z., Fu, K., Xia, Z., &Wang, Y. (2017). Models for closed-loop supply chain with trade-ins. *Omega, 66*, 308–326.

Ramani, V., & De Giovanni, P. (2017). A two-period model of product cannibalization in an atypical closed-loop supply chain with endogenous returns: The case of dellreconnect. *European Journal of Operational Research, 262*(3), 1009–1027.

Savaskan, R. C., Bhattacharya, S., & Van Wassenhove, L. N. (2004). Closed-loop supply chain models with product remanufacturing. *Management Science, 50*(2), 239–252.

Sethi, S. P. (1983). Deterministic and stochastic optimization of a dynamic advertising model. *Optimal Control Applications and Methods, 4*(2), 179–184.

Sethi, S. P., & Thompson, G. L. (2000). *Optimal control theory: Applications to management science and economics*. Berlin: Springer.

Volkswagen (2011). Corporate Social Responsibility and Sustainability Report 2011.

Zhou, W., Zheng, Y., & Huang, W. (2017). Competitive advantage of qualified WEEE recyclers through EPR legislation. *European Journal of Operational Research, 257*(2), 641–655.

# A Steady-State Game of a Net-Zero Emission Climate Regime

**Olivier Bahn and Alain Haurie**

**Abstract** In this paper we propose a very simple steady-state game model that represents schematically interactions between coalitions of countries in achieving a necessary net-zero emission of GHGs in order to stabilize climate over the long term. We start from a situation where $m$ coalitions exist and behave as $m$ players in a game of sharing a global emission budget that can only be maintained by negative emission activities. We compare a fully "cooperative" solution with a Nash equilibrium solution implemented through an international emission trading scheme. We characterize the fully cooperative and Nash equilibrium solutions for this game in a deterministic context.

**Keywords** Carbon capture and storage · Carbon dioxide removal · Climate change · Mitigation · Integrated assessment · Steady-state game

This paper is our joint contribution to Georges Zaccour Festschrift. In preparing it we decided to impose ourselves three constraints: (1) the paper should have a game theoretic flavor (Georges has dedicated his career to fostering game theory applications to business and economics); (2) the style should not be too much "high-powered maths" (so Georges could use it as a source of exercises in his game theory courses for business students); (3) the topic should be original to a point of being a little provocative (at the limit of seriousness, some would say…). On top of that there was a page limit constraint. After several exchanges we concluded that our selected topic would fit all these requirements.

O. Bahn (✉)
GERAD and Department of Decision Sciences, HEC Montréal, Montréal, QC, Canada
e-mail: olivier.bahn@gerad.ca; olivier.bahn@hec.ca

A. Haurie
ORDECSYS, Chêne-Bougeries, Switzerland

University of Geneva, Geneva, Switzerland

GERAD, HEC Montréal, Montréal, QC, Canada



P.-O. Pineau et al. (eds.), *Games in Management Science*,
International Series in Operations Research & Management Science 280,
https://doi.org/10.1007/978-3-030-19107-8_7

# 1 Introduction

In a report published in 2016 (Shell-Corp 2016), Shell Corp. declared:

> In spite of the many challenges, the practical details of providing enough energy for a better life for everyone with net-zero emissions can be envisaged, and that is reassuring, even inspiring.

This report has been updated in 2018 with an energy transition scenario where net-zero emissions are envisaged as soon as 2070 (Shell-Corp 2018).

The UNFCCC Paris Agreement, negotiated at COP-21 and signed by a majority of nations, is dedicated to limiting to less than 2 °C the surface average temperature (SAT) rise in the twenty-first century. To achieve this goal the participating countries must reduce their emissions of greenhouse gases (GHG). These emissions are mostly related to the use of fossil energy as an economic production factor. Recent research on climate policies has shown that a possible set of backstop technologies could be "negative emission technologies" (EASAC 2018) and among them, more specifically, carbon dioxide removal (CDR), which refers to technologies that reduce the level of carbon dioxide in the atmosphere. Among such technologies one finds, in particular, bio-energy with carbon capture and storage (BECCS), direct air capture (DAC), ocean fertilization, etc. (Hallegatte et al. 2016; Mathesius et al. 2015; Meadowcroft 2013; Tavoni and Socolow 2013). On the other hand, recent research on climate modeling tends to show that to limit the SAT rise to 2 °C with sufficiently high probability, one should define a global limiting carbon budget of about 1 trillion tons over the whole period starting from the Industrial Revolution to the end of the twenty-first century (Allen et al. 2009; Knutti et al. 2016). After this period the world economy should observe a "net-zero emission" regime.

In such a regime, there will still be technologies emitting GHGs, however these emissions should be offset by negative emissions obtained somewhere on the planet. In order to foster economic efficiency, an international emission trading scheme could be established where each coalition of countries has an endowment in emission rights which corresponds exactly to the level of negative emissions obtained in this coalition. Then, the coalitions will use these emission rights in a strategic way, in order to extract the most welfare benefits, taking into consideration the actions of the other countries that intervene on the international market (see Helm (2003) for a discussion of the strategic use of allowances).

In this paper we formulate a highly schematic net-zero emission climate regime model, in asymptotic steady-state, with four coalitions of countries, two types of production economy, "dirty" (carbon intensive) and "clean" (based on renewables), a technology of carbon capture and sequestration that can be applied to reduce emissions in the dirty economy and a technology of direct air capture (DAC), which represents the CDR activities. Using parameters calibrated on the RICE model (Nordhaus and Boyer 2000; Nordhaus and Yang 1996), we explore possible optimal steady-state scenarios in a sustainable world economy.

The paper is organized as follows. In Sect. 2, we recall the characterization of asymptotic steady-state in optimal economic growth models and we discuss the relevance of this concept for sustainable economic development. In Sect. 3, we propose a simple steady-state model, with two types of productive economy, and two types of emission reduction technology. In Sect. 4, we formulate the search of a fully cooperative solution. In Sect. 5, we look at a Nash equilibrium solution when the coalitions compete on an international emission trading system. In Sect. 6, we analyze different scenarios using our simple model. Finally, we conclude in Sect. 7 and propose an agenda for further research.

## 2 Asymptotic Steady-State in Optimal Economic Growth Models

In convex optimal control models with stationary dynamics, turnpike properties establish an asymptotic convergence of state trajectories toward an extremal steady state. This property is analyzed in detail in Carlson (1991). The theory has been extended to open-loop Nash equilibrium models in Carlson and Haurie (1995). The property will still hold for systems which are asymptotically stationary e.g. the DICE and RICE models for global climate change (Nordhaus 1992, 1994; Nordhaus and Boyer 2000). The relevance of the turnpike concept for sustainable development models has been discussed in several papers; see, for instance, Haurie (2002, 2003, 2005a,b) and Haurie and Moresino (2008). So, we believe it makes sense to place our analysis of a net-zero emission climate regime at the turnpike level. We further assume a zero discount rate,[1] which is also compatible with a sustainable development framework. Our analysis will therefore be developed with a long-term steady-state model.

*Remark 1* In doing so, we pay tribute to: (i) an ecological economics strand, which advocates zero growth; (ii) a theory of justice principle, which would recommend zero discount rate for issues affecting all future generations; and (iii) our desire to keep the mathematics at elementary levels.

## 3 Long-Term Steady-State Model

We propose a simple steady-state economic model with carbon emissions due to the use of fossil energy and possibility to invest in CCS (carbon capture and storage) and in a CDR technology (of a DAC type). We assume that the emission budget is shared among four coalitions of countries, which are calibrated as corresponding to: (1) European-Union (EU-28) and Switzerland; (2) USA, Japan, Canada, Australia,

[1]Notice that in the archetypal DICE and RICE models of W. Nordhaus, a hyperbolic discount rate is used which asymptotically converges to zero.

and New Zealand; (3) Brazil, Russia, India, and China (BRIC); and (4) the rest of the World (ROW). As the model is supposed to represent the result of climate negotiations, it is reasonable to assume that all regions are coalitions considered as active players.

To model the economics of energy-environment, we adapt an approach that we have introduced in Bahn et al. (2008) and exploited in several papers (Andrey et al. 2016; Bahn 2010; Bahn et al. 2012, 2015; Bahn and Haurie 2008, 2016; Bahn et al. 2009, 2017). In each coalition, economic output can come from a carbon-intensive ("dirty") production function or from a renewable-rich ("clean") one.

## *3.1 List of Parameters and Variables*

The following parameters and variables enter in the description of our model:

$i = 1$ : index of the "dirty" economy;
$i = 2$ : index of the "clean" economy;
$i = 3$ : index of the "CCS" sector;
$i = 4$ : index of the "CDR" sector;
$j$ : index of the $m$ coalitions (groups of countries);
$A_1(j)$ : total factor productivity in the "dirty" economy production function of coalition $j$;
$A_2(j)$ : total factor productivity in the "clean" economy production function of coalition $j$;
$C(j)$ : total consumption of coalition $j$, in trillions ($10^{12}$) of dollars;
$E_i(j)$ : yearly GHG emissions in the economy $i = 1, 2$ of coalition $j$, in GtC ($10^9$ tons of carbon);
$I_{K_i}(j)$ : investment in capital $i = 1, \ldots, 4$ of coalition $j$, in trillions of dollars;
$K_1(j)$ : physical stock of "dirty" productive capital of coalition $j$, in trillions of dollars;
$K_2(j)$ : physical stock of "clean" productive capital of coalition $j$, in trillions of dollars;
$K_3(j)$ : physical stock of CCS capital of coalition $j$, in trillions of dollars;
$K_4(j)$ : physical stock of CDR capital of coalition $j$, in trillions of dollars;
$L(j)$ : labor (exogenously defined world population) of coalition $j$, in millions ($10^6$) of persons;
$L_1(j)$ : labor force in the "dirty" economy of coalition $j$, in millions of persons;
$L_2(j)$ : labor force in the "clean" economy of coalition $j$, in millions of persons;
$L_4(j)$ : labor force in the "CDR" sector of coalition $j$, in millions of persons;
$p$ : price of the GHG emission rights;
$\mathrm{W}(j)$ : welfare of coalition $j$;
$Y_i(j)$ : economic output in the economy $i = 1, 2$ of coalition $j$, in trillions of dollars;

$\alpha_1(j)$ : elasticity of capital in the "dirty" economy production function of coalition $j$;
$\alpha_2(j)$ : elasticity of capital in the "clean" economy production function of coalition $j$;
$\gamma(j)$ : total factor productivity in the CDR production function of coalition $j$;
$\delta(j)$ : elasticity of CDR capital in the CDR production function of coalition $j$;
$\theta_1(j)$ : elasticity of energy in the "dirty" economy production function of coalition $j$;
$\theta_2(j)$ : elasticity of energy in the "clean" economy production function of coalition $j$;
$\pi_1(j)$ : energy price in the "dirty" economy of coalition $j$;
$\pi_2(j)$ : energy price in the "clean" economy of coalition $j$;
$\varpi(j)$ : proportion of dirty economy GHG emissions removed through CCS activities;
$\upsilon(j)$ : maximum capture rate of CCS for coalition $j$;
$\phi_1(j)$ : energy efficiency of GHG emissions in the "dirty" economy of coalition $j$;
$\phi_2(j)$ : energy efficiency of GHG emissions in the "clean" economy of coalition $j$;
$\bar{\omega}(j)$ : atmospheric carbon removed by coalition $j$ through CDR activities, in GtC.

*Remark 2* In the remainder of the paper we will put a "bar" on top of the variables to remind the reader that we are dealing with the long-term asymptotic steady-state of optimally growing economies.

## 3.2 Steady-State Production

In each coalition $j = 1, \ldots, m$, economic output ($\bar{Y}$) is obtained by the "dirty" ($i = 1$) and "clean" ($i = 2$) economies, using capital ($\bar{K}_i$), labor ($\bar{L}_i$), and energy (directly linked with emissions $\bar{E}_i$) as production factors:

$$\begin{aligned}\bar{Y}(j) = {} & \bar{A}_1(j)\,\bar{K}_1(j)^{\bar{\alpha}_1(j)}\,(\bar{\phi}_1(j)\,\bar{E}_1(j))^{\bar{\theta}_1(j)}\,\bar{L}_1(j)^{1-\bar{\alpha}_1(j)-\bar{\theta}_1(j)} \\ & + \bar{A}_2(j)\,\bar{K}_2(j)^{\bar{\alpha}_2(j)}\,(\bar{\phi}_2(j)\,\bar{E}_2(j))^{\bar{\theta}_2(j)}\,\bar{L}_2(j)^{1-\bar{\alpha}_2(j)-\bar{\theta}_2(j)}.\end{aligned} \tag{1}$$

In each coalition, total labor ($\bar{L}$) is allocated to the two types of economies and the CDR sector:

$$\bar{L}(j) = \bar{L}_1(j) + \bar{L}_2(j) + \bar{L}_4(j), \quad j = 1, \ldots, m, \tag{2}$$

**Table 1** Steady-state population level (in millions)

| $j$ | 1 | 2 | 3 | 4 | Total |
|---|---|---|---|---|---|
| $\bar{L}(j)$ | 521 | 721 | 3035 | 4422 | 8699 |

**Table 2** Productivity factor ($\bar{A}$) and elasticities ($\bar{\alpha}, \bar{\theta}$)

| $j$ | 1 | 2 | 3 | 4 |
|---|---|---|---|---|
| $\bar{A}_1(j)$ | 0.08 | 0.09 | 0.04 | 0.03 |
| $\bar{\alpha}_1(j)$ | 0.3 | 0.3 | 0.3 | 0.3 |
| $\bar{\theta}_1(j)$ | 0.05 | 0.05 | 0.05 | 0.05 |
| $\bar{A}_2(j)$ | 0.08 | 0.09 | 0.04 | 0.03 |
| $\bar{\alpha}_2(j)$ | 0.3 | 0.3 | 0.3 | 0.3 |
| $\bar{\theta}_2(j)$ | 0.05 | 0.05 | 0.05 | 0.05 |

**Table 3** Energy prices ($\bar{\pi}$) and efficiencies ($\bar{\phi}$)

| $j$ | 1 | 2 | 3 | 4 |
|---|---|---|---|---|
| $\bar{\pi}_1(j)$ | 0.425 | 0.425 | 0.425 | 0.425 |
| $\bar{\pi}_2(j)$ | 0.65 | 0.65 | 0.65 | 0.65 |
| $\bar{\phi}_1(j)$ | 0.55 | 0.55 | 0.55 | 0.55 |
| $\bar{\phi}_2(j)$ | 5 | 5 | 5 | 5 |

where the steady-state population levels $\bar{L}$ are assumed to be given as in Table 1, below. These values are representative of some optimistic projections[2] for 2100. Parameters of the production functions are next given in Table 2, below. For simplicity, we assume that elasticities are the same in all economies. Total factor productivity ($A_i$) is higher in coalitions 1 and 2, and lower in coalitions 3 and 4. What distinguishes the "clean" economy from the "dirty" one is the energy efficiency, defined as the quantity of energy obtained from one unit of emissions. We also assume that the price of energy will be higher in the "clean" economy to reflect the higher cost of renewable energy; see Eq. (8), below. These assumed values are shown below in Table 3.

## 3.3 Steady-State Capital Levels

The steady-state value of capital is defined for each coalition $j = 1, \ldots, m$ as follows, where the critical parameter is the depreciation rate ($\delta_K$) set to 10%[3] per year for all types of capital $K_i$:

$$0 = \bar{I}_{K_i}(j) - \delta_{K_i} \bar{K}_i(j), \quad i = 1, 2, 3, 4. \tag{3}$$

In other words, investment $\bar{I}_{K_i}$ must balance depreciation in a steady-state situation.

[2]Source: United Nations, World population prospects: The 2017 revision. Accessed on-line: http://esa.un.org/unpd/wpp.

[3]A sensitivity analysis with a 5% depreciation rate per year will also be performed.

## 3.4 CDR and Steady-State Emission Budget

We assume that CDR is mainly done by direct air capture (DAC). This is a capital intensive activity that we model through a production function. In coalition $j$, the quantity $\bar{\omega}$ of DAC is obtained through a combination of two factors, a dedicated capital $K_4$ and a dedicated labor $L_4$, according to the following Cobb-Douglas production function:

$$\bar{\omega}(j) = \gamma(j)\bar{K}_4(j)^{\delta(j)}\bar{L}_4(j)^{1-\delta(j)}, \tag{4}$$

where the two critical parameters have been set to the values shown in Table 4.

*Remark 3* We acknowledge the difficulty in calibrating a meaningful DAC production function. The merit of the parameter values chosen here is that they enable us to obtain numerical solutions with our model. Notice we have assumed that all countries have access to the same technology with equal factor productivity and efficiency.

## 3.5 Steady-State CCS

We assume that emissions in the "dirty" economy can be reduced through carbon capture and sequestration (CCS). This is a capital intensive activity, where in each coalition $j$ the capital $\bar{K}_3$ for CCS is an add-on to "dirty capital" $\bar{K}_1$, so we posit:

$$\bar{K}_3(j) \leq \bar{K}_1(j), \tag{5}$$

and we also assume that the investment cost for $\bar{K}_3$ is a fraction of the investment cost for $\bar{K}_1$, namely $\bar{I}_{K_3}(j) \leq \zeta_{I_3}(j)\bar{I}_{K_1}(j)$. (We have set parameter $\zeta_{I_3}$ set to 0.01 for all coalitions.) We model the fraction $\varpi_j(\bar{K}_3, \bar{K}_1)$ of emissions captured as the simple function shown below:

$$\varpi_j(\bar{K}_3(j), \bar{K}_1(j)) = \upsilon(j)\frac{\bar{K}_3(j)}{\bar{K}_1(j)}, \tag{6}$$

where the maximum capture rate parameter $\upsilon$ is set to 50% in all coalitions $j$.

**Table 4** CDR production function parameters

| $j$ | 1 | 2 | 3 | 4 |
|---|---|---|---|---|
| $\gamma(j)$ | 0.001 | 0.001 | 0.001 | 0.001 |
| $\delta(j)$ | 0.3 | 0.3 | 0.3 | 0.3 |

*Remark 4* This again corresponds to a rough estimate, given that it should apply to the entire "dirty" economy, and not only to the power generation sector say, where the capture efficiency is expected to be much higher (up to 90%).

### 3.6 Steady-State Carbon Market

Denote

$$\bar{E}(j) = \big(1 - \varpi_j(\bar{K}_3(j), \bar{K}_1(j))\big)\bar{E}_1(j) + \bar{E}_2(j) \tag{7}$$

the total emissions that have to be compensated in coalition $j = 1, \ldots, m$. One assumes that there exists an international market for emission permits. Given the quantities $\bar{\omega}(j)$ put on the market by each coalition $j$, a market clearing price $\bar{p}(\Omega)$ and the emission abatement decisions taken by the coalitions are determined by the solution of a local optimization problem, defined below:

$$\max_{\bar{E}_1(j), \bar{E}_2(j)} \left\{ \bar{Y}(j) - \sum_{i=1,2} \bar{\pi}_i(j)\bar{\phi}_i(j)\bar{E}_i(j) - \sum_{i=1}^{4} \bar{I}_{K_i}(j) + \bar{p}(\Omega)\big(\bar{\omega}(j) - \bar{E}(j)\big) \right\}. \tag{8}$$

Here $\bar{Y}$ is defined as in Eq. (1) and the permit price $\bar{p}(\Omega)$ depends on the total permit supply $\Omega = \sum_{j=1}^{m} \bar{\omega}(j)$. As the permit price is clearing the market, the following condition holds:

$$0 = \sum_{j=1}^{m} \big(\bar{\omega}(j) - \bar{E}(j; \bar{p}(\Omega))\big), \tag{9}$$

where $\bar{E}(j; \bar{p}(\Omega))$ is the emission response of coalition $j$ to the permit price. Besides, at equilibrium, the permit price is equal to the productivity of emissions in both economies for each coalition. The permit price and emission response of each coalition $j$ are thus defined by:

$$0 = \frac{\partial \bar{Y}(j)}{\partial \bar{E}_i(j, \bar{p}(\Omega))} - \bar{\pi}_i(j)\bar{\phi}_i(j) - \bar{p}(\Omega), \quad i = 1, 2 \tag{10}$$

$$0 = \sum_{j=1}^{m} \bar{E}(j; \bar{p}(\Omega)) - \Omega. \tag{11}$$

Following the same developments as in Helm (2003), we can express the marginal influence of the supply of permits on the emission levels and market price. Taking derivatives of Eqs. (10) and (11) w.r.t. $\Omega$, denoted $'$, we obtain:

$$0 = \bar{E}_i(j;\bar{p}(\Omega))' - \frac{\bar{p}(\Omega)'}{\frac{\partial^2 \bar{Y}(j)}{\partial \bar{E}_i(j;\bar{p}(\Omega))^2}}, \quad i = 1,2 \tag{12}$$

$$0 = \sum_{j=1}^{m} \Big( \big(1 - \varpi_j(\bar{K}_3(j), \bar{K}_1(j))\big)\bar{E}_1(j;\bar{p}(\Omega))' + \bar{E}_2(j;\bar{p}(\Omega))' \Big) - 1. \tag{13}$$

Combining Eqs. (12) and (13), we get:

$$0 = \sum_{j=1}^{m} \left( \left(1 - \varpi_j(\bar{K}_3(j), \bar{K}_1(j))\right) \frac{\bar{p}(\Omega)'}{\frac{\partial^2 \bar{Y}(j)}{\partial \bar{E}_1(j;\bar{p}(\Omega))^2}} + \frac{\bar{p}(\Omega)'}{\frac{\partial^2 \bar{Y}(j)}{\partial \bar{E}_2(j;\bar{p}(\Omega))^2}} \right) - 1. \tag{14}$$

Therefore, the derivatives w.r.t. $\Omega$ of the permit price is given by:

$$\frac{d}{d\Omega} p(\Omega) = \frac{1}{\sum_{j=1}^{m} \left( \frac{1-\varpi_j(\bar{K}_3(j),\bar{K}_1(j))}{\frac{\partial^2 \bar{Y}(j)}{\partial \bar{E}_1(j)^2}} + \frac{1}{\frac{\partial^2 \bar{Y}(j)}{\partial \bar{E}_2(j)^2}} \right)} \tag{15}$$

where the partial derivatives of the production function are, respectively, given for $i = 1, 2$ by:

$$\frac{\partial \bar{Y}(j)}{\partial \bar{E}_i(j)} = \bar{\theta}_i(j)\bar{A}_i(j)\, \bar{K}_i(j)^{\bar{\alpha}_i(j)} \left(\bar{\phi}_i(j)\, \bar{E}_i(j)\right)^{\bar{\theta}_i(j)-1} \bar{L}_i(j)^{1-\bar{\alpha}_i(j)-\bar{\theta}_i(j)} \tag{16}$$

and

$$\frac{\partial^2 \bar{Y}(j)}{\partial \bar{E}_i(j)^2} = \bar{\theta}_i(j)(\bar{\theta}_i(j)-1)\bar{A}_i(j)\, \bar{K}_i(j)^{\bar{\alpha}_i(j)} \left(\bar{\phi}_i(j)\, \bar{E}_i(j)\right)^{\bar{\theta}_i(j)-2} \bar{L}_i(j)^{1-\bar{\alpha}_i(j)-\bar{\theta}_i(j)}. \tag{17}$$

## 3.7 Steady-State Consumption Level and Utility Function

In each coalition $j$, consumption is what is left from output after paying for energy cost and investment, plus the revenue from selling emission permits (negative if buying):

$$\bar{C}(j) = \bar{Y}(j) - \sum_{i=1,2} \bar{\pi}_i(j)\bar{\phi}_i(j)\bar{E}_i(j) - \sum_{i=1}^{4} \bar{I}_{K_i}(j) + \bar{p}\big(\bar{\omega}(j) - \bar{E}(j)\big). \tag{18}$$

We then assume that each coalition has the following utility function:

$$\bar{W}(j) = \bar{L}(j) \log \left[ \frac{\bar{C}(j)}{\bar{L}(j)} \right]. \tag{19}$$

## 4 Fully Cooperative Solution

Introduce the weighted performance criterion defined by:

$$\bar{W} = \sum_{j=1}^{m} r(j)\bar{W}(j), \tag{20}$$

with $r(j) > 0$ and $\sum_j r(j) = 1$. The fully cooperative solution maximizes $\bar{W}$, as defined in Eq. (20), under the condition that the total emissions, net of CCS, must be compensated by DAC. That is, one has the following constraint set:

$$\begin{aligned}
\bar{W}(j) &= \bar{L}(j) \log \left[ \frac{\bar{C}(j)}{\bar{L}(j)} \right] \quad \forall j \\
\bar{C}(j) &= \bar{Y}(j) - \sum_{i=1,2} \bar{\pi}_i(j)\bar{\phi}_i(j)\bar{E}_i(j) - \sum_{i=1}^{4} \bar{I}_{K_i}(j) \quad \forall j \\
\varpi_j &= \upsilon(j)\frac{\bar{K}_3(j)}{\bar{K}_1(j)} \quad \forall j \\
\bar{K}_3(j) &\le \bar{K}_1(j) \quad \forall j \\
\bar{\omega}(j) &= \gamma(j)\bar{K}_4(j)^{\delta(j)}\bar{L}_4(j)^{1-\delta(j)} \quad \forall j \\
0 &= \bar{I}_{K_i}(j) - \delta_{K_i}\bar{K}_i(j), \quad i = 1, 2, 3, 4 \quad \forall j \\
\bar{L}(j) &= \bar{L}_1(j) + \bar{L}_2(j) + \bar{L}_4(j) \quad \forall j \\
\bar{Y}(j) &= \bar{A}_1(j)\, \bar{K}_1(j)^{\bar{\alpha}_1(j)}\, (\bar{\phi}_1(j)\, \bar{E}_1(j))^{\bar{\theta}_1(j)}\, \bar{L}_1(j)^{1-\bar{\alpha}_1(j)-\bar{\theta}_1(j)} \\
&\quad + \bar{A}_2(j)\, \bar{K}_2(j)^{\bar{\alpha}_2(j)}\, (\bar{\phi}_2(j)\, \bar{E}_2(j))^{\bar{\theta}_2(j)}\, \bar{L}_2(j)^{1-\bar{\alpha}_2(j)-\bar{\theta}_2(j)} \quad \forall j \\
\bar{E}(j) &= (1 - \varpi_j)\bar{E}_1(j) + \bar{E}_2(j) \quad \forall j \\
0 &= \sum_{j=1}^{m} \left( \bar{\omega}(j) - \bar{E}(j) \right)
\end{aligned}$$

# 5 Computing a Steady-State Nash Equilibrium

Assume that an international emission trading scheme is put in place. In a steady-state net-zero emission regime, the total supply of emission permits should be equal to the amount of negative emissions obtained through the use of DAC activities. Each coalition, considered as a "big" player $j$, may then use the supply of emission permits on the market as a strategic variable in order to maximize returns from the emission budget share they control. The other strategic variable will be the investment in CCS technologies in order to reduce the amount of emissions to offset in the "dirty" economy.

A Nash equilibrium is obtained when each coalition has chosen its strategy as its best reply to the choices made by the other coalitions, in an open-loop information structure. A strategy $\bar{s}_j$ for coalition $j$ consists more precisely of:

- a level of investments $\{\bar{I}_1(j), \bar{I}_2(j), \bar{I}_3(j), \bar{I}_4(j)\}$ in capital $K_1$, $K_2$, $K_3$, and $K_4$, respectively;
- a level of labor allocations $\{\bar{L}_1(j), \bar{L}_2(j), \bar{L}_4(j)\}$ in the two types of economies and the CDR sector, respectively;
- and an amount of emission rights $\bar{\omega}(j)$ put on the carbon market.

As indicated in Sect. 3, the supply of emission rights by coalition $j$ must be equal to the negative emissions produced by this coalition. Market clearing conditions determine the price of emission rights and the emission levels in both economies of each coalition.

To compute a Nash equilibrium, one can either use a formulation as a nonlinear complementarity problem and solve it with the help of PATH algorithm (Ferris and Munson 2000) or more simply use a cobweb approach, which in our case has always converged, although convergence is not warranted.

# 6 Scenario Analysis

We are now ready to proceed with a numerical economics exercise. We will compare the solutions obtained when assuming a fully cooperative approach (with equal weight given to the different coalitions) and a non-cooperative approach (with an international market for GHG emission rights).

## *6.1 Carbon Removed by DAC*

Table 5 gives first the amount of carbon removed through DAC activities in each coalition.

**Table 5** DAC activities

| Coalition | 1 | 2 | 3 | 4 |
|---|---|---|---|---|
| Cooperative solution | 1.94 | 3.64 | – | – |
| Nash solution | – | – | 2.13 | 2.98 |

**Table 6** GHG emissions

| | Cooperative solution | | | | Nash solution | | | |
|---|---|---|---|---|---|---|---|---|
| Coalition | 1 | 2 | 3 | 4 | 1 | 2 | 3 | 4 |
| $\bar{E}_1$ | – | – | – | – | – | – | – | – |
| $\bar{E}_2$ | 0.47 | 0.69 | 1.72 | 2.71 | 0.62 | 1.03 | 1.34 | 2.12 |

**Table 7** Capital stocks with a 10% depreciation rate

| | Cooperative solution | | | | Nash solution | | | |
|---|---|---|---|---|---|---|---|---|
| Coalition | 1 | 2 | 3 | 4 | 1 | 2 | 3 | 4 |
| $\bar{K}_1$ | – | – | – | – | – | – | – | – |
| $\bar{K}_2$ | 193.36 | 304.72 | 492.82 | 778.56 | 223.88 | 376.13 | 478.05 | 757.43 |
| $\bar{K}_3$ | – | – | – | – | – | – | – | – |
| $\bar{K}_4$ | 21.31 | 45.01 | – | – | – | – | 11.45 | 16.04 |

In the cooperative setting, the two lower-income coalitions (Coalition 3 and 4) do not enter into DAC.

In the Nash setting, conversely, the most productive coalitions (Coalition 1 and 2) do not enter into DAC, whereas Coalition 3 and 4 are engaged into DAC activities so as to supply the carbon market with emission rights (the amount of carbon removed in the atmosphere corresponds to the supply of permits).

## 6.2 Steady-State Carbon Emissions

Table 6 reports on GHG emissions for the two economies in each coalition.

In the cooperative and Nash settings, all coalitions switch to the "clean" economy, and there is thus no emission from the "dirty" economy ($\bar{E}_1 = 0$) at the steady-state.

## 6.3 Capital Stocks

Table 7 reports on capital stock levels in each coalition.

In the cooperative and Nash settings, we note that there is no need for CCS capital ($\bar{K}_3$) since only the "clean" economy is in action. Besides, in all solution settings, the accumulation of CDR capital ($\bar{K}_4$) is consistent with the DAC activities reported in Table 5.

**Table 8** Capital stocks with a 5% depreciation rate

| | Cooperative solution | | | | Nash solution | | | |
|---|---|---|---|---|---|---|---|---|
| Coalition | 1 | 2 | 3 | 4 | 1 | 2 | 3 | 4 |
| $\bar{K}_1$ | – | – | – | – | – | – | – | – |
| $\bar{K}_2$ | 532.15 | 838.72 | 1356.72 | 2000.00 | 616.29 | 1035.44 | 1315.65 | 2084.60 |
| $\bar{K}_3$ | – | – | – | – | – | – | – | – |
| $\bar{K}_4$ | 58.64 | 123.79 | – | – | – | – | 31.83 | 44.59 |

**Table 9** Labor allocations

| | Cooperative solution | | | | Nash solution | | | |
|---|---|---|---|---|---|---|---|---|
| Coalition | 1 | 2 | 3 | 4 | 1 | 2 | 3 | 4 |
| $\bar{L}_1$ | – | – | – | – | – | – | – | – |
| $\bar{L}_2$ | 423 | 563 | 2970 | 4692 | 473 | 652 | 2895 | 4587 |
| $\bar{L}_4$ | 50 | 90 | – | – | – | – | 75 | 105 |

**Table 10** Consumption: total consumption $C$ and per capita consumption $c$

| | Cooperative solution | | | | Nash solution | | | |
|---|---|---|---|---|---|---|---|---|
| Coalition | 1 | 2 | 3 | 4 | 1 | 2 | 3 | 4 |
| $\bar{C}$ | 41.47 | 64.36 | 109.42 | 172.86 | 47.43 | 77.81 | 106.93 | 168.60 |
| $\bar{c}$ | 0.08768 | 0.09871 | 0.03684 | 0.03684 | 0.10027 | 0.11934 | 0.03601 | 0.03593 |

In Table 8, we show the effect on capital stocks of lowering the capital depreciation rate from 10% to 5%. As expected a lower capital cost induces higher capital stocks.

## 6.4 *Labor Allocation*

Table 9 reports on labor allocations in each coalition.

In all solution settings, the allocation of labor to the productive economies is consistent with the fact that only the "clean" economy is active in each case.

In the cooperative setting, we note that a substantial part of the work force is allocated to the CDR sector of Coalition 1 and 2 (up to 14%). We refer to the caveat of Remark 3; probably the productivity of labor in DAC activities is assumed too low for these two coalitions. In the other cases, when active, the CDR sector requires only a small fraction (2–3%) of the total work force.

## 6.5 *Consumption*

Table 10 reports on consumption in each coalition.

In the cooperative setting, the levels of consumption per capita are quite different (by a factor 2 to 3) between Coalitions 1–2 and Coalitions 3–4. With these consumption levels, the global weighted utility criterion is equal to $-6288$.

In the Nash setting, the differences in per capita consumption levels between Coalitions 1–2 and Coalitions 3–4 are slightly higher than in the cooperative setting. In addition, per capita consumption levels for Coalition 1 and 2 are higher than in the case of full cooperation, whereas it is the opposite for Coalition 3 and 4. With these consumption levels the weighted utility criterion is equal to −9860, a lower value than in the full cooperative case, which is consistent with theory.

# 7 Conclusion

In this paper we have worked with a "toy" economic model, representing possible interactions of different coalitions of countries in managing a zero-net GHG emission regime in a steady-state. Although the extreme simplicity of the model prevents us from drawing any policy conclusion with sufficient confidence, we may, however, emphasize a few interesting outcomes of this exercise in computational economics.

1. If we believe in the Earth science that is supporting the UNFCCC Paris Agreement, as we do, we must envision a long-term future with zero-net emissions. In such a regime, all GHG emissions should be offset by negative emissions. Some countries could have a comparative advantage in harnessing direct air capture (DAC) or other carbon dioxide removal activities. Think, for instance, of the Persian Gulf countries that have deserted lands and a lot of sunshine that could be used to energize large DAC factories. Instead of selling oil and gas, these countries could sell emission rights (or both…).
2. In the computational economics experiment that we have developed, it appears that the passage through an international emission trading scheme to implement a zero-net emission regime does not lead to a Pareto optimal solution. Actually, the solution computed (with all due caveats) within a Nash setting corresponds to a lower weighted utility criterion value, with the lower income countries (our Coalitions 3–4) being worse off compared to the cooperative setting.
3. The development of this very simple model is a first step toward a deeper and more serious investigation of zero-net GHG emission regimes and exploration of the welfare implications of international agreements leading to such regimes. Further research must be done to assess the comparative advantages for the different coalitions in developing DAC or CDR activities. Similarly, one should evaluate the potential for carbon sequestration in different world regions. Then, a dynamic model, with a non-zero discount rate and long-time horizon could be envisioned. Uncertainty about the access to efficient DAC technologies could be included in the model following the approach advocated in Bahn et al. (2008, 2009). No doubt that game theory should play an important role in these investigations. Prof. Georges Zaccour could help…

**Acknowledgements** This research has been supported by the Natural Sciences and Engineering Research Council of Canada (O. Bahn) and the Qatar National Research Foundation (A. Haurie).

## References


Allen, M. R., Frame, D. J., Huntingford, C., Jones, C. D., Lowe, J. A., Meinshause, M., et al. (2009). Warming caused by cumulative carbon emissions towards the trillionth tonne. *Nature, 458*, 1163–1166.

Andrey, C., Bahn, O., & Haurie, A. (2016). Computing $\alpha$-robust equilibria in two integrated assessment models for climate change. In F. Thuijsman & F. Wagener (Eds.), *Advances in dynamic and evolutionary games* (Vol. 14, pp. 283–300). Cham: Springer.

Bahn, O. (2010). Combining adaptation and mitigation: A game theoretic approach. *INFOR, 48*, 193–201.

Bahn, O., Chesney, M., & Gheyssens, J. (2012). The effect of proactive adaptation on green investment. *Environmental Science and Policy, 18*, 9–24.

Bahn, O., Chesney, M., Gheyssens, J., Knutti, R., & Pana, A. (2015). Is there room for geoengineering in the optimal climate policy mix? *Environmental Science and Policy, 48*, 67–76.

Bahn, O., & Haurie, A. (2008). A class of games with coupled constraints to model international GHG emission agreements. *International Game Theory Review, 10*(4), 337–362.

Bahn, O., & Haurie, A. (2016). A cost-effectiveness differential game model for climate agreements. *Dynamic Games and Applications, 6*(1), 1–19.

Bahn, O., Haurie, A., & Malhamé, R. (2008). A stochastic control model for optimal timing of climate policies. *Automatica, 44*, 1545–1558.

Bahn, O., Haurie, A., & Malhamé, R. (2009). A stochastic control/game approach to the optimal timing of climate policies. In J. Filar & A. Haurie (Eds.), *Uncertainty and environmental decision making: A handbook of research and best practice*. International series in operations research & management science (Vol. 138, pp. 211–237). Boston: Springer.

Bahn, O., Haurie, A., & Malhamé, R. (2017). Limit game models for climate change negotiations. In J. Apaloo & B. Viscolani (Eds.), *Advances in dynamic and mean field games* (Vol. 15, pp. 27–48). Cham: Springer.

Carlson, D. A., & Haurie, A. (1995). A turnpike theory for infinite horizon open-loop differential games with decoupled dynamics. In G. J. Olsder (Ed.), *New trends in dynamic games and applications*. Annals of the international society of dynamic games (Vol. 3, pp. 353–376). Boston: Birkhäuser.

Carlson, D. A., Haurie, A., & Leizarowitz, A. (1991). *Infinite horizon optimal control: Deterministic and stochastic systems* (Vol. 332). New York: Springer.

EASAC. (2018). Negative emission technologies: What role in meeting Paris agreement targets? Technical report, EASAC, Secretariat Deutsche Akademie der Naturforscher Leopoldina German National Academy of Sciences Jägerberg 1 D-06108 Halle (Saale).

Ferris, M. C., & Munson, T. S. (2000). Complementarity problems in GAMS and the PATH solver. *Journal of Economic Dynamics and Control, 24*, 165–188.

Hallegatte, S., Rogelj, J., Allen, M., Clarke, L., Edenhofer, O., Field, C. B., et al. (2016). Mapping the climate change challenge. *Nature Climate Change, 6*(7), 663–668.

Haurie, A. (2002). Turnpikes in multidiscount rate environments and GCC policy evaluation. In G. Zaccour (Ed.), *Optimal control and differential games* (pp. 39–52). Dordrecht: Kluwer.

Haurie, A. (2003). Integrated assessment modeling for global climate change: An infinite horizon optimization viewpoint. *Environmental Modeling and Assessment, 8*(3), 117–132.

Haurie, A. (2005a). Hierarchical and asymptotic optimal control models for economic sustainable development. In C. Deissenberg & R. Hartl (Eds.), *Optimal control and dynamic games* (pp. 61–76). Dordrecht: Kluwer.

Haurie, A. (2005b). A multigenerational game model to analyze sustainable development. *Annals of Operations Research, 137*, 369–386.

Haurie, A., & Moresino, F. (2008). Singularly perturbed piecewise deterministic games. *SIAM Journal on Control and Optimization, 47*(1), 73–91.

Helm, C. (2003). International emissions trading with endogenous allowance choices. *Journal of Public Economics, 87*, 2737–2747.
Knutti, R., Rogelj, J., Sedlacek, J., & Fischer, E. M. (2016). A scientific critique of the two-degree climate change target. *Nature Geoscience, 9*(1), 13–18.
Mathesius, S., Hofmann, M., Caldeira, K., & Schellnhuber, H.-J. (2015). Long-term response of oceans to $CO_2$ removal from the atmosphere. *Nature Climate Change, 5*(12), 1107–1113.
Meadowcroft, J. (2013). Exploring negative territory carbon dioxide removal and climate policy initiatives. *Climatic Change, 118*(1), 137–149.
Nordhaus, W. D. (1992). An optimal path for controlling greenhouses gases. *Science, 258*, 315–1319.
Nordhaus, W. D. (1994). *Managing the global commons: The economics of climate change.* Cambridge: MIT Press.
Nordhaus, W. D., & Boyer, J. (2000). *Warming the world: Economic models of global warming.* Cambridge: MIT Press.
Nordhaus, W. D., & Yang, Z. (1996). RICE - A regional dynamic general equilibrium model of alternative climate change strategies. *American Economic Review, 86*, 741–765.
Shell-Corp. (2016). A better life with a healthy planet: Pathways to net-zero emissions. Technical report, Royal Dutch Shell.
Shell-Corp. (2018). Shell scenarios Sky: Meeting the goals of the Paris agreement. Technical report, Royal Dutch Shell.
Tavoni, M., & Socolow, R. (2013). Modeling meets science and technology: An introduction to a special issue on negative emissions. *Climatic Change, 118*(1), 1–14.

# Dynamic Models of the Firm with Green Energy and Goodwill with a Constant Size of the Output Market

**Herbert Dawid, Richard F. Hartl, and Peter M. Kort**

**Abstract** This paper analyzes a dynamic model of the firm. We focus on the effect of investment in green energy. We explicitly take into account that green energy has a positive side effect, namely that it contributes to the goodwill of the firm and thus increases demand. Different models are proposed and the solutions range from monotonic saddle point convergence to history-dependent Skiba behavior.

**Keywords** Green capital · Goodwill · Optimal investment · Skiba curve

## 1 Introduction

Georges Zaccour considerably contributed to the literature of environmental economics using dynamic models; see, e.g., the edited volumes Breton and Zaccour (1991), Carraro et al. (1994), Loulou et al. (2005), and the articles Breton et al. (2010), Jørgensen et al. (2010), André et al. (2011), Domenech et al. (2011), Masoudi and Zaccour (2013, 2014, 2017), Ben Youssef and Zaccour (2014), Eyland and Zaccour (2014), and Masoudi et al. (2016). Therefore, the topic of our paper is

H. Dawid (✉)
Department of Business Administration and Economics and Center for Mathematical Economics, Bielefeld University, Bielefeld, Germany
e-mail: hdawid@wiwi.uni-bielefeld.de

R. F. Hartl
Department of Business Administration, Production and Operations Management, University of Vienna, Vienna, Austria
e-mail: richard.hartl@univie.ac.at

P. M. Kort
Department of Econometrics and Operations Research & CentER, Tilburg University, Tilburg, The Netherlands

Department of Economics, University of Antwerp, Antwerp, Belgium
e-mail: Kort@uvt.nl



P.-O. Pineau et al. (eds.), *Games in Management Science*,
International Series in Operations Research & Management Science 280,
https://doi.org/10.1007/978-3-030-19107-8_8

green energy. A second reason for this research is that recently, more and more firms are active producers of green energy as part of their Corporate Social Responsibility (CSR) policy, e.g. Tesla or Google. We build a dynamic model of the firm in which energy is the main input of the production process. This energy can be delivered in a conventional way or by building a stock of green energy capital, e.g. by using wind mills or solar panels. This is a topical subject; see, e.g., the recent article in The Economist (2017). In addition, the firm can also reduce its energy bill using the green capital or even sell energy making use of feed-in tariffs.

Our model belongs to a stream of literature, capturing the transition from traditional to renewable resources for energy production. Wirl (1991, 2008) analyzes the investment decisions of renewable energy producers. They compete with incumbent producers with market power, which rely on production with exhaustible resources. Tsur and Zemel (2011) are closer to our setup by studying the transition from fossil fuels to solar energy within a firm. They characterize scenarios under which solar energy is adopted and finally dominates the industry. Even more close to our contribution is Amigues et al. (2015), taking into account that fossil resources are exhaustible and showing that the optimal transition may be characterized by different phases, starting with exclusive use of the non-renewable resource, exhibiting then parallel use of both resources, and a final phase in which only renewable resources are used. The investment into renewable energy may either begin before actual production of renewable energy or be delayed to wait for a sufficient increase of the energy price. Compared to Amigues et al. (2015), our paper is different in that we analyze the dynamic implications of the interplay of cost and goodwill considerations.

In fact, besides Amigues et al. (2015), another related paper in this area is Dawid et al. (2018), which is the starting point of our analysis. As in Dawid et al. (2018) we assume that the firm's stock of goodwill, created by being "green," pushes up the consumers' willingness to pay, but with a different inverse demand function than in Dawid et al. (2018). In our setting the output price is proportional to the goodwill, however, differently from Dawid et al. (2018), the market size is independent from the stock of goodwill. We consider three different models. In our first model, the firm is only able to use green energy. Numerically we find a solution with a unique saddle point stable steady state to which the firm monotonically converges. The second model analyzes a firm that, besides green energy, can also choose traditional energy obtained from the energy market (modeled as a control). For this model, we also find a unique saddle point stable steady state, but the difference with the previous model is that convergence takes place in an oscillating way. These two findings are qualitatively similar to those obtained in Dawid et al. (2018) for a different inverse demand function. The main innovative contribution of this paper compared to Dawid et al. (2018) is the consideration of the third model, in which we explicitly take into account that the green capital stock should be of certain size before the firm really can be considered as a green firm. To model this, we impose that green capital has an S-shaped effect on the development of firm goodwill. Taking into account this non-linear effect of green capital on goodwill accumulation gives rise to a history dependent solution, under which the initial level of goodwill and green capital determines whether the firm ends up in a green steady state, with positive green capital and goodwill stock, or both green capital and goodwill are fully depleted in the long run.

The content of the paper is as follows: Sect. 2 describes the general model. Section 3 analyzes the model where only green capital stock is used in the production process, whereas Sect. 4 looks at the extension where the production process has two possible inputs, green capital and traditional energy. In Sect. 5 we analyze the implications of the S-shaped effect of green capital. Finally, Sect. 6 concludes.

## 2 The Model

In Dawid et al. (2018), we analyze a dynamic model of the firm with a production process that has energy as input. Energy can be generated by either traditional energy or green energy. The usage rate of traditional energy, e.g. oil or gas, is modeled as a control variable, $X$. An alternative way of obtaining energy is to build up a stock of green capital, $K$. Green capital stock can be represented by solar panels or windmills. The evolution of the green capital stock over time follows the traditional capital accumulation equation

$$\dot{K} = I - \delta K, \qquad K(0) = K_0$$

in which $I$ is investment in the green capital stock, whereas $\delta$ is the depreciation rate.

The firm's own energy infrastructure, $K$, generates a flow of green energy, $h(K)$, that adds to the energy bought on the energy grid/market, $X$, yielding $E := X + h(K)$. We assume that $h(K) = K$. Hence, the production factor energy, $E$, is given by

$$E = K + X,$$

generating an output $q(E)$.

Investing in and using green energy makes that the firm exposes itself as a "green firm." This helps to create goodwill, $G$. Goodwill in turn helps to enhance the effect of advertising, $a$. We get that the goodwill dynamics becomes

$$\dot{G} = f(K)a - \delta_G G, \qquad G(0) = G_0. \tag{1}$$

where the effectiveness of advertising, $f(K)$, is an increasing function of the green capital stock, and $\delta_G$ is the depreciation rate of goodwill.

The firm's output can be sold on the market. The output price increases in goodwill and decreases in quantity, i.e.

$$p = p(G, q(E)), \quad p_G > 0, p_q < 0.$$

where $p$ is the output price.

The firm incurs some costs. First, there are advertising costs denoted by $C_a(a)$. Further we have (green energy) investment costs $C_s(I)$, and a constant unit cost of traditional energy that equals $p_X$.

The resulting dynamic model of the profit maximizing firm then equals

$$\max_{I,a,X} \int_0^{\infty} e^{-rt} \{p(G, q(E)) q(E) - C_a(a) - C_s(I) - p_X X\} dt, \tag{2}$$

$$\dot{G} = f(K) a - \delta_G G, \qquad G(0) = G_0 \tag{3}$$

$$\dot{K} = I - \delta K, \qquad K(0) = K_0 \tag{4}$$

$$E = K + X. \tag{5}$$

We take into account that advertising is non-negative, i.e. $a \geq 0$. On the other hand, we allow disinvestments, but under the restriction that the stock of green capital satisfies $K \geq 0$. Note that the consumption of conventional energy, $X$, can be positive or negative. Indeed, if the firm has a high green capital stock, $K$, and/or a small production rate (proportional to $E$), then $X = E - K$ can be negative and the excess energy $|E - K|$ can be sold making use of a feed-in tariff, which is assumed to be equal to $p_X$.

We depart from Dawid et al. (2018) in two different ways. First, where in Dawid et al. (2018) we employ the linear inverse demand function $p = \max[g(G) - \alpha q(E), 0]$, here we adopt the more multiplicative formulation

$$p = g(G) \max[\beta - \alpha q(E), 0].$$

where $\alpha$ and $\beta$ are positive constants, and $g(.)$ is an increasing function of goodwill $G$. We chose this formulation because in this way the willingness to pay goes up with goodwill, but the overall market size stays the same.[1]

Second, in Dawid et al. (2018), the effectiveness of advertising function $f(K)$ is linear in $K$. We also study the model with this functional form in the next two sections. However, Sect. 5 contains an analysis where $f(K)$ is S-shaped, the motivation of which is that green energy capital should have reached some threshold level before a firm is really recognized being a green firm.

[1]In the following analysis we will implicitly assume that the price is positive and drop the maximum operator in the inverse demand. We have checked that along optimal paths we consider in the numerical analysis this assumption is indeed correct.

## 3 Analysis of the Model with Only Clean Input

Before we analyze the general model, we first look at a simplified variant, where the traditional energy use is abolished, i.e., $X = 0$. Setting $X = 0$, the dynamic model of the firm turns into

$$\max_{I,a} \int_0^\infty e^{-rt} \{[g(G)(\beta - \alpha q(K))] q(K) - C_a(a) - C_s(I)\} dt$$

$$\dot{G} = f(K)a - \delta_G G, \qquad G(0) = G_0$$

$$\dot{K} = I - \delta K, \qquad K(0) = K_0, \ K \geq 0.$$

Following Grass et al. (2008), and Feichtinger and Hartl (1986), the current value Hamiltonian equals

$$H = [g(G)(\beta - \alpha q(K))] q(K) - C_a(a) - C_s(I) + \lambda (f(K)a - \delta_G G)$$
$$+\mu (I - \delta K) + \nu K.$$

This yields the following necessary optimality conditions of the maximum principle:

$$H_I = 0 = -C_s'(I) + \mu,$$
$$H_a = 0 = -C_a'(a) + \lambda f(K),$$
$$\dot{\lambda} = (r + \delta_G)\lambda - g'(G)(\beta - \alpha q(K)) q(K),$$
$$\dot{\mu} = (r + \delta)\mu - g(G) q'(K)(\beta - 2\alpha q(K)) - \lambda a f'(K) - \nu.$$

To proceed with the analysis we use the following functional forms:

$$C_a(a) = \frac{\varphi}{2} a^2, \tag{6}$$

$$C_s(I) = \frac{\gamma}{2} I^2, \tag{7}$$

$$f(K) = K, \tag{8}$$

$$q(K) = \eta K, \tag{9}$$

$$g(G) = \theta G^{\rho}, \qquad \text{with} \theta > 0, 0 < \rho < 1. \tag{10}$$

Combining these with the necessary optimality conditions, we get:

$$H_I = 0 = -\gamma I + \mu,$$
$$H_a = 0 = -\varphi a + \lambda K,$$

$$\dot{\lambda} = (r + \delta_G)\lambda - \rho\theta G^{\rho-1}\eta K(\beta - \alpha\eta K),$$
$$\dot{\mu} = (r + \delta)\mu - \theta G^{\rho}\eta(\beta - 2\alpha\eta K) - \lambda a - \nu,$$

with $\nu \geq 0$ and $\nu K = 0$. The optimal investment rate and advertising rate are given by

$$I = \frac{\mu}{\gamma},$$
$$a = \frac{\lambda K}{\varphi}.$$

Combining this with the dynamic equations for states and co-states, we obtain the canonical system

$$\dot{\lambda} = (r + \delta_G)\lambda - \rho\theta G^{\rho-1}\eta K(\beta - \alpha\eta K), \tag{11}$$
$$\dot{\mu} = (r + \delta)\mu - \theta G^{\rho}\eta(\beta - 2\alpha\eta K) - \lambda^2\frac{K}{\varphi} - \nu, \tag{12}$$
$$\dot{G} = \frac{\lambda K^2}{\varphi} - \delta_G G, \qquad G(0) = G_0, \tag{13}$$
$$\dot{K} = \frac{\mu}{\gamma} - \delta K, \qquad K(0) = K_0. \tag{14}$$

Analyzing this canonical system leads to the following result:

**Proposition 1** *The canonical system* (11)–(14) *need not have a unique steady state.*

*Proof* Assuming that $K > 0$ we set $\nu = 0$. Hence, from (11), (13), and (14) we obtain that in a possible steady state it must hold that

$$\hat{G} = \frac{\lambda}{\varphi\delta_G}\left(\frac{\mu}{\gamma\delta}\right)^2, \tag{15}$$
$$\hat{K} = \frac{\mu}{\gamma\delta}, \tag{16}$$
$$\hat{\lambda} = \frac{\rho\theta\eta}{\gamma\delta(r + \delta_G)}\left(\frac{\hat{\lambda}}{\varphi\delta_G}\left(\frac{\mu}{\gamma\delta}\right)^2\right)^{\rho-1}\mu\left(\beta - \alpha\eta\frac{\mu}{\gamma\delta}\right). \tag{17}$$

From (17) we can explicitly solve for $\lambda$:

$$\hat{\lambda} = \left(\frac{\rho\theta\eta(\gamma\delta)^{1-2\rho}(\varphi\delta_G)^{1-\rho}}{(r + \delta_G)}\right)^{\frac{1}{2-\rho}}\mu^{\frac{2\rho-1}{2-\rho}}\left(\beta - \alpha\eta\frac{\mu}{\gamma\delta}\right)^{\frac{1}{2-\rho}}. \tag{18}$$

We now substitute (15), (16), and (18) into (12) and setting $\nu = 0$, $\dot{\mu} = 0$. We eventually obtain that

$$(r+\delta)\,\mu^{\frac{2-4\rho}{2-\rho}} = \left[\theta\eta\left(\frac{1}{\varphi\delta_G\gamma^2\delta^2}\right)^{\rho}\left(\frac{\rho\theta\eta\,(\gamma\delta)^{1-2\rho}\,(\varphi\delta_G)^{1-\rho}}{(r+\delta_G)}\right)^{\frac{\rho}{2-\rho}}\right]$$
$$\times\left(\beta-\alpha\eta\frac{\mu}{\gamma\delta}\right)^{\frac{\rho}{2-\rho}}\left(\beta-2\alpha\eta\frac{\mu}{\gamma\delta}\right)$$
$$+\left[\frac{1}{\gamma\delta\varphi}\left(\frac{\rho\theta\eta\,(\gamma\delta)^{1-2\rho}\,(\varphi\delta_G)^{1-\rho}}{(r+\delta_G)}\right)^{\frac{2}{2-\rho}}\right]\left(\beta-\alpha\eta\frac{\mu}{\gamma\delta}\right)^{\frac{2}{2-\rho}} \tag{19}$$

This is an equation involving several nonlinear terms in $\mu$, which possibly gives rise to multiple steady states. ■

Indeed, for the numerical values

$$\begin{aligned}&\beta = 1;\ \gamma = 5;\ \eta = 1;\ \theta = 0.6;\ \rho = 0.6;\\&\alpha = 0.15;\ \delta_G = 0.2;\ \delta = 0.2;\ r = 0.02;\ \varphi = 30,\end{aligned} \tag{20}$$

we found two steady states with $\mu_1 = 0.048$ and $\mu_2 = 2.469$. The steady state values of states and controls are given by

$$\begin{aligned}G_1^* &= 0.0006, & G_2^* &= 1.971,\\ K_1^* &= 0.0481, & K_2^* &= 2.469,\\ a_1^* &= 0.0025, & a_2^* &= 0.160,\\ I_1^* &= 0.0096, & I_2^* &= 0.493.\end{aligned} \tag{21}$$

We now perform a stability analysis, in which we will show that only the larger steady state, $(G_2, K_2)$, can be a stable equilibrium. To do so, we apply the formulas of Dockner (1985) and the theory described in Feichtinger et al. (1994). This involves determining the determinant of the Jacobian of the dynamic system and Dockner's $K$. This expression, which we denote by $\kappa$, is defined as

$$\kappa = \det\begin{pmatrix}\frac{\partial\dot{G}}{\partial G} & \frac{\partial\dot{G}}{\partial\lambda}\\ \frac{\partial\dot{\lambda}}{\partial G} & \frac{\partial\dot{\lambda}}{\partial\lambda}\end{pmatrix} + \det\begin{pmatrix}\frac{\partial\dot{K}}{\partial K} & \frac{\partial\dot{K}}{\partial\mu}\\ \frac{\partial\dot{\mu}}{\partial K} & \frac{\partial\dot{\mu}}{\partial\mu}\end{pmatrix} + 2\det\begin{pmatrix}\frac{\partial\dot{G}}{\partial K} & \frac{\partial\dot{G}}{\partial\mu}\\ \frac{\partial\dot{\lambda}}{\partial K} & \frac{\partial\dot{\lambda}}{\partial\mu}\end{pmatrix}. \tag{22}$$

After a long derivation, we obtain that the determinant of the Jacobian of the canonical system (11)–(14) in a steady state is given by

$$\det J = \delta_G\delta\,(r+\delta_G)\,(r+\delta)$$

$$+\delta_G 2\alpha\eta^2\theta\,(r+\delta_G)\,\frac{1}{\gamma}\left(\frac{1}{\varphi\delta_G}\left(\frac{\mu}{\gamma\delta}\right)^2\right)^{\rho}\lambda^{\rho}$$

$$-\eta\rho\theta\left(\frac{1}{\varphi\delta_G}\left(\frac{\mu}{\gamma\delta}\right)^2\right)^{\rho-1}\left(\beta-2\alpha\eta\frac{\mu}{\gamma\delta}\right)\frac{2\frac{\mu}{\gamma\delta}}{\gamma\varphi}\,(r+2\delta_G)\,\lambda^{\rho}$$

$$+3\,(1-\rho)\,\rho\theta\eta\left(\frac{1}{\varphi\delta_G}\left(\frac{\mu}{\gamma\delta}\right)^2\right)^{\rho-2}\left(\beta-\alpha\eta\frac{\mu}{\gamma\delta}\right)\frac{\left(\frac{\mu}{\gamma\delta}\right)^3}{\gamma\varphi^2}\lambda^{\rho}$$

$$+\,(1-\rho)\,\rho\theta\eta\,(r+\delta)\left(\frac{1}{\varphi\delta_G}\left(\frac{\mu}{\gamma\delta}\right)^2\right)^{\rho-2}\left(\beta-\alpha\eta\frac{\mu}{\gamma\delta}\right)\delta\frac{\left(\frac{\mu}{\gamma\delta}\right)^3}{\varphi}\lambda^{\rho-2}$$

$$+\,(1-\rho)\,\rho\theta\eta\left(\frac{1}{\varphi\delta_G}\left(\frac{\mu}{\gamma\delta}\right)^2\right)^{2\rho-2}\left(\beta-\alpha\eta\frac{\mu}{\gamma\delta}\right)2\alpha\eta^2\theta\frac{\left(\frac{\mu}{\gamma\delta}\right)^3}{\varphi\gamma}\lambda^{2\rho-2}$$

$$-\,(\eta\rho\theta)^2\left(\frac{1}{\varphi\delta_G}\left(\frac{\mu}{\gamma\delta}\right)^2\right)^{2\rho-2}\left(\beta-2\alpha\eta\frac{\mu}{\gamma\delta}\right)^2\frac{\left(\frac{\mu}{\gamma\delta}\right)^2}{\varphi}\frac{1}{\gamma}\lambda^{2\rho-2}$$

$$-\delta_G\,(r+\delta_G)\,\frac{1}{\gamma\varphi}\lambda^2.$$

For $\kappa$ from (22) we obtain in the steady state

$$\begin{aligned}\kappa &= -\delta\,(r+\delta)-(2-\rho)\,(r+\delta_G)\,\delta_G\\ &\quad-\frac{2\alpha\eta^2\theta}{\gamma}\left(\frac{1}{\gamma\delta}\right)^{\frac{3\rho}{2-\rho}}\left(\frac{\rho\theta\eta}{\varphi\delta_G\,(r+\delta_G)}\right)^{\frac{\rho}{2-\rho}}\left(\beta-\alpha\eta\frac{\mu}{\gamma\delta}\right)^{\frac{\rho}{2-\rho}}\mu^{\frac{3\rho}{2-\rho}}\\ &\quad+\frac{1}{\gamma\varphi}\left(\frac{\rho\theta\eta\,(\gamma\delta)^{1-2\rho}\,(\varphi\delta_G)^{1-\rho}}{(r+\delta_G)}\right)^{\frac{2}{2-\rho}}\left(\beta-\alpha\eta\frac{\mu}{\gamma\delta}\right)^{\frac{2}{2-\rho}}\mu^{\frac{4\rho-2}{2-\rho}}.\end{aligned}\tag{23}$$

For the *larger* steady state $(G_2, K_2)$ in (21), we obtain the following values for the stability indicators:

$$\det J_2 = 0.00415 \tag{24}$$

$$\kappa_2 = -0.1346 \tag{25}$$

$$\det J_2-(\kappa_2/2)^2 = -0.00038. \tag{26}$$

From Table 1 (respectively Figure 1) in Feichtinger et al. (1994) which builds on Dockner (1985), we conclude that the steady state is a (locally) stable equilibrium with monotonic convergence to the steady state. On the other hand, the determinant in the smaller steady state $(G_1, K_1)$ in (21) is negative:

$$\det J_1 = -0.00073.$$

From Feichtinger et al. (1994) and from Dockner (1985), we conclude that this steady state is unstable and cannot constitute an equilibrium.

## 4 Analysis of the Complete Model with Clean and Conventional Input

In the complete model (2)–(5), the firm can use both, the conventional input, $X$, and the green capital stock, $K$. To derive the optimality conditions we first define the Hamiltonian

$$\begin{aligned} H &= g(G)(\beta - \alpha q(K+X))\, q(K+X) - C_a(a) - C_s(I) - p_X X \\ &\quad + \lambda (f(K)\, a - \delta_G G) + \mu (I - \delta K) + \nu K. \end{aligned}$$

The necessary optimality conditions are

$$\begin{aligned} H_I &= 0 = -C_s'(I) + \mu, \\ H_a &= 0 = -C_a'(a) + \lambda f(K), \\ H_X &= 0 = g(G)\, q'(K+X)(\beta - 2\alpha q(K+X)) - p_X, \\ \dot{\lambda} &= (r + \delta_G)\lambda - g'(G)(\beta - \alpha q(K+X))\, q(K+X), \\ \dot{\mu} &= (r+\delta)\mu - g(G)\, q'(K+X)(\beta - 2\alpha q(K+X)) - \lambda a f'(K) - \nu. \end{aligned}$$

We again employ the special functions (6)–(10), to obtain

$$\begin{aligned} H_I &= 0 = -\gamma I + \mu, \\ H_a &= 0 = -\varphi a + \lambda K, \\ H_X &= 0 = \theta G^{\rho} \eta (\beta - 2\alpha\eta (K+X)) - p_X, \\ \dot{\lambda} &= (r+\delta_G)\lambda - \rho\theta G^{\rho-1}(\beta - \alpha\eta(K+X))\,\eta(K+X), \\ \dot{\mu} &= (r+\delta)\mu - \theta G^{\rho}\eta(\beta - 2\alpha\eta(K+X)) - \lambda a - \nu. \end{aligned}$$

It follows that the optimal controls are given by

$$I = \frac{\mu}{\gamma},$$

$$a = \frac{\lambda K}{\varphi},$$

$$X = \frac{\eta\beta - \frac{p_X}{\theta G^\rho}}{2\alpha\eta^2} - K. \tag{27}$$

The steady state values of the $G$ and $K$ are again given by (15) and (16). Since an analytic approach of the stability analysis does not lead to anything, from now on we use a numerical analysis. Solving the system of equations $\dot{\lambda} = 0$ and $\dot{\mu} = 0$ allows to determine the steady states and the stability indicators also for this extended model. We keep the same parameter values as in (20) and in addition, we impose

$$p_X = 0.1. \tag{28}$$

Again a unique steady state in the relevant state space with the required saddle-point property of the state/co-state dynamics can be identified[2]:

$$\mu^* = 2.016$$
$$\lambda^* = 2.26$$
$$G^* = 1.532$$
$$K^* = 2.016$$
$$a^* = 0.152$$
$$I^* = 0.403$$
$$X^* = 0.89.$$

Concerning the stability indicators of this steady state, we obtain:

$$\det J = 0.00288$$
$$\kappa = -0.07$$
$$\det J - (\kappa/2)^2 = 0.0016$$

Using again Table 1 (respectively Figure 1) in Feichtinger et al. (1994), we conclude that the steady state is again locally stable. However, we now have complex eigenvalues, implying that transient oscillations in the state dynamics occur before convergence to the steady state. The result that the use of conventional input by the firm induces non-monotone convergence towards the steady state is consistent with findings obtained in Dawid et al. (2018) for a different inverse demand function.

[2]There exists also a steady state of the state/co-state dynamics with $K^* > 0$ and $X^* < 0$. However for this steady state $\det J < 0$ and it is not saddle point stable.

## 5 Convex-Concave Effects of Green Capital on Goodwill

In our analysis so far we have assumed that the effect of green capital of the firm on the effectiveness of advertisement for the build-up of goodwill is linear. However, arguably below a certain minimal level of green capital stock the consumers hardly recognize the activities of the firm and this does not generate substantial increase of goodwill. Furthermore, a firm already perceived as being very environmentally friendly can hardly improve its reputation by increasing its level of green capital. An S-shaped form of the function $f(K)$ determining the effectiveness of advertisement for the increase of the goodwill stock (see (1)) is more appropriate to capture such a situation than the linear form used so far. Therefore, in this section we analyze our model (in the version with clean and conventional input) for the following form of the function $f(K)$

$$f(K) = \frac{\zeta_1}{1 + Exp[-\zeta_2(K - \tilde{K})]}.$$

which has an S-shaped form for appropriate values of the parameters $\zeta_1, \zeta_2, \tilde{K}$. In Fig. 1 we depict the function $f(K)$ for

$$\zeta_1 = 2, \quad \zeta_2 = 6.6 \quad \tilde{K} = 2,$$

which will be the default values in our following analysis.

Taking into account that under this formulation there exists an upper bound for the returns on investment in the goodwill stock and that the marginal effort needed to sustain a certain goodwill stock is increasing with respect to the level, we adapt the parameters given in (20) such that the effect of the goodwill stock on demand is now linear. In particular, in the following analysis we assume now $\rho = 1, \gamma = 9, p_X = 0.175$, and otherwise stick to values in (20). In this scenario three steady

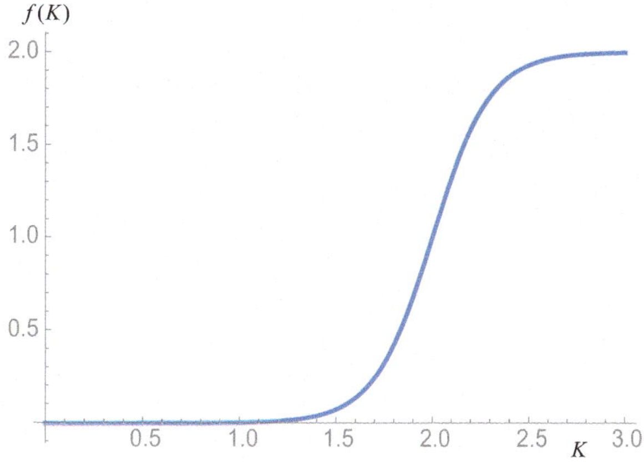


**Fig. 1** The effect of green capital on the increase of the goodwill stock

states of the canonical system can be found. Two of them are saddle points, namely

$$\begin{array}{ll} \mu_1^* = 0 & \mu_2^* = 4.4 \\ \lambda_1^* = 0 & \lambda_2^* = 4.49 \\ G_1^* = 0 & G_2^* = 2.7 \\ K_1^* = 0 & K_2^* = 2.44 \\ X_1^* = 0 & X_2^* = 0.526 \end{array}$$

The first steady state corresponds to a scenario in which the firm stops producing and essentially leaves the market, whereas the firm is an active producer in the second steady state. Between these two there is an unstable node with $\mu_3^* = 3.35, \lambda_3^* = 3.66, G_3^* = 0.198, K_3^* = 1.86, X_3^* = 0$. In order to obtain deeper insights concerning the global dynamics under optimal investment, in particular about the basins of attraction of the two locally stable fixed points, we rely on a global numerical analysis.

In particular, we determine two "local value functions" around the two saddle points of the canonical system, i.e. value functions resulting from trajectories of the state and control variables that are optimal, conditional on the assumption that these trajectories converge to the upper or the lower steady state. To determine these optimal value functions (and the corresponding investment functions) we rely on a collocation method, in which an approximation based on Chebyshev polynomials of the value function is determined such that the Hamilton-Jacobi-Bellman equation associated with the control problem is solved on a suitably determined grid in the state space. The optimal investment functions are then determined from the first order conditions using this approximate value function for the problem. A more detailed description of the method can be found, for example, in Dawid et al. (2015).

Using this approach we obtain the two local value functions shown in Fig. 2a, where the blue function is the local value function corresponding to trajectories converging to the lower steady state at zero and the orange function to that of the upper steady state. It can be clearly seen that for low initial values of green capital the firm's value function is larger if it chooses a trajectory converging to the low steady state. Convergence to the upper steady state is optimal for initial values above the Skiba curve given by the intersection of the two local value functions. It can be clearly seen that a certain minimal initial stock of green capital is necessary to make convergence to the upper "green" steady state optimal and that this level increases as the initial stock of goodwill becomes larger.

In Fig. 2b, two trajectories are depicted that both depart from one point on the Skiba curve, namely, $G(0) = 1.5$, $K(0) = 1.3$. The red trajectory converges to the origin, in which the firm gradually leaves the business. On the green trajectory, we see that the green capital stock has to reach a sufficiently high level of green capital before the firm's goodwill starts increasing. Here, the firm stays in business forever, because the trajectory converges to the steady state where both, green capital and goodwill are positive.

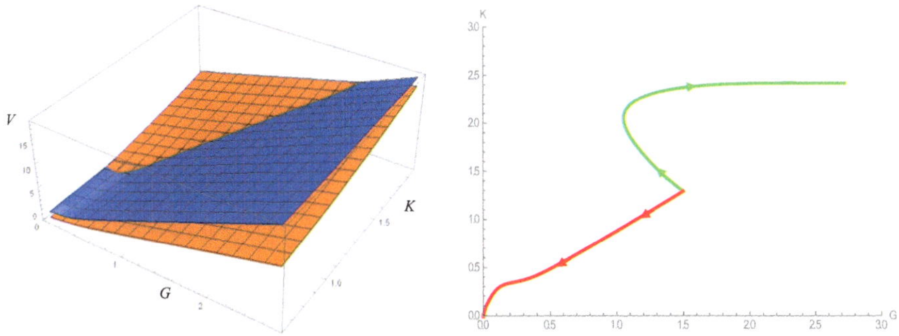


(a) (b)

**Fig. 2** (**a**) "Local value functions" corresponding to trajectories converging to the lower (blue) and upper (yellow) steady state. The intersection between the two value functions is a Skiba curve along which the firm is indifferent between choosing controls inducing convergence to the upper and the lower steady state; (**b**) the two optimal trajectories for an initial point on the Skiba curve

Figure 3 takes a closer look at these two trajectories, distinguishing between the dynamics of states, controls, and instantaneous profits. If the firm chooses the trajectory converging to the lower steady state (red lines), it does not invest in goodwill ($a$=0) and only initially invests small amounts ($I$) in green capital. Both of these stocks ($K$, $G$) monotonically decrease to zero. At the same time initially the firm uses relatively large amounts of conventional energy ($X$) for production in order to satisfy the relatively large demand induced by the high initial goodwill stock. Since goodwill decreases over time, the same holds for the demand and, therefore, the use of conventional energy quickly converges to zero. Due to the decrease of demand over time, instantaneous profits ($F$) are high in the first periods, but then become zero in the long run. Intuitively, the firm exploits its high initial goodwill stock using cost minimizing input factors without trying to keep the high goodwill in the long run.

The same objective value can be obtained by the firm by choosing the green trajectory converging to the high steady state with a positive long run level of green capital and goodwill. Also in this scenario, the goodwill stock initially decreases, because the initial level of green capital is so low that the firm's advertising activities have essentially no positive effect on the dynamics of its goodwill stock (see also Fig. 1). Once the firm has built up sufficient green capital, the returns of investment in the goodwill stock are sufficiently large such that this stock starts to increase and converges to a relatively high positive level. For this trajectory, initial profits are negative (when the firm invests in the build-up of the stock), but positive in the long run. Also, it should be observed that a positive amount of conventional energy is used in the long run under our parametrization.

Comparing these two trajectories highlights that under these initial conditions the goodwill stock initially decreases no matter whether or not the firm plans to build up green capital and keep a positive goodwill stock in the long run. If the firm

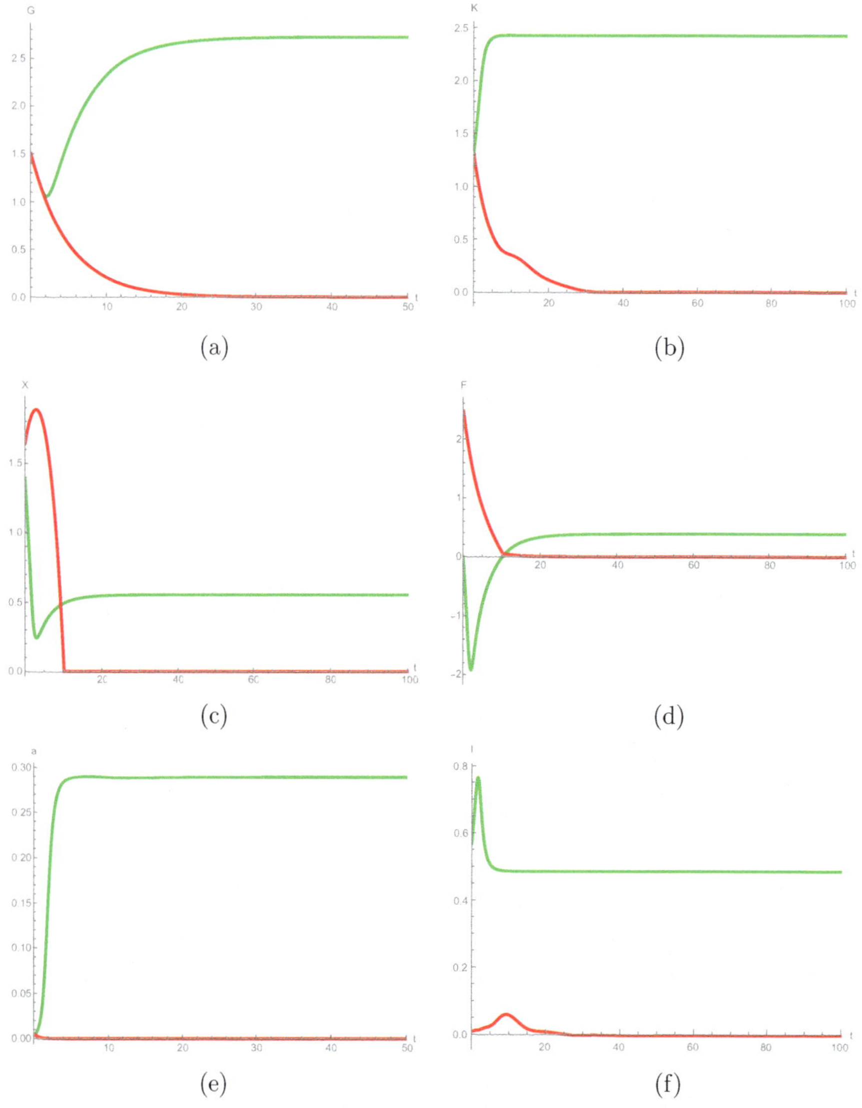


**Fig. 3** Dynamics under the optimal trajectories converging to the low steady state (red) and high steady state (green) for $G(0) = 1.5$, $K(0) = 1.3$: (**a**) goodwill stock; (**b**) green capital stock; (**c**) conventional input; (**d**) instantaneous profit; (**e**) investment in goodwill; (**f**) investment in green capital

targets the positive steady state it has to invest heavily in green capital during the early periods in order to reach a part of the S-shaped curve $f(K)$, determining the effectiveness of advertising, that allows it to prevent through advertising activities the goodwill stock from falling. If the initial stock of green capital is too low, then the investments needed to push the stock of green capital to such a minimal level are too costly. For such initial conditions it is always optimal for the firm to exploit its initial stock of goodwill (mainly using conventional input). This explains why in Fig. 2a the area with $K \leq 1$ is completely below the Skiba curve, implying that for such low levels of initial green capital the optimal trajectory leads to the lower steady state no matter how large the initial goodwill stock is. Furthermore, it is quite intuitive that the marginal value of an additional unit of green capital is more valuable for a firm targeting the upper steady state compared to a firm targeting the lower steady state. This is, because for a firm targeting the upper steady state the higher level of green capital increases the future effectiveness of its advertising activities, in addition to providing production input, whereas for a firm following a trajectory to the lower steady state (without advertising) only the second of these two effects is present. Hence, the "local value function" corresponding to the upper steady state increases more steeply with respect to the level of green capital, which explains that for high levels of green capital it is optimal for the firm to converge to the positive steady state (see again Fig. 2a).

Qualitatively similar observations about the properties of optimal trajectories to those discussed above can be made when considering initial conditions in the interior of the basins of attraction of the two locally stable steady states rather than the one on the Skiba curve used here.

## 6 Conclusions

In this paper we consider a dynamic model of the firm whose main production factor is energy. These days, firms have the choice between using conventional energy and employing green energy by investing in windmills and/or solar panels. We explicitly take into account that green energy has a positive side effect, namely that it contributes to the goodwill of the firm and thus increases demand. The latter happens only after the green capital stock has reached a significantly large size. We show that this feature can lead to history-dependent Skiba behavior. We plan to do future work in this area where, among others, we want to investigate the effect of government intervention in the form of subsidies and investment grants.

**Acknowledgements** The authors are grateful for helpful comments from two anonymous referees.

## References

Amigues, J.-P., Ayong Le Kama, A., & Moreaux, M. (2015). Equilibrium transitions from non-renewable energy to renewable energy under capacity constraints. *Journal of Economic Dynamics and Control, 55*, 89–112.

André, F., Sokri, A., & Zaccour, G. (2011). Public disclosure programs vs. traditional approaches for environmental regulation: Green goodwill and the policies of the firm. *European Journal of Operational Research, 212*, 199–212.

Ben Youssef, S., & Zaccour, G. (2014). Absorptive capacity, R&D spillovers, emissions taxes and R&D subsidies. *Strategic Behavior and the Environment, 4*(1), 41–58.

Breton, M., Sbragia, L., & Zaccour, G. (2010). Dynamic models for international environmental agreements. *Environmental and Resource Economics, 45*, 25–48.

Breton, M., & Zaccour, G. (Eds.). (1991). *Advances in operations research in the oil and gas industry*. Paris: Éditions Technip.

Carraro, C., Haurie, A., & Zaccour, G. (Eds.). (1994). *Environmental management in a transition to market economy: A challenge to governments and business*. Paris: Éditions Technip.

Dawid, H., Hartl, R. F., & Kort, P. M. (2018). Dynamic models of the firm with green energy and goodwill. In G. Feichtinger, R. Kovacevic, & G. Tragler (Eds.), *Control systems and mathematical methods in economics - Essays in honor of Vladimir M. Veliov* (pp. 279–296). Berlin: Springer.

Dawid, H., Keoula, M., Kopel, M., & Kort, P. M. (2015). Product innovation incentives by an incumbent firm: A dynamic analysis. *Journal of Economic Behavior and Organization, 117*, 411–438.

Dockner, E. J. (1985). Local stability analysis in optimal control problems with two state variables. In G. Feichtinger (Ed.), *Optimal control theory and economic analysis* (Vol. 2, pp. 89–103). Amsterdam: North-Holland.

Domenech, P.-A., Saint-Pierre, P., & Zaccour, G. (2011). Forest conservation and CO2 emissions: A viable approach. *Environmental Modeling & Assessment, 16*(6), 519–539.

Eyland, T., & Zaccour, G. (2014). Carbon tariffs and cooperative outcomes. *Energy Policy, 65*, 718–728.

Feichtinger, G., & Hartl, R. F. (1986). *Optimale Kontrolle ökonomischer Prozesse: Anwendungen des Maximumprinzips in den Wirtschaftswisssenschaften*. Berlin: de Gruyter.

Feichtinger, G., Novak, A., & Wirl, F. (1994). Limit cycles in intertemporal adjustment models: Theory and applications. *Journal of Economic Dynamics and Control, 18*, 353–380.

Grass, D., Caulkins, J. P., Feichtinger, G., Tragler, G., & Behrens, D. A. (2008). *Optimal control of nonlinear processes*. Berlin: Springer.

Jørgensen, S., Martín-Herrán, G., & Zaccour, G. (2010). Dynamic games in the economics and management of pollution. *Environmental Modeling & Assessment, 15*(6), 433–446.

Loulou, R., Waaub, J.-P., & Zaccour, G. (Eds.). (2005). *Energy and environment*. Berlin: Springer.

Masoudi, N., Santugini, M., & Zaccour, G. (2016). A dynamic game of emissions pollution with uncertainty and learning. *Environmental and Resource Economics, 64*(3), 349–372.

Masoudi, N., & Zaccour, G. (2013). A differential game of international pollution control with evolving environmental costs. *Environment and Development Economics, 18*(6), 680–700.

Masoudi, N., & Zaccour, G. (2014). Emissions control policies under uncertainty and rational learning in a linear-state dynamic model. *Automatica, 50*, 719–726.

Masoudi, N., & Zaccour, G. (2017). Adapting to climate change: Is cooperation good for the environment? *Economics Letters, 153*, 1–5.

The Economist. (2017). Renewable energy: A world turned upside down. Feb 25 2017.

Tsur, Y., & Zemel, A. (2011). On the dynamics of competing energy sources. *Automatica, 47*(1), 1357–1365.

Wirl, F. (1991). (Monopolistic) resource extraction and limit pricing: The market penetration of competitively produced synfuels. *Environmental Resource Economics, 1*, 157–178.

Wirl, F. (2008). Intertemporal investments into synfuels. *Natural Resource Modeling, 21*(3), 466–488.

# A Review of Experiments on Dynamic Games in Environmental and Resource Economics

**Dina Tasneem and Hassan Benchekroun**

**Abstract** This chapter reviews the existing experimental literature on the behavioral outcomes in dynamic common pool resource games. We categorize the contributions in three sections. The first group of contributions compares the observed behavior to different cooperative and noncooperative theoretical benchmarks. The second group covers contributions that design experiments that aim to find behavioral support for the use of specific types of strategies. Both the first and second groups exclusively use a discrete time framework. Finally, we discuss the exceptions in this literature that study behavior in continuous time.

**Keywords** Resource economics · Environmental economics · Experiments · Dynamic games · Differential games · Common pool resources · Tragedy of the common

## 1 Introduction

Environmental and natural resource economics have been fertile areas of applications of dynamic games (see, e.g., Jorgensen et al. 2010; Long 2011 or Benchekroun and Long 2012 for surveys). Inefficiencies in the management of common pool resources, such as the fisheries, forestry, grazing, water, or the climate, have been extensively used to highlight the role of absent or ill-defined property rights in the failure of the welfare theorems in economics. The payoff of each economic agent typically depends on the actions of all the other agents involved in the economic activity. Game theory is then a natural tool to model situations where the resource

D. Tasneem
American University of Sharjah, Sharjah, United Arab Emirates
e-mail: dtasneem@aus.edu

H. Benchekroun (✉)
McGill University, Montréal, QC, Canada
e-mail: hassan.benchekroun@mcgill.ca



P.-O. Pineau et al. (eds.), *Games in Management Science*,
International Series in Operations Research & Management Science 280,
https://doi.org/10.1007/978-3-030-19107-8_9

is exploited by a fixed number of strategic agents such as firms, municipalities, regions, or countries. Natural resource and environmental games often share another important characteristic: actions at a given date impact the future states of the game. Clearly, extracting one barrel of oil today changes the states of the future reserves of oil. On the climate change front, whether the global temperature will stay below the critical thresholds as prescribed by the UN International Panel on Climate Change (IPCC) will depend on the contemporaneous mitigation efforts undertaken by all countries. Taking these intertemporal constraints into account requires the use of a dynamic framework that captures the evolution of the states of the game and its dependence on current actions.

While there is an extensive literature on dynamic resource or environmental games, the experimental analysis of strategic behavior explicitly taking into account the intertemporal dimension of the game is scarce. We review in this chapter contributions that have examined agents behavior when facing a resource dynamics.

Game theoretic modeling makes assumptions regarding economic agent's fundamental decision-making behavior, e.g., selfish behavior ignoring externalities caused to others. The first group of the contributions we review examines, within specific common pool resource dynamic games or dynamic pollution games, whether the observed behavior supports cooperation, typically corresponding to the Pareto optimal/first best outcome, or selfish behavior, typically corresponding to the outcome of a noncooperative equilibrium of the game. Some of these experimental studies also examine different institutional and environmental factors that can induce a cooperative behavior that mitigates the "tragedy of the commons." In Sect. 2 we cover those studies that compare behavior to different cooperative and noncooperative benchmarks.

Another important specification of a game is the definition of the strategy set available to the players. In dynamic games, strategies can be history dependent allowing retaliation or punishment such as trigger strategies, or can consist of an action path (open-loop strategy) chosen at the start of the game or can consist of state contingent plans such as Markovian strategies (Basar and Olsder 1999; Dockner et al. 2000; Haurie et al. 2012; Basar and Zaccour 2018a,b). The equilibrium outcome of the game may vary, sometimes substantially, depending on the strategy space considered. In Sect. 3 we summarize the findings of experiments that examine behavioral support for the use of specific types of strategies.

The papers reviewed in Sects. 2 and 3 model the dynamics of the common property resource game using a discrete time framework (difference games), with the dynamics of the system described with difference equations. In Sect. 4 we cover in more detail the much less studied case where the dynamics of the resource is modeled using a continuous time framework. The implication on the experimental design is that in continuous time events evolve in an asynchronous fashion (differential games), and the dynamics are modeled with differential equations. In Sect. 4 we present experiments implemented in continuous time involving dynamic externality within the context of a common property renewable resource. Concluding remarks are offered in Sect. 5.

## 2 Equilibrium Prediction vs Behavior

In this section we discuss few experimental studies that compare behavior to the predictions of noncooperative game theory in the presence of dynamic externalities.

Herr et al. (1997) is one of the early experimental studies to examine behavior in the laboratory in a common-pool resource (CPR) game with and without dynamic appropriation externalities. Time-dependent externalities arise when appropriation of the resource by any player increases the cost of appropriation for all players in the current *and* future periods, whereas under time-independent externalities the appropriation by a player at a given time increases the cost of all players in that moment only, the appropriation externality is of a static nature. The authors compare behavior in these two settings. They consider behavioral observations to three benchmark outcomes of the game, namely the Pareto optimal, the sub-game perfect Nash equilibrium, and the outcome of myopic strategy. The myopic case corresponds to the outcome of a game where agents fail to consider the impact of their extraction on their own future costs of appropriation. Herr et al. (1997) illustrate their results using a groundwater basin as the CPR, and consider a finite time horizon dynamic where the depth-to-water is the unique state variable of the game and for each player the cost of extraction depends on the depth of the aquifer. The marginal cost of pumping a unit of water is assumed to be a linear increasing function of the depth-to-water: it increases at a constant rate with each resource unit extracted. When the depth-to-water is reset in each period the inter-temporal externality feature of the problem is muted. The myopic outcome obtains when players fail to take into account that current extraction impacts the future values of depth-to-water, this corresponds to a game where agents completely discount future payoffs. Such myopic behavior may be justified on the grounds of the difficulty to compute the solution to the dynamic optimization problem facing an agent.[1]

In the experiment the resource extractors are subjects ordering "tokens." A subject's monetary earnings depend on her token order and her cost is based on her token order as well the group token order. The experiments delivered two important findings:

- The cooperative outcomes are poor predictors of behavior relative to the non-cooperative benchmark outcomes. This is true for the time-dependent and the time-independent settings.
- In the time-dependent designs, the observed payoffs are significantly lower than those in the time-independent designs. This is partly explained by the presence of myopic behavior. The presence of myopic players might exacerbate the aggressive extraction of non-myopic players and result in a more severe tragedy of the commons.

[1]The possibility of myopic playing has been explicitly included in transboundary pollution differential games (see, e.g., Benchekroun and Martín-Herrán 2016).

Addressing myopic behavior could be an important part of policy interventions in the management of CPRs and represents a promising line of future research.

Mason and Phillips (1997) also compared behavior under static vs dynamic externality as part of their experimental study of common pool extraction. An important specificity of their model with respect to the vast CPR literature is that in addition to sharing access of an input, players, i.e., firms also share the market of output: firms are oligopolists in the market of output. In this context Cornes et al. (1986) show that the socially optimal industry size is larger than one but finite. This feature is particularly relevant for resources that have no close substitutes and therefore a substantial markup is to be expected, e.g., Gulf Coast oysters versus Pacific Coast oysters, or West Coast refiners preference for Alaskan crude oil over crude oil from the Middle East. The objective of Mason and Phillips (1997) is to study the effect of industry size on behavior in the commons, in particular the tendency of firms to collude. They ran experiments with markets (groups) of two, three, four, or five harvesters, and examined collusive behavior under a static externality only and in the presence of both a static and dynamic externality. A benchmark scenario is the optimal industry size when the players compete à la Cournot. Under a static externality only, they observed that firms tend to collude, i.e. produce less than the Cournot equilibrium quantity, and that for the benchmark case considered the empirically optimal industry size is four. However when a dynamic externality is present the observed behavior of markets is no longer consistent with a collusive behavior or cooperation between players. Then, the optimal industry size is three, the same optimal size obtained when firms play à la Cournot.

The two papers discussed above compare behavior in the presence of static externality and both static and dynamic externalities (Herr et al. 1997) or only dynamic externality (Mason and Phillips 1997). Both studies intend to test the hypothesis that dynamic externalities exacerbate the tragedy of the commons and their findings support this hypothesis. Giordana et al. (2010) also address a similar concern of whether the delayed, rather than an immediate, realization of the consequences of over exploitation exacerbates the problem of the commons. However they formulate their question differently: they test whether adding immediacy of a static externality in an otherwise dynamic environment can help mitigating the tragedy of the commons, by increasing the salience of the problem to the players. Therefore, they compare behavior in the presence of dynamic externality vs. static and dynamic externalities. The common pool resources they examine are coastal groundwater reservoirs where excessive pumping from the reservoirs increases the risk of natural seawater intrusion into the aquifers, thereby rendering them useless for agricultural and human consumption. Giordana et al. (2010) use a dynamic game with two substitutable common pool resources with different exploitation costs. In one treatment the common-pool resources generate only a dynamic externality, while in the other treatment the common-pool resources generate both static and dynamic externality. They compare behavior against three extraction paths, the sub-game perfect, myopic and Pareto optimal. Their experimental design is such that sub-game perfect and myopic benchmark extraction paths are same under both treatment. As stated above, they hypothesize that users who faces both static and dynamic externality of their exploitation of the resource are likely to be more conservative in their exploitation than users who experience only dynamic externalities. They assume that the immediacy of the static externality may enhance awareness of the consequences of their actions and encourage more socially beneficial behavior. The experimental observations do not support this hypothesis. They also find in both of their treatments behavior follow the myopic prediction more closely.

Noussair et al. (2015) test the canonical renewable resource model (Hardin 1968) in a framed field experiment where experienced recreational fishers make decision on their individual per period catch from an allowable catch for a four member group. Given that the fishermen involved are experienced, they are well aware of the negative externalities inflicted on the group when they choose to overfish. Sixteen fishermen were assigned to groups of four with fixed membership. The game repeats for four periods and each period lasts 1 h. The experiment poses a social dilemma along three dimensions: duration of the game since the game stops if the fish stock is exhausted, the number of fish caught, and the monetary benefit associated with the catch. The allowable catch for the group was affected negatively by the total catch of the group in the previous periods. The authors find no evidence of cooperation and the results of this field experiment are consistent with the predictions of noncooperative game theory which assumes selfish agents.

There have been a very few experimental studies specifically designed to represent the climate change game. The main concern of these studies has been the inherently dynamic nature of the problem and its effect on the ability of the parties to cooperate. The principal effect of the accumulation of greenhouse gases will be felt in the future and in some cases by different generations. In this context one of the behavioral concerns is that decision-makers are myopic, therefore, much worse at processing the future consequences of their actions than immediate ones (Calzolari et al. 2016). Calzolari et al. (2016) compare cooperation in three different environments with different degrees of persistence of greenhouse gas emissions. They report similar levels of cooperation (in terms of average emission) in all environments. But interestingly in the dynamic environment they find cooperation levels deteriorate for high stocks of pollution. In this dynamic externality treatments emission strategies seemed to be increasing in the stock of pollution. Based on this result the authors warn that successful climate policy may require starting early mitigation efforts while pollution stock is low enough. Otherwise too high of a pollution stock may itself work against any possible cooperation among the parties.

Sherstyuk et al. (2016) compare behavior in intragenerational and intergenerational dynamic games of climate change with pollution generating production and external cost from accumulated pollutant over time. In the intragenerational treatment, the dynamic game is played throughout by the same group of subjects, while in the intergenerational treatments, the dynamic game is played by several groups (generations) of subjects. The authors designed the later treatment to describe reality more closely where the countries' decision-makers and citizens may care more about their immediate welfare and care only partially about the welfare of the future generations. They find while in the intragenerational treatment a significant fraction of the groups show sign of cooperation and ability to approach socially optimal outcome, behavior in the intergenerational treatment, even when caring for future generations is incentivized, resembles the noncooperative outcome of the game. They speculate, this result arises from additional strategic uncertainty imposed by different generations.

Pevnitskaya and Ryvkin (2013) present the result of a laboratory experiment on a public bad dynamic game. In this game private production generates pollutant that accumulates over time and imposes cost on all producers. This simple game captures the basic structure of many social dilemmas such as the problems of local or global pollution, and renewable or nonrenewable common pool resource exploitation. The study compares the collective behavior to the Markov perfect equilibrium and Pareto optimal solution of the game. In the laboratory the average accumulation of the public bad is less than that of the Markov Perfect Nash equilibrium though it remains above the social optimum. The authors report the effect of framing the problem within an environmental context, that is framing the public bad as pollution. In one of their treatment settings they find significant decrease in production decision therefore lower pollution when the problem is explicitly framed as a pollution game. They argue that the environmental context may have activated pro-environmental behavior in the laboratory as well as to some extent worked as a proxy for experience.

In considering the noncooperative equilibrium of a CPR game, very little to no attention is actually given to the space of strategies considered. However in dynamic game theory it is well known that the set of strategies considered can have important implications on the equilibrium of the game (see, e.g., Dockner et al. 2000 or Haurie et al. 2012). The most important sets considered in the dynamic game theory are the set of time (only)- dependent strategies (open-loop strategies), state-dependent strategies (Markovian strategies), history-dependent strategies (e.g., trigger strategies that allow for punishments depending on the history of play). There are a very few experimental studies that focus on behavioral assumption of the strategy types used to solve dynamic games. In the next section we cover contributions intended to address this issue.

## 3 Behavior and Strategy Types

In many occasions the set of sub-game perfect equilibria in a dynamic game can be large and varied in nature (Vespa 2011). Experimental studies can play important role in studying the issue of equilibrium selection in such games. For example, Vespa (2011) presented their experiment participants with a dynamic common pool game of two players who share a common resource that grows at an exogenous rate. The efficient outcome of the game requires the players to let the resource grow such that eventually they receive a large return from it. The game has both Markov perfect equilibria and history- dependent grim-trigger strategy equilibrium in its set of sub-game perfect Nash equilibria. While in their game the efficient outcome cannot be supported by a Markov equilibrium, it can be supported by a sub-game perfect grim-trigger strategy. They use a large state space, but to make the environment simple enough for the laboratory participants, they restrict the choice space to maximum three points. Comparing behavior with a list of possible strategies they conclude that Markov strategies can be a reasonable assumption for behavior in this environment. The modal behavior in their experiment mostly agrees with a sub-game perfect Markov equilibrium strategy. Though, if they increase the incentives for cooperation the Markov-perfect equilibrium strategy loses its popularity to some extent.

Battaglini et al. (2016) study behavior in a dynamic public good game where the public good accumulates over time. In these games again the set of sub-game perfect Nash equilibria includes both stationary Markov perfect equilibrium and non-Markovian equilibria. As Markovian strategies depend on the accumulated level of the public good, an increase in current investment by one agent results in reduction in future investment by all agents and there by leads to under provision of the public good. Cooperative outcome of the game can be achieved using some form of carrot-and-stick equilibrium strategies. The framework can be applied to issues such as pollution abatement as well as many others. Their experiment includes the case where players can make both positive or negative (reversible) contribution to the public good account as well as the case where the contribution have to be non-negative (irreversible). Theoretically in their model investment and therefore steady state public good stock is higher in the irreversible case under the assumption of symmetric Markov perfect strategy. Though under the assumption of history-dependent strategies, involving punishments and rewards for past actions, they prove that in case of reversibility the optimal investments can be achieved as a sub-game perfect equilibrium. Considering the most efficient history- dependent sub-game perfect Nash equilibrium in each case, they find that the investment and therefore steady state public good stock is higher in the reversible case than irreversible case. This stark difference in the comparative static predictions arising from the behavioral assumption in their model allows them to design an experiment to test the behavioral relevance of Markovian and history-dependent strategies. The result of the experiment at the aggregate level supports the comparative statics of the Markov perfect equilibrium that irreversible investment leads to higher public good production than reversible investment. They show that the Markov perfect equilibrium strategy they consider does not do such good job with the finer details of the individual investing behavior. Though, they find evidence that their subjects' investment choice responds to the evolution of the stock of the public good.

## 4 Continuous Time Games

In all the experimental studies we discussed till now and in most of the literature, each participant takes her action on a period by period basis, as everyone has made their decision in one period, the time moves on to the next, that is, decision making happens in discrete time. In reality, decision making in dynamic settings not necessarily happens in such orderly fashion (Janssen et al. 2010). Natural social systems unfold in real time, when decisions are made in an "asynchronous fashion" with continuous updates of states and information (Huberman and Glance 1993). Implementing laboratory experiments in continuous time is quite recent and one of the focus of these studies is to compare behavior under continuous time interaction and discrete time interaction in the lab (Friedman and Oprea 2012; Calford and Oprea 2017; Oprea et al. 2014; Bigoni et al. 2015; Horstmann et al. 2016) . It is quite rare to find experimental studies studying behavior in real time/continuous time in an environment with dynamic externality. Janssen et al. (2010) and Tasneem et al. (2017) are two exceptions.

The experimental design in Janssen et al. (2010) is based on field research on governance of social-ecological systems and particularly keen to approximate the field settings. For this purpose they present the experimental subjects (in group of fives) with a 29-by-29 computer-simulated grid of cells with a shared renewable resource (experimental tokens). The players harvest tokens from this simulated resource field in real time for 4 min, where the resource's renewal rate depends on the density of the resource and therefore the players face both spatial and temporal resource dynamics. The best collective outcome of this game can be achieved if the players thoughtfully and patiently decide where and when to harvest. Some of the experimental treatments allow costly punishment and/or written communication among the players. Without any punishment or communication possible the study replicates "tragedy of the commons" with fast resource depletion. Only availability of costly punishment could not improve welfare of the players in the game. Communication and communication paired with costly punishment improves harvesting decision significantly. One of the interesting findings of this paper is that communication by itself has a long-lasting effect on cooperative behavior but as they pair communication with punishment that cooperative behavior dissipates when communication and costly punishment is taken away.

The experimental design in Tasneem et al. (2017) is based on an infinite horizon linear quadratic differential game. As it has been presented in the theoretical literature the game admits a linear Markov-perfect equilibrium as well as a continuum of equilibria with strategies that are nonlinear functions of the state variable (see Dockner and Van Long (1993) in the context of a transboundary pollution game, Fujiwara (2008), Colombo and Labrecciosa (2013a,b), Lambertini and Mantovani (2014, 2016) or Bisceglia (2018) in the case of a renewable resource oligopoly and Kossioris et al. (2011) in the case of a shallow lake problem and Dockner and Wagener (2014) for a more general treatment). The objective of this study is to examine the empirical relevance of these linear and nonlinear equilibria in a two-player common property renewable resource game. Given that this is the first experimental study of a differential game, we describe it into a bit more detail.

The linear quadratic differential game framework is a workhorse model in economics differential game literature because of its analytical tractability. In a linear quadratic differential game the objective functions of the players are quadratic and the system of state equations are linear. In this paper two identical agents $i$, $j$ share access to a renewable resource stock, denoted by $S$, with natural reproduction function of the stock given by $F(S)$. The instantaneous payoff function of each player is given by

$$u_i(q_i(t)) = q_i(t) - \frac{q_i(t)^2}{2}.$$

where $q_i(t)$ is the extraction by player $i$ at time $t$ and the stock dynamics is given by

$$\dot{S}(t) = F(S(t)) - q_i(t) - q_j(t).$$

For the sake of simplicity they consider the range of stock where reproduction function of the resource is given by $F(S) = \delta S$, where $\delta$ represents the intrinsic growth rate of the resource. The instantaneous payoff function reaches its maximum when $q = 1$. Therefore the cooperative strategy should support extraction of 1 for each player at the steady state.

Note that in this problem there is no static externality per say, there is only a dynamic externality. Nonlinear equilibria in the presence of static externalities coming from an oligopolistic behavior (as in Mason and Phillips (1997) above) were shown to exist in this game as well (see Fujiwara (2008), Colombo and Labrecciosa (2013a,b), Lambertini and Mantovani (2014, 2016) or Bisceglia (2018)). Since the focus is to investigate behavioral support for equilibria with nonlinear strategies, the study is done in the simplest framework that generates those equilibria, i.e. without static externalities.

To solve the game they focus on the set of stationary Markovian strategies, such that at any point in time, the extraction decision of a player depends only on the state of the stock at that moment. Each player $i$ takes the other player's strategy as given and chooses a stationary Markovian strategy that maximizes the discounted sum of his instantaneous payoff over infinite horizon. The game admits a piecewise linear Markov-perfect equilibrium as well locally defined continuum of nonlinear Markov-perfect equilibria.[2] These Markovian equilibrium strategies vary in the extent of aggressiveness of exploitation of the resource and therefore support infinitely many stable steady states, varying from the best possible steady state (linear Markovian strategy) to very low steady states portraying the tragedy of the commons phenomena. We reproduce Fig. 1 from the paper that presents several examples of equilibrium strategies in this game for the parameter values chosen for the experiment. In the figure, the line called "Steady State" represents steady state extractions at different stock levels. The intersection of this line with a strategy represents the steady state corresponding to the strategy. The strategy labeled "Linear" is the noncooperative linear strategy. In this game, in case of the linear strategy, extraction of the players reaches the best possible steady state.

[2] A global Markov-perfect equilibrium strategy is defined over the whole state space. A local Markov-perfect Nash equilibrium strategy is defined over an interval strictly included in the state space (Tasneem et al. 2017).

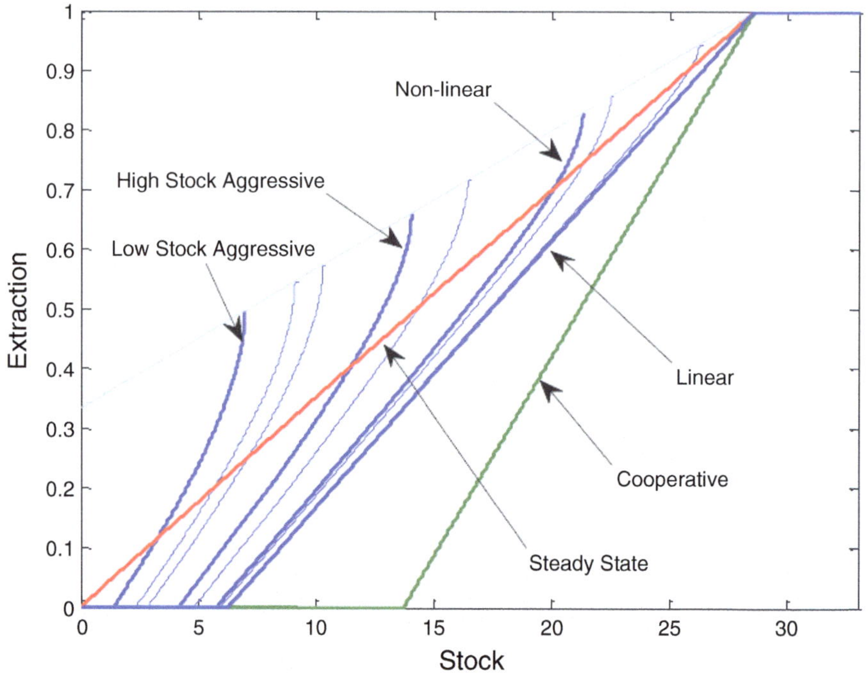


**Fig. 1** Representative equilibrium strategies

The locally defined curves represent different nonlinear equilibrium strategies. The nonlinear Markovian strategies may sustain many possible steady states, including those that result from resource depletion. The graph also shows the cooperative linear strategy called "Cooperative."

The experiment is implemented in real/continuous time. That is all information are updated every second and players can change their decision any time they want as many times they want. The laboratory subjects are paired randomly in a group at the beginning of the experimental session. Each group plays several rounds of simulated common pool resource games.[3] The experimenters set the initial extraction rate and the starting stock level for each simulation. Within a session there are four games for practice and six games for pay. For the practice game the initial extraction rates are set according to the linear Markov perfect strategy for both players. For the six games for pay, the first two games are set with linear strategy initial extraction rate. For next two games, the initial extraction rate is set according

[3]They implemented the discount rate by applying it to instantaneous payoffs every second. To implement infinite horizon each simulation of the game end with a continuation payoff for each player computed as the discounted sum of payoffs for the player out to infinity. This computation assumed that the extraction rate forever stayed the same as it was at the end of the simulation, and it took into account whether the stock level would ever go to zero.

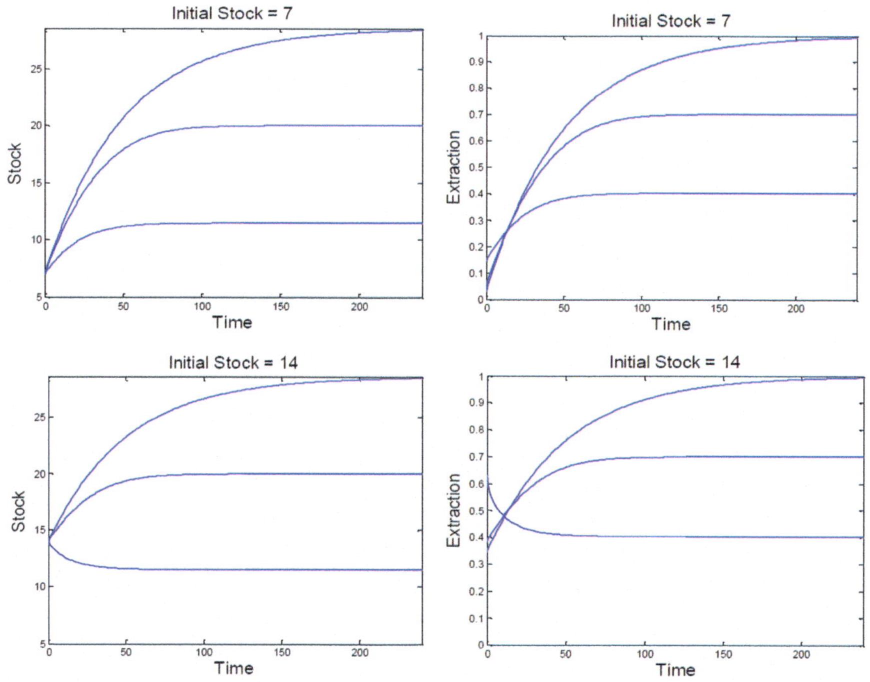


**Fig. 2** Predicted time paths for experimental parameters

to the nonlinear strategy labeled "Nonlinear" in Fig. 1. And for the last two the initial extraction rate is set according to the nonlinear strategy labeled "High Stock Aggressive." A session either belongs to the low initial stock treatment ($S_0 = 7$) or high initial stock treatment ($S_0 = 14$) . The theoretical time path predictions for these strategies are shown in their Fig. 3 (reproduced here as Fig. 2).

The paper first tests the empirical relevance of the Markovian equilibrium strategies by looking into whether any of the steady-states supported by the equilibrium strategies has been frequently reached by the players. They use an algorithm called MSER-5 commonly used in computer simulations literature to identify the time of convergence of a process for the purpose of characterizing steady states in their experimental data. For each group in each simulation of the game the algorithm identifies if a steady state exists, if so the algorithm identifies the extraction rates at the steady state. Table 3 in the paper (reproduced as Table 1 here) summarizes the findings.

According to their table, in every game in each treatment, the majority of behaviors resulted in a steady state stock management. The table categorizes the estimated steady state total extractions into those similar to linear equilibrium steady state, the steady states with total extraction approximately 2, or those similar to any of the steady states supported by nonlinear equilibrium strategies, the steady states

**Table 1** Percentage of games reaching a steady state

| | Both treatments | | | | Treatment 1 $S_0 = 7$ | | | | Treatment 2 $S_0 = 14$ | | | |
|---|---|---|---|---|---|---|---|---|---|---|---|---|
| | Range | All $SS$ (%) | $SS_L$ (%) | $SS_{NL}$ (%) | Range | All $SS$ (%) | $SS_L$ (%) | $SS_{NL}$ (%) | Range | All $SS$ (%) | $SS_L$ (%) | $SS_{NL}$ (%) |
| Game 5 | 0.04–2.99 | 67.16 | 16.42 | 43.28 | 0.04–2.16 | 69.70 | 21.21 | 42.42 | 0.40–2.99 | 64.71 | 11.76 | 44.12 |
| Game 6 | 0.02–3.00 | 71.64 | 17.91 | 49.25 | 0.02–2.06 | 78.79 | 24.24 | 51.51 | 0.29–3.00 | 61.76 | 8.82 | 47.06 |
| Game 7 | 0.00–2.13 | 67.16 | 17.91 | 47.76 | 0.00–2.01 | 63.64 | 27.27 | 36.36 | 0.54–2.13 | 70.59 | 8.82 | 58.82 |
| Game 8 | 0.25–3.32 | 67.16 | 22.39 | 35.82 | 0.25–2.41 | 69.70 | 30.30 | 33.33 | 0.31–3.32 | 64.71 | 14.71 | 38.24 |
| Game 9 | 0.01–2.35 | 61.19 | 19.40 | 35.82 | 0.01–2.13 | 60.60 | 24.24 | 33.33 | 0.08–2.36 | 61.76 | 14.71 | 38.24 |
| Game 10 | 0.05–2.80 | 68.66 | 20.90 | 37.31 | 0.05 -2.37 | 66.67 | 30.30 | 30.30 | 0.14–2.80 | 70.59 | 11.76 | 44.12 |
| All games (Mean) | 0.00–3.32 | 67.16 (1.43) | 19.15 (2.00) | 41.54 (1.03) | 0.00–2.41 | 68.18 (1.36) | 26.26 (2.00) | 37.88 (0.82) | 0.08–3.32 | 65.69 (1.50) | 11.76 (2.00) | 45.10 (1.20) |

$SS$ denotes steady states reached by the players

$SS_L$ denotes the steady states reached by the players that can be theoretically supported by the linear equilibrium strategy. Given the possibility of human error they consider any steady state, to be similar to the steady state supported by the linear equilibrium strategy, if it deviates by less than 5% of the linear strategy steady state total extraction of 2

$SS_{NL}$ denotes the steady states reached by the players that can be theoretically supported by some nonlinear equilibrium strategy. In addition there are some steady states reached by the players that cannot be supported by either linear or nonlinear strategies

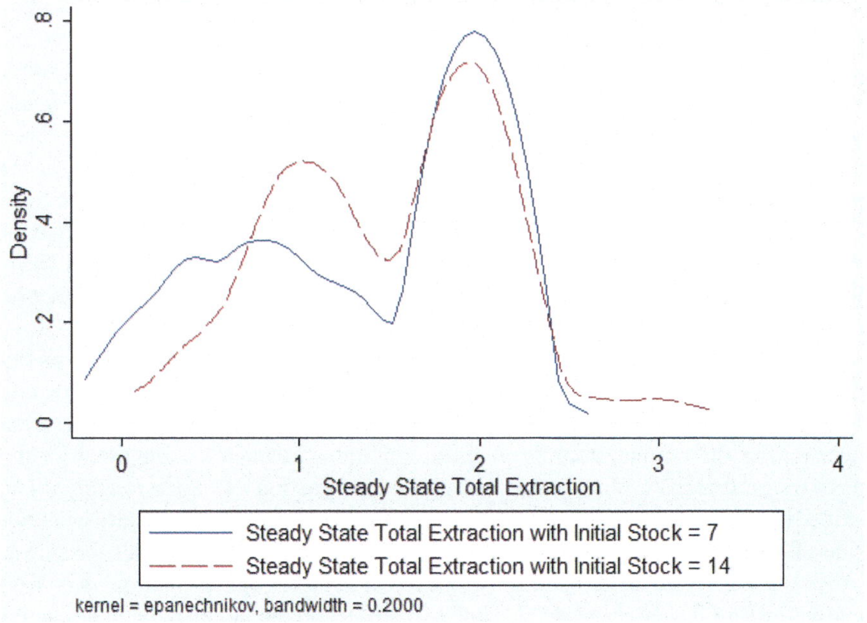


**Fig. 3** Distribution of steady states by treatment

with total extraction between 0 and 2. They also find some steady states that are not consistent with either linear or nonlinear equilibrium strategies, for example, the steady states that are significantly greater than 2.

To test the treatment effect they compare the distribution of steady state total extraction rates under initial sock of 7 and 14. Their Figure 8 (reproduced as Fig. 3 here) presents a density estimate of the distributions of steady state total extraction rates for these two experimental treatments. Theoretically, increasing the initial stock level eliminates a set of equilibrium strategies that exist at the lower initial stock level. As the authors demonstrate in Fig. 1, when starting stock is 7, one of the most aggressive strategies available is labeled "Low Stock Aggressive," but if the starting stock is 14, one of the most aggressive strategies available is "High Stock Aggressive." Therefore, the higher starting stock of 14 eliminates the nonlinear strategies between those two strategies. When they compare the empirical steady state distributions they find them significantly different from each other. A two-sample Kolmogorov Smirnov test indicates that steady state total extraction distribution in Treatment 2 ($S_0 = 14$) contains larger values than that of Treatment 1 ($S_0 = 7$). They conclude that starting with a healthy stock appears to improve extraction behavior in terms of the steady states achieved. Given that symmetry of the equilibrium strategies is a common assumption in the theoretical literature, they look into symmetry of extraction rates at the steady states. Out of the 270 games that reached steady state, 144 were symmetric by their criteria. The study does not find any significant effect of different initial resource extraction rates on behavior.

Next they compare the extraction behavior of the players against Markov equilibrium strategies. The theoretical strategies suggest the more aggressive the strategies are, the faster is the increase in extraction with increase in stock. Also the more and more aggressive strategies get the smaller and smaller stock levels will be subject to positive extraction. In their data actual extraction behavior varies widely and deviates quite far from theoretical suggestions. A large fraction of the play shows sign of over extraction of the resource at low level compared to even the most aggressive theoretical strategy. A good share of actual extraction behavior shows qualitative similarity with nonlinear strategies in terms of raising extraction more and more as the stock grows. Also the players seem to adjust their extraction later downward (upward) following initial over (under) extraction.

The authors also use general-to-specific modeling to find the empirical model that best fit extraction behavior of each player in each play. One of the interesting findings of this analysis is that, though most of the players condition their extraction decision on the current stock level (in a continuous manner as suggested by the Markovian strategies), in a small but significant fraction of cases, roughly 14%, extraction did not condition on the stock level at all. The authors call these behavior rule-of-thumb, as their investigation reveals that in most of these cases the player extracts at a low rate, or zero, until the stock is built up enough for maximum or high extraction rate. The average steady state extraction corresponding to these strategies is significantly higher than that in other categories. Out of the plays that reached the best steady states about 50% includes at least one player with rule-of-thumb behavior. These rule-of-thumb behavior turned out to be quite efficient as it adjusted quickly when the stock reached the level to support best extraction.

## 5 Conclusion

The chapter reviewed the relatively scarce literature on experiments examining the behavioral outcomes in dynamic common pool resource games. The main finding is that when the dynamics is taken into account there is even less evidence of cooperation than in settings where the dynamics of the resource is muted. There is support of myopic behavior where agents simply ignore the intertemporal constraints. This myopic behavior tends to exacerbate the tragedy of the commons. In the cases of resource extraction there is evidence that agents do cooperate more when the resource stocks, with even agents refraining from extraction until the stock reaches a certain threshold. In a differential game where the experiments allow for continuous time settings, there is support for the use of rule-of-thumb strategies, that is strategies that consist of refraining from extraction until the stock reaches a certain threshold and extract at a constant rate. These strategies share the simplicity of myopic strategies. However they do not imply an exacerbation of the tragedy of the commons, and even result in outcomes that are closer to cooperative outcomes than the noncooperative equilibrium outcome. The framing of the experimental setting can also influence the outcome of the game.

More research is needed to understand the implication of the resource dynamics in the tragedy of the commons within full-fledged continuous time experiments, including the case of more sophisticated games where agents exploit several species (Vardar and Zaccour 2018). The lessons learned from these experiments are relevant for policy making and the priorities that regulators need to have. Regulators should pay particular attention to incentives that combat myopic behavior of agents facing intertemporal constraints. Policies that set moratoriums on extraction until some stock is reached bare obvious similarities with rule-of-thumb strategies which are shown to be more conducive to cooperation. Future experiments that examine how the design of these thresholds and the duration of a harvesting season impact the cooperative behavior among players and how this impact depends on group sizes would bring valuable insights into natural resource policy modeling.

Mastering the translation of differential games in laboratory or field experiments will deliver important insights into the behavior of subjects within the context of competition and intertemporal constraints in general. Building this know-how will be very fruitful in addressing important questions in dynamic games in general such as the behavior of subjects in the presence of a leader (Stackelberg differential games), the (dis)advantage of a regulator in taking the lead, the impact of dissemination of information regarding the states of the game. The behavioral lessons learned in these important class of games will guide regulators and players in general to design institutions that are more conducive to cooperation and that reduce existing inefficiencies.

An experimental approach can be particularly useful when the theoretical analysis of strategic interactions yields inconclusive outcomes. Such situation arises when the game admits multiple equilibria, or when different behavioral assumptions in defining strategies or objective functions of the players lead to different outcomes in the game. For example, Tasneem et al. (2017) present a differential common pool resource game with multiple Markov Perfect equilibria that are Pareto ranked. Further insight into the outcome of this game can only be gained through an experiment designed to study equilibrium selection. The equilibrium outcome of a differential game typically depends on the space of strategies considered. These contexts warrant experimental studies to test the relevance of the underlying behavioral assumptions of different strategy types. Vespa (2011) and Battaglini et al. (2016) are two early examples of such attempts. Another promising line for future research is to examine the role of social status in resource games. Benchekroun and Long (2016) present a common pool renewable resource oligopoly incorporating social status in the objective function of the players. Their analysis suggests existence of social status based on relative harvest exacerbates the tragedy of the commons, while social status based on relative profit can have an opposite effect in certain conditions. It is difficult to know whether social status or relative output/payoff plays any role in common pool resource exploitation decision in real life and if it does, through what channel, output or profit. Experimental studies can shed light on the relative importance of each channel (output versus profits) and by the same token provide insights into the relevant policy interventions in the presence of status concerned harvesters. The lessons learned from the response of strategic agents facing joint intertemporal constraints can be valuable in dynamic oligopolistic games in general (Lambertini 2018; Basar and Zaccour 2018a,b), such as investment games (e.g., Huisman and Kort 2015), competition under price stickiness, branding decisions in marketing (Crettez et al. 2018; Pnevmatikos et al. 2018). These lessons can help narrow down the set of available strategies considered in a theoretical framework, based on empirical observations, which in turn can make the theoretical analysis more salient.

## References

Basar, T., & Olsder, G. J. (1999). *Dynamic noncooperative game theory* (2nd ed.). Philadelphia: SIAM.
Basar, T., & Zaccour, G. (2018a). *Handbook of dynamic game theory* (Vol. 1, 670 pp.). Berlin: Springer.
Basar, T., & Zaccour, G. (2018b). *Handbook of dynamic game theory* (Vol. 2, 670 pp.). Berlin: Springer.
Battaglini, M., Nunnari, S., & Palfrey, T. R. (2016). The dynamic free rider problem: A laboratory study. *American Economic Journal: Microeconomics, 8*(4), 268–308.
Benchekroun, H., & Martín-Herrán, G. (2016). The impact of foresight in a transboundary pollution game. *European Journal of Operational Research, 251*(1), 300–309.
Benchekroun, H., & Van Long, N. (2012). Collaborative environmental management: A review of the literature. *International Game Theory Review, 14*(4), 1240002.
Benchekroun, H., & Van Long, N. (2016). Status concern and the exploitation of common pool renewable resources. *Ecological Economics, 125*, 70–82.
Bigoni, M., Casari, M., Skrzypacz, A., & Spagnolo, G. (2015). Time horizon and cooperation in continuous time. *Econometrica, 83*(2), 587–616.
Bisceglia, M. (2018). *Optimal Taxation in a Common Resource Oligopoly Game*. Working Paper.
Calford, E., & Oprea, R. (2017). Continuity, inertia, and strategic uncertainty: A test of the theory of continuous time games. *Econometrica, 85*(3), 915–935.
Calzolari, G., Casari, M., & Ghidoni, R. (2016). *Carbon is forever: A climate change experiment on cooperation* (No. 1065). Quaderni-Working Paper DSE.
Colombo, L., & Labrecciosa, P. (2013a). Oligopoly exploitation of a private property productive asset. *Journal of Economic Dynamics and Control, 37*(4), 838–853.
Colombo, L., & Labrecciosa, P. (2013b). On the convergence to the Cournot equilibrium in a productive asset oligopoly. *Journal of Mathematical Economics, 49*(6), 441–445.
Cornes, R., Mason, C. F., & Sandler, T. (1986). The commons and the optimal number of firms. *The Quarterly Journal of Economics, 101*(3), 641–646.
Crettez, B., Hayek, N., & Zaccour, G. (2018). Brand imitation: A dynamic-game approach. *International Journal of Production Economics, 205*, 139–155.
Dockner, E. J., Jorgensen, S., Van Long, N., & Sorger, G. (2000). *Differential games in economics and management science*. Cambridge: Cambridge University Press.
Dockner, E. J., & Van Long, N. (1993). International pollution control: Cooperative versus noncooperative strategies. *Journal of Environmental Economics and Management, 25*(1), 13–29.
Dockner, E. J., & Wagener, F. (2014). Markov perfect Nash equilibria in models with a single capital stock. *Economic Theory, 56*(3), 585–625.
Friedman, D., & Oprea, R. (2012). A continuous dilemma. *American Economic Review, 102*(1), 337–363.

Fujiwara, K. (2008). Duopoly can be more anti-competitive than monopoly. *Economics Letters, 101*(3), 217–219.
Giordana, G. A., Montginoul, M., & Willinger, M. (2010). Do static externalities offset dynamic externalities? An experimental study of the exploitation of substitutable common-pool resources. *Agricultural and Resource Economics Review, 39*(2), 305–323.
Hardin, G. (1968). The tragedy of the commons. *Science, 162*, 1243–1248.
Haurie, A., Krawczyzk, J. B., & Zaccour, G. (2012). *Games and dynamic games*. Singapore: World Scientific.
Herr, A., Gardner, R., & Walker, J. M. (1997). An experimental study of time-independent and time-dependent externalities in the commons. *Games and Economic Behavior, 19*(1), 77–96.
Horstmann, N., Kraemer, J., & Schnurr, D. (2016). Oligopoly competition in continuous time. Available at SSRN 2630664.
Huberman, B. A., & Glance, N. S. (1993). Evolutionary games and computer simulations. *Proceedings of the National Academy of Sciences, 90*(16), 7716–7718.
Huisman, K. J. M., & Kort, P. M. (2015). Strategic capacity investment under uncertainty. *The RAND Journal of Economics, 46*(2), 376–408.
Janssen, M. A., Holahan, R., Lee, A., & Ostrom, E. (2010). Lab experiments for the study of social-ecological systems. *Science, 328*(5978), 613–617.
Jorgensen, S., Martin-Herran, G., & Zaccour, G. (2010). Dynamic games in the economics and management of pollution. *Environmental Modeling & Assessment, 15*(6), 433–467.
Kossioris, G., Plexousakis, M., Xepapadeas, A., & de Zeeuw, A. (2011). On the optimal taxation of common-pool resources. *Journal of Economic Dynamics and Control, 35*(11), 1868–1879.
Lambertini, L. (2018). *Differential games in industrial economics.* Cambridge: Cambridge University Press. https://doi.org/10.1017/9781316691175
Lambertini, L., & Mantovani, A. (2014). Feedback equilibria in a dynamic renewable resource oligopoly: Pre-emption, voracity and exhaustion. *Journal of Economic Dynamics and Control, 47*(C), 115–122.
Lambertini, L., & Mantovani, A. (2016). On the (in) stability of nonlinear feedback solutions in a dynamic duopoly with renewable resource exploitation. *Economics Letters, 143*, 9–12.
Long, N. V. (2011). Dynamic games in the economics of natural resources: A survey. *Dynamic Games and Applications, 1*(1), 115–148.
Mason, C. F., & Phillips, O. R. (1997). Mitigating the tragedy of the commons through cooperation: An experimental evaluation. *Journal of Environmental Economics and Management, 34*(2), 148–172.
Noussair, C. N., van Soest, D., & Stoop, J. (2015). Cooperation in a dynamic fishing game: A framed field experiment. *The American Economic Review, 105*(5), 408.
Oprea, R., Charness, G., & Friedman, D. (2014). Continuous time and communication in a public-goods experiment. *Journal of Economic Behavior & Organization, 108*, 212–223.
Pevnitskaya, S., & Ryvkin, D. (2013). Environmental context and termination uncertainty in games with a dynamic public bad. *Environment and Development Economics, 18*(1), 27–49.
Pnevmatikos, N., Vardar, B., & Zaccour G. (2018). When should a retailer invest in brand advertising? *European Journal of Operational Research, 267*(2), 754–764.
Sherstyuk, K., Tarui, N., Ravago, M. L. V., & Saijo, T. (2016). Intergenerational games with dynamic externalities and climate change experiments. *Journal of the Association of Environmental and Resource Economists, 3*(2), 247–281.
Tasneem, D., Engle-Warnick, J., & Benchekroun, H. (2017). An experimental study of a common property renewable resource game in continuous time. *Journal of Economic Behavior & Organization, 140*, 91–119.
Vardar, B., & Zaccour, G. (2018). Exploitation of a productive asset in the presence of strategic behavior and pollution externalities, G-2018-43, Cahiers du GERAD.
Vespa, E. (2011). An experimental investigation of strategies in the dynamic common pool game. Available at SSRN 1961450.

# Managerial Incentives and Polluting Inputs Under Imperfect Competition

**Denis Claude and Mabel Tidball**

**Abstract** This paper explores the link between upstream input pricing and downstream strategic delegation decisions. It complements earlier contributions by studying how environmental emissions and tax payments alter the incentives business owners have to divert their managers from profit maximization in favor of sales revenue generation. Two scenarios are compared depending on whether the upstream supplier precommits to a fixed input price or adopts a flexible price strategy. Corresponding Subgame-Perfect Nash-Equilibria are characterized and elements of comparative statics analysis are presented. The analysis confirms that previous results—showing that a price precommitment makes the upstream supplier better off and downstream firms worse off—carry over to situations in which production generates pollution.

**Keywords** Precommitment · Externality · Delegation · Vertical relations · Managerial incentives

## 1 Introduction

This chapter bridges two fields of research in which Georges Zaccour has been active: the analysis of *vertical relations* and *environmental and resource economics*. The former studies relations between firms that intervene successively along the value chain whereas the latter deals with the relations between the economy and the environment. We can trace back his interest for vertical relations to his early contributions on the analysis of energy markets (Zaccour 1983, 1987; Breton et al.

D. Claude (✉)
LEDi, Université de Bourgogne-Franche Comté, Dijon, France
e-mail: denis.claude@u-bourgogne.fr

M. Tidball
CEE-M, Université de Montpellier, Inra, Cnrs, SupAgro, Montpellier, France
e-mail: mabel.tidball@inra.fr



P.-O. Pineau et al. (eds.), *Games in Management Science*,
International Series in Operations Research & Management Science 280,
https://doi.org/10.1007/978-3-030-19107-8_10

1990). Today this interest is mainly manifested in his work on *marketing channels*.[1] But, it also comes up tangentially in a variety of contributions ranging from environmental economics—where sustainable tourism development may require tourism destinations to delegate expenditures in environmental remediation to a regional authority (Claude and Zaccour 2009)—to *institutional economics*—where good institutions are produced by the strategic precommitment of civil society to fight corruption (Ngendakuriyo and Zaccour 2013).

Strategies of "delegation" and "precommitment" are at the heart of the literature on *strategic delegation* to which this chapter contributes.[2] Starting with Vickers (1985), Fershtman and Judd (1987), and Sklivas (1987), this literature has examined the incentives owners have to delegate production decisions. By hiring a manager, the owner of a firm can *credibly* commit to pursue a goal that differs from maximizing profit. The managerial compensation contract will then be designed to convey the appropriate managerial incentives to guide the manager in his day-to-day decisions. Its terms will provide for a variable *target-based bonus* that rewards the manager's performance in achieving some alternative goal to (pure) profit maximization. This variable part may be based on any one or a combination of the following criteria: profit, sales volume or revenue, market share, and corporate social responsibility or environmental objectives. Since rational managers respond to financial incentives conveyed by the variable part of their remuneration, they will be encouraged to deviate from profit maximization.

As is well known, by choosing to reward sales revenue rather than profit, the owner encourages the manager to adopt a more aggressive market behavior. Namely, the managerial firm will produce more (for any level of production of its competitors) than a profit-maximizing owner-managed firm. Financial disclosure rules usually ensure that the incentives embedded in managerial compensation contracts are common knowledge.[3] Any change in the performance criteria presiding over managerial compensation will then affect the expectations of competing firms. This opens the door to a *strategic manipulation* of compensation contracts: each owner attempting to alter the expectations of rival firms to its own advantage.

But deviating from profit maximization is only profitable when the deviation is unilateral. And, since all owners face similar incentives to deviate, widespread deviations are to be expected and excessive output supply will result into lower profits for all. Hence, the opportunity to strategically delegate day-to-day production decisions to a manager closes as a trap on firm owners who actually find themselves confronted with a *Prisoner's Dilemma*.

Strategic delegation provides a much needed rationale for observed deviations from profit maximization. This rationale, however, assumes that firms are vertically integrated and produce their own inputs. When this assumption is relaxed, the vertical externality linked to input pricing appears to have a disciplining role on

[1] See Jorgensen and Zaccour (2004) and the references therein.

[2] For literature surveys, see Lambertini (2017), Kopel and Pezzino (2018), and Sengul et al. (2012).

[3] See Vural (2018) for the example of the ball bearing company SKF.

downstream firms' behavior. If duopolists buy in inputs from a common monopolist supplier, Park (2002) shows that strategic delegation becomes unprofitable for business owners. Wang and Wang (2010) reach the same conclusion by assuming that managerial incentive contract rewards a combination of profit and market share (rather than profit and sale revenues). On the contrary, Liao (2008, 2010) shows that strategic delegation retains its strategic value if the input supplier precommits to a fixed input price. This conclusion is backed by Wang (2015), who proves it true when compensation contracts reward the manager's performance relative to peers.[4] Finally, Claude (2018) shows that Park (2002)'s main results no longer holds if we consider a downstream market consisting of more than two—but a finite number of—firms.

This paper re-examines the link between vertical externalities and strategic delegation decisions. We consider an extended version of Park (2002)'s model in which downstream firms generate pollution emissions when they process the intermediate product into a final good. Specifically, we assume that the emission of a firm, per unit of output produced, is inversely related to that firm's productivity in processing inputs. Since pollution emissions are assumed to be taxed, downstream firms have an incentive to internalize them, at least in part.

Two papers have investigated the consequences of strategic delegation for environmental policy-making. In a Cournot duopoly with homogeneous goods and pollution emissions, Barcena-Ruiz and Garzon (2002) find that strategic delegation is profitable. At the equilibrium, managerial firms produce and emit more than profit-maximizing owner-managed firms. Consequently, the optimal environmental tax is higher than that required to regulate a standard Cournot market. Pal (2012) generalizes this result to differentiated industries. However, none of these papers has investigated how factor market imperfections alter managerial incentives in downstream markets, which is the main purpose of our paper. Since our model encompasses those of Park (2002) and Claude (2018), we check the robustness of their results.

The paper is organized as follows. Section 2 describes the model. Section 3 solves the managerial sub-game. Sections 4 and 5 characterize the sub game perfect Nash equilibrium (SPNE) depending on whether the upstream monopolist precommits to a price. SPNE outcomes are compared in Sect. 6. The last section concludes.

## 2 The Model

We extend the model of Park (2002) to account for the existence of pollution (or waste). Consider a vertical market structure with upstream monopoly and downstream quantity competition. The single upstream monopolist (indexed by *up*) produces at no cost a homogeneous input $x$ that it sells at a non-discriminatory

[4]This form of strategic delegation was first studied by Fumas (1992) and Miller and Pazgal (2002).

price $k > 0$. Let $x_i$ denote the input consumption of firm $i$ ($i = 1, 2, \ldots, n$) and $X = \sum_{i=1}^{n} x_i$ the aggregate input demand. The upstream supplier's profit function is simply $\pi_{\text{up}} = k\,X$.

Downstream firms rely on the same technology to turn the intermediate product $x$ into a final good $y$. Let $y_i$ denote the output of Firm $i$ ($i = 1, 2, \ldots, n$) and $Y = \sum_{i=1}^{n} y_i$ the aggregate downstream output. Firm $i$'s production function is $y_i = \varepsilon\, x_i$, where $\varepsilon \in (0, 1]$ is a parameter measuring Firm $i$'s productivity.

Pollution emissions come as a by-product of production. More precisely, they are inversely related to Firm $i$'s productivity and given by $e_i = (1 - \varepsilon)\, x_i$.[5] We assume that the government levies a tax on pollution emissions at a rate $\tau \geq 0$. Obviously, the more inefficient the firm is (the lower the value of $\varepsilon$), the higher is the quantity of pollution emitted for each unit of the final good $y$ produced and the higher is the firm's tax bill for a given level of output. Conversely, if $\varepsilon$ is assumed equal to 1, then the firm no longer emits pollution and its environmental tax bill is zero. Without loss of generality, we assume that downstream firms face no other production cost than that associated with input purchase. Demand for the final good $y$ is represented by the inverse demand function $P(Y) = a - bY$ with $a, b > 0$. Under the above assumptions, Firm $i$'s profit function is given by

$$\pi_i = P(Y)y_i - kx_i - (1 - \varepsilon)\tau x_i. \tag{1}$$

Using the production function, the above can be expressed in terms of $y$ only:

$$\pi_i = P(Y)y_i - A\,(k, \tau, \varepsilon)\, y_i, \qquad \text{where} \qquad A\,(k, \tau, \varepsilon) = \frac{k + \tau}{\varepsilon} - \tau, \tag{2}$$

denotes firm i's *effective* marginal cost of production; namely, the sum of the firm's input expenditure $(k/\varepsilon)$ and environmental tax bill $(\tau\,(1 - \varepsilon)\,/\varepsilon)$. Observe that $\partial A/\partial \varepsilon < 0$, $\partial A/\partial \tau > 0$ and $\partial A/\partial k > 0$, which is in accordance with intuition. Indeed, an increase in the price of factors ($\tau$ for emissions, and $k$ for the intermediate product) leads to a corresponding increase in marginal cost, whereas an increase in the productivity parameter $\varepsilon$ translates into a reduced (effective) marginal cost.

The owners of downstream firms may hire managers to run their firms on their behalf. The decision to hire a manager leads to the so-called *divorce between*

---

[5]We assume that the more efficient the firm is, the less input is used per unit of output and thus the lower the level of waste or emission generated. This assumption conforms to empirical findings by Shadbegian and Gray (2003). Examining the determinants of environmental performance at paper mills, they found that high productivity plants pollute less. More precisely, a 10% higher productivity level is associated with a 2.5% lower emission per unit of output. Furthermore, they found that unexpectedly high productivity levels are associated with unexpectedly low levels of emissions per unit of output. Shadbegian and Gray (2003) advance two main explanations for their results. On the one hand, newer production plants may be more efficient in production but also designed so as to reduce pollution emissions or waste. On the other hand, older, more inefficient, firms may face less regulatory pressure and retrofitting their facilities may be extremely difficult so that abatement possibilities are reduced.

*ownership and control.* Indeed, when management and control functions are divorced, agency problems arise if the manager's interests diverge from those of the owner (Sengul et al. 2012). In the remainder of this paper, we assume that all agency problems have been resolved[6] to focus our attention on the strategic value of delegation decisions. Following Vickers (1985), Sklivas (1987), and Fershtman and Judd (1987), we assume that owners hire managers in order to credibly commit their businesses to objectives that differ from profit maximization. Here, we assume that this objective is the maximization of a weighted sum of profits and sales revenue; i.e.,

$$F_i = \alpha_i \pi_i + (1 - \alpha_i) [P(Y) y_i] . \tag{3}$$

The profit weight $\alpha_i$ is chosen strategically by the owner of firm $i$ in order to manipulate the anticipations of rival firms to its own advantage (i.e., in a profitable way). The corresponding departure from profit maximization is credible since the inclusion of a target-based bonus in managerial compensation contracts ensures that appropriate incentives are conveyed to managers.[7] Depending on the value selected for $\alpha_i$, owners can encourage a wide range of behaviors. To see this, let us rewrite Eq. (3) as:

$$F_i = P(Y) y_i - \alpha_i A(k, \tau, \varepsilon) y_i . \tag{4}$$

Then the performance measure has straightforward interpretation. By choosing $\alpha_i = 1$, firm owners encourage pure profit maximization. However, if $\alpha_i$ is set lower than one, they direct their managers to pursue revenue generation at the expense of profits and if $\alpha_i$ is set greater than one they direct their managers to pursue cost minimization at the expense of profit generation. We let $\vec{\alpha} = (\alpha_1, \alpha_2, \ldots, \alpha_n)$ denote the profile of profit weights chosen by downstream owners.

[6]After delegation, the manager may not act in the owner's interest and engage in opportunistic practices and other self-serving behaviors. Managerial opportunism arises from two main sources (Eisenhardt 1989) : (a) the objectives of the owner and the manager may conflict and (b) the owner may not be able to observe the behavior of the manager. The first is of little relevance in our context. Indeed, we assume that the manager is offered a performance-based bonus which reduces conflicts of interests and ensures that the manager will adhere to the owner's supply strategy. However, appropriate monitoring and governance mechanisms should be put in place in order to treat the second source of managerial opportunism. Indeed, unobservable behaviors may result in expropriation of the company funds (Shleifer and Vishny 1997). Common examples include the excessive consumption of perquisites such as luxurious offices and company jets. More broadly, managerial discretion may result in the allocation of company funds to the pursuit of pet projects or that of an irrational expansion of the firm.

[7]Consider a two-part compensation contract $w_i = w_i^F + w_i^V F_i$ where $w_i^F \geq 0$ denotes the *base salary* and $w_i^V > 0$ denotes the *bonus rate that rewards performance* as measured by $F_i$. Given this compensation contract, it is equivalent for manager $i$ to maximize the compensation $w_i$ or the target $F_i$. For more on this point, see Kopel and Pezzino (2018).

We consider two scenarios depending on whether the upstream monopolist is able to precommit to a fixed input price. In the first scenario, we assume that price is difficult to change and communicate to customers. In this case, price commitments are credible. The order of moves is then as follows. The upstream monopolist moves first and sets the input price. After observing this price, downstream owners design their managerial compensation contracts simultaneously and independently. Obviously, the latter simply means setting the value of the profit weight $\alpha_i$. At the last stage of the game, managers compete in quantities so as to maximize their compensation, taking into account their target-based bonus and the input price.

In the second scenario, we assume that prices are easy to change and communicate and the upstream monopolist is unable (or unwilling) to make a price precommitment. Without price commitment the upstream monopolist retains the opportunity to adjust the input price to changes in the behavior of downstream firms. The order of moves is then as follows. In the first stage of the game, firm owners simultaneously and independently choose the compensation contracts that will be offered to hired managers (i.e., they set $\alpha_i$, $i = 1, 2, \ldots, n$). In the second stage, when the profile of managerial incentives $\vec{\alpha}$ is known and common knowledge, the upstream monopolist sets the input price $k$. Then, in the last stage, managers compete in quantities so as to maximize their compensation taking into account their target-based bonuses and the input price.

The difference between the timing of moves in the above two scenarios is illustrated in Fig. 1. Both three stage games are solved by backward induction. Since the last stage is common to both games, it will be analyzed separately in the next section.

Throughout the paper, we make the following assumption:

**Assumption 1** *We assume that the efficiency of firms is sufficiently high,* $1 \geq \varepsilon > \frac{\tau}{a+\tau}$, *(or the price of energy is sufficiently low,* $\frac{a\varepsilon}{(1-\varepsilon)} > \tau > 0$*) to ensure that all firms are active at the equilibrium.*

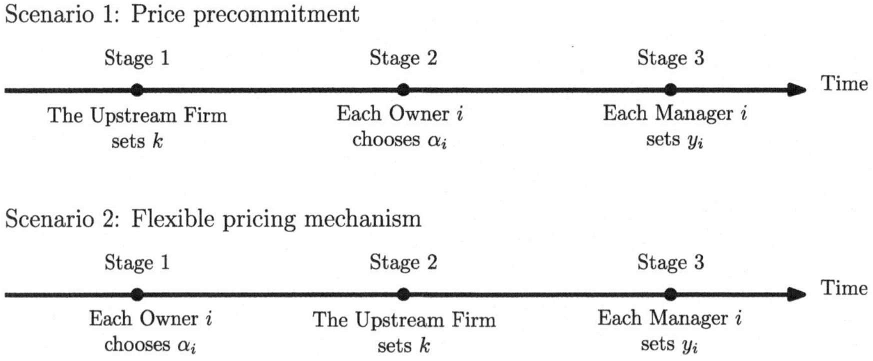


**Fig. 1** Timing of the games: precommitment vs. flexible pricing

## 3 The Managerial Subgame

At stage 3, given $\vec{\alpha}$ and $k$, each manager $i$ simultaneously and independently chooses a quantity $y_i$ so as to maximize $F_i$, $i = 1, 2, \ldots, n$. Assuming an interior solution, the resulting system of first-order conditions is

$$P(Y) + P'(Y)y_i = \alpha_i A\,(k, \tau, \varepsilon)\,, \quad i = 1, 2, \ldots, n. \tag{5}$$

Obviously, these conditions repeat the standard provision that marginal revenue must equal the marginal cost of production. However, when the manager strikes the balance between marginal costs and revenue, he does not consider the firm's actual marginal cost of production $A\,(k, \tau, \varepsilon)$, but its depreciated (or inflated) value by the weight factor $\alpha_i$. By solving the i-th equation in (5) for $y_i$ we obtain:

$$y_i = \frac{\alpha_i A\,(k, \tau, \varepsilon) - P(Y)}{P'(Y)}, \quad i = 1, 2, \ldots, n. \tag{6}$$

If $\alpha_i < 1$, observe that the manager $i$ is induced to produce more than a profit-maximizing firm owner (for each output choice of its competitors). By choosing profit weights that are lower than one, owners induce their managers to be more aggressive on the output market.

Summing Eq. (6) over $i$ results in:

$$Y = \frac{A\,(k, \tau, \varepsilon) \sum_{i=1}^{n} \alpha_i}{P'(Y)} - n\frac{P(Y)}{P'(Y)}, \tag{7}$$

or, equivalently,

$$Y P'(Y) + n P(Y) = A\,(k, \tau, \varepsilon) \sum_{i=1}^{n} \alpha_i. \tag{8}$$

This last equation implicitly defines the equilibrium industry output $Y^\star$ as a function of a weighted sum of the marginal costs of production incurred by the firms making up the industry. This result repeats the observation by Bergstrom and Varian (1985a,b) that Cournot-Nash quantities depend solely on the sum of the firms' characteristics and, especially, are independent of how those characteristics are distributed. By substituting the inverse demand function $P(Y) = a - bY$ into the fixed-point equation (8) and solving for the equilibrium industry output level, we obtain:

$$Y^\star = \frac{na - A(k, \tau, \varepsilon) \sum_{i=1}^{n} \alpha_i}{b(n+1)}. \tag{9}$$

Now, by plugging this quantity back into Firm $i$'s (inclusive) reaction function (6), we obtain:

$$y_i^\star = \frac{a + A(k,\tau,\varepsilon)\left(\sum_{i=1}^n \alpha_i - (n+1)\alpha_i\right)}{(n+1)b}. \tag{10}$$

The implied equilibrium price is:

$$P^\star := P(Y^\star) = \frac{a + A(k,\tau,\varepsilon)\sum_{i=1}^n \alpha_i}{(n+1)}. \tag{11}$$

# 4 Price Precommitment

In this first scenario, the upstream monopolist precommits to an input price $k$. So, at the time when owners design their managers' compensation contracts, the pricing policy of the upstream monopolist is known and common knowledge. The timing of the game is depicted in Fig. 1. At Stage 2, each owner simultaneously and independently chooses the incentives to provide to management (i.e., the profit weight $\alpha_i$). At stage 1, the upstream supplier sets the input price $k$. Now, we resume the backward induction procedure starting with the design of managerial incentive contracts.

## 4.1 Choice of Managerial Incentive Contracts

Each owner $i$ sets the profit weight $\alpha_i$ so as to maximize its profit taking the input price $k$ as given; i.e., owner $i$ solves

$$\max_{\alpha_i} \pi_i = \left[P(Y^\star) - A\left(k,\tau,\varepsilon\right)\right] y_i^\star, \quad i = 1, 2, \ldots, n. \tag{12}$$

Assuming an interior solution, we obtain the following system of $n$ first-order conditions for profit maximization:

$$P'(Y^\star)\frac{\partial Y^\star}{\partial \alpha_i} y_i^\star + \frac{\partial y_i^\star}{\partial \alpha_i}\left[P(Y^\star) - A\left(k,\tau,\varepsilon\right)\right] = 0, \quad i = 1, 2, \ldots, n. \tag{13}$$

The first term on the left-hand side corresponds to a gain in sales revenue linked to the price increase resulting from the reduction in total industry output. The second term corresponds to a profit loss linked to the reduction in firm $i$'s supply. Conditions (13) indicate that managerial incentive contracts should be designed so as to balance these two countervailing effects.

From Eqs. (9) and (10), we obtain (See Appendix 1):

$$\partial Y^\star/\partial \alpha_i = -A/\left(b\left(n+1\right)\right), \qquad \partial y_i^\star/\partial \alpha_i = -nA/\left(b\left(n+1\right)\right)). \tag{14}$$

Replacing in Eq. (13), $P(Y)$, $P'(Y)$, $\partial Y/\partial\alpha_i$, $\partial y_i/\partial\alpha_i$ and $y_i$ by their respective expressions, after straightforward calculations, we obtain:

$$-(n-1)\left(a + A\sum_{i=1}^{n}\alpha_i\right) - (n+1)\,A\,(\alpha_i - n) = 0, \quad i = 1, 2, \ldots, n. \tag{15}$$

Solving each of the above equations, we find the expression of $\alpha_i$ as a function of $\sum_{i=1}^{n}\alpha_i$:

$$\alpha_i = n - \tfrac{(n-1)}{(n+1)A}\left(a + A\sum_{i=1}^{n}\alpha_i\right), \quad i = 1, 2, \ldots, n. \tag{16}$$

Summing Eq. (15) over $i = 1, 2, \ldots, n$, results in

$$-(n-1)\,na - \left(1+n^2\right)A\sum_{i=1}^{n}\alpha_i + n^2\,(n+1)\,A = 0, \tag{17}$$

from which we extract the expression of the weighted sum of managerial incentives

$$\sum_{i=1}^{n}\alpha_i = \tfrac{n((n+1)nA-(n-1)a)}{(n^2+1)A}. \tag{18}$$

Plugging this sum back into Eq. (16) we find:

$$\alpha_i^c = n - \tfrac{(n-1)}{(1+n^2)A}\left(a + n^2 A\right), \quad i = 1, 2, \ldots, n. \tag{19}$$

Plugging $\alpha_i^c$ back into Eqs. (9)–(11) yields:

$$P\left(Y^c\right) = \tfrac{a+n^2A}{(1+n^2)}, \quad Y^c = \tfrac{n^2(a-A)}{b(1+n^2)}, \quad y_i^c = \tfrac{n(a-A)}{b(1+n^2)}, \quad \pi_i^c = \tfrac{n(a-A)^2}{b(1+n^2)^2}, \tag{20}$$

$i = 1, 2, \ldots, n$. For the time being, let us assume that $k$ is given and fixed. Then, equilibrium values are presented in (20). As expected the solution is symmetric. Furthermore, the solution is admissible (i.e., prices and quantities are strictly positive) provided that $a > A$. From Eq. (19), we obtain:

$$\alpha_i^c = \alpha^c = 1 - \tfrac{(n-1)}{(1+n^2)}\,(a/A - 1)\,, \quad i = 1, 2, \ldots, n. \tag{21}$$

Since admissibility implies $a > A$, it follows that $(\frac{a}{A} - 1) > 0$ and $\alpha^c < 1$. In other words, equilibrium managerial compensation contracts provide that managers will be rewarded for deviations from profit maximization that favor sales revenue generation. Now, observe that $A(k, \tau, \varepsilon) > A\,(k, \tau, 1) = k$. Since $\partial\alpha^c/\partial A = a\,(n-1)\,/\left(A^2\left(1+n^2\right)\right) > 0$, we find that the equilibrium profit weight $\alpha^c$ reaches a minimum when production generates no pollution (i.e., when $\varepsilon = 1$):

$$\alpha^c|_{\varepsilon=1} = 1 - \tfrac{(n-1)}{(1+n^2)}\,(a/k - 1)\,. \tag{22}$$

This is the same expression as found by Claude (2018). Finally, still assuming that $k$ is given, we cannot exclude that $\alpha^c$ might be negative for some parameter values. Indeed, this could be the case for $a > (n(n+1)A)/(n-1)$. However, in the next section it is shown that $\alpha^c$ ranges in the interval $[0, 1]$ when evaluated at the SPNE price $k^c$.

## 4.2 Monopoly Pricing

At stage 1, the upstream supplier sets the input price $k$ so as to maximize its profit; i.e., the monopolist solves

$$\max_k \pi_{\text{up}}^c = kn\,(y^c/\varepsilon)\,.$$

Assuming an interior solution exists, the first-order condition for profit maximization is

$$y^c + k\left(\frac{\partial y^c}{\partial k}\right) = 0. \tag{23}$$

Solving the above equation for $k$, after straightforward computations, we obtain the following proposition.

**Proposition 1** *Suppose that the upstream monopolist precommits to a fixed input price. There exists a unique Subgame-Perfect Nash-Equilibrium. Equilibrium outcomes are, respectively,*

$$k^c = \frac{a\varepsilon - \tau(1-\varepsilon)}{2}, \qquad \alpha^c = 1 - \frac{2(n-1)\,k^c}{(n^2+1)(a\varepsilon+\tau(1-\varepsilon))}, \qquad y^c = \frac{n\,k^c}{b(n^2+1)\varepsilon}, \tag{24}$$

$$P^c = \frac{a\varepsilon\left(n^2+2\right)+n^2(\tau(1-\varepsilon))}{2\left(n^2+1\right)\varepsilon}, \quad \pi_{\text{up}}^c = \frac{n^2(k^c)^2}{b\left(n^2+1\right)\varepsilon^2}, \qquad \pi_i^c = \frac{n(k^c)^2}{b\left(n^2+1\right)^2\varepsilon^2}. \tag{25}$$

Assumption 1 ensures that all the quantities given in the above proposition are nonnegative at the SPNE.

*Remark 1* If $\varepsilon = 1$, we obtain the managerial incentive as in Claude (2018),

$$\alpha^c|_{\varepsilon=1} = 1 - \frac{(n-1)}{\left(1+n^2\right)},$$

and the same equilibrium values. Moreover, as $\varepsilon$ tends to $\tau/(a+\tau)$ the input price $k^c$ tends to zero, $\alpha^c$ tends to one, and both upstream and downstream profits tend to zero.

We obtain the following comparative statics results:

$$\frac{\partial k^c}{\partial \varepsilon} > 0, \qquad \frac{\partial y^c}{\partial \varepsilon} > 0, \qquad \frac{\partial \alpha^c}{\partial \varepsilon} < 0, \qquad \frac{\partial \pi_i^c}{\partial \varepsilon} > 0, \qquad \frac{\partial \pi_{\text{up}}^c}{\partial \varepsilon} > 0. \tag{26}$$

A reduction in pollution emissions or, equivalently in our model, an increase in the productivity of downstream firms, allows the upstream supplier to charge a higher input price $k$. Also, it provides firm owners with incentives to assign a lower weight to profit maximization in managerial compensation contracts. This lower weight encourages managers to adopt a more aggressive market behavior, which results in higher firm and industry output levels. As a result, both upstream and downstream profits rise. The opposite comparative statics hold for an increase in the tax rate $\tau$. This should come as no surprise given the inverse relationship between productivity and pollution emissions in our model.

Finally, we have $\alpha^c \in [0, 1]$, $\lim_{n \to +\infty} \alpha^c = 1$, and $\frac{\partial \alpha^c}{\partial n} > 0$, $\forall n > 2$. We conclude that $\alpha^c$ is monotonically increasing in $n$ and converges to one as the number of downstream firms becomes arbitrarily large. Indeed, as the number of firms increases, the downstream market becomes more competitive implying that aggregate output rises and market price falls. This in turn leads to a reduction in marginal revenue that must be compensated by lower production volumes. Firm owners achieve the required production cut by assigning a lower weight $(1 - \alpha)$ to sales in managerial compensation schemes.

## 5 Flexible Pricing Mechanism

We now turn to the second scenario in which the order of firms' moves is reversed. At the time when the upstream monopolist sets the input price, it is assumed that the terms of managerial compensation contracts are known and common knowledge. The timing of the game is depicted in Fig. 1. In stage 1, firm owners choose managerial incentives. In stage 2, the upstream monopolist sets the input price. We now are in a position to resume the backward induction procedure starting with the resolution of the monopolist pricing problem.

### *5.1 Input Pricing*

In stage 2, the upstream monopolist sets the input price $k$ so as to maximize its profit: $\pi_{\text{up}}^\star = k \sum_{i=1}^n x_i^\star$. Recall that $y_i^\star = \varepsilon x_i^\star$ so that $x_i^\star = y_i^\star / \varepsilon^\star$. The decision problem of the upstream monopolist then writes as

$$\max_k \pi_{\text{up}}^\star = k \sum_{i=1}^n \left( y_i^\star / \varepsilon^\star \right).$$

Assuming an interior solution exists, the optimal input price solves

$$\sum_{i=1}^{n} y_i^\star + k\left[\frac{\partial}{\partial k}\left(\sum_{i=1}^{n} y_i^\star\right)\right] = 0. \tag{27}$$

Replacing $y_i^\star$ by its value from (10) and solving for the optimal input price $k$ gives:

$$k^f = \frac{n a \varepsilon - \tau (1-\varepsilon)\sum_{i=1}^{n}\alpha_i}{2\sum_{i=1}^{n}\alpha_i}. \tag{28}$$

As we shall see below, Assumption 1 ensures that $k^f$ is positive for the SPNE value of $\alpha_i$.

Note that

$$\frac{\partial k^f}{\partial \alpha_i} = -\frac{n a \varepsilon}{2\left(\sum_{i=1}^{n}\alpha_i\right)^2} < 0. \tag{29}$$

An increase in the weight downstream firms place on profits results in a decrease in the output price. Alternatively, a greater sale orientation causes a reduction in the input price. The economic intuition behind this result is straightforward. As the profit weight decreases, downstream managers are induced to behave more aggressively, producing more. The increase in input demand then explains the increase in input price.

## 5.2 Strategic Delegation

In stage 1, each owner $i$ simultaneously and independently sets the profit weight $\alpha_i$ so as to maximize its profit. In other words, each owner $i$ solves the following problem:

$$\max_{\alpha_i} \pi_i^f = \left[P(Y^f) - A\left(k^f,\tau,\varepsilon\right)\right] y_i^f, \quad i = 1,2,\ldots,n.$$

First-order conditions for profit maximization are:

$$\left(b\frac{\partial Y^f}{\partial \alpha_i} + \frac{1}{\varepsilon}\frac{\partial k^f}{\partial \alpha_i}\right)\left(P(Y^f) - \alpha_i^f A^f\right) + \left(A^f + \frac{\alpha_i^f}{\varepsilon}\frac{\partial k^f}{\partial \alpha_i} + b\frac{\partial Y^f}{\partial \alpha_i}\right)\left(P(Y^f) - A^f\right) = 0, \tag{30}$$

$i = 1,2,\ldots,n$. Assuming a symmetric equilibrium, it is shown in Appendix 2 that condition (30) reduces to

$$g(\alpha) := \omega_0 + \omega_1\alpha + \omega_2\alpha^2 + \omega_3\alpha^3 = 0, \tag{31}$$

where

$$\omega_0 = a^2 n^2 (n+1)\varepsilon^2, \quad \omega_2 = -n\tau(1-\varepsilon)(n(n+1)(a\varepsilon+\phi) - 2a\varepsilon), \tag{32}$$

$$\omega_1 = -a\left(n^2-1\right)\varepsilon((n+2)\phi - n\tau(1-\epsilon)), \quad \omega_3 = -n\left(n^2+1\right)\tau^2(\varepsilon-1)^2, \tag{33}$$

with $\phi = (a\varepsilon - \tau(1-\varepsilon))$.

The following proposition provides a characterization of equilibrium managerial incentives:

**Proposition 2 (Equilibrium Managerial Incentives)** *Suppose that the upstream monopolist uses a flexible pricing mechanism. Then, there exists a unique symmetric SPNE characterized as follows.*

1. *The equilibrium profit weight takes values on the interval* $(0, 1]$*;*
2. *If production generates no pollution emission, then it is given by* $\alpha^f\big|_{\varepsilon=1} = \frac{n^2}{(n-1)(n+2)}$*;*
3. *It is equal to one—implying pure profit maximization—in two cases:*
   (a) *if the downstream market consists in a duopoly and production generates no pollution emission* ($n = 2, \varepsilon = 1$) *and,*
   (b) *in the limit case where downstream firms are so inefficient that they prefer to be inactive at the equilibrium* ($\varepsilon = \tau/(a+\tau)$).

*Proof* See Appendix 3. □

We remark that Points 3($a$) and (b) in Proposition 2 include as special cases results by Claude (2018) which correspond to the model without pollution ($\varepsilon = 1$).

Assuming that $\alpha_i = \alpha^f$ for all $i = 1, 2, \ldots, n$, and plugging (28) into Eqs. (9)–(11) gives the following proposition:

**Proposition 3** *Equilibrium prices, quantities, and profit levels are, respectively, given by*

$$k^f = \frac{a\varepsilon - \tau(1-\varepsilon)\alpha^f}{2\alpha^f}, \quad P^f = \frac{a\varepsilon(n+2) + n\tau(1-\varepsilon)\alpha^f}{2(n+1)\varepsilon}, \quad y^f = \frac{\alpha^f k^f}{b(n+1)\varepsilon}, \tag{34}$$

$$\pi^f_{\text{up}} = \frac{n\alpha^f\left(k^f\right)^2}{b(n+1)\varepsilon^2}, \quad \pi^f = \frac{\alpha k^f((n+1)(a\epsilon + \tau(\epsilon-1)) - k^f(\alpha^f n + n + 1))}{b(n+1)^2\epsilon^2}. \tag{35}$$

Under Assumption 1, it is straightforward to show that the prices $k^f$ and $P^f$ and the quantities $y^f$ and $Y^f$ are strictly positive.

# 6 Comparing Equilibrium Outcomes

This section attempts to compare equilibrium outcomes depending on whether the upstream monopolist makes a price precommitment. A major difficulty in doing this is due to the complexity of the expression for $\alpha^f$. To begin with, we analyze how the decision to precommit alters managerial incentives in the downstream market. We are able to state the following proposition.

**Proposition 4 (Comparison of Managerial Incentives)**

1. *The equilibrium profit weight is lower when the upstream monopolist precommits to a fixed input price; i.e.,* $\alpha^c < \alpha^f$.
2. *As the number of downstream firms becomes arbitrarily large and irrespective of the pricing strategy adopted by the upstream monopolist, the equilibrium profit weight converges to* 1*, implying that the behavior of managerial firms is eventually identical to that of profit-maximizing owner-managed firms; i.e.,* $\lim_{n\to\infty} \alpha^h = 1, \quad h = c, f$.

*Proof* See Appendix 4. □

The above proposition extends previous results by Claude (2018) that were established under the assumption that there is no pollution ($\varepsilon = 1$). It states that downstream owners offer compensation schemes that favor sale orientation over profit maximization if the upstream monopolist makes a price precommitment.

The intuition for this result can be traced back to the difference in the timing of moves between the two games that we consider. In the precommitment scenario, downstream owners set managerial incentives when the input price is known and has become common knowledge. Accordingly, in their decision-making process, they take the input price as fixed and given. By contrast, in the other scenario, the upstream monopolist adjusts the input price to the observed degree of competition on the downstream market (as proxied by managerial incentives). Then, downstream owners recognize that the price they pay for the input $x$ depends on the managerial incentives they give to their managers. Specifically, they know that a greater sale orientation (a lower value for $\alpha^f$) results in higher total downstream production, which, in turn, implies an increase in both input consumption and input price. The implied surge in production costs reduces the extent to which firm owners find it profitable to divert managers away from profit maximization; i.e., the value of $(1 - \alpha^f)$.

Finally, as the number of firms rises, the downstream market becomes increasingly competitive so that the strategic value of delegation vanishes. Then, the behavior of managerial firms converges to that of profit-maximizing owner-managed firms.

Next, we compare equilibrium input prices between the two scenarios. When the downstream market structure is a duopoly and no environmental externality exists, Liao (2008, 2010) showed that the upstream monopolist sets the same equilibrium price irrespective of whether a price commitment was made. However, Claude (2018) proved that precommitment results in a lower input price if more than two firms operate on the downstream market. The following proposition extends this result to more general contexts in which production is polluting the environment:

**Proposition 5 (Comparison of Input Prices)** *If the downstream industry (i) is a duopoly which generates no pollution emission or (ii) consists of infinitely many firms, then the upstream monopolist sets the same price in both scenarios (commitment or no commitment). Otherwise, precommitment results in a lower equilibrium input price.*

*Proof* From Eqs. (24) and (34), we obtain

$$k^c - k^f = -\frac{a\varepsilon\left(1-\alpha^f\right)}{2\alpha^f} \leq 0.$$

If $n = 2$ and $\varepsilon = 1$, then $\alpha^f = 1$ and $k^c = k^f = a/2$. Moreover, since $\alpha^f$ tends to 1 as the number of downstream firms becomes arbitrarily large, it follows that $\lim_{n\to\infty} k^c - k^f = 0$. □

Turning to the comparison of production levels, we are able to prove the following proposition:

**Proposition 6 (Comparison of Downstream Production Levels)** *Downstream managers are encouraged to choose higher output levels if the upstream monopolist precommits to a fixed input price; i.e.,* $y^c - y^f > 0$. *Consequently, precommitment results in more pollution.*

*Proof* We have

$$\Delta y = y^c - y^f = \frac{a\varepsilon\,(n-1) - \left(n^2\left(1-\alpha^f\right) + n - \alpha^f\right)(1-\varepsilon)\,\tau}{2b\,(n+1)\left(n^2+1\right)\varepsilon}. \tag{36}$$

It is easy to see that $y^c - y^f > 0$ if and only if $\alpha^f > \tilde{\alpha}$ where

$$\tilde{\alpha} = 1 - \frac{(n-1)(a\varepsilon - \tau(1-\varepsilon))}{(n^2+1)\tau(1-\varepsilon)}. \tag{37}$$

From Proposition 4, recall that $\alpha^f > \alpha^c$. A little algebra shows that $\alpha^c > \tilde{\alpha}$. Since $\alpha^f > \tilde{\alpha}$, it follows that $y^c > y^f$. □

The intuition for this result can be understood by considering Propositions 4 and 5 above. The former shows that, in the precommitment scenario, firm owners design compensation schemes so as to encourage a greater sale orientation (or, equivalently, a more aggressive market behavior). Under the same assumption, the latter proves that the upstream monopolist sets a lower input price. This, in turn, implies that the final good $y$ becomes less costly to produce. Then, the greater sale orientation combines with reduced production costs to encourage greater production and emissions of pollutant.

Next, we turn to the comparison of downstream and upstream profit levels. In the absence of environmental externalities, Claude (2018) showed the upstream monopolist is strictly better off when committing to a fixed input price. The exact opposite holds for downstream firms: if the upstream supplier engages in fixed price contracts, they earn lower profits.

These result might seem surprising at first sight, since the upstream monopolist charges a lower input price in the precommitment scenario. However, the basic intuition for this result is simple. If the upstream monopolist adopts a flexible pricing mechanism, downstream owners anticipate that a more aggressive market behavior will result in a higher input price. Then, each owner provides his management with lower sales incentives. In other words, each owner assigns a lower weight $(1 - \alpha)$ to sales in managerial compensation schemes. This alleviates the problem of "over-competition" among firms arising from strategic delegation.

However, by waiving the right to adjust the input price to changes in managerial incentives, the upstream monopolist place firm owners back into their initial "prisoner's dilemma" situation. This strategic move creates the conditions for an overproduction that is unprofitable only for downstream owners. Indeed, precommitment makes the upstream monopolist better off since the surge in input consumption implied by overproduction more than compensates for the loss in revenue due to a reduced input price.

Unfortunately, if we relax the assumption that production causes no pollution, it becomes difficult to sign the difference between profits in the two scenarios for arbitrary parameter values. With that said, we are still able to shed some light on how profits compare and offer interesting insights on this issue. Let us recall that the admissible values of $\varepsilon$ lie in the range $\tau/(a + \tau) < \varepsilon < 1$. By evaluating upstream and downstream profits at both extremities, we obtain the following proposition.

**Proposition 7 (Comparison of Profits)**

1. *If production causes no pollution, then the upstream monopolist is better off committing to a fixed input price:*

$$\left(\pi_{\text{up}}^{c} - \pi_{\text{up}}^{f}\right)|_{\varepsilon=1} > 0.$$

*However, the precommitment decision of the upstream supplier is detrimental to downstream firms:*

$$\left(\pi_i^c - \pi_i^f\right)\Big|_{\varepsilon=1} < 0.$$

2. *For a sufficiently high level of emission per unit of the final good produced (or a sufficiently high environmental tax rate), downstream firms stop producing implying zero-profits for all:*

$$\pi_{\text{up}}^{c}|_{\varepsilon=\frac{\tau}{a+\tau}} = \pi_{\text{up}}^{f}|_{\varepsilon=\frac{\tau}{a+\tau}} = 0 = \pi_i^c|_{\varepsilon=\frac{\tau}{a+\tau}} = \pi_i^f|_{\varepsilon=\frac{\tau}{a+\tau}}, \quad i = 1, 2, \ldots, n.$$

*Proof* By plugging the value of $\varepsilon$ into the corresponding expressions. □

We retrieve the same results as in Claude (2018). As a work around for the difficulty in signing differences in upstream and downstream profits, we ran numerical simulations. Despite extensive efforts, we found it impossible to find even one numerical example that is admissible and reverses the ranking between profits in Proposition 7. If production generates few pollution emissions ($\varepsilon$ is close to 1), numerical simulations confirm that the upstream firm is better-off when committing to a fixed input price (i.e., $\pi^c_{\text{up}} > \pi^f_{\text{up}}$). The opposite (i.e., $\pi^c_{\text{up}} < \pi^f_{\text{up}}$) was obtained only for so low values of $\varepsilon$ that downstream firms produce nothing. Finally, we confirmed numerically that downstream firms make lower profits in the precommitment scenario. We conclude that previous results by Claude (2018) are robust to the introduction of pollution emissions from productive activities.

## 7 Conclusion

Recent advances in the strategic delegation literature emphasize that factor market imperfections reduce the incentive business owners have to manipulate the structure of incentives embedded in managerial compensation contracts so as to encourage managers to deviate from profit maximization. The purpose of this paper was to re-examine this issue by allowing for pollution emissions and related environmental tax payments.

Two scenarios were considered depending on whether the upstream monopolist supplies the intermediate product through fixed price contracts or relies on a flexible pricing scheme. The corresponding two games were solved by backward induction. In both cases, we proved the existence of a unique Subgame-Perfect Nash-Equilibrium (SPNE). Furthermore, we showed that the value of the equilibrium profit weight is restricted to the range from 0 to 1. It was shown that equilibrium managerial incentives encourage pure profit maximization only in limit cases. Hence, non-profit managerial incentives are expected to be the norm rather than the exception. Furthermore, except in limit cases, the following results hold. A price precommitment results in a lower input price which encourages greater production of the final good and greater pollution emissions. Moreover, it makes the upstream supplier better off and downstream firms worse off. We conclude that upstream suppliers will choose to sign fixed price contracts with their customers.

This paper has limitations. For tractability reason, we assumed identical firms. Relaxing this assumption offers interesting challenges for future research. Furthermore, our analysis has focused exclusively on how precommitment alters managerial incentives in downstream market. Accordingly, the rate of environmental taxation was regarded as exogenous and the welfare consequences of precommitment were not investigated. Future research could examine the normative question of optimal environmental policy.

**Acknowledgements** We would like to thanks the two anonymous referees for their useful comments and suggestions. This work was supported by the ANR GREEN-Econ research project (ANR-16-CE03-0005).

## Appendix 1: Proof of Eq. (14)

Recall that Firm $i$'s equilibrium output level is

$$y_i^\star = \frac{(a+A(k,\tau,\varepsilon)(\sum_{i=1}^n \alpha_i-(n+1)\alpha_i))}{(n+1)b}, \tag{38}$$

whereas that of Firm $j$ $(j \neq i)$ is

$$y_j^\star = \frac{(a+A(k,\tau,\varepsilon)(\sum_{i=1}^n \alpha_i-(n+1)\alpha_j))}{(n+1)b}. \tag{39}$$

Differentiating Eqs. (38) and (39) yields :

$$\partial y_i^\star/\partial\alpha_i = \frac{A(1-(n+1))}{(n+1)b} = -\frac{nA}{(n+1)b}, \tag{40}$$

$$\partial y_j^\star/\partial\alpha_i = \frac{A}{(n+1)b}, \tag{41}$$

and

$$\sum_{j\neq i} \frac{\partial y_j}{\partial\alpha_i} = \frac{(n-1)A}{(n+1)b}. \tag{42}$$

Finally, we obtain:

$$\frac{\partial Y}{\partial\alpha_i} = \sum_{j=1}^n \frac{\partial y_j}{\partial\alpha_i} = \sum_{j\neq i} \frac{\partial y_j}{\partial\alpha_i} + \frac{\partial y_i}{\partial\alpha_i}, \tag{43}$$

$$= \frac{((n-1)A-nA)}{(n+1)b} = -\frac{A}{(n+1)b}. \tag{44}$$

## Appendix 2: Proof of Eq. (31)

Let us consider the first-order conditions for profit maximization which are given by:

$$-\left(b\frac{\partial Y}{\partial\alpha_i} + \frac{1}{\varepsilon}\frac{\partial k}{\partial\alpha_i}\right) + \frac{\partial y_i}{\partial\alpha_i}(P(Y)-A) = 0, \quad i = 1, 2, \ldots, n. \tag{45}$$

From Eq. (7), observe that

$$y_i = \frac{a-\alpha_i A}{b} - Y, \tag{46}$$

so that

$$\frac{\partial y_i}{\partial \alpha_i} = -\left[\frac{1}{b}\left(A + \frac{\alpha_i}{\varepsilon}\frac{\partial k}{\partial \alpha_i}\right) + \frac{\partial Y}{\partial \alpha_i}\right]. \tag{47}$$

Plugging $y_i$ and $\partial y_i/\partial \alpha_i$ back into (45) results in

$$\left[b\frac{\partial Y}{\partial \alpha_i} + \frac{\partial A_i}{\partial \alpha_i}\right]\left(\frac{a-\alpha_i A}{b} - Y\right) + \left(\frac{1}{b}\left(A + \frac{\alpha_i}{\varepsilon}\frac{\partial k}{\partial \alpha_i}\right) + \frac{\partial Y}{\partial \alpha_i}\right)(P(Y) - A) = 0. \tag{48}$$

Given that

$$b\left(\frac{a-\alpha_i A}{b} - Y\right) = P(Y) - \alpha_i A, \tag{49}$$

we obtain:

$$\frac{1}{b}\left(b\frac{\partial Y}{\partial \alpha_i} + \frac{\partial A}{\partial \alpha_i}\right)\left(b\left(\frac{a-\alpha_i A}{b} - Y\right)\right) + \frac{1}{b}\left(\left(A + \frac{\alpha_i}{\varepsilon}\frac{\partial k}{\partial \alpha_i}\right) + b\frac{\partial Y}{\partial \alpha_i}\right)(P(Y) - A) = 0, \tag{50}$$

and, finally,

$$\left(b\frac{\partial Y}{\partial \alpha_i} + \frac{1}{\varepsilon}\frac{\partial k}{\partial \alpha_i}\right)(P(Y) - \alpha_i A) + \left(\left(A + \frac{\alpha_i}{\varepsilon}\frac{\partial k}{\partial \alpha_i}\right) + b\frac{\partial Y}{\partial \alpha_i}\right)(P(Y) - A) = 0. \tag{51}$$

Differentiating Eqs. (9) and (10) with respect to $\alpha_i$ yields

$$\frac{\partial Y}{\partial \alpha_i} = -\frac{1}{b(n+1)}\left[A + \sum_i \frac{\alpha_i}{\epsilon}\frac{\partial k}{\partial \alpha_i}\right], \tag{52}$$

$$\frac{\partial y_i}{\partial \alpha_i} = -\frac{1}{(n+1)b}\left(nA + (n+1)\left(\frac{\alpha_i}{\varepsilon}\frac{\partial k}{\partial \alpha_i}\right) - \sum_{i=1}^{n}\frac{\alpha_i}{\varepsilon}\frac{\partial k}{\partial \alpha_i}\right). \tag{53}$$

Plugging the following quantities

$$Y|_{\text{sym}} = \frac{n(a-\alpha A)}{b(n+1)}, \qquad y_i|_{\text{sym}} = \frac{(a-\alpha A)}{b(n+1)}, \qquad k|_{\text{sym}} = \frac{a\varepsilon - \tau(1-\varepsilon)\alpha}{2\alpha}, \tag{54}$$

$$\left.\frac{\partial Y}{\partial \alpha_i}\right|_{\text{sym}} = -\frac{(a-\alpha A)}{b(n+1)}, \qquad \left.\frac{\partial y_i}{\partial \alpha_i}\right|_{\text{sym}} = -\frac{(na+\alpha A)}{b(n+1)}, \qquad \left.\frac{\partial k}{\partial \alpha_i}\right|_{\text{sym}} = -\frac{a\varepsilon}{2n\alpha^2}. \tag{55}$$

into

$$\left(b\frac{\partial Y}{\partial \alpha} + \frac{1}{\varepsilon}\frac{\partial k}{\partial \alpha}\right)(P(Y) - \alpha A) + \left(\left(A + \frac{\alpha}{\varepsilon}\frac{\partial k}{\partial \alpha}\right) + b\frac{\partial Y}{\partial \alpha}\right)(P(Y) - A) = 0, \tag{56}$$

yields the cubic equation (31).

## Appendix 3: Proof of Proposition 2

Let us consider the existence problem first. Let $\delta_0 = \omega_2^2 - 3\omega_3\omega_1$ and $\delta_1 = 2\omega_2^3 - 9\omega_3\omega_2\omega_1 + 27\omega_3^2\omega_0$. The discriminant of the third-degree polynomial $g(\alpha)$ can be expressed as $\Delta = -\left(\delta_1^2 - 4\delta_0^3\right) / \left(27\omega_3^2\right)$. Let us show that $\Delta < 0$ so that Eq. (31) has a unique real solution $\alpha^f$. To this end, it is convenient to express the discriminant as $\Delta = \eta\,\Gamma$ where

$$\eta = -4a^2 n(n+1)\tau^2(\epsilon - 1)^2\epsilon^2 < 0 \quad \text{and} \tag{57}$$

$$\Gamma = \beta_0 + \beta_1 z + \beta_2 z^2 + \beta_3 z^3 + \beta_4 z^4 \quad \text{with} \quad z = (\varepsilon - 1) < 0 \quad \text{and} \tag{58}$$

$$\beta_0 = a^4\left(n^2 + n - 2\right)^2\left((n-1)n(n+2)^2 - 2\right)\epsilon^4, \tag{59}$$

$$\beta_1 = -2a^3(n(n(n(n(n(n(3n(n+2)-5)+5)+19)-31)-29)+16)+12)\tau\epsilon^3, \tag{60}$$

$$\beta_2 = a^2(n+1)(n(n(n(n(n(n(11n-14)+31)-26)+13)+64)-16)-24)\tau^2\epsilon^2, \tag{61}$$

$$\beta_3 = -2a(n+1)^2\left(2n^2 - 1\right)\left(n^2(n(2n-3)+2) - 4\right)\tau^3\epsilon, \tag{62}$$

$$\beta_4 = n^3(n+1)^3\left(2n^2 - 1\right)\tau^4. \tag{63}$$

Examining Eqs. (59)–(63), we find that $\beta_j$ is positive (resp., negative) if $j$ is even (resp., odd), for all $j = 0, \ldots, 4$. Since $z < 0$, it follows that $\Gamma > 0$ and thus $\Delta < 0$.

We proceed by showing that $\alpha^f$ takes values on the interval (0, 1]. Since $g(0) = \omega_0 > 0$, it follows that $\alpha^f$ cannot be equal to zero. Moreover, when $\frac{\tau}{a+\tau} < \varepsilon \leq 1$ we find that

$$g(1) = \sum_{i=0}^{3}\omega_i = -(a\varepsilon - \tau\,(1-\varepsilon))\,(a\,(n-2)\,(n+1)\varepsilon + (n-1)(1-\varepsilon)n\tau) < 0. \tag{64}$$

It follows that $g(\alpha)$ has a sign change on the interval (0, 1). Finally, if $\varepsilon = \tau/(a+\tau)$ recall that $(a\varepsilon - \tau\,(1-\varepsilon)) = 0$, so that $g(1) = 0$. Hence, the unique real root $\alpha^f$ takes values on (0, 1].

## Appendix 4: Proof of Proposition 4

1. We proved that $g(\alpha)$ has a sign change on the interval $(0, 1)$. Observe that $g(\alpha) > 0$ implies that $\alpha < \alpha^f$. It follows that $\alpha^c < \alpha^f$ if $g(\alpha^c) > 0$. Then, the proof reduces to showing that $g(\alpha^c) > 0$.

   Straightforward computations yield $g(\alpha^c) = \Phi h(t)$ where $\Phi = \frac{2a\varepsilon(a\varepsilon-\tau(1-\varepsilon))}{(n^2+1)^2(a\varepsilon+\tau(1-\varepsilon))^3} > 0$ and $h(t) = h_0 + h_1\tau + h_2\tau^2 + h_3\tau^3$, with

$$h_0 = a^3\varepsilon^3(n+1)\left(n^2+1\right)\left(n^2-2n+2\right) > 0,$$
$$h_1 = a^2(2n+1)((n-1)n+2)^2(1-\varepsilon)\varepsilon^2 > 0, \tag{65}$$
$$h_2 = a\left(n\left(n\left(n^3+2n+4\right)+3\right)+2\right)(1-\varepsilon)^2\varepsilon > 0,$$
$$h_3 = n(n+1)^2(1-\varepsilon)^3 > 0. \tag{66}$$

   Let $\Delta(h)$ denote the discriminant of $h(\tau)$. By direct computation, we find that

$$\Delta(h) = -a^6(n-1)^6 n(n+1)^2(\varepsilon-1)^6\varepsilon^6\Gamma, \quad \text{where} \tag{67}$$
$$\Gamma = \left(8n^{10}-n^9+6n^8+59n^7-48n^6+156n^5+64n^4+28n^3+96n^2-68n-64\right). \tag{68}$$

   Since $\Gamma$ is strictly positive, it follows that $\Delta(h) < 0$. Hence, the cubic equation $h(\tau)$ has a unique real solution. Since

$$h(\tau)|_{\tau=0} = a^3(n+1)\left(n^2+1\right)\left(n^2-2n+2\right)\varepsilon^3 > 0,$$

   and $h'(t) > 0$, it follows that $h(\tau) > 0$ for all $\tau \in [0, \frac{a\varepsilon}{1-\varepsilon}]$. Then $\Phi>0$ implies that $g(\alpha^c) > 0$ so that $\alpha_c < \alpha_f$.
2. Since $\lim_{n\to\infty}\alpha^c(n) = 1$ and $\alpha^c(n) < \alpha^f(n) \leq 1$, it follows that $\lim_{n\to\infty}\alpha^f(n) = 1$.

## References

Barcena-Ruiz, J. C., & Garzon, M. B. (2002). Environmental taxes and strategic delegation. *Spanish Economic Review, 4*(4), 301–310.
Bergstrom, T. C., & Varian, H. R. (1985a). Two remarks on Cournot equilibria. *Economics Letters, 19*(1), 5–8.
Bergstrom, T. C., & Varian, H. R. (1985b). When are Nash equilibria independent of the distribution of agents' characteristics? *The Review of Economic Studies, 52*(4), 715–718.

Breton, M., Haurie, A., & Georges, Z. (1990). Methods in the analysis of multistage commodity markets. In C. T. Leondes (Ed.), *Advances in large scale systems dynamics and control: Vol. 36. Control and dynamic systems* (pp. 75–105). Cambridge: Academic.
Claude, D. (2018). A reconsideration of the link between vertical externality and managerial incentives. *Managerial and Decision Economics, 39*(5), 526–534.
Claude, D., & Zaccour, G. (2009). Investment in tourism market and reputation. *Journal of Public Economic Theory, 11*(5), 797–817.
Eisenhardt, K. M. (1989). Agency theory: An assessment and review. *The Academy of Management Review, 14*(1), 57–74.
Fershtman, C., & Judd, K. L. (1987). Equilibrium incentives in oligopoly. *The American Economic Review, 77*(5), 927–940.
Fumas, V. S. (1992). Relative performance evaluation of management. *International Journal of Industrial Organization, 10*(3), 473–489.
Jorgensen, S., & Zaccour, G. (2004). Models of marketing channels. In *Differential games in marketing. International series in quantitative marketing* (Vol. 15). Boston: Springer.
Kopel, M., & Pezzino, M. (2018). Strategic delegation in oligopoly (Chap. 24). In L. Corchon & M. Marini (Eds.), *Handbook of game theory and industrial organization*. Cheltenham: Edward Elgar.
Lambertini, L. (2017). *An economic theory of managerial firms: Strategic delegation in oligopoly. Routledge studies in the economics of business and industry*. Abingdon: Routledge.
Liao, P.-C. (2008). Input price contracts and strategic delegation with downstream duopoly. *Academia Economic Papers, 36*(4), 387–404.
Liao, P.-C. (2010). Discriminatory input pricing and strategic delegation. *Managerial and Decision Economics, 31*(4), 263–276.
Miller, N., & Pazgal, A. (2002). Relative performance as a strategic commitment mechanism. *Managerial and Decision Economics, 23*(2), 51–68.
Ngendakuriyo, F., & Zaccour, G. (2013). Fighting corruption: To precommit or not? *Economics Letters, 120*(2), 149–154.
Pal, R. (2012). Delegation and emission tax in a differentiated oligopoly. *The Manchester School, 80*(6), 650–670.
Park, E.-S. (2002). Vertical externality and strategic delegation. *Managerial and Decision Economics, 23*(3), 137–141.
Sengul, M., Gimeno, J., & Dial, J. (2012). Strategic delegation a review, theoretical integration, and research agenda. *Journal of Management, 38*(1), 375–414.
Shadbegian, R., & Gray, W. (2003). What determines environmental performance at paper mills? The roles of abatement spending, regulation, and efficiency. *Topics in Economic Analysis & Policy, 3*(1), 1144–1144.
Shleifer, A., & Vishny, R. W. (1997). A survey of corporate governance. *The Journal of Finance, 52*(2), 737–783.
Sklivas, S. D. (1987). The strategic choice of managerial incentives. *The RAND Journal of Economics, 18*(3), 452–458.
Vickers, J. (1985). Delegation and the theory of the firm. *The Economic Journal, 95*(Supplement: Conference Papers), 138–147.
Vural, D. (2018). Disclosure practices by family firms: Evidence from Swedish publicly listed firms. *Accounting in Europe, 15*(3), 347–373.
Wang, L. F. S., & Wang, Y.-C. (2010). Input pricing and market share delegation in a vertically related market: Is the timing order relevant? *International Journal of the Economics of Business, 17*(2), 207–221.
Wang, Z. (2015). Delegation and vertical externalities. *Economics Bulletin, 35*(2), 1128–1135.
Zaccour, G. (1983). *Calcul des pseudo-équilibres sur des réseaux représentant des marchés séquentiels*. M.Sc. thesis, H.E.C, Montreal.
Zaccour, G. (1987). *Théorie des jeux et marchés énergétiques: marché européen du gaz naturel et échanges d'éléctricité*. Ph.D. thesis, HEC Montréal.

# Non-linear Incentive Equilibrium Strategies for a Transboundary Pollution Differential Game

**Javier de Frutos and Guiomar Martín-Herrán**

**Abstract** In this paper we apply non-linear incentive strategies to sustain over time an agreement. We illustrate the use of these strategies in a linear-quadratic transboundary pollution differential game. The incentive strategies are constructed in such a way that in the long run the pollution stock (the state variable) is close to the steady state of the pollution stock under the cooperative mode of play. The non-linear incentive functions depend on the emission rates (control variables) of both players and on the current value of the pollution stock. The credibility of the incentive equilibrium strategies is analyzed and the performance of open-loop and feedback incentive strategies is compared in their role of helping to sustain an agreement over time. We present numerical experiments to illustrate the results.

**Keywords** Incentive equilibria · Differential games · Credibility · Environmental Economics

## 1 Introduction

This paper revisits one of the mechanisms already proposed in the literature to ensure the sustainability over time of an agreement reached at the starting date of a game. An agreement will last for its whole intended duration if, at any intermediate instant of time, each player stands to receive a greater payoff being part of the agreement rather than leaving it. A first approach proposed in the literature to sustain

J. de Frutos
IMUVA and Dpto. Matemática Aplicada, Facultad de Ciencias, Universidad de Valladolid, Valladolid, Spain
e-mail: frutos@mac.uva.es

G. Martín-Herrán (✉)
IMUVA and Dpto. Economía Aplicada, Facultad de Ciencias Económicas y Empresariales, Universidad de Valladolid, Valladolid, Spain
e-mail: guiomar@eco.uva.es



P.-O. Pineau et al. (eds.), *Games in Management Science*,
International Series in Operations Research & Management Science 280,
https://doi.org/10.1007/978-3-030-19107-8_11

over time an agreement is the design of an agreement which is time-consistent or agreeable. In a time-consistent agreement the coordinated payoffs-to-go are greater than the non-cooperative ones along the cooperative state trajectory, and hence, no player finds it optimal to switch to his non-cooperative control at any intermediate instant of time. An agreement is agreeable if the comparison condition holds along any state trajectory. Time-consistent agreements are analyzed, among others, in Petrosjan (1997), Petrosjan and Zenkevich (1996), Petrosjan and Zaccour (2003), and Jørgensen and Zaccour (2001a, 2002); and agreeable agreements are studied, for example, in Kaitala and Pohjola (1990, 1995) and Jørgensen et al. (2003, 2005).

The second approach that can be generally called equilibrium approach is to embody the cooperative solution with an equilibrium property such that, by definition each player will find individually rational to stick to his part of the coordinated solution. One option to build a cooperative equilibrium is to use the so-called trigger strategies (Tolwinski et al. 1986; Haurie and Pohjola 1987). These are strategies based on the past actions in the game and they include a threat to punish, credibly and effectively, any player who cheats on the agreement. These strategies are non-Markovian because they are based on all past information of the game evolution to the current time.

This paper examines the use of non-linear incentive strategies as another option to implement cooperative solutions by means of non-cooperative play. Ehtamo and Hämäläinen (1986, 1989, 1993, 1995), Jørgensen and Zaccour (2001b), Martín-Herrán and Zaccour (2005, 2009), and De Frutos and Martín-Herrán (2015), among others, propose incentive strategies to support the cooperative solution in two-player differential games. The incentive is designed in such a way that a coordinated outcome becomes a Nash equilibrium. Incentive strategies are functions of the possible deviation of the other player and recommend to each player to implement his part of the coordinated or agreed solution whenever the other player is doing so. One important characteristic of the incentive strategies is their credibility. The credibility property of incentive strategies requires that each player sticks to the agreed-upon incentive strategy and does not revert to the cooperative solution, even when the other player chooses to break the agreement. If the incentive strategies are credible no player will be tempted to unilaterally deviate from the agreed decision.

Incentive strategies have been extensively used in the differential games literature in different areas, especially environmental economics (Jørgensen and Zaccour 2001b; Martín-Herrán and Zaccour 2005, 2009; Breton et al. 2008; De Frutos and Martín-Herrán 2015) and marketing (Jørgensen and Zaccour 2003; Martín-Herrán and Taboubi 2005; Jørgensen et al. 2006; Buratto and Zaccour 2009; De Giovanni et al. 2016; Taboubi 2019; De Giovanni 2018). Many of these works have not addressed the analysis of the credibility of the incentive strategies. In general, this property cannot be studied analytically, even for games that belong to the class of solvable games such as linear-state and linear-quadratic differential games. All the papers previously cited, except De Frutos and Martín-Herrán (2015), study games belonging to these classes and the incentive strategies are constructed in such a way that the incentive equilibrium is the cooperative solution. Furthermore, the strategies are assumed to be linear and decision-dependent, i.e., each player makes his current decision contingent on the current decision of his opponent. When the credibility property is analyzed the common conclusion is that credibility is assured only for sufficiently small values of the deviation from the agreed solution.

This result led us to consider whether the definition of less restrictive strategies could help to guarantee the sustainability of an agreement over time. In De Frutos and Martín-Herrán (2015) we consider state-dependent and decision-dependent equilibrium strategies defined as non-linear functions of the control variables of both players and the current value of the state variable. More importantly, we look for an incentive strategy equilibrium such that the steady state of the optimal state trajectory is close enough but not necessarily identical to the steady state of the state variable under the cooperative mode of play. We show that the incentive equilibrium is credible in a larger region than the one associated with the usual linear incentive strategies.

Incentive equilibrium strategies can be seen as the implementation of an implicit social norm[1] defined by the following stylized features. First, each player sticks to the agreement if the other player does the same; second, a unilateral deviation by one of the players is followed by a deviation of the other player. Furthermore, in the approach proposed by De Frutos and Martín-Herrán (2015) the response to a deviation is required only if the state variable is far from the cooperative solution and becomes more stringent as the observed deviation from the cooperative solution increases.

As far as we know De Frutos and Martín-Herrán (2015) is the first paper in which a weaker concept of incentive strategies has been used. It is worth noticing that this weaker concept of incentive strategies that implies more flexibility leads to more credibility. The focus of this follow-up paper is to analyze non-linear incentive strategies if the players use open-loop strategies instead of stationary Markovian strategies as previously assumed. We compare the performance of open-loop and feedback incentive strategies when maintaining an agreement over time and we study the credibility of the incentive strategies when one of the players deviates from the incentive equilibrium. The two information structures are compared for the well-known linear-quadratic transboundary pollution differential game proposed in Van der Ploeg and De Zeeuw (1992) and Dockner and Long (1993). We present numerical experiments to illustrate the results.

The rest of the paper is organized as follows. In Sect. 2 we briefly recall the formulation of the linear-quadratic transboundary pollution differential game, its cooperative solution, the open-loop non-cooperative Nash strategies, as well as the steady-state pollution stocks under cooperative and non-cooperative modes of play. In Sect. 3 we define the incentive strategies and equilibrium, and in Sect. 4 we analyze their credibility. Section 5 concludes.

---

[1] We are indebted to an anonymous reviewer for bringing this interpretation to our attention.

# 2 A Linear-Quadratic Transboundary Pollution Differential Game

For simplicity in the exposition and with the objective of comparing our results with those obtained in De Frutos and Martín-Herrán (2015), we focus on a particular linear-quadratic model that has been extensively studied in the environmental economics literature. The formulation is borrowed from Van der Ploeg and De Zeeuw (1992) and Dockner and Long (1993). Let player $i$'s optimization problem be given by[2]:

$$\max_{u_i} \left\{ W_i(u_1, u_2, x_0) := \int_0^\infty \left[ u_i \left( A_i - \frac{1}{2} u_i \right) - \frac{1}{2} \varphi_i x^2 \right] e^{-\rho t} \, dt \right\} \tag{1}$$

$$\text{s.t.} : \dot{x} = \beta(u_1 + u_2) - \alpha x, \quad x(0) = x_0, \tag{2}$$

where $\beta$, $A_i$, and $\varphi_i$, $i \in \{1, 2\}$ are positive parameters and $0 < \alpha < 1$. The control variable $u_i$ is the emissions of player (country) $i$ and the state variable $x$ represents the accumulated stock of pollution and its dynamics is defined by the linear ordinary differential equation (2), where parameter $\alpha$ denotes the natural absorption rate. The state dynamics says that the variation in the pollution stock level is the sum of emissions, scaled by parameter $\beta$, minus what is absorbed by nature. Assuming that emissions are a proportional by-product of industrial activities, the objective of player $i$ is given by the difference between revenues from industrial activities and pollution damage costs. Function $u_i \left( A_i - \frac{1}{2} u_i \right)$ represents the concave revenue function of player $i$. Pollution induces damage costs, given by $\frac{1}{2} \varphi_i x^2$, assumed to depend on accumulated pollution. Parameter $\rho$ is a positive constant discount rate.

If countries (players) use open-loop strategies they choose a time profile of actions at the beginning of the game and commit themselves to retain these preannounced profiles from the rest of the game. If countries choose state-dependent decision rules as their strategies, they choose emission strategies that are functions of the pollution stock. State-dependent Markovian strategies imply that, whenever country $i$ makes a decision that results in a change in the pollution stock, country $j$ immediately reacts. This action and reaction pattern implies more competitive behavior and the outcome of the game is further from the cooperative level.

In this paper we assume that the players restrict themselves to open-loop strategies (Haurie et al. 2012), meaning that the players base their decisions only on time and an initial condition. An open-loop strategy selects the control action according to a decision rule $\mu_i$, which is a function of the initial state $x_0$: $u_i(t) = \mu_i(x_0, t)$. Because the initial state is fixed, there is no need to distinguish between $u_i(t)$ and $\mu_i(x_0, t)$. Using an open-loop strategy means that the player commits, at the initial time, to a fixed time path for his control, i.e., his choice of control at each instant of time is predetermined. More precisely, the set of admissible controls for

[2]To simplify the notation we will drop the explicit dependence on the time variable when no confusion can arise.

Player $i$, $i = 1, 2$, $\mathscr{U}_i$ is defined as the set of nonnegative absolutely continuous functions $u_i = u_i(t)$ defined in $\mathbb{R}_+ = [0, +\infty)$ with values in $\mathbb{R}_+$ such that if $(u_1, u_2) \in \mathscr{U} = \mathscr{U}_1 \times \mathscr{U}_2$, the initial value problem (2) possesses a unique solution defined in $\mathbb{R}_+$. The pair $(u_1^N, u_2^N) \in \mathscr{U}$ is an open-loop Nash equilibrium for the differential game (1)–(2) if

$$W_1(u_1^N, u_2^N) \geq W_1(u_1, u_2^N), \quad W_2(u_1^N, u_2^N) \geq W_2(u_1^N, u_2)$$

for all $u_1$ and $u_2$ such that $(u_1, u_2^N) \in \mathscr{U}$ and $(u_1^N, u_2) \in \mathscr{U}$.

If players agree to cooperate they solve an optimal control problem in which they jointly maximize the aggregate payoff

$$W_1 + W_2 = \int_0^\infty \sum_{i=1}^{2} \left[ u_i \left( A_i - \frac{1}{2} u_i \right) - \frac{1}{2} \varphi_i x^2 \right] e^{-\rho t} \, dt,$$

subject to dynamics (2). Martín-Herrán and Zaccour (2009) proved that the cooperative optimal controls read $u_i^c = A_i + \beta(a^c x + b^c)$ where superscript $c$ stands for cooperation and coefficients $a^c$ and $b^c$ are the quadratic and linear coefficients of the quadratic value function. These coefficients can be found in Martín-Herrán and Zaccour (2009) (p. 272) and allow us to compute the steady-state pollution stock under cooperation:

$$x_{ss}^c = \frac{\beta(\rho + \alpha)(A_1 + A_2)}{(\rho + \alpha)\alpha + 2\beta^2(\varphi_1 + \varphi_2)}. \tag{3}$$

The following proposition characterizes the Nash equilibrium if the players do not cooperate and use open-loop strategies.

**Proposition 1** *Assuming interior solutions, the pair $(u_1^N(t), u_2^N(t))$ is an open-loop Nash equilibrium of the differential game (1)–(2), where*

$$u_i^N(t) = A_i - \frac{\varphi_i x_{ss}^N}{\rho + \alpha} - \beta(x_0 - x_{ss}^N) \frac{\varphi_i}{\rho + \alpha - \xi} e^{\xi t},$$

$$\xi = \frac{\rho - \sqrt{(\rho + 2\alpha)^2 + (\varphi_1 + \varphi_2)\beta^2}}{2},$$

*and superscript $N$ stands for Nash equilibrium.*

*The optimal state trajectory is*

$$x^N(t) = (x_0 - x_{ss}^N) e^{\xi t} + x_{ss}^N,$$

*where $x_{ss}^N$ denotes the steady state given by:*

$$x_{ss}^N = \frac{(\rho + \alpha)\beta(A_1 + A_2)}{(\rho + \alpha)\alpha + \beta^2(\varphi_1 + \varphi_2)}. \tag{4}$$

*Proof* We define the current-value Hamiltonian of player $i$

$$H^i(x, u_1, u_2, \lambda_i) = u_i\left(A_i - \frac{1}{2}u_i\right) - \frac{1}{2}\varphi_i x^2 + \lambda_i(\beta(u_1+u_2) - \alpha x),$$

where $\lambda_i$ is the $i$-th player costate variable associated with the state variable, $x$.

Assuming interior solution, the sufficient conditions for optimality derived from the Pontryagin maximum principle include

$$\frac{\partial H^i}{\partial u_i} = A_i - u_i + \beta\lambda_i = 0,$$
$$\dot{x} = \beta(u_1+u_2) - \alpha x, \quad x(0) = x_0,$$
$$\dot{\lambda}_i = \rho\lambda_i - \frac{\partial H^i}{\partial x} = (\rho+\alpha)\lambda_i + \varphi_i x, \quad \lim_{t\to\infty} e^{-\rho t}\lambda_i(t) = 0.$$

Solving the system of linear ordinary differential equations and taking into account the initial and transversality conditions, the expressions of $u_i^N(t)$ and $x^N(t)$ in the statement can be easily derived. □

As usual in this kind of models the non-cooperative solution leads to emission levels greater than those prescribed by the cooperative solution. The comparison of (3) and (4) clearly shows that the steady state of the pollution stock is lower if players cooperate than if they do not. Then, one can assume that if one player deviates from the cooperative solution, he is choosing an emission level greater than that corresponding to the cooperative solution. The general main objective of this paper is the design of an incentive strategy implying that the players will not depart importantly from their part of the coordinated solution. Specifically, the long-run pollution stock will be near the long-run pollution stock attained under cooperation.

## 3 Incentive Equilibria

For the sake of completeness, let us recall the definition of an incentive equilibrium. The admissible incentive strategies for Player $i$ are functions $\psi_i$ defined in $\mathbb{R}_+ \times \mathbb{R}_+ \times \mathbb{R}_+$ such that for every $(u_1, u_2) \in \mathscr{U}$, $(\Psi_1(\cdot), \Psi_2(\cdot)) \in \mathscr{U}$ where $\Psi_i(t) = \psi_i(u_1(t), u_2(t), x(t))$, $i = 1, 2$, with $x(\cdot)$ the solution of (2). We denote by $\Gamma_i$ the set of admissible strategies for Player $i$.

**Definition 1** A pair $\psi_1(v_1, v_2, x)$, $\psi_2(v_1, v_2, x)$ with $\psi_i \in \Gamma_i$, $i = 1, 2$, is an incentive equilibrium at $(u_1^*, u_2^*) \in \mathscr{U} = \mathscr{U}_1 \times \mathscr{U}_2$ when players use open-loop information structure iff for all $u_1 \in \mathscr{U}_1$ and $u_2 \in \mathscr{U}_2$,

$$W_1(u_1^*, u_2^*) \geq W_1(u_1, \psi_2(u_1, u_2^*, \hat{x})), \quad W_2(u_1^*, u_2^*) \geq W_2(\psi_1(u_1^*, u_2, \check{x}), u_2),$$

where $\hat{x}$ and $\check{x}$ satisfy, for almost all $t \geq 0$, $\dot{\hat{x}}(t) = \beta(u_1(t)+\psi_2(u_1(t), u_2^*(t), \hat{x}(t)))-\alpha\hat{x}(t)$, and $\dot{\check{x}}(t) = \beta(\psi_1(u_1^*(t), u_2(t), \check{x}(t)) + u_2(t)) - \alpha\check{x}(t)$, respectively, with $\hat{x}(0) = \check{x}(0) = x_0$. Furthermore, $u_1^*(t) = \psi_1(u_1^*(t), u_2^*(t), x^*(t))$, $u_2^*(t) = \psi_2(u_1^*(t), u_2^*(t), x^*(t))$, where $\dot{x}^*(t) = \beta(u_1^*(t) + u_2^*(t)) - \alpha x^*(t)$, with $x^*(0) = x_0$.

From now on and for simplicity in the notation we use the term open-loop (feedback) incentive equilibrium strategies to denote the incentive equilibrium strategies when the players use open-loop (feedback) information structure.

An incentive equilibrium is thus characterized by the following pair of optimal control problems

$$\max_{u_i \in \mathscr{U}_i} W_i(u_i, u_j^*) = \int_0^\infty \left(u_i(t)\left(A_i - \frac{1}{2}u_i(t)\right) - \frac{1}{2}\varphi_i x^2(t)\right) e^{-\rho t}\, dt, \tag{5}$$

$$\text{s.t. } :\dot{x}(t) = \beta(u_i(t) + \psi_j(u_i(t), u_j^*(t), x(t))) - \alpha x(t), \quad x(0) = x_0, \tag{6}$$

with $u_i^* = \arg\max_{u_i} W_i(u_i, u_j^*)$, $i, j = 1, 2$, $i \neq j$. The equilibrium condition $u_i^* = \psi_i(u_i^*, u_j^*, x^*)$, $i, j = 1, 2$, $i \neq j$, has to be satisfied.

We would like to remark that the players use open-loop strategies, meaning that each player looks for a function $t \mapsto u_i^*(t)$ satisfying $u_i^* = \arg\max_{u_i} W_i(u_i, u_j^*)$ and the equilibrium condition $u_i^*(t) = \psi_i(u_i^*(t), u_j^*(t), x^*(t))$, $i, j = 1, 2$, $i \neq j$. In consequence, even if the incentive strategies explicitly depend on the state variable, at $t = 0$ the players commit to the entire time-path of the controls. Conversely, in De Frutos and Martín-Herrán (2015) the players use feedback strategies, that is, they look for functions $x \mapsto u_i^*(x)$ with $x$ denoting the state variable. In this paper results under both information structures are compared.[3]

The linear incentive strategies previously proposed in the literature are a particular case of Definition 1. This literature (except De Frutos and Martín-Herrán 2015) assumes that the incentive equilibrium is the cooperative solution, $(u_1^c, u_2^c)$, and the following affine function has been usually proposed as incentive strategy (see, for example, Ehtamo and Hämäläinen 1986; Martín-Herrán and Zaccour 2005, 2009):

$$\psi_j(u_i, u_j, x) = \psi_j(u_i) = u_j^c + D_j(u_i - u_i^c), \quad i, j = 1, 2, \ i \neq j, \tag{7}$$

with $D_j$, $j = 1, 2$, denoting an appropriate non-zero constant.

However, in this paper, as in De Frutos and Martín-Herrán (2015), we look for an incentive strategy equilibrium $(u_1^*, u_2^*)$ such that the steady state of the pollution stock of the system when the incentive strategies are used, $x_{ss}^*$, is greater but close to this value under cooperation, $x_{ss}^c$, and lower than the steady-state value under non-cooperation, $x_{ss}^N$.

[3]The pollution stock and emission time-paths for the Nash and incentive equilibria as well as the payoffs when the players use feedback strategies presented later in this paper have been taken from De Frutos and Martín-Herrán (2015).

For the sake of completeness, let us recall the form of the incentive functions $\psi_j$, $j = 1, 2$, in Definition 1 we choose to attain the purpose:

$$\psi_j(u_i, u_j, x) = (u_j^c + D_j(u_i - u_i^c))\phi(x - x_{ss}^c, \varepsilon) + u_j(1 - \phi(x - x_{ss}^c, \varepsilon)), \quad (8)$$

where $\varepsilon > 0$ is a small positive parameter and $\phi(x, \varepsilon)$ is a smooth function satisfying

$$\phi(x, \varepsilon) = 0, \text{ if } x \leq \varepsilon; \quad \phi(x, \varepsilon) = 1, \text{ if } x \geq 2\varepsilon. \quad (9)$$

The definition of the non-linear incentive in (8) and the cutoff function in (9) show that the incentive strategy is exclusively implemented if one player deviates from the cooperative outcome (and emits at a greater level) and, therefore, at some time $t$ the trajectory $x(t)$ is above $x_{ss}^c$. If $x(t)$ is far from $x_{ss}^c$, then the linear incentive in (7) applies and pushes the players' emissions in such a way that the pollution path returns close to the steady-state value under cooperation, $x_{ss}^c$. Conversely, the non-linear incentive strategy (8) when $x(t)$ is close enough to $x_{ss}^c$ (the distance measured by parameter $\varepsilon$) allows the players to choose any time path.

We observe that in the limit $\varepsilon \to 0$ the non-linear incentive strategy (8) reduces to the linear incentive strategy defined in (7). In this case, $x_{ss}^*$ is equal to $x_{ss}^c$. Furthermore, $u_i^* = u_i^c$, $i = 1, 2$.

The following proposition characterizes the incentive equilibrium if the non-linear incentive in (8) is used and the players restrict themselves to open-loop strategies.

**Proposition 2** *If the incentive strategy is defined by (8), then the open-loop interior incentive equilibrium $(u_1^*(t), u_2^*(t))$ satisfies the equilibrium conditions $u_i^* = \psi_i(u_i^*, u_j^*, x^*)$, $i, j = 1, 2, i \neq j$ together with the following set of optimality conditions:*

$$\begin{aligned} u_i &= A_i + \beta(1 + D_j\phi(x - x_{ss}^c, \varepsilon))\xi_i, \\ \dot{x} &= \beta(u_i + \psi_j(u_i, u_j^*, x)) - \alpha x, \quad x(0) = x_0, \\ \dot{\xi}_i &= \left(\rho + \alpha - \beta\frac{\partial \psi_j}{\partial x}(u_i, u_j^*, x)\right)\xi_i + \varphi_i x, \ \lim_{t\to\infty} e^{-\rho t}\xi_i(t) = 0, \ i = 1, 2, i \neq j, \end{aligned} \quad (10)$$

*where $\xi_i$ denotes the costate variable of player $i$.*

*Proof* Let us define the Hamiltonian for player $i$ as

$$H_i(u_i, u_j, \xi_i, x) = u_i\left(A_i - \frac{1}{2}u_i\right) - \frac{1}{2}\varphi_i x^2 + \xi_i(\beta(u_i + \psi_j(u_i, u_j, x)) - \alpha x).$$

Assuming interior strategies, the maximum principle optimality conditions read:

$$\frac{\partial H_i}{\partial u_i}(u_i, u_j, \xi_i, x) = 0,$$

$$\dot{\xi}_i = \rho\xi_i - \frac{\partial H_i}{\partial x}, \quad \lim_{t\to\infty} e^{-\rho t}\xi_i(t) = 0, \quad i = 1, 2, i \neq j,$$

$$\dot{x} = \beta(u_i + \psi_j(u_i, u_j, x)) - \alpha x, \quad x(0) = x_0.$$

From these conditions, those in the statement of the proposition immediately follow. □

The non-linear incentive equilibrium cannot be analytically characterized and numerical methods are required for the analysis of these incentive strategies. To numerically solve the system of optimality conditions (10) we first introduce a large $T > 0$ and substitute the transversality condition $\lim_{t\to\infty} e^{-\rho t}\xi_i(t) = 0$ by the approximate boundary condition $\xi_i(T) = 0$, $i = 1, 2$. The resulting boundary value problem is solved by means of a collocation method implemented in the MATLAB subroutine `bvp4c.m`, see Kierzenka and Shampine (2001). The procedure is repeated with a larger $T$ until no differences between approximate solutions are found.

For illustration purposes we consider a symmetric example and fix the following values of the parameters: $A_1 = A_2 = 0.5$, $\varphi_1 = \varphi_2 = 1$, $\alpha = 0.2$, $\beta = 1$, $\rho = 0.1$. The threshold $\varepsilon$ in the cutoff function in (9) is set to $\varepsilon = 0.025$ and the initial pollution stock was set to $x_0 = 0$. The parameter $D_j$ in (8) was set to $D_j = 1$, $j = 1, 2$ as in Martín-Herrán and Zaccour (2009).

Using the above parameters values, we have represented in Fig. 1 the phase diagram of the optimality system (10) in the symmetric case. The variables are the pollution stock in the abscissas axis and emissions in the ordinates axis. The system

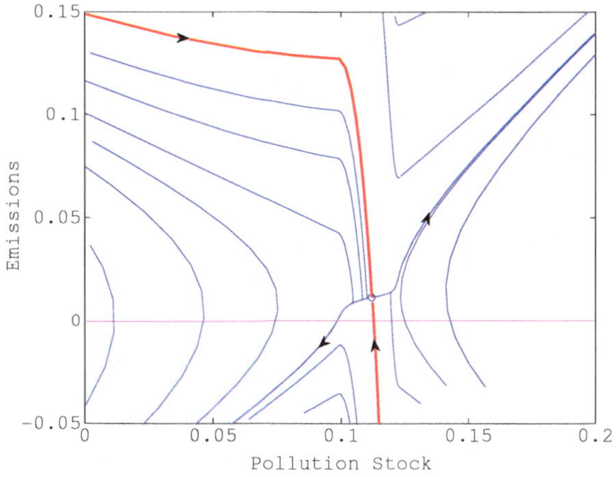


**Fig. 1** Pollution stock-emissions phase diagram in the symmetric case. In red the stable variety of the unique equilibrium of (10). Parameters values $A_1 = A_2 = 0.5$, $\varphi_1 = \varphi_2 = 1$, $\alpha = 0.2$, $\beta = 1$, $\rho = 0.1$, $\varepsilon = 0.025$, $D_j = 1$, $j = 1, 2$

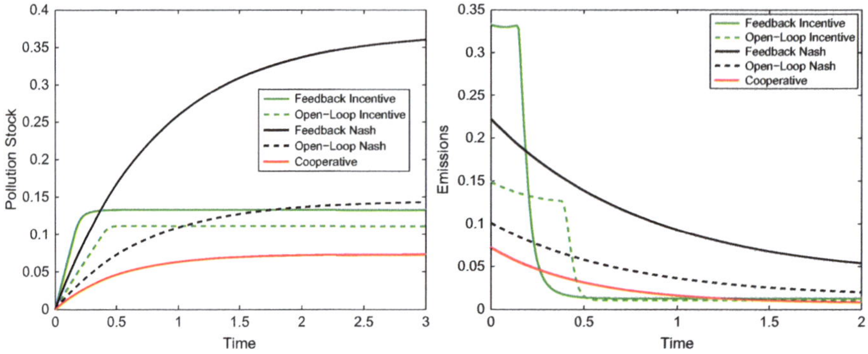


**Fig. 2** Nash, incentive, and cooperative pollution stock and emission time-paths

possesses a unique steady state (represented by a circle in the figure) which is a saddle point. In the figure the stable variety is highlighted in red color. This curve represents, for a given initial condition, the unique non-linear symmetric open-loop incentive equilibrium. Note that, for simplicity, in the figure the positivity conditions on emissions have not been imposed.

Figure 2 shows the optimal pollution and emission time-paths for five different modes of play. The solid red lines represent the cooperative state (left) and control (right) optimal time-paths. The optimal time-paths corresponding to the non-cooperative Nash equilibrium are represented using solid (black) and dashed (black) lines for the feedback and open-loop information structures, respectively. Finally, solid (green) and dashed (green) lines are the optimal time-paths for the feedback and open-loop incentive equilibrium strategies, respectively. A first message from Fig. 2 (left) is that the incentive strategies attain their objective of approaching the long-run cooperative level of the pollution stock. Furthermore, for the same threshold $\varepsilon$ in the cutoff function the open-loop incentive equilibrium is closer to the cooperative optimal time-paths than the feedback incentive equilibrium. The transition to the steady state in the open-loop case is smoother than in the feedback case. This effect is clearer in the case of emissions (right chart). The main difference is in the short run. The emissions in the feedback incentive equilibrium are initially very high, while they are much lower in the open-loop case. However, after a very short period of time, the first ones decrease sharply, while the second ones begin their fall later. The times at which the emissions start to decline towards their stationary levels depend on the value of parameter $\varepsilon$ in the cutoff function. The smaller $\varepsilon$, the earlier the emissions should begin their fall in order to attain a steady state closer to the cooperative steady state. In the long run the emissions in both open-loop and feedback incentive equilibria are very similar to the cooperative emissions.

## 4 Credibility

In this section we focus on the study of the credibility of the incentive strategies. The incentive strategies are credible if player $j$ deviates unilaterally from his incentive equilibrium action, $u_j = u_j^*(\cdot)$, then, it will be more beneficial for player $i$ to follow the incentive strategy, rather than to stick to $u_i = u_i^*(\cdot)$. Hence, if the credibility property is satisfied, there will not be any temptation for unilateral deviation from the pair $u_j = u_j^*(\cdot)$, $j = 1, 2$. It is understood that the function $u_i^*(\cdot)$ is either a function of the time variable $t$ if players use open-loop strategies or a function of the state variable $x$ if players use feedback strategies.

**Definition 2** A pair $\psi_1(v_1, v_2, x)$, $\psi_2(v_1, v_2, x)$, with $\psi_i \in \Gamma_i$, $i = 1, 2$ of incentive equilibrium strategies at $(u_1^*, u_2^*)$ is credible in a set $U_1 \times U_2 \subset \mathscr{U}_1 \times \mathscr{U}_2$ iff given $u_1 \in U_1$ and $u_2 \in U_2$ there exist $\hat{u}_1 \in \mathscr{U}_1$ and $\check{u}_2 \in \mathscr{U}_2$ such that

$$W_1(\psi_1(\hat{u}_1, u_2, \hat{x}), u_2) \geq W_1(u_1^*, u_2), \quad W_2(u_1, \psi_2(u_1, \check{u}_2, \check{x})) \geq W_2(u_1, u_2^*), \tag{11}$$

where $\hat{x}$ and $\check{x}$ satisfy $\dot{\hat{x}} = \beta(\psi_1(\hat{u}_1, u_2, \hat{x}) + u_2) - \alpha\hat{x}$ and $\dot{\check{x}} = \beta(u_1 + \psi_2(u_1, \check{u}_2, \check{x})) - \alpha\check{x}$, respectively, with $\hat{x}(0) = \check{x}(0) = x_0$.

A sufficient, although obviously not necessary, condition for credibility is that for all $u_1 \in U_1$ and $u_2 \in U_2$

$$W_1(\psi_1(u_1^*, u_2, \hat{x}), u_2) \geq W_1(u_1^*, u_2); \quad W_2(u_1, \psi_2(u_1, u_2^*, \check{x})) \geq W_2(u_1, u_2^*), \tag{12}$$

where $\hat{x}$ and $\check{x}$ are defined as in Definition 2, with $\hat{u}_1 = u_1^*$ and $\check{u}_2 = u_2^*$. The stricter condition (12) is not always the best possible option as a response to a deviation even if the players restrict themselves to the use of incentive strategies of the form $\psi_i$. The use of non-linear strategies allows the players to optimize their response to deviations from the agreement while maintaining their commitment to the implementation of incentive strategies. The examples below present some of the practical consequences of the use of the weaker Definition 2.

Note that, in the case of linear incentive strategies, $\psi_j$ is given by (7). Then, Definition 2 reduces to $W_i(\psi_i(u_j), u_j) \geq W_i(u_i^*, u_j)$, $\forall u_j \in U_j$ with $u_i^* = u_i^c$, for $i = 1, 2$, which is the credibility definition usually proposed in the literature (see, for example, Jørgensen and Zaccour 2003; Martín-Herrán and Zaccour 2005, 2009). A set of general conditions ensuring the credibility of linear incentive strategies for the model described by (1)–(2) has been characterized in Martín-Herrán and Zaccour (2009). A comparison between the sets of deviations for which credibility can be obtained under linear and non-linear incentive equilibrium strategies can be found in De Frutos and Martín-Herrán (2015). In this last paper it is shown that the region of credibility attained with the non-linear incentive contains and is considerably greater than that reported for the linear incentive in Martín-Herrán and Zaccour (2009). This result allows us to conclude that the introduction of flexibility can be a useful device to facilitate the sustainability of an agreement over time.

**Table 1** Players' payoffs under open-loop and feedback strategies

| | $W_1^{ol}$ | $W_2^{ol}$ | $W_1^f$ | $W_2^f$ |
|---|---|---|---|---|
| $U_1 = u_1^N, U_2 = u_2^N$ | $3.14 \times 10^{-3}$ | $3.14 \times 10^{-3}$ | $-3.68 \times 10^{-1}$ | $-3.68 \times 10^{-1}$ |
| $U_1 = u_1^*, U_2 = u_2^*$ | $1.68 \times 10^{-2}$ | $1.68 \times 10^{-2}$ | $3.91 \times 10^{-4}$ | $3.91 \times 10^{-4}$ |
| $U_1 = u_1^*, U_2 = u_2^N$ | $8.30 \times 10^{-4}$ | $2.16 \times 10^{-2}$ | $-5.81 \times 10^{-1}$ | $3.74 \times 10^{-1}$ |
| $U_1 = \psi_1(u_1^*, u_2^N, x), U_2 = u_2^N$ | $2.16 \times 10^{-3}$ | $2.39 \times 10^{-3}$ | $-3.72 \times 10^{-1}$ | $-3.79 \times 10^{-1}$ |

Player 2 deviates to $u_2^N$

Definition 2 requires conditions (11) to be checked in some subset of admissible controls $U_1 \times U_2 \subset \mathscr{U}_1 \times \mathscr{U}_2$. In order to be able to analyze the credibility properties of the non-linear incentive strategies we assume that the set of possible deviations is restricted to convex combinations of the cooperative control $u_i^c$ and the non-cooperative Nash equilibrium $u_i^N$: $U_i = \{u_i = \theta u_i^c + (1-\theta)u_i^N\}$. The following tables allow us to illustrate the credibility of the non-linear incentive strategies and to compare this property for the open-loop and feedback information structures.

Tables 1, 2, 3 and 4 show the players' payoffs when they play either open-loop (first two columns) or feedback (last two columns) strategies (superscripts *ol* and $f$, respectively). Each row presents the different strategies used by the players in each case.

The first row in Table 1 shows the players' payoffs when they do not cooperate and play the open-loop or the feedback Nash equilibrium. The second row presents these payoffs when they follow their open-loop or feedback incentive strategies. Both players under both information structures improve their payoffs with respect to the non-cooperative Nash levels. In the third row the payoffs are no longer identical for both players, when player 1 continues to play his part of the incentive equilibrium, while player 2 deviates to his part of the non-cooperative Nash equilibrium. Let us note that it could be considered irrational that player 2 implements a strategy that would provide him a smaller payoff than that associated with the incentive strategy. Table 1 shows that the deviation from the incentive equilibrium to $u_2^N$ provides a greater payoff to the deviating player, i.e. $W_2(u_1^*, u_2^N) > W_2(u_1^*, u_2^*)$. The credibility of the incentive equilibrium requires the existence of a feasible $u_1$ such that $W_1(\psi_1(u_1, u_2^N, x), u_2^N) > W_1(u_1^*, u_2^N)$. The fourth row in Table 1 shows that last inequality is satisfied in particular for $u_1 = u_1^*$. Furthermore, the deviating player (player 2) is penalized because $W_2(\psi_1(u_1^*, u_2^N, x), u_2^N) < W_2(u_1^*, u_2^*)$. In fact, deviating from the incentive strategy player 2 even worsens his payoff compared to his payoff in the non-cooperative case: $W_2(\psi_1(u_1^*, u_2^N, x), u_2^N) < W_2(u_1^N, u_2^N)$. All these results apply when the players use either open-loop or feedback strategies.

Table 1 also allows us to compare the relative improvement of the players' payoffs when moving from their non-cooperative mode of play to the incentive equilibrium as well as the relative loss with respect to the cooperative solution both under open-loop and feedback strategies. Because we are analyzing a completely symmetric game, each player under cooperation receives half of the total cooperative payoff. The individual cooperative payoff is given by $W_1^c = W_2^c = 2.64 \times 10^{-2}$. Let us denote by $W_i^{Nol}$ and $W_i^{Nf}$ the players' payoffs when the open-loop and feedback Nash equilibrium, respectively, is played (first row in Table 1). Equivalently, let us denote by $W_i^{*ol}$ and $W_i^{*f}$ the players' payoffs when the open-loop and feedback incentive equilibrium, respectively, is implemented (second row in Table 1).

The comparison of the Nash equilibrium payoffs $W_i^{Nol}$ and $W_i^{Nf}$ with the individual cooperative payoff $W_i^c$ shows that if the players implement the feedback Nash equilibrium the payoffs are much smaller than if the open-loop Nash equilibrium is played, which is, already, an 88.1% lower than the cooperative ones. This loss of welfare is a well-known consequence of the non-cooperative mode of play that can be mitigated if the players agree to follow incentive strategies (either open-loop or feedback), as can be seen in the second row of Table 1. The payoffs when the incentive equilibrium is played compared to the payoffs under cooperation account for a fall of a 36.7% and a 98.5%, under open-loop and feedback strategies, respectively. If the incentive equilibrium payoffs are compared to the non-cooperative Nash payoffs, they show an increase of more than one order of magnitude in both cases, being four times greater in the open-loop than in the feedback case. These results clearly show that the players can find neatly advantageous the use of non-linear incentive strategies.

Tables 2, 3 and 4 present the different payoffs when player 2 deviates from the incentive equilibrium to different convex combinations of his cooperative control, $u_2^c$, and his part of the non-cooperative Nash equilibrium, $u_2^N$. In Tables 2, 3, and 4 the weight assigned to the non-cooperative part decreases from 0.75 to 0.5 and to 0.25, respectively. Concerning the rationality property of player 2, these tables show that it is rational that player 2 deviates from the incentive equilibrium when he changes to a strategy in which no-cooperation is weighted at least as cooperation. However, Table 4 shows that it is irrational that player 2 deviates to $u_2^{(3)} = 0.75u_2^c + 0.25u_2^N$ or to $u_2^{(3)} = u_2^c$ not shown in the table. These results are applicable both for the open-loop and feedback incentive strategies.

The last two rows in Tables 2, 3 and 4 allow to analyze the credibility of the incentive strategies when player 2 deviates to $u_2^{(i)}$, $i = 1, 2, 3$. Player 1 implements

**Table 2** Players' payoffs under open-loop and feedback strategies

| | $W_1^{ol}$ | $W_2^{ol}$ | $W_1^f$ | $W_2^f$ |
|---|---|---|---|---|
| $U_1 = u_1^*, U_2 = u_2^*$ | $1.68 \times 10^{-2}$ | $1.68 \times 10^{-2}$ | $3.91 \times 10^{-4}$ | $3.91 \times 10^{-4}$ |
| $U_1 = u_1^*, U_2 = u_2^{(1)}$ | $1.08 \times 10^{-2}$ | $2.01 \times 10^{-2}$ | $-3.58 \times 10^{-1}$ | $2.70 \times 10^{-1}$ |
| $U_1 = \psi_1(u_1^*, u_2^{(1)}, x), U_2 = u_2^{(1)}$ | $1.23 \times 10^{-2}$ | $1.22 \times 10^{-2}$ | $-1.14 \times 10^{-1}$ | $-1.54 \times 10^{-1}$ |
| $U_1 = \psi_1(u_1^N, u_2^{(1)}, x), U_2 = u_2^{(1)}$ | $1.36 \times 10^{-2}$ | $1.01 \times 10^{-2}$ | $-9.22 \times 10^{-2}$ | $-3.92 \times 10^{-1}$ |

Player 2 deviates to $u_2^{(1)} = 0.25u_2^c + 0.75u_2^N$

**Table 3** Players' payoffs under open-loop and feedback strategies

| | $W_1^{ol}$ | $W_2^{ol}$ | $W_1^f$ | $W_2^f$ |
|---|---|---|---|---|
| $U_1 = u_1^*, U_2 = u_2^*$ | $1.68 \times 10^{-2}$ | $1.68 \times 10^{-2}$ | $3.91 \times 10^{-4}$ | $3.91 \times 10^{-4}$ |
| $U_1 = u_1^*, U_2 = u_2^{(2)}$ | $2.01 \times 10^{-2}$ | $1.76 \times 10^{-2}$ | $-1.54 \times 10^{-1}$ | $1.34 \times 10^{-1}$ |
| $U_1 = \psi_1(u_1^*, u_2^{(2)}, x), U_2 = u_2^{(2)}$ | $2.02 \times 10^{-2}$ | $1.75 \times 10^{-2}$ | $-3.02 \times 10^{-2}$ | $-4.75 \times 10^{-2}$ |
| $U_1 = \psi_1(u_1^N, u_2^{(2)}, x), U_2 = u_2^{(2)}$ | $2.26 \times 10^{-2}$ | $5.01 \times 10^{-3}$ | $1.23 \times 10^{-1}$ | $-4.46 \times 10^{-1}$ |

Player 2 deviates to $u_2^{(2)} = 0.5u_2^c + 0.5u_2^N$

**Table 4** Players' payoffs under open-loop and feedback strategies

| | $W_1^{ol}$ | $W_2^{ol}$ | $W_1^f$ | $W_2^f$ |
|---|---|---|---|---|
| $U_1 = u_1^*, U_2 = u_2^*$ | $1.68 \times 10^{-2}$ | $1.68 \times 10^{-2}$ | $3.91 \times 10^{-4}$ | $3.91 \times 10^{-4}$ |
| $U_1 = u_1^*, U_2 = u_2^{(3)}$ | $2.88 \times 10^{-2}$ | $1.44 \times 10^{-2}$ | $5.29 \times 10^{-2}$ | $-4.87 \times 10^{-2}$ |
| $U_1 = \psi_1(u_1^*, u_2^{(3)}, x), U_2 = u_2^{(3)}$ | $2.87 \times 10^{-2}$ | $1.45 \times 10^{-2}$ | $2.08 \times 10^{-2}$ | $-6.00 \times 10^{-3}$ |
| $U_1 = \psi_1(u_1^N, u_2^{(3)}, x), U_2 = u_2^{(3)}$ | $3.12 \times 10^{-2}$ | $-1.41 \times 10^{-4}$ | $2.91 \times 10^{-1}$ | $-5.19 \times 10^{-1}$ |

Player 2 deviates to $u_2^{(3)} = 0.75u_2^c + 0.25u_2^N$

$U_1 = \psi_1(u_1, u_2^{(i)}, x)$ with either $u_1 = u_1^*$ (third row) or $u_1 = u_1^N$ (fourth row). Comparing the entries in the third (fourth) row with those in the second, it can be deduced that the incentive equilibrium is credible in the same scenarios when it is rational for player 2 to deviate (Tables 2 and 3) and is not credible when the deviation is irrational for player 2 (Table 4). The results are qualitatively similar for open-loop and feedback strategies.

Finally, from the comparison of $W_2(\psi_1(u_1, u_2^{(i)}, x), u_2^{(i)})$ and $W_2(u_1^*, u_2^*)$ for $u_1 \in \{u_1^*, u_1^N\}$ in Tables 2 and 3 we can deduce that deviating from the incentive strategy player 2 always worsens his payoff compared to his payoff in the incentive equilibrium. More precisely, if $u_2 = u_2^{(1)} = 0.25u_2^c + 0.75u_2^N$ both choices for player 1, $u_1 = u_1^*$ or $u_1 = u_1^N$ lead to a payoff for player 2 lower than $W_2(u_1^*, u_2^*)$ regardless of whether players use open-loop or feedback strategies. If $u_2 = u_2^{(2)} = 0.5u_2^c + 0.5u_2^N$, the same comment applies for the feedback case. In the open-loop case the choice $u_1 = u_1^*$ for player 1, although being credible, it does not penalize enough player 2 with respect to his payoff in the incentive equilibrium. However, if player 1 optimizes his choice by moving to $u_1 = u_1^N$, then $W_1(\psi_1(u_1^N, u_2^{(2)}, x), u_2^{(2)}) > W_1(\psi_1(u_1^*, u_2^{(2)}, x), u_2^{(2)}) > W_1(u_1^*, u_2^*)$ and $W_2(\psi_1(u_1^N, u_2^{(2)}, x), u_2^{(2)}) < W_2(u_1^*, u_2^*)$.

Figure 3 for the open-loop strategies and Fig. 4 for the feedback strategies show the time paths of the emission rates and the pollution stock for the non-cooperative, cooperative, and incentive strategies used by the players, as well as the strategies when player 2 deviates to his part of the non-cooperative equilibrium, while player

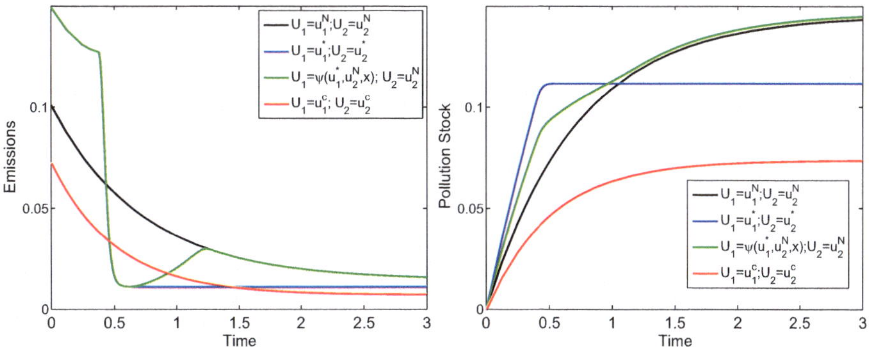


**Fig. 3** Credibility open-loop incentive strategies

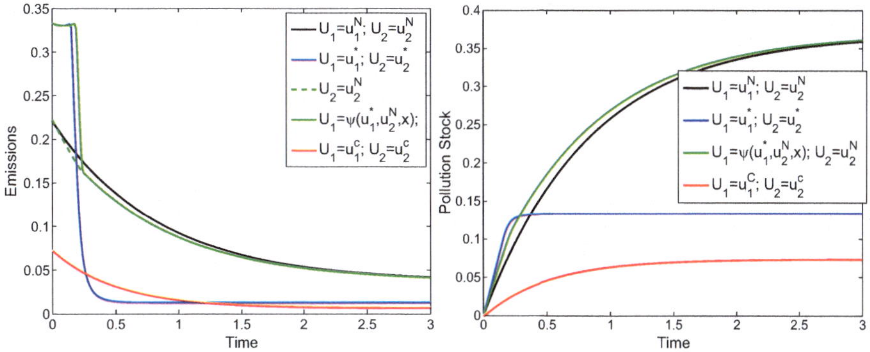


**Fig. 4** Credibility feedback incentive strategies

1 sticks to his part of the incentive equilibrium. These time paths allow us to analyze the credibility of the incentive strategies. The optimal paths when the non-cooperative Nash equilibrium or the cooperative solution is played are represented using solid black and red lines, respectively. Those associated with the incentive equilibrium are depicted using solid blue line. Finally, the solid green line shows the time paths when player 2 deviates while player 1 plays the incentive strategy.

Figure 3 shows that when player 2 deviates from the incentive equilibrium and follows his part of the non-cooperative Nash equilibrium the emission time-paths are described by the black solid line starting around 0.1. When player 1 responds using the incentive strategy $U_1 = \psi(u_1^*, u_2^N, x)$ his emission time-path initially starts around 0.15 and follows the incentive equilibrium strategy, decreasing up to a minimum level. When player 1 realizes that the pollution stock is far from the pollution stock under cooperation departs from his part of the incentive equilibrium and increases the emissions up to a level similar to the non-cooperative level. From this point on the emission time-paths run very closely. This behavior translates into a pollution stock time-path associated with the strategies $U_1 = \psi(u_1^*, u_2^N, x)$ and

$U_2 = u_2^N$ that initially evolves between the time-paths for the non-cooperative Nash equilibrium and the incentive equilibrium. The time path initially increases at a speed greater than that corresponding to the Nash equilibrium. From a time on the incentive strategy used by player 1, $U_1 = \psi(u_1^*, u_2^N, x)$, allows to approximate the pollution stock time-path under non-cooperation.

Figure 4 collects the different feedback strategies. The deviating player (player 2) initially follows the discontinuous green line starting around 0.22, corresponding to the emission time-path for $U_1 = \psi(u_1^*, u_2^N, x)$ and $U_2 = u_2^N$. This time path evolves quite close to the non-cooperative emission time-path. Let us note that although in Fig. 4 we display the time paths associated with the different strategies, the players are using feedback strategies, and as such they are taking their optimal decisions depending on the value of the state variable (the pollution stock). Hence, the discontinuous green line does not coincide with the continuous black line because player 1 is playing $U_1 = \psi(u_1^*, u_2^N, x)$ instead of $u_1^N$, implying a different value of the pollution stock. Player 1's emissions (continuous green line) start around 0.33 and sharply decrease imitating the incentive equilibrium time-path up to a point in time where the trajectory reaches the discontinuous green line. From this time on the emission time-paths for both players coincide and evolve close to the non-cooperative emission time-path. The main difference with respect to the previous case is that in the case of open-loop strategies player 1 reduces his emissions too sharply and then he has to raise them during an intermediate period of time to follow a trajectory similar to the non-cooperative case. However, in the case of feedback, the decrease in emissions towards values close to those of the non-cooperative scenario is monotonous and smoother. As the right chart in Fig. 4 shows the pollution stock when $U_1 = \psi(u_1^*, u_2^N, x)$ and $U_2 = u_2^N$ is not far away from the non-cooperative pollution stock even in the short term.

## 5 Concluding Remarks

This paper examines the use of non-linear incentive strategies as another option to sustain over time an agreement by means of non-cooperative play. Incentive strategies have been extensively used in the differential games literature in different areas and have been proposed to support the cooperative solution in two-player differential games. The incentive is designed in such a way that a coordinated outcome becomes a Nash equilibrium. If the incentive strategies are credible, no player will be tempted to unilaterally deviate from the agreed decision. As far as we know all the previous literature on incentive strategies, except De Frutos and Martín-Herrán (2015), study games belonging to the linear-state or linear-quadratic classes and the incentive strategies are constructed in such a way that the incentive equilibrium is the cooperative solution. Furthermore, the strategies are assumed to be linear and decision-dependent, i.e., each player makes his current decision contingent on the current decision of his opponent. Most of the previous works have not addressed the analysis of the credibility of the incentive strategies.

This paper is a follow-up of De Frutos and Martín-Herrán (2015) and assumes state-dependent and decision-dependent equilibrium strategies defined as non-linear functions of the control variables of both players and the current value of the state variable. We relax the definition of incentive equilibrium in the sense that we look for an incentive strategy equilibrium such that the steady state of the optimal state trajectory is close enough, but not necessarily identical, to the steady state of the state variable under cooperation. We show that the definition of less restrictive strategies helps to guarantee the sustainability of an agreement over time. We analyze a well-known linear-quadratic transboundary pollution differential game and present numerical experiments to illustrate the results. We compare the incentive equilibrium strategies, its credibility and the players' payoff when players use open-loop strategies and when they focus on stationary Markovian strategies.

The consideration of the incentive strategies as a social norm raises the question of whether the use of evolutionary game concepts could give a new insight about the use of incentive equilibrium strategies. One interesting idea could be to see whether a more flexible social norm (non-linear incentive strategies) could survive when confronted to a more efficient social norm (linear incentive strategies). We postpone this study for further research.[4]

**Acknowledgements** We are grateful to two anonymous reviewers for valuable comments and suggestions in an earlier draft of this paper. This research is partially supported by MINECO under projects MTM2016-78995-P (AEI) (Javier de Frutos) and ECO2014-52343-P and ECO2017-82227-P (AEI) (Guiomar Martín-Herrán) and by Junta de Castilla y León VA024P17 and VA105G18 co-financed by FEDER funds (EU).

## References

Breton, M., Sokri, A., & Zaccour, G. (2008). Incentive equilibrium in an overlapping-generations environmental game. *European Journal of Operational Research, 185*, 687–699.

Buratto, A., & Zaccour, G. (2009). Coordination of advertising strategies in a fashion licensing contract. *Journal of Optimization Theory and Applications, 142*, 31–53.

De Frutos, J., & Martín-Herrán, G. (2015). Does flexibility facilitate sustainability of cooperation over time? A case study from environmental economics. *Journal of Optimization Theory and Applications, 165*, 657–677.

De Giovanni, P. (2018). A joint maximization incentive in closed-loop supply chains with competing retailers: The case of spent-battery recycling. *European Journal of Operational Research, 268*, 128–147.

De Giovanni, P., Reddy, P. V., & Zaccour, G. (2016). Incentive strategies for an optimal recovery program in a closed-loop supply chain. *European Journal of Operational Research, 249*, 605–617.

Dockner, E. J., & Long, N. V. (1993). International pollution control: Cooperative versus noncooperative strategies. *Journal of Environmental Economics and Management, 25*, 13–29.

Ehtamo, H., & Hämäläinen, R. P. (1986). On affine incentives for dynamic decision problems. In T. Basar (Ed.), *Dynamic games and applications in economics* (pp. 47–63). Berlin: Springer.

[4]We thank an anonymous reviewer for pointing out this question.

Ehtamo, H., & Hämäläinen, R. P. (1989). Incentive strategies and equilibria for dynamic games with delayed information. *Journal of Optimization Theory and Applications, 63*, 355–369.

Ehtamo, H., & Hämäläinen, R. P. (1993). Cooperative incentive equilibrium for a resource management problem. *Journal of Economic Dynamics and Control, 17*, 659–678.

Ehtamo, H., & Hämäläinen, R. P. (1995). Credibility of linear equilibrium strategies in a discrete time fishery management game. *Group Decision and Negotiation, 4*, 27–37.

Haurie, A., Krawczyk, J. B., & Zaccour, G. (2012). *Games and dynamic games*. Singapore: World Scientific.

Haurie, A., & Pohjola, M. (1987). Efficient equilibria in a differential game of capitalism. *Journal of Economic Dynamics and Control, 11*, 65–78.

Jørgensen, S., Martín-Herrán, G., & Zaccour, G. (2003). Agreeability and time-consistency in linear-state differential games. *Journal of Optimization Theory and Applications, 119*, 49–63.

Jørgensen, S., Martín-Herrán, G., & Zaccour, G. (2005). Sustainability of cooperation overtime in linear-quadratic differential games. *International Game Theory Review, 7*, 395–406.

Jørgensen, S., Taboubi, S., & Zaccour, G. (2006). Incentives for retailer promotion in a marketing channel. In A. Haurie, S. Muto, L. A. Petrosjan, & T. E. S. Raghavan (Eds.), *Annals of the international society of dynamic games* (Vol. 8, pp. 365–378). Boston: Birkhäuser.

Jørgensen, S., & Zaccour, G. (2001a). Time consistent side payments in a dynamic game of downstream pollution. *Journal of Economic Dynamics and Control, 25*, 1973–1987.

Jørgensen, S., & Zaccour, G. (2001b). Incentive equilibrium strategies and welfare allocation in a dynamic game of pollution control. *Automatica, 37*, 29–36.

Jørgensen, S., & Zaccour, G. (2002). Time consistency in cooperative differential games. In G. Zaccour (Ed.), *Decision and control in management science (in Honor of Alain Haurie)* (pp. 349–366). Dordrecht: Kluwer Academic Publishers.

Jørgensen, S., & Zaccour, G. (2003). Channel coordination over time: Incentive equilibria and credibility. *Journal of Economic Dynamics and Control, 27*, 801–822.

Kaitala, V., & Pohjola, M. (1990). Economic development and agreeable redistribution in capitalism: Efficient game equilibria in a two-class neoclassical growth model. *International Economic Review, 31*, 421–437.

Kaitala, V., & Pohjola, M. (1995). Sustainable international agreements on greenhouse warming: A game theory study. In C. Carraro & J. A. Filar (Eds.), *Annals of the international society of dynamic games* (Vol. 2, pp. 67–87). Boston: Birkhäuser.

Kierzenka, J., & Shampine, L. F. (2001). A BVP solver based on residual control and the MATLAB PSE. *ACM TOMS, 27*, 299–316.

Martín-Herrán, G., & Taboubi, S. (2005). Incentive strategies for shelf-space allocation in duopolies. In A. Haurie & G. Zaccour (Eds.), *Dynamic games: Theory and applications. GERAD 25th anniversary series* (pp. 231–253). New York: Springer.

Martín-Herrán, G., & Zaccour, G. (2005). Credibility of incentive equilibrium strategies in linear-state differential games. *Journal of Optimization Theory and Applications, 126*, 1–23.

Martín-Herrán, G., & Zaccour, G. (2009). Credible linear incentive equilibrium strategies in linear-quadratic differential games. In P. Bernhard, V. Gaitsgory, & O. Pourtallier (Eds.), *Annals of the international society of dynamic games* (Vol. 10, pp. 261–291). Boston: Birkhäuser.

Petrosjan, L. A. (1997). Agreeable solutions in differential games. *International Journal of Mathematics, Game Theory and Algebra, 7*, 165–177.

Petrosjan, L. A., & Zaccour, G. (2003). Time-consistent Shapley value allocation of pollution cost control. *Journal of Economic Dynamics and Control, 27*, 381–398.

Petrosjan, L. A., & Zenkevich, N. A. (1996). *Game theory*. Singapore: World Scientific.

Taboubi, S. (2019). Incentive mechanisms for price and advertising coordination in dynamic marketing channels. *International Transactions in Operational Research, 26*, 2281–2304.

Tolwinski, B., Haurie, A., & Leitmann, G. (1986). Cooperative equilibria in differential games. *Journal of Mathematical Analysis and Applications, 119*, 192–202.

Van der Ploeg, F., & De Zeeuw, A. J. (1992). International aspects of pollution control. *Environmental and Resource Economics, 2*, 117–139.

# Strategic Interaction Among Firms in Output and Emission Markets: A Unified Framework

**Francisco J. André and Luis M. de Castro**

**Abstract** Cap-and-trade (CAT) programs are nowadays a common tool used by authorities to regulate polluting emissions and tackle environmental problems such as Climate Change. In this chapter, we analyze the implications of firm's strategic behavior in product and emission permit markets for the success of these policies. We survey the related literature focusing on the relevance of market structure and firms' competition. We develop a simple but unifying setting to revisit some of the main academic results on the subject.

**Keywords** Emissions trading · Oligopoly · Market interaction · Cournot model · Stackelberg model

## 1 Introduction and Literature Review

There are basically two ways of introducing a price that incorporates environmental externalities in the markets: price-based and quantity-based regulations. A carbon tax is a paradigmatic example of price-based regulation, while the quantity-based approach usually takes the form of a cap-and-trade (CAT) program. Such a system involves setting an aggregate cap for emissions that is lower than the business-as-usual level and creating tradable emission rights that are distributed among the polluters.

CAT programs represent nowadays an important tool commonly used by authorities to regulate pollution emissions. These programs are in place for several pollutants at national and international levels. In the 1990s, the US established one

F. J. André (✉)
Department of Economic Analysis and ICEI, Universidad Complutense de Madrid, Madrid, Spain
e-mail: andre@ccee.ucm.es

L. M. de Castro
ICEI, Universidad Complutense de Madrid, Madrid, Spain



P.-O. Pineau et al. (eds.), *Games in Management Science*,
International Series in Operations Research & Management Science 280,
https://doi.org/10.1007/978-3-030-19107-8_12

of the best known and more successful CAT systems in the world for sulfur-dioxide in the framework of the Clean Air Act (see, e.g., Ellerman 2000). The EU ETS (European Union Emission Trading System) was implemented in 2005 for the entire EU $CO_2$ emissions, and it is nowadays considered the most important emissions market in the world.

CAT programs change the nature of the regulatory process with respect to traditional command-and-control policies by shifting the task of identifying the appropriate pollution control strategies from the environmental authority to the polluters. This is attractive for individual firms because they enjoy more flexibility to comply with the regulation, either reducing emissions or acquiring permits from other firms. More importantly, it is also attractive from a social perspective because, as long as marginal abatement costs differ across polluters, the market can play a positive role in achieving a specified environmental target in an efficient way.

Two of the most important aspects that have been addressed in the literature are the existence of market imperfections and the different mechanisms to make the initial allocation of permits. This work mainly focuses on the first one, imperfect competition. Since a detailed analysis of the initial permit allocation is beyond the scope of this study, we restrict ourselves to the allocation method that has been traditionally more used in practice, which is called grandfathering and consists of allocating the permits for free among firms based on business-as-usual emissions.[1]

At least two different issues regarding grandfathering have been investigated in the literature: the effect of sequential versus simultaneous allocation and the impact of related imperfect output markets. The sequential version typically takes places in schemes with sovereign governments that are allowed to announce their permit allocations at different dates. Such was the case in the early phases of the EU ETS, when domestic permit allocations were often announced at separate times. MacKenzie (2011) concludes that this option may result in strategic behavior of the different countries involved, giving rise to an aggregate welfare loss.

One of the central results in the literature shows that a CAT system is cost-effective under a set of assumptions. The first approach to the theory behind emissions trading was formalized by Baumol and Oates (1971), who proposed a set of arbitrary standards and charges on emissions sufficient to attain these standards. Although they admitted that the system does not generally produce a Pareto-efficient allocation of resources, it does result in a cost-effective result (i.e., achieving a specified reduction in pollution levels at a minimum cost). In a static framework, Montgomery (1972) showed that cost-effectiveness is achieved regardless of the allocation rule chosen, as the initial allocation of emission permits does not affect the market equilibrium. In a dynamic setting, Cronshaw and Kruse (1996) showed that (static and dynamic) cost-effectiveness holds when permits are bankable in a

[1]Permit auctioning is progressively gaining more relevance as an alternative allocation method. Alvarez and André (2015, 2016) and Alvarez et al. (2019) address the efficiency issues regarding this method.

competitive permit market with perfect foresight, but only if the firms are not subject to profit regulations.

The desirable cost-effectiveness property of CAT systems crucially relies on the permit market being perfect. If this is not the case, cost-effectiveness is not guaranteed anymore. Moreover, the allocation method is not innocuous either. Imperfect competition allows firms to pass pollution costs on to consumers. If firms receive permits for free, they essentially get reimbursed for costs they never had to incur, which is commonly known as windfall profits. According to Hintermann (2011), existing firms favor freely allocated tradable permits not only because they convey rents that represent a wealth transfer from consumers to firms but also for the fact that it sets entry barriers, as long as the newcomers, unlike the incumbents, have to purchase permits to operate.

Unfortunately, the perfect market assumption rarely holds in practice as permit markets typically cover large firms (or countries) operating in highly concentrated markets with market power, like electricity, cement or refining.[2] When the markets are not perfectly competitive, the strategic behavior of firms becomes a key issue because the environmental policy aims at reducing pollution emissions at the minimum cost while firms aim at maximizing profits, and both objectives are not necessarily compatible. The most obvious consequences of the lack of perfect competition are distorted prices and inefficient equilibrium allocations. Game theory plays a central role in this area for studying, on the one hand, the interactions among firms and their strategic behavior and, on the other hand, the interplay between the product market and the permit market.

Market power has been studied in the literature under different approaches, including static and dynamic settings and different market structures, such as Cournot or Stackelberg.[3] The results depend on whether there is market power in the output market, in the permit market or in both. Therefore, we can split the related literature in three different lines. In what follows, we present a short overview of this literature.

[2] One rationale behind this fact is that operating a permit market with a large number of small firms would involve an unbearable amount of transaction costs associated with monitoring, reporting, and verifying emissions (see, e.g., Cason and Gangadharan 2003 or Montero 1997). So, markets with a large number of firms have traditionally been regulated by means of taxes and emission markets tend to be used in markets with a relatively small number of (typically large) participants.

[3] Note that the term "Stackelberg model" is used with, at least, two slightly different meanings. In classic microeconomic theory, it refers to a very specific model with two firms in which one of them acts as a leader in output quantity. In the differential games literature, on the other hand, the term "Stackelberg game" is typically used to denote a sequential game whether there is one or several leaders or dominant firms, one or many followers or price-taking firms and the strategic variable can be either output or price.

## 1.1 Market Power Only in the Permit Market

The ground-breaking paper in this line is Hahn (1984). Considering a single dominant firm and a price-taking fringe, he stated that the efficiency loss due to market power depends on the initial allocation of permits, and the permit price is an increasing function of the leader's allocation. The dominant firm will manipulate the price (upwards if it is a seller and downwards if it is a buyer) unless the initial allocation equals the cost-effective one, which requires a perfectly informed regulator.[4]

Hagem and Westskog (1998) extended the Hahn setting in a dynamic two-period model considering oligopolists and a competitive fringe in a Cournot-type game, and found a non-optimal distribution of abatement in an imperfectly competitive market with banking and borrowing.

## 1.2 Market Power Only in the Output Market

Different authors, such as Sartzetakis (2004), have shown that perfect competition in the permit market is not sufficient to render a cost-effective outcome if the product market is not competitive. Within the framework of a Cournot duopoly, Sartzetakis (1997b) compares the efficiency of a competitive emissions market to a command-and-control regulation. He shows that emissions trading modify the allocation of emissions among firms and hence their production choices, and there is an output redistribution effect that favors the less efficient firm.

Meunier (2011) analyzes the efficiency of permit trading between two imperfectly competitive markets and concludes that the integration of permit markets (even if they are perfectly competitive) can decrease welfare because of imperfect competition in product markets. Theoretically, if markets are perfectly competitive, a unique global permit market that covers all polluting activities would be more efficient, but under market power, several permit markets may be more efficient than an integrated one.

Ehrhart et al. (2008) show that, if the output market is not perfectly competitive, firms may have incentives to collude in the permit market since an increase in the permit prices may decease output and lead to higher profits.

André and de Castro (2017) analyzed if the existence of scarcity rents can make the firms agree on a more stringent policy and concluded that this event is more likely to happen under Cournot than under Stackelberg competition, and the chances increase if the firms are endowed with a large initial amount of permits.

---

[4]André and Arguedas (2018) extended the Hahn model to consider technology adoption and showed that the initial distribution of permits (in particular, the amount of permits initially given to the dominant firm) is crucial in determining, not only the allocation of emissions and abatement but also the existence of over- or underinvestment in technology.

## 1.3 Market Power in the Permit and the Output Market

Market interaction implies that one firm's actions will influence, not only its productions cost but also the costs of other firms and the market equilibrium. One dominant firm may increase its profits by increasing the industry costs. In the industrial organization literature, this strategy is known as "raising rivals' costs."

Misiolek and Elder (1989) concluded that a single dominant firm can manipulate the permit market to drive up the fringe firm's cost in the product market, what they call exclusionary manipulation. The dominant firm will buy more (or sell fewer) permits to increase its market share and profits relative to the fringe.

Eshel (2005) presents a model with a competitive fringe and a dominant firm that simultaneously selects the set prices in both markets. The permit price is set above the dominant firm's marginal abatement cost if the profit decrease in the permit market is overweighed by the profit increase in the product market. When the dominant firm is not able to set a price mark-up in the product market and is a buyer of permits, it sets the price below its marginal abatement cost.

Hintermann (2011) found that a firm that has a dominant position in both markets will manipulate both prices to increase its profits at the expense of consumers and tax payers, and overall efficiency cannot be achieved by means of permit allocation alone. While Hahn (1984) stated that efficiency in the permit market can be achieved when the initial allocation to the dominant firm equals the cost-effective one, Hintermann shows that this is not the case when there is interaction between the product and the permit market. Hintermann (2017), in a simplified version of his 2011 model, shows that it is not the presence of "double" market power per se, but simply the transmission of input costs into output prices which leads to a failure of Hahn's prescription of full free allocation.

The result by De Feo et al. (2013) contrasts with the one by Eshel (2005). They model the interaction between the tradable emissions permits market (upstream) and the output market (downstream) by considering a three-stage game: in the first stage, a dominant firm sets the price of permits, in the second stage permits are traded and, in the third stage, firms compete "a la Cournot" in the output market. They found that the dominant firm may set a permit price above its marginal abatement cost, even when it is a net buyer of permits and cannot set the product price.

Within a two-stage game, Sartzetakis (1997a) considers a Cournot duopoly where one of the two Cournot players has price setting power in the permits market. In the first stage, the leader choses the permit price and, in the second one, both firms make their output and abatement decisions taking permit price as given.

The main purpose of this paper is to present a unified view of the literature on emissions trading under market power. It brings together different strands and highlights some of the main results. For this purpose, in Sect. 2, we set up a simple two-period model where two firms compete in the output and the emissions markets. Within this model, we replicate some relevant results in the related literature, paying more attention to those results concerning the cross-market links. In Sect. 3, we set up a particular model to analyze the relationship between the level of permit price and the degree of competition in both markets. In Sect. 4, we set up a slightly simplified version of this particular model to explore different market structures depending on which is the mainstream market: output or permits. Finally Sect. 5 concludes.

## 2 A Unified Model

In this section, we develop a two-period partial equilibrium model with two markets. The permit market allows for full banking and borrowing and can be linked to one or several output sectors. We focus on one of those sectors, that is composed by two firms producing a final good (energy, for instance) and emitting a global pollutant. We consider different structures in both markets. Our aim is to offer a somewhat general model that can give rise (with slight modifications) to some of the basic results of the literature under different market structures.

### 2.1 *Basic Elements*

We consider two firms labeled as $i = 1, 2$ and two periods, $j = 1, 2$. We denote the variables with two subscripts where the first subscript refers to the firm ($i$) and the second to the period ($j$). The variables that are not firm-specific are denoted with a single subscript for the period. Let $x_{ij}$ be the output of firm $i$ in period $j$ and $X_j$ the total output in period $j$, with $X_j = x_{1j} + x_{2j}$. In period $i$ firms face the inverse demand function $P_i(X)$, with $P_i^{'}(X) < 0$. Let $e_{ij}$ denote the emissions, net of abatement, of firm $i$ in period $j$. Both firms have the same cost function so that the cost of firm $i$ is given by $C(x_i, e_i)$, which depends on output and emissions and is continuous and twice differentiable in both arguments with the following properties:

$$\frac{\partial C}{\partial x_i} > 0, \quad \frac{\partial^2 C}{\partial x_i^2} \geq 0, \quad \frac{\partial C}{\partial e_i} < 0, \quad \frac{\partial^2 C}{\partial e_i^2} > 0, \quad \frac{\partial^2 C}{\partial x_i \partial e_i} < 0. \tag{1}$$

These common assumptions mean that the cost is increasing and convex with respect to output and decreasing and convex with respect to emissions, i.e., producing dirty is cheaper than producing clean. They can also be thought of as implying a positive and increasing marginal abatement cost. The fifth condition implies that not only total cost but also marginal output cost is decreasing in emissions.

We assume that the firms are subject to a CAT scheme so that they cannot emit more than their holdings of permits, and emissions can be perfectly monitored without cost by the regulatory authorities. Thus, emissions become a factor of production that has to be paid for. As highlighted by Hahn (1984), the assumption that marginal abatement costs are increasing is equivalent to the assumption of downward sloping demand curves for emission permits. This assumption implies that the firms attain a regular minimum in solving its profit maximization problem.

The firms can receive an initial endowment of permits in each period, and they can purchase additional required permits or sell the remaining ones in the market. Let $S_{ij}$ be the initial endowment of permits of firm $i$ at period $j$ and $S_j$ the total number of permits issued in period $j$, so that $S_j = S_{1j} + S_{2j}$. Denote as $y_{ij}$ the purchases of permits by firm $i$ at period $j$, where a positive value corresponds to a purchase and a negative value to a sale. Unused permits may be sold or banked to be used in the next period. Let $B_{ij}$ be the amount permits banked by firm $i$ at period $j$. By constructions, the following balance condition must hold at every period:

$$B_{ij} = S_{ij} + y_{ij} - e_{ij} \tag{2}$$

which simply states that the amount of permits banked equals the difference between the total amount of permits held by the firm (initial endowment plus purchases) and its net emissions.

To have a flexible enough framework, output competition is modeled using the conjectural variations approach.[5] In oligopoly theory, a conjecture is a firm's belief about the rival's response to its own strategy, i.e., firm $i$'s conjecture about firm $j$'s reaction can be formally defined as the expectation formed by firm $i$ about $\partial x_j^R(x_i)/\partial x_i$, where $x_j^R(x_i)$ is the reaction function (or best reply function) of firm $j$. Many of the standard economic models implicitly assume that this conjecture is fixed. The conjectural variations literature investigates how the market equilibrium changes when firms' conjectures change or vary across different markets.

If we assume that both firms have constant and identical conjectural variations equal to δ, it follows:

$$\delta = \frac{\partial x_2}{\partial x_1} = \frac{\partial x_1}{\partial x_2}; \qquad \frac{dX}{dx_1} = \frac{dx_1}{dx_1} + \frac{dx_2}{dx_1} = 1 + \delta \tag{3}$$

which means that, when one firm increases its own production by one unit, it conjectures that total output in the market will increase by $1 + \delta$ units.

The attractiveness of this approach is that it can be seen as a general framework to include different well-known market structures as particular cases. The Cournot equilibrium is obtained when $\delta = 0$. The competitive (or Bertrand) model corresponds to $\delta = -1$ and collusion is obtained when $\delta = 1$.

We can also assume that firms have constant but different conjectural variations:

[5]The conjectural variation approach is typically viewed as a reduced form approximation to a repeated dynamic game. The conjectural variations model includes monopolistic and competitive behavior as special cases. This model is discussed by Bresnahan (1981) and Seade (1980) and is surveyed by Dixit (1986).

$$\delta_i = \left(\frac{\partial x_k}{\partial x_i}\right); \qquad \frac{dX}{dx_i} = \frac{dx_i}{dx_i} + \frac{dx_k}{dx_i} = 1 + \delta_i \qquad i \neq k \tag{4}$$

The Stackelberg model with firm 1 being a leader and firm 2 a follower corresponds to $\delta_1 = 0.5$, $\delta_2 = 0$.

Both firms maximize their discounted profit in the two-period horizon with $\beta$ being the discount factor. To have a full description of the model, we need to determine whether the output and the permit markets are competitive or not. In what follows we study each case separately.

## 2.2 Competitive Output Market

Let us study first the case in which the output market is competitive, which, in turn, gives raise to two different scenarios depending on whether the permit market is competitive or not.

### 2.2.1 Competitive Permit Market

Each firm faces the following two-period profit maximization problem:

$$\begin{aligned} &\underset{\{x,e,y\}}{\text{Max}}\ P_1(X)x_{i1} - C(x_{i1}, e_{i1}) - p_1 y_{i1} + \beta\left[P_2(X)x_{i2} - C(x_{i2}, e_{i2})\, c x_{i2} - p_2 y_{i2}\right] \\ &s.t. \qquad B_{i1} = S_{i1} - e_{i1} + y_{i1} \\ &\qquad\quad\ \ B_{i2} = S_{i2} - e_{i2} + y_{i2} + B_{i1} \\ &\qquad\quad\ \ B_{i1} \geq 0 \\ &\qquad\quad\ \ B_{i2} \geq 0 \end{aligned} \tag{5}$$

Let $\lambda_j$ and $\mu_j$ ($j = 1, 2$) be the multipliers of the equality and inequality constraints, respectively, corresponding to both periods. For the time being, we restrict our analysis to the first-order conditions (FOCs) related to the emissions permit market:

$$FOC\,(y_{i1}) \Rightarrow \lambda_1 = p_1 \tag{6}$$

$$FOC\,(e_{i1}) \Rightarrow -\frac{\partial C}{\partial e_{i1}} = \lambda_1 \Rightarrow e_{i1} = f\,(p_1) \tag{7}$$

$$FOC\,(y_{i2}) \Rightarrow \lambda_2 = \beta p_2 \tag{8}$$

$$FOC\,(e_{i2}) \Rightarrow -\beta\frac{\partial C}{\partial e_{i2}} = \lambda_2 \Rightarrow e_{i2} = f\,(p_2) \tag{9}$$

$$FOC\,(B_{i1}) \Rightarrow -\lambda_1 + \lambda_2 + \mu_1 = 0 \tag{10}$$

$$FOC\,(B_{i2}) \Rightarrow -\lambda_2 + \mu_2 = 0 \tag{11}$$

$$B_{i1} \geq 0; \quad \mu_1 \geq 0; \quad B_{i1}\mu_1 = 0 \tag{12}$$

$$B_{i2} = 0; \mu_2 \geq 0 \tag{13}$$

Since the marginal abatement cost is positive and the model has only two periods, clearly it is optimal for the firms not to bank any permit at period 2, so in equilibrium, $B_{i2} = 0, \quad i = 1, 2$. That is the reason why condition (13) differs from (12).

As we know from Montgomery (1972), in a competitive market, cost-effectiveness is achieved regardless of the allocation rule chosen. In our case, this result straightforwardly follows from Eqs. (7) and (9), which are the standard conditions according to which the marginal cost of abatement equals the permit price and therefore the marginal cost is equalized across firms for any initial allocation of the permits.

Using Eqs. (6) and (8) into Eq. (10) and taking into account that the multiplier $\mu$ is non-negative, we conclude that if the problem has a unique solution, the present value permit prices are not decreasing over time. Formally,

$$p_1 - \beta p_2 = \mu_1 \geq 0$$

In a similar way, it can be established by using Eqs. (7) and (9) into (10) that the marginal abatement cost is not increasing over time in present value:

$$\frac{\partial C}{\partial e_{i1}} - \beta\frac{\partial C}{\partial e_{i2}} = \mu_1 \geq 0$$

According to Eqs. (12) and (13), the amount of banked permits in a period can be positive only if the multiplier $\mu$ for that period is zero. This is in line with one of the results in Cronshaw and Kruse (1996). They stated "Suppose that one of the firms is not subject to profit regulation. Then, that firm is only willing to bank permits if the futures price is the same in the two periods, or equivalently if the spot price rises with the rate of interest" (Page 185). It should be also noted that in this case, the present value of marginal abatement cost is also equated among periods.

### 2.2.2 Imperfect Competition in the Permit Market

Now we assume that, although the output market is competitive, firm 1 acts as a dominant firm and firm 2 as a follower in the permit market.[6] The dominant firm solves the following problem with regard to the emissions permit market:

$$\begin{aligned} &\underset{\{e_{11},y_{11},e_{12},y_{12},p_1,p_2\}}{\text{Min}} \quad C\left(x_{11},e_{11}\right)+p_1 y_{11}+\beta\left[C\left(x_{12},e_{12}\right)+p_2 y_{12}\right] \\ &s.t. \quad B_{11}=S_{11}+y_{11}-e_{11} \\ &\qquad B_{12}=S_{12}+y_{12}-e_{12}+B_{11} \\ &\qquad y_{11}=-y_{21} \\ &\qquad y_{12}=-y_{22} \end{aligned} \tag{14}$$

The static one-period model can be seen as a particular case by considering $\beta = 0$ and not allowing for banking or borrowing. In this way, we can adapt our model to represent Hahn's (1984) static framework. The follower FOCs are Eqs. (6) and (7). Now the problem of firm 1 can be stated as

$$\begin{aligned} &\underset{\{p_1\}}{\text{Min}}\ C\left(x_{11},e_{11}\right)+p_1 y_{11} \\ &s.t. \qquad y_{11}=e_{11}-S_{11}=S_{21}-e_{21}\left(p_1\right) \end{aligned} \tag{15}$$

The FOC with respect to $p$ yields

$$\frac{\partial C}{\partial e_{11}}\frac{\partial e_{11}}{\partial e_{21}}\frac{\partial e_{21}}{\partial p_1}+\frac{\partial y_{11}}{\partial e_{11}}\frac{\partial e_{11}}{\partial e_{21}}\frac{\partial e_{21}}{\partial p_1}p_1+y_{11}=0 \tag{16}$$

This is the first result by Hahn (1984). He stated that if the dominant firm does not receive an amount of permit equal to the number that it holds in equilibrium, total abatement cost will exceed the cost-minimizing solution (Proposition 1, page 756).

Simple manipulation of Eq. (16) makes Hahn's statement clear in our model:

$$\frac{\partial e_{11}}{\partial e_{21}}=-1\Rightarrow\left(-\frac{\partial C}{\partial e_{11}}-p_1\right)\frac{\partial e_{21}}{\partial p}+y_{11}=0$$

Note that the dominant firm equals its marginal abatement cost to the permit price only when there is no trade ($y_{11} = 0$).

The second Hahn's result says that if a regular interior minimum exists, a transfer of permits from any of the price takers to the dominant firm will increase the permit price. This result can be immediately shown by differentiating the FOC (14).

[6] Qualitatively similar results can be obtained if firm 2 is replaced by a fringe of competitive firms.

$$\frac{\partial p}{\partial S_{11}} = \left[\left(-\frac{\partial C}{\partial e_{11}} - p_1\right)\left(\frac{\partial^2 e_{21}}{\partial p_1^2}\right) + \left(\frac{\partial^2 C}{\partial e_{11}^2}\right)\left(\frac{\partial^2 e_{21}}{\partial p_1^2}\right) - 2\left(\frac{\partial e_{21}}{\partial p_1}\right)\right]^{-1} > 0$$

The expression in brackets is just the second-order condition and must be positive to attain a minimum of problem (15). An immediate implication is that the number of permits that the dominant firm demands increases with its initial allocation of permits.

As we know from Montgomery (1972), when the permit market is competitive, the distribution of permits is merely an equity issue without relevance for the sake of efficiency or cost-effectiveness. But as soon as we relax the perfect competition assumption, the distribution of permits matters, with regard not only to equity considerations but also to the overall cost of the system.

Let us now consider a two-period framework with banking and borrowing. For convenience, denote gross emissions (i.e., emissions in the absence of abatement activities) as $Z$ and the amount of emissions abated as $q$. Then, net emissions equal the difference between gross emissions and abatement:

$$e_{ij} = Z_{ij} - q_{ij} \tag{17}$$

Taking into account that it is optimal for both firms not to keep permits at the end of the second period, firm 2 faces the following constraint:

$$q_{21} + q_{22} = Z_{21} + Z_{22} - S_{21} - S_{22} - (y_{21} + y_{22}) \tag{18}$$

We consider the situation where firm 2 is a net buyer of permits, which implies $y_{21} + y_{22} > 0$. Firm 2 solves the following problem:

$$\begin{array}{ll} \underset{\{q_{21}, y_{21}, q_{22}, y_{22}\}}{\text{Min}} & C(x_{21}, q_{21}) + p_1 y_{21} + \beta [C(x_{22}, q_{22}) + p_2 y_{22}] \\ s.t. & Eq.(16) \end{array} \tag{19}$$

The FOCs of this problem are as follows:

$$FOC(y_{21}) \Rightarrow \lambda = p_1 \tag{20}$$

$$FOC(q_{21}) \Rightarrow \frac{\partial C}{\partial q_{21}} = \lambda \tag{21}$$

$$FOC(y_{22}) \Rightarrow \lambda = \beta p_2 \tag{22}$$

$$FOC(q_{22}) \Rightarrow \beta \frac{\partial C}{\partial q_{22}} = \lambda \tag{23}$$

Equations (20) and (22) imply that the present value of the permit price must be constant over time in equilibrium. Equations (21) and (23) show that it is optimal for firm 2 in equilibrium to abate emissions until the present value of marginal abatement costs equals the present value price of permits. These are again the results by Cronshaw and Kruse (1996). From (20) to (23), the permit prices can be expressed as a function of the correspondent abated quantities.

The dominant firm faces a constraint similar to that of the follower:

$$q_{11} + q_{12} = Z_{11} + Z_{12} - S_{11} - S_{12} - (y_{11} + y_{12}) \tag{24}$$

The leader minimizes the cost of abatement in both periods minus the income from selling permits.

$$\begin{aligned} &\mathrm{Min}_{\{q_{11}, y_{11}, q_{12}, y_{12}\}}\ C\,(x_{11}, q_{11}) - p_1\,(y_{21})\,y_{11} + \beta\,[C\,(x_{12}, q_{12}) + p_2\,(y_{22})\,y_{12}] \\ &s.t. \qquad Eq.(24) \\ &\qquad\quad y_{1j} = -y_{2j} \end{aligned} \tag{25}$$

From the first-order conditions, we get

$$\frac{\partial C}{\partial q_{11}} = \frac{\partial p_1\,(y_{21})}{\partial y_{21}} \frac{\partial y_{21}}{\partial y_{11}} y_{11} + p_1\,(y_{21}) \tag{26}$$

$$\frac{\partial C}{\partial q_{12}} = \frac{\partial p_2\,(y_{22})}{\partial y_{22}} \frac{\partial y_{22}}{\partial y_{12}} y_{12} + p_2\,(y_{22}) \tag{27}$$

According to these conditions, it is optimal for the dominant firm to abate emissions until the present value of marginal abatement cost in each period equals the present value of the marginal revenue from selling permits.

The follower's marginal abatement costs will exceed the marginal abatement costs of the dominant firm in each period. The dominant firm sells too few permits and hence abates too little as compared to a cost-effective distribution of abatement across agents. This result is equivalent to the one by Hagem and Westskog (1998). According to them, "in the banking and borrowing system the monopolist extracts the full monopoly rent from the total sale of permits over both periods" (Page 95).

## 2.3 *Imperfect Competition in the Output Market*

Consider now that the output market is imperfectly competitive. For the sake of tractability, in this section we focus on the case in which the permit market is competitive although the output market is not.

### 2.3.1 Competitive Permit Market

Each firm solves (5). If the permit market is competitive, Eqs. (6)–(13) must hold in equilibrium. With regard to the output market, we consider conjectural variations in line with Eq. (3). The FOCs with respect to output result in

$$FOC\left(x_{i1}\right) \Rightarrow P_1 + x_{i1}\frac{dP_1}{dX_1}\left(1+\delta\right) - \frac{\partial C}{\partial x_{i1}} = 0 \tag{28}$$

$$FOC\left(x_{i2}\right) \Rightarrow P_2 + x_{i2}\frac{dP_2}{dX_2}\left(1+\delta\right) - \frac{\partial C}{\partial x_{i2}} = 0 \tag{29}$$

Several authors have stated that the output market structure matters when considering the efficiency of CAT policies. Focusing on the EU ETS, Meunier (2011) concluded that even if firms are price takers in permit markets, the integration of different permit markets can decrease welfare because of imperfect competition in product markets. A similar result can be obtained in our framework. To this aim, let us consider that, instead of two individual firms, $i = 1, 2$ represent two different emission markets with $n_i$ different firms in each of them, denoting by $X_i$ the aggregate quantity of goods produced at market $i$ and by $P_i(X_i)$ the inverse demand function, which is assumed to satisfy the conditions required to ensure existence and uniqueness of a Cournot equilibrium. Particularly the price function is not too convex and quantities are strategic substitutes (see Meunier 2011 for details). To replicate the Cournot case, we set $\delta = 0$.

In each market, a CAT is implemented with the local price of emissions denoted as $p_i$. Let us assume the static version of our canonical model with $\beta = 0$ and assume that there is no an initial free allocation of permits (i.e., $S_i = 0$). Then (5) becomes

$$\begin{aligned} &\text{Max}_{\{x_i, e_i\}} P_i\left(X_i\right) x_i - C\left(x_i, e_i\right) - p_i y_i \\ &e_i = y_i \end{aligned} \tag{30}$$

If we consider the same cost function for all agents, the equilibrium is symmetric in the sense that output and emissions are equally distributed among firms on each output market. Therefore, individual quantities are $X_i/n_i$ and individual emissions are $E_i/n_i$ where $E_i$ is the overall quantity of emissions in market $i$.

Total quantity produced can be written as a function of total emissions, say $X_i^*(E_i)$, as the unique solution of the following equation:

$$P_i + P_i'\frac{X_i}{n_i} = \frac{\partial C_i}{\partial x_i}\left({}^{X_i}\!/_{n_i}, {}^{E_i}\!/_{n_i}\right) \tag{31}$$

The demand for permits of each firm in market $i$ as a function of the permit price, say $E_i^*(p_i)$, is implicitly determined by the following condition:

$$p_i = -\frac{\partial C_i}{\partial e_i}\left(X_i^*\left(E_i^*(p_i)\right)\Big/_{n_i}, E_i^*(p_i)\Big/_{n_i}\right), \quad i = 1, 2 \tag{32}$$

The welfare implications of the interaction of a competitive market for emission permits with an oligopolistic product market can be analyzed with the introduction of the following welfare function:

$$W(X_1, X_2, E_1, E_2) = \sum_i Z(X_i) - n_i C_i\left(X_i\Big/_{n_i}, E_i\Big/_{n_i}\right) \tag{33}$$

Welfare is the sum of surpluses net of production cost. Gross surplus from consumption is $V(X)$ with $dV/dX = P(X)$. The optimal allocation of emissions denoted $(E_{1*}, E_{2*})$ solves the following problem:

$$\begin{aligned} &\underset{\{E_1, E_2\}}{\text{Max}}\ W(X_1(E_1), X_2(E_2), E_1, E_2) \\ &s.t. \\ &\qquad E_1 + E_2 = \overline{E} \end{aligned} \tag{34}$$

On each market, an additional unit of emissions increases the local net surplus by

$$\frac{dW_i}{dE_i} = \left(P_i - \partial C_i\Big/_{\partial x_i}\right)\partial X_i^*\Big/_{\partial E_i} - \partial C_i\Big/_{\partial e_i} \tag{35}$$

The first term is the market power effect. An additional permit increases production and, due to the existence of market power, this has a positive effect on welfare.

If the optimal allocation of emissions is interior, it satisfies the first-order condition:

$$\left(P_1 - \partial C_1\Big/_{\partial x_1}\right)\partial X_1^*\Big/_{\partial E_1} - \partial C_1\Big/_{\partial e_1} = \left(P_2 - \partial C_2\Big/_{\partial x_2}\right)\partial X_2^*\Big/_{\partial E_2} - \partial C_2\Big/_{\partial e_2} \tag{36}$$

With an integrated permit market, local permit prices are equalized and the marginal costs of emissions for each firm are equalized across output markets. If the difference between the product price and the marginal cost is not the same in both markets, then the market allocation does not satisfy Eq. (36) and welfare is lower than in Eq. (34). This is the result of Meunier (2011), which says that the integration of markets does not increase welfare in general. The inefficiency of an integrated permit market arises from the divergence between price and marginal cost and the sensitivity of production to emissions.

Ehrhart et al. (2008) investigate the effect of a permit price increase on firm's profit and consumer surplus under imperfect competition in the product market in a symmetric setting (i.e., $\delta_1 = \delta_2 = \delta$). The influence of a permit price increase on firms' profit is ambiguous because there are two counteracting effects. A negative effect is due to more expensive permit purchasing costs and a positive effect is due to higher revenues. The latter is related to the imperfect competition product market because a permit price increase will lead to a decreasing output level, and to an increase in the output price, implying a revenue increase. Under certain conditions, a higher permit price will increase firms' profits but decrease social welfare.

To replicate these findings, we solve problem (30) in two stages by backward induction. In the second stage, the cost minimization problem for firm $i$ is

$$\underset{\{e_i\}}{\text{Min}} C\left(x_i, e_i\right) + pe_i \tag{37}$$

The FOC states that the marginal cost of emissions equals the permit price. Based on Eq. (1) conditions, the cost function is convex in emissions and the second-order condition is always fulfilled. Let us denote as $e_{i^*}$ the amount of emissions that minimizes abatement costs.

In the first stage, we solve the profit maximization problem given by

$$\underset{\{x_i, x_{-i}, p\}}{\text{Max}} \; P\left(x_i + x_{-i}\right) x_i - TC\left(x_i, e_i^*\left(x_i, p\right)\right) \tag{38}$$

The FOC for maximization is

$$P\left(x_i + x_{-i}\right) + (1+\delta)\frac{\partial P\left(x_i + x_{-i}\right)}{\partial X} x_i - \frac{\partial TC\left(x_i, p\right)}{\partial x_i} = 0 \tag{39}$$

Due to symmetry, in equilibrium we have $x_i = x_{-i} = x^*$. Differentiating the profit function and taking (39) into account yields

$$\frac{d}{dp}\left(P\left(2x^*\right)x^* - TC^*\left(x^*, p\right)\right) = (1-\delta)\, x^* \frac{\partial P\left(2x^*\right)}{\partial X}\frac{\partial x^*}{\partial p} - \frac{\partial TC^*}{\partial p} \tag{40}$$

And the sign is ambiguous. The condition for Eq. (40) being positive is as follows:

$$\left[\frac{(1-\delta)\, x^* {}^{\partial P\left(2x^*\right)}\big/_{\partial X}}{(3+\delta)\, {}^{\partial P\left(2x^*\right)}\big/_{\partial X} - {}^{\partial TC^*\left(x^*, p\right)}\big/_{\partial x_i \partial x_i}}\right]\frac{\partial e_i^*}{\partial x_i} - e_i^* > 0$$

This is Ehrhart et al.'s (2008) result. They stated that "under certain parameter ranges, a higher permits price induces higher firm profits for all types of expected competition, with the exception of a monopoly scenario ($\delta = 1$)" (Page 352).

As it was defined in (33), welfare is the sum of surpluses net of production cost.

$$W\left(x_1, x_2, p\right) = \int_0^{2x^*} P(z)dz - 2TC^*\left(x^*, p\right)$$

Independent of the effects on firms' profits, an increasing permit price never leads to an increase in social welfare because the negative effect on the consumers' surplus always outweighs the possibly increasing profits. Analytically,

$$\frac{\partial W(x_1, x_2, p)}{\partial p} = \left[2P\left(2x^*\right) - 2\frac{\partial TC^*(x^*, p)}{\partial x^*}\right]\frac{dx^*}{dp} - 2\frac{\partial TC^*(x^*, p)}{\partial p} \leq 0$$

This is the second result by Ehrhart et al. (2008).

## 3 A Particular Model with Endogenous Permit Prices

In order to investigate a framework in which there is market power both in the permit and in the output market, we need to introduce some more specific structures in the model. To this end, we assume some specific production and abatement functions. We adopt linear and linear-quadratic specifications for the sake of tractability. A similar model can be found by Sartzetakis (1997b) or André and de Castro (2017).

In the output market we keep a duopolistic framework, with linear demand. The inverse demand function is $P = a - bX$. The firms' cost can be linearly separated into output cost and abatement cost. On the production side, the firms face a constant marginal cost, $c$. Gross emissions are assumed to be proportional to the firms' output ($e_i = rx_i$), $r$ being the pollution intensity, which is common for both firms. Total abatement cost is given by the linear-quadratic function $(d + tq_i)q_i$, where $q_i$ is total abatement by firm $i$.[7] In this section, we abstract from market power in the emission market, but, in the next one, we use a simplified version of this model to consider simultaneous market power in the output and in the emission market.

The profit maximization problem considered in (5) becomes

$$\begin{aligned} \underset{\{y_{ij}, x_{ij}, q_{ij}, B_{ij}\}}{\text{Max}} \quad & (a - bx_{i1} - bx_{-i1})\,x_{i1} - cx_{i1} - \left(d + tq_{i1}\right)q_{i1} - p_1 y_{i1} + \\ & + \beta\left[P_2\left(x_{i2} + x_{-i2}\right)x_{i2} - cx_{i2} - \left(d + tq_{i2}\right)q_{i2} - p_2 y_{i2}\right] \\ s.t. \quad & B_{i1} = S_{i1} - rx_{i1} + q_{i1} + y_{i1} \\ & B_{i2} = S_{i2} - rx_{i2} + q_{i2} + y_{i2} + B_{i1} \\ & B_{i1} \geq 0 \\ & B_{i2} \geq 0 \end{aligned} \tag{41}$$

The corresponding Kuhn Tucker conditions from Eqs. (6)–(13), (25), and (26) take the following form:

[7] This cost function allows us to keep simple analytical expressions while, at the same time, second-order conditions are guaranteed. As mentioned below, it is widely used in the literature, and it is also very convenient for comparison purposes.

$$FOC\left(y_{i1}\right) \Rightarrow \lambda_1 = p_1 \tag{42}$$

$$FOC\left(q_{i1}\right) \Rightarrow d + 2tq_{i1} = p_1 \Rightarrow q_{i1} = \frac{p_1 - d}{2t} \tag{43}$$

$$FOC\left(x_{i1}\right) \Rightarrow P_1 - bx_{i1}\left(1+\delta\right) = c + rp_1 \tag{44}$$

$$FOC\left(y_{i2}\right) \Rightarrow \lambda_2 = \beta p_2 \tag{45}$$

$$FOC\left(q_{i2}\right) \Rightarrow -\beta\left(d + 2tq_{i2}\right) = \lambda_2 \Rightarrow q_{i2} = \frac{p_2 - d}{2t} \tag{46}$$

$$FOC\left(x_{i2}\right) \Rightarrow P_2 - bx_{i2}\left(1+\delta\right) = c + rp_2 \tag{47}$$

$$FOC\left(B_{i1}\right) \Rightarrow -\lambda_1 + \lambda_2 + \mu_1 = 0 \tag{48}$$

$$FOC\left(B_{i2}\right) \Rightarrow -\lambda_2 + \mu_2 = 0 \tag{49}$$

$$B_{i1} \geq 0;\ \mu_1 \geq 0;\ B_{i1}\mu_1 = 0 \tag{50}$$

$$B_{i2} = 0;\ \mu_2 > 0 \tag{51}$$

From Eqs. (44) and (47), it is trivial to see that the product price is increasing in the permit price, as a simple algebraic manipulation yields the following values for the output market equilibrium:

$$X_1 = 2 \cdot \frac{P_1 - c - rp_1}{b\left(1+\delta\right)} \Rightarrow P_1 = \frac{\left(1+\delta\right)a + 2\left(c + rp_1\right)}{3+\delta} \tag{52}$$

As usual, the market is closed when total emissions net of abatement equal the total amount of permits distributed in the market:

$$r\left(X_1 + X_2\right) - \left(q_{11} + q_{21}\right) - \left(q_{12} + q_{22}\right) = S_1 + S_2 \tag{53}$$

Considering that total output in each period is

$$X_j = \frac{2\left(a - c - rp_j\right)}{b\left(3 + \delta\right)} \tag{54}$$

From the first-order conditions, abatement is a function of the permit price, and using (40) and (43) into (50) and (51) leads us to the permit price.

$$p_2 = \frac{8rt^2\left(a - c\right) - 2b\left(3 + \delta\right)\left[\left(S_1 + S_2\right)t^2 - 2dt\right]}{\left(1 + \beta\right)\left[4r^2t^2 + 2bt\left(3 + \delta\right)\right]} \tag{55}$$

From this equation, we can analyze and compare the endogenous permit price in the different market structures, represented by parameter $\delta$. From (54), we know that total output is decreasing in both the permit price and the conjectural variation parameter.

**Lemma 1** *The quantities abated and the permit price decrease at a decreasing rate as $\delta$ increases.*

An increase in $\delta$ can be interpreted as a reduction in market competition. Therefore, Lemma 1 states that, as the output market becomes less competitive, firms tend to abate less and also to demand less permits, which causes the permit price to decrease. Proposition 1 allows us to compare two important cases such as the Stackelberg model and a collusive market.

**Proposition 1** *The price of permits is higher when the product market is controlled by a Stackelberg leader than in the case of a cartel. Moreover, it is higher than in the case of a Cournot duopoly. But it is lower than if the product market is competitive.*

The permit price level under different market structures follows the same behavior as the total quantities produced in each type of market. The rationale for this result is that a higher output tends to generate more emissions and a higher demand for permits, which pushes the permit price up. It is also worth noting that the permit price depends on the output technology (parameter $c$) and the abatement technology (parameters $d$ and $t$), which are assumed to be the same for both firms.

## 4 Market Power in Both Markets

In this section, we address simultaneous power market in the output and permit markets. For the sake of tractability, we focus on a version of the model introduced in Sect. 3 with some additional simplifications in line with De Feo et al. (2013). In this section, we aim to underline the relevance of the game timing in the role of the permit market. Specifically, we show that qualitatively different results follow depending on which market (output or emissions) is considered as the mainstream market.

Following De Feo et al. (2013), we assume that the inverse demand product function is $P = 1 - X = 1 - x_1 - x_2$ and normalize the marginal output cost to zero and the pollution intensity to 1. Therefore emissions equal product quantity. The abatement cost function of firm $i$ is now $\frac{1}{2}t_i q_i{}^2$ so that the marginal abatement cost is $t_i q_i$. We assume that firm 1 is the leader in the permit market and it has the most efficient abatement technology. In particular, we set $t_1 = 1; t_2 > 1$.

First, we assume that the leading mainstream market is the output one. We do so by using a timing of the game in which firms interact first in the output market and then in the permit market. As usual, in the last stage firm 2 solves its minimum cost problem in terms of abatement and obtains the familiar first-order condition that equates marginal abatement costs to the permit price. The dominant firm minimizes its own costs anticipating the reaction of the follower and taking into consideration the permit market clearing condition. In this case, we obtain the following values for firm 2's optimal abatement and demand for permits in equilibrium as a function of output and the initial allocation of permits:

$$q_2 = \frac{x_1 + (1+t)\,x_2 - (1+t)\,S_2 - S_1}{1+2t} \tag{56}$$

$$y_2 = -y_1 = \frac{(S_1 - tS_2) - (x_1 - tx_2)}{1+2t} = \frac{1}{t}\left(tq_2 - q_1\right) = \frac{1}{t}\left(p - q_1\right) \tag{57}$$

Equation (54) shows that the dominant firm is a net buyer of permits ($y_1 > 0$) when the permit price is lower than its marginal abatement cost ($p < q_1$). Similarly, the dominant firm is a net seller of permits when the price exceeds its marginal abatement cost. In both cases (whatever the output market structure), firm 1 will use its dominant position to reduce its cost and increase its profit. Therefore, only the dominant firm can take advantage of its strategic position in the permit market while the competitive firm simply acts as a price taker.

De Feo et al. (2013) set the game in a different order. Firms decide first in the permit market and then take output decisions based on the equilibrium permit price. This timing gives the permit market more relevance as we show below. Using backward induction and considering a Cournot model of oligopoly, firm 2 (solves the following problem in stage 4):

$$\begin{aligned} &\underset{x_2}{\text{Max}} \quad (1 - x_1 - x_2)\,x_2 - \tfrac{1}{2}tq_2^2 - py_2 \\ &s.t. \qquad q_2 = x_2 - y_2 - S_2 \end{aligned} \tag{58}$$

In stage 3, the leader solves

$$\begin{aligned} &\underset{x_1}{\text{Max}} \quad (1 - x_1 - x_2)\,x_1 - \tfrac{1}{2}q_1^2 - py_1 \\ &s.t. \qquad q_1 = x_1 + y_2 - S_1 \end{aligned} \tag{59}$$

The reactions functions turn out to be

$$\begin{aligned} x_2 &= \tfrac{1-x_1+ty_2+tS_2}{2+t} \\ x_1 &= \tfrac{1-x_2-y_2+S_1}{3} \end{aligned} \tag{60}$$

and equilibrium output is

$$\begin{aligned} x_2 &= \tfrac{2+(3t+1)y_2+3tS_2-S_1}{3t+5} \\ x_1 &= \tfrac{1+t-(2t+2)y_2-tS_2+S_1(2+t)}{3t+5} \end{aligned} \tag{61}$$

In stage 2 the follower decides its demand of permits as a price taker, solving the problem:

$$\begin{aligned} &\operatorname*{Max}_{y_2} \left(1 - x_1^* - x_2^*\right) x_2^* - \tfrac{1}{2}tq_2^2 - py_2 \\ &s.t. \qquad q_2 = x_2 - y_2 - S_2 \end{aligned} \tag{62}$$

From the FOC we find

$$y_2 = \frac{4 + 14t + 6t^2 - p(3t+5)^2 - S_2\left(9t^2 + 19t\right) - S_1\left(3t^2 + 7t + 2\right)}{6t\,(2+t) - 2} \tag{63}$$

In stage 1 the dominant firm sets the permit price by solving the following problem:

$$\begin{aligned} &\operatorname*{Max}_{p} \left(1 - x_1^* - x_2^*\right) x_1^* - \tfrac{1}{2}q_1^2 + py_2 \\ &s.t. \qquad y_2 = \tfrac{4+14t+6t^2-p(3t+5)^2-S_2\left(9t^2+19t\right)-S_1\left(3t^2+7t+2\right)}{6t(2+t)-2} \end{aligned} \tag{64}$$

Finally the permit price is obtained:

$$\begin{aligned} p = {} & \tfrac{54t^4+294t^3+498t^2+274t+32}{3(t+3)(3t+1)(3t+5)^2} - S_2\tfrac{63t^4+327t^3+501t^2+197t}{3(t+3)(3t+1)(3t+5)^2} - \\ & - S_1\tfrac{27t^4+183t^3+381t^2+245t-4}{3(t+3)(3t+1)(3t+5)^2} \end{aligned} \tag{65}$$

Equations (63) and (65) correspond to Eqs. (10) and (12) in De Feo et al. (2013).[8] From these equations, they demonstrate in their Proposition 2 that the leader always marks up the price of permits above its equilibrium marginal abatement cost, whether it is a net seller or buyer of permits.[9]

[8]Apart from other differences in notation, $S_1$ and $S_2$ correspond to $\alpha S$ and $(1-\alpha)S$, respectively, in DeFeo et al.

[9]The proof is quite cumbersome and is omitted here for simplicity of exposition.

As indicated by De Feo et al. (2013), the above result is in contrast with Eshel (2005), who considered a leader and a price-taking competitive fringe instead of a Cournot duopoly in the output market and a dominant firm in the permits market as De Feo et al. (2013) did. Nevertheless, we kept the same market structure model but changed the mainstream market (by altering the timing) and also got a different result. The reason of this difference is the fact that not only the leader but also the follower adopts a "raise the rival's cost" strategy in the permits market. In simple terms, the strategic interaction in the output market generates additional distortions in the permits market.

## 5 Conclusions

This chapter has described the role played by market competition within cap-and-trade programs, paying special attention to the interaction between the output market and the emission permits market. We have adapted a simple unifying model to replicate some of the results in the related literature. The conjectural variations approach provides such a unifying framework to make comparisons between different oligopolistic structures.

We have analyzed how market power in the output market affects the price in the emissions market. Our analysis shows that the permit price level under different market structures tends to follow the same behavior as the quantities produced in the output market. We have analyzed four different structures that we enumerate from higher to lower equilibrium output quantities (and hence, from higher to lower price of permits): a competitive market, a Stackelberg oligopoly, a Cournot oligopoly, and a cartel.

We have also shown in a simplified model how the firm's strategic behavior changes as we change the timing of the game (which, according to our interpretation, corresponds to changing the leading market). This result has been shown previously in the literature using different market structures, but in this paper we show that also keeping the same market structure the equilibrium permit price is different as we set the game in a different timing, that is to say, when we change the roles of markets as "the main market" and the secondary one.

**Acknowledgments** Francisco J. André and Luis M. de Castro acknowledge support from the Spanish Ministry of Economy, Industry and Competitiveness and Development Fund, ERDF (grant ECO2015-70349-P). F.J. André also thanks Universidad Complutense de Madrid and Banco Santander (grant PR26/16-15B-4) and the European Commission (project INNOVEC'EAU, SOE1/P1/F0173).

## Appendix

**Proof of Lemma 1** Consider the following changes of variables to simplify the expression of the emissions permits price in period 2.

$$\begin{aligned} A &= 8r(a-c)t^2 \\ B &= 2b\left[(S_1+S_2)t^2 - 2dt\right] \\ C &= (1+\beta)4r^2t^2 \\ D &= (1+\beta)2bt \end{aligned}$$

where $A$ must be positive to have a positive output in equilibrium (which requires $a - c - rp_2 > 0$). The sign of $B$ is ambiguous while $C$ and $D$ are clearly positive.

Using this notation, (52) takes the form

$$p_2 = \frac{8r(a-c)t^2 - 2b(3+\delta)\left[(S_1+S_2)t^2 + 2dt\right]}{(1+\beta)\left[4r^2t^2 + 2bt(3+\delta)\right]} \Rightarrow p_2 = \frac{A-(3+\delta)B}{C+(3+\delta)D} \tag{66}$$

Differentiation with respect to $\delta$ gives

$$\frac{\partial p_2}{\partial \delta} = \frac{-B\left[C+(3+\delta)D\right] - \left[A-(3+\delta)B\right]D}{\left[C+(3+\delta)D\right]^2} = \frac{-BC-AD}{\left[C+(3+\delta)D\right]^2} \tag{67}$$

Using this value we can compute

$$q_{12} = \frac{p_2-d}{2t} \Rightarrow \frac{\partial q_{12}}{\partial \delta} = \frac{1}{2t}\frac{\partial p_2}{\partial \delta} \tag{68}$$

Assume (67) is positive. Therefore B must be negative and $-BC > AD$. If this is the case, we have the following inequality:

$$\begin{aligned} &-2b\left[(S_1+S_2)t^2 - 2dt\right](1+\beta)4r^2t^2 > 8r(a-c)t^2(1+\beta)2bt \\ &-\left[(S_1+S_2)t^2 - 2dt\right]r > (a-c)2t \\ &2rdt - (a-c)2t > (S_1+S_2)rt^2 \Rightarrow rd - (a-c) > 0 \end{aligned}$$

But the last expression must be negative if product quantity is positive because

$$a-c-rp_2 > 0 \Rightarrow a-c-r\left(d+2tq_{i2}\right) > 0 \Rightarrow a-c-rd > 0 \Rightarrow rd-(a-c) < 0$$

Therefore, it must be $AD > -BC$ and so (67) must be negative.

To see that is decreasing at a decreasing rate, we just take the second-order derivative:

$$\frac{\partial^2 p_2}{\partial \delta^2} = \frac{2D(BC+AD)}{\left[C+(3+\delta)D\right]^3} > 0$$

**Proof of Proposition 1** Let us consider firm 1 as the Stackelberg leader. The conjectural variations are $\delta_2 = 0$ and $\delta_1 = -1/2$. The total output for the first period is

$$X_1 = x_{11} + x_{21} = \frac{2\left(P_1 - c - rp_1\right)}{b} + \frac{\left(P_1 - c - rp_1\right)}{b} = \frac{3\left(P_1 - c - rp_1\right)}{b} \tag{69}$$

And the corresponding product price is

$$P_1 = a - bX_1 = \frac{a + 3\left(c + rp_1\right)}{4}$$

leading to the standard Stackelberg result where the leader is producing a double quantity than the follower as long as the marginal product cost is the same.

$$X_1 = x_{11} + x_{21} = \frac{a - c - rp_1}{2b} + \frac{a - c - rp_1}{4b} = \frac{3\left(a - c - rp_1\right)}{4b} \tag{70}$$

The market clearing condition becomes

$$r\left(\frac{3\left(a - c - rp_1\right)}{4b} + \frac{3\left(a - c - rp_2\right)}{4b}\right) - S_1 - S_2 - \frac{p_1 - d}{t} - \frac{p_2 - d}{t} = 0$$

And taking into account that $p_1 = \beta p_2$, the last expression leads us to

$$r\left(\frac{6\,(a - c)}{4b} - \frac{3rp_2\,(1 + \beta)}{4b}\right) - S_1 - S_2 - \frac{\left(p_2\,(1 + \beta) - 2d\right)}{t} = 0$$

Therefore, in the second period, the permit price is

$$p_2 = \frac{6rt\,(a-c) - 4b\left[(S_1+S_2)\,t - 2d\right]}{(1+\beta)\left[3r^2t + 4b\right]} = \frac{6rt^2\,(a-c) - 4b\left[(S_1+S_2)\,t^2 - 2dt\right]}{(1+\beta)\left[3r^2t^2 + 4bt\right]} \tag{71}$$

Consider the following changes of variable to simplify the expression of the emissions permit price in period 2.

$$\begin{aligned} A &= rt^2\,(a - c) \\ B &= b\left[(S_1 + S_2)\,t^2 - 2dt\right] \\ C &= (1 + \beta)\,r^2t^2 \\ D &= (1 + \beta)\,2bt \end{aligned}$$

where $A$ must be positive to have a positive output in equilibrium (which requires $a - c - rp_2 > 0$). The sign of $B$ is ambiguous while $C$ and $D$ are clearly positive. Using this notation, (71) becomes

$$p_2^S = \frac{6A - 4B}{(1 + \beta)\,[3C + 2D]} \tag{72}$$

Now we proceed to compare the Stackelberg permit price with other oligopolistic structures. The conjectural variation for a Cournot competition takes the value $\delta = 0$ and the permit price simplified form is

$$p_2^C = \frac{8A - 6B}{(1 + \beta)\,[4C + 3D]} \tag{73}$$

The comparison shows that the Stackelberg permit price is higher:

$$\begin{aligned}&(6A - 4B)\,[4C + 3D] = 24AC + 18AD - 16BC - 12BD\\&(8A - 6B)\,[3C + 2D] = 24AC + 16AD - 18BC - 12BD\\&B > 0 \Rightarrow 18AD - 16BC > 16AD - 18BC\end{aligned}$$

The result is also valid if $B < 0$ because we have $AD > -BC$, which implies

$$AD > -BC \Rightarrow 2AD > -2BC \Rightarrow 18AD - 16BC > 16AD - 18BC$$

We apply a similar procedure to prove the rest of comparisons in the proposition.

## References


Alvarez, F., & André, F. J. (2015). Auctioning versus grandfathering in cap-and-trade systems with market power and incomplete information. *Environmental and Resource Economics, 62*, 873–906.

Alvarez, F., & André, F. J. (2016). Auctioning emission permits with market power. *The BE Journal of Economic Analysis and Policy, 16*(4). https://doi.org/10.1515/bejeap-2015-0041.

Alvarez, F., Mazón, C., & André, F. J. (2019). Assigning pollution permits: are uniform auctions efficient? *Economic Theory, 67*(1), 211–248.

André, F. J., & Arguedas, C. (2018). Technology adoption in emission trading programs with market power. *The Energy Journal, 39*, 145–174.

André, F. J., & de Castro, L. M. (2017). Scarcity climate rents under a carbon price with oligopoly competition. *Oxford Economic Papers, 69*(2), 470–495.

Baumol, W. J., & Oates, W. E. (1971). The use of standards and prices for protection of the environment. *The Swedish Journal of Economics, 73*(1), 42–54.

Bresnahan, T. F. (1981). Duopoly models with consistent conjectures. *The American Economic Review, 71*(5), 934–945.

Cason, T., & Gangadharan, L. (2003). Transactions cost in tradable permit markets: An experimental study of pollution markets designs. *Journal of Regulatory Economics, 23*, 145–165.

Cronshaw, M. B., & Kruse, J. B. (1996). Regulated firms in pollution permit markets. *Journal of Regulatory Economics, 9*, 179–189.
De Feo, G., Resende, J., & Sanin, M. E. (2013). Emission permit trading and downstream strategic market interaction. *The Manchester School, 81*, 780–802.
Dixit, A. (1986). Comparative statics for oligopoly. *International Economic Review, 27*, 107–122.
Ehrhart, K. M., Hoppe, C., & Löschel, R. (2008). Abuse of EU emissions trading for tacit collusion. *Environmental and Resource Economics, 41*, 347–361.
Ellerman, A. D. (2000). *Markets for clean air: The US acid rain program*. Cambridge: Cambridge University Press.
Eshel, D. M. D. (2005). Optimal allocation of tradable pollution rights and market structures. *Journal of Regulatory Economics, 28*(2), 205–223.
Hagem, C., & Westskog, H. (1998). The design of a dynamic tradeable quota system under market imperfections. *Journal of Environmental Economics and Management, 36*, 89–107.
Hahn, R. (1984). Market power and transferable property rights. *Quarterly Journal of Economics, 99*, 753–765.
Hintermann, B. (2011). Market power, permit allocation and efficiency in emission permit markets. *Environmental and Resource Economics, 49*, 327–349.
Hintermann, B. (2017). Market power in emission permit markets: Theory and evidence from the EU ETS. *Environmental and Resource Economics, 66*, 89–112.
MacKenzie, I. (2011). Tradable permits allocations and sequential choice. *Resource and Energy Economics, 33*(1), 268–278.
Meunier, G. (2011). Emission permit trading between imperfectly competitive product markets. *Environmental and Resource Economics, 50*, 347–264.
Misiolek, W., & Elder, H. (1989). Exclusionary manipulation of markets for pollution rights. *Journal of Environmental Economics and Management, 16*, 156–166.
Montero, J. P. (1997). Marketable pollution permits with uncertainty and transaction costs. *Resource and Energy Economics, 20*, 27–50.
Montgomery, W. D. (1972). Markets in licenses and efficient pollution control programs. *Journal of Economic Theory, 5*, 395–418.
Sartzetakis, E. (1997a). Raising rivals' costs strategies via emission permits markets. *Review of Industrial Organization, 12*, 751–765.
Sartzetakis, E. S. (1997b). Tradeable emission permits regulations in the presence of imperfectly competitive product markets: Welfare implications. *Environmental and Resource Economics, 9*, 65–81.
Sartzetakis, E. S. (2004). On the efficiency of competitive markets for emission permits. *Environmental and Resource Economics, 27*, 1–19.
Seade, J. (1980). On the effects of entry. *Econometrica, 48*(2), 479–489.

# Human vs River: Cooperation in Environmental Games Through Environmental Personhood

**Michèle Breton and Suzanne Zaccour**

**Abstract** This chapter opens a conversation between law and game theory on the personhood status of environmental entities. Specifically, we consider the granting of personhood status to a river that suffers from the production activity of a firm creating economic value, but also pollution emissions as a by-product. If no one lives downstream, for instance, traditional responses to the environmental problem are unsatisfactory. We show that environmental personhood can help achieve efficient bargaining solutions between polluters and environmental entities. We also report on various approaches that have been taken in some countries to endow environmental persons with means to protect their rights.

**Keywords** Environment · Game theory · Law · Personhood

## 1 Introduction

On October 18th, 1929, the Privy Council overturned the Supreme Court of Canada's decision that women were not "persons" for the purpose of the Constitution. In November of 2016, a court in Argentina ruled that Cecilia, a chimpanzee detained at the zoo of Mendoza, was not a thing but a legal person, with legal capacity. In 2017, New Zealand's Whanganui River was declared to be a legal person. It has been decades since the legal personhood of corporations has ceased to be controversial. Since 1985, the Canada Business Corporations Act grants corporations the rights and privileges of a natural person.

M. Breton (✉)
HEC Montréal, Montréal, QC, Canada
e-mail: michele.breton@hec.ca

S. Zaccour
Oxford University, Oxford, UK
e-mail: suzanne.zaccour@mail.mcgill.ca

 
P.-O. Pineau et al. (eds.), *Games in Management Science*,
International Series in Operations Research & Management Science 280,
https://doi.org/10.1007/978-3-030-19107-8_13

In law, "person" and "human being" are not synonymous. Human beings can be considered property, and objects can gain personhood status. In this article, we open a conversation between law and game theory on the personhood status of environmental entities. Specifically, we consider the granting of personhood status to a river that is being polluted by one (or more) firms. While the literature on environmental personhood has mostly focused on standing (the capacity to sue), we conceptualize legal personhood as a vehicle to facilitate cooperation between the environment and the polluters. We show that cooperation with environmental "persons" may be preferable to alternative solutions such as laisser-faire, Government regulation, and non-cooperative or cooperative solutions involving interested parties.

The inspiration for this work comes from Jørgensen and Zaccour (2001), which considers two countries producing along a river and generating pollution. One of the countries, being downstream, suffers the costs of both countries' emissions. Similar settings have also been considered in Fernandez (2009), Fanokoa et al. (2011), and Shi et al. (2016), among others, where various cooperative game solutions are analyzed in the context of downstream pollution. A first question that comes to mind is the following: What if there is nobody downstream—should the upstream polluter continue to pollute with impunity? Recent developments granting legal status to rivers and other environmental entities give rise to a second question: Can we envision a bargaining solution between the polluter and the river?

As pointed out in Sigman (2002), rivers should represent a good case for cooperation, as they typically involve a limited number of players and well-defined costs. Using water-quality data, the author however finds evidence of international downstream pollution spillovers and concludes that cooperation is not effective between countries sharing rivers.

The aim of this paper is twofold. First, we explore the recent literature on environmental personhood. Second, we use a stylized model to illustrate cooperative and regulatory approaches that could be taken to protect an environmental entity. Our objective is to illustrate how environmental personhood can help resolve a number of issues in the context of protection of the environment, namely:

- Legal standing
- Tragedy of the commons and inefficiency of non-cooperative solutions
- Enforcement of cooperative solutions.

It is worth mentioning that an important issue in the literature on downstream pollution is the time consistency of cooperative solutions, as it is the case in Jørgensen and Zaccour (2001). Here, for the sake of simplicity, we consider a static game model, and do not address the time-consistency issue.

The rest of the paper is organized as follows: Sect. 2 presents a stylized model involving a river, a polluting firm, a downstream resident, and a regulator, and discusses the limitations of various approaches that can be taken to protect the river. Section 3 introduces the concept of environmental personhood, and Sect. 4 discusses practical implementation issues. Section 5 briefly concludes.

# 2 The Model and the Players

## 2.1 *The Polluting Firm*

The problem we are concerned with starts with a polluting firm and a river. The firm undertakes a production activity that creates not only economic value but also pollution emissions as a by-product. We normalize the production technology in such a way that the pollution level corresponding to the optimal production activity of the firm when there is no environmental concern is equal to 1.

We assume that the firm may suffer from adverse consequences of its pollution emissions and can use various mitigation mechanisms to reduce this environmental cost (e.g., reduction in the production activity or filters). Let $m_1$ represent the level of mitigation by the firm. Environmental damage and mitigation costs are assumed to be increasing and convex functions. We use a stylized model with a quadratic specification, so that the cost incurred by the firm is

$$z_1 = \frac{D_1}{2}E^2 + \frac{M_1}{2}m_1^2,$$

where $D_1 \geq 0$, $M_1 > 0$, $E = 1-m_1$ is the total pollution (after mitigation). The first term of the right-hand side represents the environmental damage cost of pollution and the second term is the cost of mitigation. Accordingly, the mitigation level that minimizes the firm's total cost is

$$m_1^F = \frac{D_1}{M_1 + D_1}$$

and the corresponding cost and pollution level are then

$$z_1^F = \frac{1}{2}\frac{M_1 D_1}{D_1 + M_1}$$

$$E^F = \frac{M_1}{D_1 + M_1}.$$

Note that when the firm does not suffer damages from pollution ($D_1 = 0$), the optimal mitigation level is 0.

## 2.2 *The Downstream Resident*

Now suppose that concerned citizens living downstream of the firm suffer damages from its polluting emissions (e.g., loss of enjoyment of their property, damages to their land, harm to their health). To simplify, we consider a single player (Player 2) representing the group of downstream residents and assume that the environmental damage cost suffered by Player 2 is given by $D_2E^2$, where $D_2 \geq 0$. Different avenues can be envisioned to reduce the pollution in the river and the cost borne by Player 2, among which traditional tort and property law, undertaking of additional mitigation activities, and public law regulation.

### 2.2.1 Tort Law

One of the traditional legal responses to a polluted river is for the downstream resident to sue the polluting firm under the law of tort (or extracontractual obligations). A court can make the firm pay damages for the harm caused to the resident. An injunction can even be granted in certain circumstances, forcing the firm to change or stop its activities. Assuming that the firm expects to have to pay the entirety of the damages borne by the downstream residents and that $D_2 > 0$, the optimal mitigation level of the firm becomes

$$m_1^T = \frac{D_1 + D_2}{D_1 + D_2 + M_1} > m_1^F,$$

with the corresponding cost and pollution levels

$$\begin{aligned} z_1^T &= \frac{M_1}{2} \frac{D_1 + D_2}{D_1 + D_2 + M_1} > z_1^F \\ E^T &= \frac{M_1}{D_1 + D_2 + M_1} < E^F. \end{aligned}$$

This solution, in which the firm internalizes the damage cost of its downstream neighbors, results in a higher mitigation level (less pollution) and a higher cost for the firm (not including the legal costs).

However, tort and property law responses to environmental damage face important limitations, including litigation costs, issues of standing and enforcement problems.

In some cases, the costs that the residents are entitled to recover, that is

$$z_2^T = \frac{D_2}{2} \left( \frac{M_1}{D_1 + D_2 + M_1} \right)^2,$$

may not be sufficient to justify litigation. More importantly, these costs may result in an underestimation of the environmental impact of polluting activities (O'Donnell and Talbot-Jones 2018). For example, the plaintiffs may be able to recover the costs necessary to improve the water quality to the relevant standard, but not those needed to restore the river's ecosystem. In any case, there is no guarantee that the plaintiffs will use the amount they recover to undertake mitigation activities.

The standing and underestimation problems are exacerbated if no one lives downstream. In that case, no human being is directly harmed by the deterioration of the river (other than by living on a more polluted planet). The ones to suffer are the river, fish, and other non-human animals, which do not currently have rights or the capacity to sue. Who, then, could sue the firm, and for how much?

We should note that many jurisdictions grant standing to concerned citizens or organizations to bring suit in the public interest, even if they are not directly affected by the dispute. This form of public interest litigation has been used for environmental purposes, and several jurisdictions even allow for wide interpretations of the public interest (see, e.g., Bélanger and Halley (2017) on the Québec case). Even then, however, practical and enforcement problems remain.

An important limitation to environmental lawsuits by concerned citizens is the cost and uncertainty of tort litigation. This kind of action represents important risks for a person who is not directly affected by the environmental degradation, especially in jurisdictions where the losing party is liable for the costs incurred by the winning party. In addition, a limited range of legal actions and remedies is possible. While someone may demand an injunction to have the firm stop or modify its activities, or contest the legality of a permit granted to the firm, one cannot demand a financial reparation for the harm caused to the river. Finally, the lack of pre-established representatives to take action to protect the river (the problem of the commons) represents another hurdle. There is no guarantee that any concerned citizen will be prompted to take on the battle personally.

### 2.2.2 Mitigation

Alternatively, suppose that the downstream resident decides to undertake mitigation activities (e.g., adaptive measures, cleaning),[1] so that the pollution level after mitigation by both parties becomes

$$E = 1 - m_1 - m_2.$$

We assume that the mitigation costs for Player 2 have the same functional form as for Player 1, so that the total cost for the downstream resident is

$$z_2 = \frac{D_2}{2}E^2 + \frac{M_2}{2}m_2^2,$$

where $M_2 > 0$.

The best response of each player to the mitigation level of the other is then

[1]Using a representative player is equivalent to considering that the downstream residents agree to coordinate their mitigation decisions and seek a cooperative solution, which minimizes the total cost borne by the group. A noncooperative solution among $n$ downstream residents can also be envisioned.

$$m_1 = D_1 \frac{1 - m_2}{D_1 + M_1} \tag{1}$$

$$m_2 = D_2 \frac{1 - m_1}{M_2 + D_2}, \tag{2}$$

where it is apparent that the mitigation levels of the firm and of the downstream resident are strategic substitutes, and that no mitigation activity will be undertaken by Player 2 if $D_2 = 0$. Assuming that $D_2 > 0$, the equilibrium solution is then

$$\begin{aligned} m_1^N &= \frac{D_1 M_2}{N} < m_1^F < m_1^T \\ m_2^N &= \frac{D_2 M_1}{N} > 0 \\ N &= D_1 M_2 + D_2 M_1 + M_1 M_2 \end{aligned}$$

and the corresponding pollution level and costs incurred by the players are

$$E^N = \frac{M_1 M_2}{N} < E^F \tag{3}$$

$$z_1^N = \frac{1}{2} D_1 M_1 M_2^2 \frac{D_1 + M_1}{N^2} < z_1^F < z_1^T \tag{4}$$

$$z_2^N = \frac{1}{2} D_2 M_1^2 M_2 \frac{D_2 + M_2}{N^2} \tag{5}$$

$$\begin{aligned} Z^N &\equiv z_1^N + z_2^N \\ &= \frac{1}{2} M_1 M_2 \frac{D_1^2 M_2 + D_2^2 M_1 + D_1 M_1 M_2 + D_2 M_1 M_2}{N^2}. \end{aligned} \tag{6}$$

This solution results in a lower pollution level and lower mitigation level and cost for the polluting firm than the laissez-faire scenario, while the cost for the downstream resident and the total cost can be higher or lower, depending on the value of the parameters.

It is well known however that this non-cooperative equilibrium solution is unsatisfactory, as it leads to over-exploitation of the environmental resources (the so-called *tragedy of the commons* presented in the seminal paper by Hardin (1968)) and is generally not Pareto efficient. Indeed, it is easy to show that any increase $\theta$ in the mitigation level of each player satisfying

$$0 < \theta < \min_{j=1,2} \left\{ \frac{2 D_j M_1 M_2}{N \left(4 D_j + M_j\right)} \right\}$$

will lead to a decrease in the total costs of both players.

In Jørgensen and Zaccour (2001), a cooperative solution is proposed, where the downstream player is better off compensating the polluting firm for some of its mitigation costs, rather than acting noncooperatively. However, the underestimation problem remains: there is no reason for the cooperating players (the polluting firm and the downstream resident) to consider environmental costs that are not affecting them directly (e.g., the degradation of the river). Moreover, no one is specifically habilitated to take decisions or negotiate with the firm on behalf of the river. For sure, citizens and organizations can sign contracts or enter into settlement negotiations with polluting firms, but this scenario is unlikely when, for instance, no one lives downstream. Additionally, any contractual engagement will be undertaken toward the party voluntarily attempting to protect the river, rather than towards the river itself.

## 2.3 *The Regulator*

We now consider a third possibility to reduce the global environmental cost and/or the pollution level of the river, that is, public law regulation (zoning laws, permit granting, emissions taxes). In this scenario, we still assume that directly concerned citizens interact noncooperatively with the polluting firm by undertaking mitigation activities when $D_2 > 0$.

Consider, for instance, the simplest approach to environmental legislation, that is, a tax on polluting emissions.[2] Assuming a tax rate of $\tau$, the firm's total cost becomes

$$z_1 = \frac{D_1}{2}E^2 + \frac{M_1}{2}m_1^2 + \tau\,(1 - m_1)$$

and the mitigation level that minimizes the firm's cost is then

$$m_1(\tau, m_2) = \frac{\tau + D_1\,(1 - m_2)}{D_1 + M_1}.$$

When $D_2 > 0$, using Eq. (2), the equilibrium solution is then, for $0 < \tau < M_1$,

$$m_1^R(\tau) = \frac{\tau\,(D_2 + M_2) + D_1 M_2}{N} > m_1^N$$

$$m_2^R\,(\tau) = \frac{D_2\,(M_1 - \tau)}{N} < m_2^N$$

[2]Other policy instruments could be modeled similarly, for instance, subsidizing mitigation activities undertaken to reduce the environmental damage to the river.

$$E^R(\tau) = \frac{M_2(M_1 - \tau)}{N} < E^N. \tag{7}$$

The impact of an emission tax on the noncooperative solution is an increase in the mitigation level of the polluting firm and a decrease in that of the downstream resident, resulting in an overall lower level of pollution. The corresponding environmental costs are then

$$z_1^R(\tau) = z_1^N + \frac{1}{2}\tau\left(D_1M_2(2D_2 + M_2) + M_1(D_2 + M_2)^2\right)\frac{2M_1 - \tau}{N^2} \tag{8}$$

$$z_2^R(\tau) = z_2^N - \frac{1}{2}\tau D_2M_2(D_2 + M_2)\frac{2M_1 - \tau}{N^2}. \tag{9}$$

A feasible environmental tax will result in a higher cost for the firm and a lower cost for the downstream resident.

There are many possibilities for deciding on an environmental tax rate. Let us assume that the legislator selects a tax rate minimizing the sum of environmental and mitigation costs, given by

$$Z(\tau) = \frac{D_3}{2}\left(E^R(\tau)\right)^2 + \frac{M_1}{2}\left(m_1^R(\tau)\right)^2 + \frac{M_2}{2}\left(m_2^R(\tau)\right)^2,$$

where we assume that the environmental damage cost coefficient $D_3$ used by the legislator accounts for a large spectrum of social and environmental costs, including those borne by the firm and the downstream residents, so that $D_3 > D_1 + D_2$.

The optimal tax rate is then given by

$$\tau^O = \begin{cases} M_1 - M_1(D_2 + M_2)\frac{D_1M_2 + M_1(D_2 + M_2)}{O} & \text{if } D_1 < D_2 + \frac{M_2(D_3 - D_1)}{D_2} \\ 0 & \text{otherwise,} \end{cases}$$

where

$$O = M_2^2(D_3 + M_1) + D_2^2(M_1 + M_2) + 2D_2M_1M_2.$$

Under an environmental tax designed to minimize the total environmental damage and mitigation costs, the equilibrium solution is then

$$m_1^O = M_2\frac{D_2^2 + D_3M_2}{O} > m_1^N$$

$$m_2^O = D_2M_1\frac{D_2 + M_2}{O} < m_2^N$$

$$E^O = M_1M_2\frac{D_2 + M_2}{O} < E^N$$

$$Z^O = \frac{1}{2} M_1 M_2 \frac{D_2^2 + D_3 M_2}{O}.$$

This solution results in a reduction of the total pollution and of the total environmental damage and mitigation costs with respect to the non-cooperative scenario.

However, environmental legislation famously faces considerable compliance problems (see, e.g., Heyes 2000; Weiss and Jacobson 2000; Russell 2001; Wang et al. 2003; Van Rooij 2006; Sparovek et al. 2010; Kim 2015). Moreover, as seen in Eq. (8), an environmental tax is not Pareto-improving with respect to the non-cooperative equilibrium. Since the regulator should consider all interested parties' competing interests, we can wonder if this solution can be improved, in particular by using cooperative game theory approaches. Finally, we show in Sect. 3.2.2 that the first-best solution can generally not be attained using a constant tax rate.

This is where environmental personhood intervenes.[3]

# 3 Environmental Personhood

We contend that environmental personhood can solve the standing problem and prompt efficient cooperative solutions. In this section, we first explore the legal arguments for environmental personhood, and then show how environmental personhood allows to model the river as a player in a cooperative game.

Let us start by defining that a "legal person" is an entity capable of having rights and obligations. As such, personhood can be characterized as merely a rights-container into which different rights can be poured (Wise 2013). Certain persons, such as adult human beings, possess the right to vote and a large array of fundamental rights. Others, such as New Zealand's Whanganui River, possess more limited rights, such as the ownership of a riverbed (Hutchison 2014). Only legal persons can appear in court and vindicate their rights. If an environmental entity has rights, the standing problem disappears. A part of the environment that has become a person can also presumably enter into contracts and out-of-court settlements.

But does it make (legal) sense to speak of a river as a person? Actually, personhood is merely a policy decision. Personhood is the vehicle that the law uses to identify whose' interests are worthy of recognition, who counts (Fagundes 2001). As such, enslaved people were denied personhood status despite being human beings because they did not count in the eyes of the law. Thus, determinations of personhood, be it for corporations, animals, fetuses, or rivers, "are strongly result driven" (Fagundes 2001). For the cynical, "'person' might legally mean whatever the law makes it mean" (Dewey 1926). When scholars ponder the personhood status of fetuses or animals, they are speaking in terms of policy, not biology.

[3]Animal personhood, a hot topic in the law of persons, is another possible avenue for addressing the problem under consideration.

Likewise, deciding what environmental entities are deserving of personhood, how their boundaries are to be defined, and what rights they will be granted is a matter of policy. River rights will develop differently from mountain rights, for example (Berry 2010). Precisely what kinds of rights a given entity would have is likely to depend on context, ecology, and political compromise. As it is noted in Stone (2010), "to say that the environment should have rights is not to say that it should have every right we can imagine, or even the same body of rights as human beings have. Nor is it to say that everything in the environment should have the same rights as every other thing in the environment." The Whanganui River, for instance, can no longer be owned in its entirety; however, the parts of the river that were privately owned before the change in its status remained so. Moreover, existing public access rights and navigation rights were preserved, such that the river does not have the power to exclude the public from accessing it (Hutchison 2014). Personhood, then, is a tool to recognize interests that may not be adequately protected otherwise.

## *3.1 Intrinsic Interests*

If personhood is a policy question, then the policy reasons for protecting the environment abound. Pollution is one of our biggest global killers. Every year, more than three million children under age five die from environmental factors (WHO 2018).

Water pollution alone costs Canadians $300 million annually in health expenses (Government of Canada 2018). Reserving personhood status to humans and corporations constructs non-human animals and nature as resources, as property for humans to consume (Hutchison 2014). This anthropocentric and capitalist framework is put under increasing pressure by environmentalist and animal rights scholars who wish to allow non-humans to object to the abuse and degradation to which they are routinely subjected.

Granting personhood status to a natural feature can serve as a recognition of its intrinsic interests, that is, the river is protected independently of who lives downstream. There is scholarly debate regarding whether the protection of the environment should be subordinated to the interests of human beings (either living humans or future generations), or whether the environment has independent interests (Shelton 2015; Berg 2007; Feinberg 1984; Gordon 2018). More generally, determinations of personhood can rest on the interests of the newly recognized entity or on those of currently recognized persons. Some argue that an entity that does not have interests evidenced by sentience, ability to form relationships and biological life (for example) can only obtain legal personhood based on the interests of others. This question can have practical interest: legal personhood based on the protection of the interests of others is generally more limited. Tribe (1973) notes that "the best interests of individual persons (and even of future human generations) are not demonstrably congruent with those of the natural order as a whole."

In any case, long-term environmental wellbeing, whether for its sake or for the sake of future generations, routinely conflicts with short-term interests of human agents. States may not act in the best interest of nature or of future generations when they compete for economic activities by attempting to pass the laxest environmental legislation. According to Shelton (2015), "While rules governing standing may already permit a government official or agency to represent a natural object as trustee on behalf of the public trust, this representation may not serve to protect the intrinsic value of the environment, especially where the government's short term interests conflict with more long term ecological interests." More generally, governments must balance a wide range of competing interests and cannot give priority to those of the environment or a group of people. As for the protection of nature by individuals, even if they can be granted standing, they must argue in anthropocentric terms, for example showing that they suffer harm from being unable to see a rare animal if its habitat is destroyed. Thus, nature-focused rights approaches differ from purely human-centered duties.

## *3.2 A Cooperative Solution*

We have seen that environmental personhood allows the river to protect its own rights independently of humans' short-term interests. We will return to the practical question of how a river can make decisions in the next section. For now, let us identify the river with Player 2 and the polluting firm with Player 1 and see what happens if, instead of one suing the other or relying on regulation, the two players agree to coordinate their mitigation efforts in order to avoid legal costs and to minimize the total environmental cost.[4] We may assume that the damage cost parameter $D_2$ includes damages to the downstream residents, but also damages to the river, non-human animals, etc. The cooperative solution minimizes

$$Z = \sum_{j=1}^{2} \frac{D_j}{2} E^2 + \frac{M_j}{2} m_j^2$$

and is given by

$$\begin{aligned} m_1^C &= M_2 \frac{D_1 + D_2}{C} \\ m_2^C &= M_1 \frac{D_1 + D_2}{C}, \end{aligned}$$

[4]To simplify, we assume that downstream residents do not participate in the coordination process or in the mitigation effort.

where

$$C = (D_1 + D_2)(M_1 + M_2) + M_1 M_2.$$

The corresponding pollution level and the costs incurred by the players are then

$$\begin{aligned}
E^C &= \frac{M_1 M_2}{C} \\
z_1^C &= \frac{1}{2} M_1 M_2^2 \frac{(D_1 + D_2)^2 + D_1 M_1}{C^2} \\
z_2^C &= \frac{1}{2} M_1^2 M_2 \frac{(D_1 + D_2)^2 + D_2 M_2}{C^2} \\
Z^C &= \frac{1}{2} M_1 M_2 \frac{D_1 + D_2}{C}.
\end{aligned}$$

It is obvious that the cooperative solution results in a lower global cost than the non-cooperative solution in Eqs. (3)–(6) (this is due to the convexity of the mitigation costs). The *benefit of cooperation* is given by

$$Z^N - Z^C = \frac{1}{2} M_1^2 M_2^2 \frac{(D_1 + D_2)(D_1 - D_2)^2 + D_1^2 M_1 + D_2^2 M_2}{N^2 C^2} > 0.$$

### 3.2.1 Individual Rationality and Side Payments

Now, it is not necessarily the case that the polluting firm benefits from adhering to the cooperative solution when $D_1 << D_2$. In particular, when $D_1 = 0$, the firm does not suffer from pollution and does not mitigate in the non-cooperative solution ($z_1^N = 0$), so that all the environmental cost is borne by the river, while in the cooperative solution, the firm abates and incurs mitigation costs.

The usual mechanism used to ensure that a cooperative solution is individually rational consists of allowing for side payments or transfers between the cooperating players. Various solution concepts have been proposed to distribute the benefits of cooperation among players. For instance, *bargaining solutions* propose acceptable ways (according to various desirable properties) in which two players can share the benefits of cooperation (for instance, the Nash (1950), Kalai and Smorodinsky (1975) and egalitarian (Kalai 1977) bargaining solutions).

As an illustration, assume that

$$\begin{aligned}
z_1^C - z_1^N &> 0 \\
z_2^C - z_2^N &< 0,
\end{aligned}$$

that is, the environmental cost for the polluting firm is higher under the cooperative solution than under the non-cooperative one (and the reverse is true for the river).

The egalitarian bargaining solution involves a side payment

$$SP = \frac{z_1^C - z_1^N + z_2^N - z_2^C}{2} > 0$$

paid by the river to the firm. As a result, the total environmental cost for each player becomes

$$z_1^C - SP = z_1^N + \frac{1}{2}\left(Z^C - Z^N\right) < z_1^N$$

$$z_2^C + SP = z_2^N + \frac{1}{2}\left(Z^C - Z^N\right) < z_2^N.$$

### 3.2.2 Efficiency with Respect to Environmental Tax

It is worth mentioning that the cooperative solution, which optimizes the global welfare, cannot in general be attained using an environmental tax. In our example, for instance, setting $D_3 = D_1 + D_2$, it is not possible to find a solution to the system

$$m_1^R(\tau) = m_1^C$$

$$m_2^R(\tau) = m_2^C,$$

except when $D_1 = 0$. When $D_1 = 0$, then the optimal tax

$$\tau^O = \frac{D_2 M_1 M_2}{D_2 (M_1 + M_2) + M_1 M_2}$$

yields the cooperative solution. When $D_1 > 0$, the total cost under an optimal environmental tax is higher:

$$Z^O - Z^C = \frac{M_1^2 M_2^2 D_1^2}{2CO}.$$

### 3.2.3 Enforcement

While a cooperative solution with a side payment reduces the environmental cost for both players, it is not self-enforcing; for instance, using (1), the best response of the polluting firm to the river's cooperative mitigation level is

$$m_1 = D_1 \frac{1 - m_2^C}{D_1 + M_1}$$

$$= m_1^C - \frac{D_2 M_1 M_2}{C (D_1 + M_1)},$$

that is, a reduction in its mitigation level. In addition, there is nothing preventing the firm to take the side payment before reverting to a lower mitigation level.

Environmental personhood provides a plausible solution to the enforcement problem. As we have seen, if the river is a person, then it has rights and obligations, and it can sue and be sued. Domestic contract law can serve as an enforcement mechanism.

The river-person can enter into contracts. If a capable river undertook a contractual obligation to pay mitigation costs, for instance, such an obligation would be enforceable directly against the river. We can likewise imagine mutual obligations between neighboring capable rivers, who cooperate in conservation efforts. By contrast, the polluting firm cannot conclude a contract with an object. It would be possible for the firm to enter into contracts with environmental groups or the government, but as we have already discussed, this scenario is unlikely, particularly when no one lives downstream. Why would random citizens go to the trouble of engaging their personal responsibility to convince a firm to pollute less a river whose state does not affect them personally?

Another issue is that even if such a contract were signed, it would have to be signed between two legal persons: the firm and an organization or human people. What if the organization closes? What if its direction and purpose change? What if the group of citizens dissolves or its members go bankrupt? There are legal solutions to these problems, but the lack of personhood status of the river is a source of legal risk and reduces the likelihood that cooperative solutions will emerge. The river, by contrast, is (hopefully) permanent. Its rights and obligations will survive a change of guardians. Assuming that the government has endowed the river with some financial resources to undertake mitigation strategies and negotiate deals (or that it has resources from other sources such as previous lawsuits against polluters), dealing directly with the river is preferable for the firm.

## 4 Implementation

We have now seen that environmental personhood solves three practical problems: the standing issue, the difficulty in protecting environmental interests when they conflict with human interests, and the enforcement problem. There still remains the problem of implementing environmental personhood in practice.

With or without standing, how can the river stand in court without any legs, sign a contract without any hands, vindicate its rights without the ability to speak? This problem is easy to solve, and indeed has already been solved. Infants, people with severe cognitive disabilities, and patients in a coma all possess rights even as they require help to vindicate them. Parents, family members and guardians decide on their health care, manage their property, and make other decisions in their best interest. Similarly for an environmental entity, guardians can be appointed to make sure that its rights are respected, and, more importantly for our purposes, to engage in negotiation and mutually beneficial cooperation with polluters.

Observing the development of environmental personhood in New Zealand, India, Ecuador, Bolivia, the United States and Columbia, two main approaches can be discerned regarding actions taken by humans in the name of the environment. In some instances, one or more specific guardians are appointed. In New Zealand, for example, two people, appointed from nominations made by the State and by Māori groups (iwi) with interests in the Whanganui River, become the "human face" of the river that can speak on its behalf. The guardians of the river owe their responsibilities to the river, not the appointors.

Another approach is to let any person intervene to protect an environmental entity. The constitution of Ecuador, the first country to recognize nature as a subject of rights, states that "All persons, communities, peoples and nations can call upon public authorities to enforce the rights of nature" (article 71). The State also permits "any natural person or legal entity, human community or group, to file legal proceedings and resort to judicial and administrative bodies without detriment to their direct interest, to obtain from them effective custody in environmental matters, including the possibility of requesting precautionary measures that would make it possible to end the threat or the environmental damage that is the object of the litigation" (article 397). Citizens of Bolivia can also sue individuals and groups to protect Mother Earth's rights to life, to diversity of life, to water, to clean air, to equilibrium, to restoration, and to live free of contamination (Ley de Derechos de la Madre Tierra, Ley 071, 2010).

It is not necessary for our purpose to decide that one approach is superior to the other. However, we can observe that for the protection of nature as a whole, all citizens may be granted the right to take action, while for the protection of specific natural features, such as a river, the practice seems to be to place this responsibility within the hands of a small group of people. The choice of guardians for a river will necessarily be context-specific. Granting decision-making abilities to a few heads should facilitate negotiations with polluters (having polluters negotiate with the whole of the country is impractical and brings us back to the problems of lawsuits for the public interest) as well as avoid the risk of opening the floodgates to excessive litigation. Note that the appointment of guardians can also have the added benefit of involving indigenous groups in the protection of the environment and recognizing specific indigenous nations' connection with a given natural feature.

Now, there are two reasons why the situation of appointed guardians is superior to relying on concerned citizens or organizations to step up as Player 2 and voluntarily embrace the protection of the river. First, by assigning the responsibility to defend the river's interest to a specific person or group of people, a tragedy of the commons scenario—where everyone expects someone else to undertake mitigation activities—can be avoided. Second, a guardian would have fiduciary obligations towards the river. That is, it would have actual enforceable legal obligations to act in the river's best interest. Of course, if all of the river's guardians decide not to fulfill these obligations, still a human being is needed to bring the situation to the attention of the courts or the relevant regulatory body, making environmental personhood but an imperfect solution.

## 5 Conclusion

Game theory is not partial to the identity of players. By staging a dialogue between law and game theory, we have proposed a fresh perspective on environmental negotiation. Indeed, we have shown that environmental personhood can help in achieving efficient "bargaining" solutions taking into account the economic interest of production facilities generating pollution as a by-product along with the interest of environmental entities, such as rivers, in preserving their integrity. We also reported on various approaches that have been taken in some countries to endow environmental persons with means to protect their rights.

Determinations of personhood, however, go well beyond the capacity to negotiate. The project of environmental personhood will have implications for our society and the extra-legal norms it values. Thus, while some scholars warn that environmental rights could crowd out social norms of protection of the environment (Gordon 2018) or legitimize the imperialistic project of further domesticating nature (Livingston 2004), an alternative consequence is that environmental personhood will strengthen our commitment towards its protection. Legal change has symbolic and expressive value. In shaping "what society thinks of as human," (Hutchison 2014), the law may raise the costs of disregard for the environment on the part of polluters (e.g., through shareholder activism or customer retaliation), further favoring cooperative solutions. While the environmental crisis is a political problem (Cole 1992)—not an economic or a legal one—we propose that legal change can, by altering the rules of the game, participate in shaping its outcome.

## References


Bélanger, M., & Halley, P. (2017). Accès à la justice pour protéger l'environnement au Québec: Réflexions sur la capacité à agir des particuliers et des groupes environnementaux. *McGill Law Journal, 62*(3), 603–632.

Berg, J. (2007). Elephants and embryos: A proposed framework for legal personhood, *Hastings Law Journal, 59*, 369–406.

Berry, T. (2010). *Evening thoughts*. Berkeley, CA: Counterpoint.

Cole, L. W. (1992). Empowerment as the key to environmental protection: The need for environmental poverty law. *Ecology Law Quarterly, 19*, 619–683.

Dewey, J. (1926). The historic background of corporate legal personality. *Yale Law Journal, 35*(6), 655–673.

Fagundes, D. (2001). What we talk about when we talk about persons: The language of a legal fiction. *Harvard Law Review, 114* (6).

Fanokoa, P. S., Telahigue, I., & Zaccour, G. (2011). Buying cooperation in an asymmetric environmental differential game. *Journal of Economic Dynamics and Control, 35*(6), 935–946.

Feinberg, J. (1984). *The moral limits of the criminal law: Harm to others*. Oxford: Oxford University Press.

Fernandez, L. (2009). Wastewater pollution abatement across an international border. *Environment and Development Economics, 14*(1), 67–88.

Gordon, G. J. (2018). Environmental personhood. *Columbia Journal of Environmental Law, 43*(1), 49–91.

Government of Canada. (2018). *Water in Canada*. Retrieved August 20, 2018, from https://www.canada.ca/en/environment-climate-change/services/water-overview/publications/water-in-canada.html

Hardin, G. (1968). The tragedy of the commons. *Science, 162*, 1243–1248.

Heyes, A. (2000). Implementing environmental regulation: Enforcement and compliance. *Journal of Regulatory Economics, 17*(2), 107–129.

Hutchison, A. (2014). The Whanganui river as a legal person. *Alternative Law Journal, 39*(3), 179–182.

Jørgensen, S., & Zaccour, G. (2001). Time consistent side payments in a dynamic game of downstream pollution. *Journal of Economic Dynamics and Control, 25*(12), 1973–1987.

Kalai, E. (1977). Proportional solutions to bargaining situations: Intertemporal utility comparisons. *Econometrica, 45*(7), 1623–1630.

Kalai, E., & Smorodinsky, M. (1975). Other solutions to Nash's bargaining problem. *Econometrica, 43*(3), 513–518.

Kim, S. H. (2015). Time to come clean? Disclosure and inspection policies for green production. *Operations Research, 63*(1), 1–20.

Livingston, J. A. (2004). John A. Livingston. In J. Derrick (Ed.), *Listening to the land: Conversations about nature, culture, and eros* (pp. 53–66). White River Junction, VT: Chelsea Green Publishing.

Nash, J. (1950). The bargaining problem. *Econometrica, 18*(2), 155–162.

O'Donnell, E., & Talbot-Jones, J. (2018). Creating legal rights for rivers: Lessons from Australia, New Zealand, and India. *Ecology and Society, 23*(1), 7.

Russell, C. (2001). Monitoring, enforcement, and the choice of environmental policy instruments. *Regional Environmental Change, 2*(2), 73–76.

Shelton, D. (2015). Nature as a legal person. *VertigO–la revue électronique en sciences de l'environnement*, Hors-série 22.

Shi, G. M., Wang, J. N., Zhang, B., Zhang, Z., & Zhang, Y. L. (2016). Pollution control costs of a transboundary river basin: Empirical tests of the fairness and stability of cost allocation mechanisms using game theory. *Journal of Environmental Management, 177*, 145–152.

Sigman, H. (2002). International spillovers and water quality in rivers: Do countries free ride? *American Economic Review, 92*(4), 1152–1159.

Sparovek, G., Berndes, G., Klug, I. L., & Barretto, A. G. (2010). Brazilian agriculture and environmental legislation: Status and future challenges. *Environmental Science and Technology, 44*(16), 6046–6053.

Stone, C. D. (2010). *Should trees have standing?: Law, morality, and the environment*. Oxford: Oxford University Press.

Tribe, L. H. (1973). Ways not to think about plastic trees: New foundations for environmental law. *Yale Law Journal, 83*, 1315–1348.

Van Rooij, B. (2006). Implementation of Chinese environmental law: Regular enforcement and political campaigns. *Development and Change, 37*(1), 57–74.

Wang, H., Mamingi, N., Laplante, B., & Dasgupta, S. (2003). Incomplete enforcement of pollution regulation: Bargaining power of Chinese factories. *Environmental and Resource Economics, 24*(3), 245–262.

Weiss, E. B., & Jacobson, H. K. (Eds.). (2000). *Engaging countries: Strengthening compliance with international environmental accords*. Cambridge, MA: MIT Press.

Wise, S. M. (2013). Nonhuman rights to personhood. *Pace Environmental Law Review, 30*, 1278–1281.

World Health Organization. (2018). *Children's environmental health*. Retrieved August 20, 2018, from http://www.who.int/ceh/publications/factsheets/fs284/en/

# A Dynamic Game with Interaction Between Kantian Players and Nashian Players

**Ngo Van Long**

**Abstract** This paper defines the concept of feedback Kant-Nash equilibrium for a discrete-time model of resource exploitation by infinitely-lived Kantian and Nashian players, where we define Kantian agents as those who act in accordance with the categorical imperative. We revisit a well-known dynamic model of the tragedy of the commons and ask what would happen if not all agents are solely motivated by self interest. We establish that even without external punishment of violation of social norms, if a sufficiently large fraction of the population consists of Kantian agents, the tragedy of the commons can be substantially mitigated.

**Keywords** Kantian equilibrium · Rule of behavior · Categorical imperative

## 1 Introduction

Even though the theory of the tragedy of the commons (Gordon 1954; Hardin 1968) has issued a stern warning against the regime of resource management under common access, economists have become increasingly acquainted with *the Ostrom facts*: many communities have been able to manage their common property resources in a sustainable way (Ostrom 1990). The key mechanism behind these successful communities is the operation of social norms. There are a number of dynamic models of common property resources where some subset of agents observe social norms. This literature includes the interesting contributions of Sethi and Somanathan (1996) and Breton et al. (2010). The former paper assumes that agents are myopic, while the latter paper considers far-sighted agents. A common feature of models with social norms is that some subset of agents is endowed with the propensity to punish community members who violate norms.

N. V. Long (✉)
McGill University, Montréal, QC, Canada
e-mail: ngo.long@mcgill.ca

 
P.-O. Pineau et al. (eds.), *Games in Management Science*,
International Series in Operations Research & Management Science 280,
https://doi.org/10.1007/978-3-030-19107-8_14

This paper takes a different approach. We introduce into a model of common property resource a subset of players called Kantian agents and we enquire whether even without punishment against violation, a society that has a sufficiently large number of Kantians can attenuate the tragedy of the commons. For this purpose, we define the concept of feedback Kant-Nash equilibrium in a discrete-time model of resource exploitation by infinitely-lived Kantian and Nashian players.

Our research question is: In a dynamic game of exploitation of a common property resource, does the presence of a group of Kantian agents lead to a higher steady-state welfare and environmental quality? Using an adaptation of the fish-war model of Levhari and Mirman (1980), we show that if Kantian agents constitute a large share of the population, the resource stock can attain a steady state that is sufficiently close to the social optimal.

## 2 A Brief Review of the Literature on the Role of Morality and Kantian Behavior in Economics

The words "Kantian economics" first appeared in the title of an influential paper by Laffont (1975). He asks "Why is it that (at least in some countries) people do not leave their beer cans on the beaches?" This question is difficult to answer using the Standard Model of Economic Behavior. The impact one's own "welfare" from leaving one's beer cans on the beach is certainly negligible while the effort to properly dispose of them is not. Yet many people would make the required effort. Laffont's explanation is very simple, yet compelling: "Every economic action takes place in the framework of a moral or ethics." He refers to Kant's categorical imperative. Kant wrote that "There is only one categorical imperative, and it is this: Act only on the maxim by which you can at the same time will that it should become a universal law" (Kant 1785; translated by Hill and Zweig 2002, p. 222). Other eminent economists have also alluded to Kantian behavior in economics (Arrow 1973; Sen 1977).

Many economists have pointed out that the standard axiom of homo œconomicus is clearly inadequate to explain economic behavior. In fact, as Smith (2003, p. 465) pointed out, "the values to which people respond are not confined to those… based on the narrowly defined canons of rationality." This quoted sentence has its roots in the work of Smith (1790), where the role of natural sympathies in human activities was discussed at length.[1] Smith (2003, p. 466) elaborates on this points[2]:

> Research in economic psychology has prominently reported examples where "fairness" considerations are said to contradict the rationality assumptions of the standard socioeconomic science model. But experimental economics have reported mixed results on rationality: people are often better (e.g., in two-person anonymous interactions), in agree-

[1] Smith (2003) emphasizes these roots and the importance of Adam Smith's moral philosophy.

[2] I thank a reviewer for drawing my attention to the article of Smith (2003), and the relevant quote.

> ment with (e.g., in flow supply and demand markets), or worse (e.g., in asset trading), in achieving gains for themselves and others than is predicted by rational analysis. Patterns of these contradictions and confirmations provide important clues to the implicit rules or norms that people may follow, and can motivate new theoretical hypotheses for examination in both the field and the laboratory. The pattern of results greatly modifies the prevailing, and I believe misguided, rational SSSM, and richly modernizes the unadulterated message of the Scottish philosophers.[3]

For the analysis of certain economic activities, Roemer (2010, 2015) has proposed a useful mathematical formulation of the Kantian rule of behavior, admitting the possibilities that agents have different cost functions or profit functions. This formulation may be briefly described as follows. Consider an activity that yields negative or positive externalities, such as playing loud music, or keeping the side walk in front of your house clean and safe. Roemer suggests that, as a Kantian, your current activity level $x > 0$ is morally appropriate if and only if any scaling up or scaling down of that activity level by a factor $\lambda \neq 1$ would make you worse off, were *everyone else* to scale up or down their activity levels by the same proportion. Clearly, Kantian agents are not optimizing in the standard economic sense. They are acting according to a moral norm. As Roemer (2015) puts it, a Kantian agent would explain her behavior as follows:

> I hold a norm that says: "If I want to deviate from a contemplated action profile (of my community's members), then I may do so only if I would have all others deviate in like manner." (Roemer 2015, p. 46)

Is such a behavioral rule rational? Harsanyi (1980) gives an affirmative answer. It is as if socially responsible individuals made a rational commitment to a comprehensive joint strategy. According to Harsanyi (1980, p. 130), "behavior based on a rational commitment must be classified as truly rational behavior." Harsanyi's concept of rule-utilitarianism (1980) is similar in spirit to Roemer's concept of Kantian equilibrium, even though in philosophy the Kantian doctrine is opposed to the consequentialism that utilitarians advocate (Russell 1945).

In Laffont (1975) and Roemer (2010, 2015), all individuals are Kantians. This assumption must be relaxed in order to model real-world situations, where Kantians and Non-Kantians interact. Papers dealing with such issues include Long (2016, 2017) and Grafton et al. (2017). The present paper belongs to this stream of literature. Its main contribution is to provide an analysis of Kant-Nash equilibrium in a discrete-time framework, where agents use feedback strategies. Specifically, we use here the concept of a generalized Kant-Nash equilibrium. This concept was defined in Long (2017) so that the two extreme cases (called exclusive Kant-Nash equilibrium and inclusive Kant-Nash equilibrium) are special cases of this more general concept. While this paper is not a place for a detailed philosophical discussion, we feel it necessary to expand a bit more on these concepts.

The Kantian categorical imperative (CI for short), "Act only on the maxim by which you can at the same time will that it should become a universal law," seems

[3]For a fully articulate exposition of Adam Smith's philosophical views, see Muller (1993).

to suppose that one should do what one would wish everyone else to do. Therefore, it could be argued that the most fundamental property of the CI is its universality. The demand for universality is consistent with the notion of "inclusive Kantians" introduced in Long (2017), where Kantians test the appropriateness of a proposed action level by asking themselves: "what would the world be like if every human being would deviate from this action level in the same way?" (Please refer to Roemer's concept of scaling up, or scaling down an activity level by a scalar $\lambda > 0$, as mentioned above.)

At the same time, from a practical viewpoint, it would seem more realistic to ask: "what would this community (at this time and this place) be like if all members of the community were to deviate from the proposed action level in the same way?" In asking this question (and bearing in mind that the words "this community" are not unambiguous) it seems that certain subset of humanity or of the current society is being excluded from consideration. This practical argument seems to be in line with the notion of "exclusive Kantians" which was mentioned in Long (2017).

If one agrees that both notions of "inclusive Kantians" and "exclusive Kantians" have certain merit (depending on the scope of application), it would seem natural to encompass both notions in a generalized formulation. Thus, Long (2017) proposes the concept of a generalized Kant-Nash equilibrium, in which Kantians would ask themselves the following question: If I were to deviate from the proposed action level $x$ by scaling it up or down by a factor $\lambda > 0$, what would this community (at this place and this time) be like if some members of society would deviate by the same factor $\lambda$, while other members would deviate by a factor $\mu$, where $\mu = (\lambda - 1)\tau + 1$? Clearly, if $\tau = 0$, this means that these members were supposed to stay put (the exclusive case), and if $\tau = 1$, all members are included in the thought experiment (the inclusive case). Then by restricting $\tau$ to be in the interval [0, 1], the generalized Kant-Nash equilibrium admits the exclusive Kant-Nash equilibrium and the inclusive Kant-Nash equilibrium as special cases.[4] While this formulation clearly departs from the pure Kantian doctrine, it seems that one can find some partial support among moral philosophers for not adhering to the pure Kantian doctrine. The following paragraph from Johnson and Cureton (2018) may shed some light on this issue[5]:

> All specific moral requirements, according to Kant, are justified by this principle, which means that all immoral actions are irrational because they violate the CI. Other philosophers, such as Hobbes, Locke, and Aquinas, had also argued that moral requirements are based on standards of rationality. However, these standards are either instrumental principles of rationality for satisfying one's desires, as in Hobbes, or external rational principles that are discoverable by reason, as in Locke and Aquinas. Kant agreed with many of his

[4]A reviewer rightly points out that there is an issue about observability. How does a Kantian know who is a Nashian and who is a Kantian? A partial reply to this criticism would be that, in a model of common property resource exploitation, a Kantian needs to only know the population share of Nashians. In this simple model, there is no need to know if a specific individual one meets is Kantian or Nashian.

[5]I thank a reviewer for this quote.

predecessors that an analysis of practical reason reveals the requirement that rational agents must conform to instrumental principles. Yet he also argued that conformity to the CI (a non-instrumental principle), and hence to moral requirements themselves, can nevertheless be shown to be essential to rational agency.

## 3 A Dynamic Game with Kantian and Nashian Players

Consider a community consisting of $m$ infinitely-lived individuals. Let $M = \{1, 2, \ldots, m\}$ denote the set of individuals. Assume that a subset $K$ of these individuals behave according to the Kantian norm. Without loss, let $K = \{1, 2, \ldots, k\}$. The complementary set, denoted by $N = \{k+1, k+2, \ldots, k+n\}$, where $n = m - k$, consists of members that behave in a Nashian fashion.

Let $S_t$ denote the stock of a natural asset (e.g., $S_t$ is the biomass in the community's fishing ground). Let $Q_t$ denote the community's aggregate exploitation from the biomass for consumption, where $Q_t \leq S_t$. The dynamics of the biomass is given by

$$S_{t+1} = F(S_t, Q_t)$$

where $F_S > 0$ and $F_Q < 0$.

Let $x_{it}$ denote the resource exploitation effort by Kantian agent $i$ in period $t$ and $y_{jt}$ the exploitation effort by Nashian agent $j$ in period $t$. Define

$$X_t = \sum_{i \in K} x_{it}, X_{-i,t} = X - x_{it}, Y_t = \sum_{j \in N} y_{jt}, Y_{-j,t} = Y_t - y_{jt}$$

and

$$Q_t = X_t + Y_t$$

The utility level of Kantian agent $i$ in period $t$ is $u_i(x_{it})$, where $u_i(.)$ is a strictly concave and increasing function. Furthermore, we assume that

$$\lim_{x_{it} \to 0} u'(x_{it}) = \infty$$

This ensures that the agent always wants to achieve a strictly positive level of consumption, as long as $S_t > 0$. The same assumption is made for the utility function $u_j(.)$ of Nashian agents.

At each date $z = 1, 2, 3, \ldots$, the Nashian agent $j$ seeks to maximize her remaining life-time payoff starting from time $z$ (denoted by $\Omega_{jz}$), where

$$\Omega_{jz} = \sum_{t=z}^{\infty} \beta^{t-z} u_j(y_{jt})$$

where $\beta \in (0, 1)$ is the discount factor. In solving her problem, she takes as given (her conjectures of) the feedback extraction rules $\psi_h(.)$ of all other Nashian agents $h \in N - \{j\}$, where

$$y_{ht} = \psi_h(S_t)$$

and the feedback extraction rules $\theta_f(.)$ of Kantian agents $f \in K$, where

$$x_{ft} = \theta_f(S_t)$$

(We assume that their conjectures are correct). Her optimal solution must satisfy the Bellman equation

$$V_{Nj}(S_t) = \max_{y_{jt}} \left\{ u_j(y_{jt}) + \beta V_{Nj}(S_{t+1}) \right\} \tag{1}$$

where $V_{Nj}(S)$ is her value function, and

$$S_{t+1} = F\left(S_t, y_{jt} + \sum_{h \in N-\{j\}} \psi_h(S_t) + \sum_{f \in K} \theta_f(S_t)\right)$$

Kantian agents behave differently. In deciding whether she should choose an exploitation level $x_{it}^* > 0$ or a different level, a Kantian agent $i$ would ask herself the following question: If I deviate from $x_{it}^*$ by choosing some $x_{it} = \lambda x_{it}^*$, where $\lambda > 0$ and $\lambda \neq 1$, what would happen to my payoff, *assuming all other Kantians would deviate in the same way*?[6] Then $x_{it}^*$ is her correct action level if and only if any $\lambda \neq 1$ would result in a lower life-time payoff. That is, $x_{it}^*$ must satisfy the following condition:

$$1 = \arg\max_{\lambda} \left\{ u_i(\lambda x_{it}^*) + \beta V_{Ki}(S_{t+1}(\lambda)) \right\} \tag{2}$$

where $V_{Ki}(S)$ is her value function, and

$$S_{t+1}(\lambda) \equiv F\left(S_t, \lambda x_{it}^* + \sum_{h \in N} \psi_h(S_t) + \sum_{f \in K-\{i\}} \lambda \theta_f(S_t)\right)$$

[6]In this section, for the sake of expositional simplicity, we are assuming that Kantians are exclusive, in the sense explained at the end of Sect. 2. In the next section, we will consider a slightly more general hypothesis about the Kantians.

Condition (2) yields the Kantian choice of exploitation level, $x_{it}^* = \theta_i(S_t)$. Then the following equation holds for Kantians:

$$V_{Ki}(S_t) = u_i(\phi_i(S_t)) + \beta V_{Ki}(S_{t+1}) \tag{3}$$

A Kant-Nash equilibrium is a strategy profile $(\theta_1, \ldots, \theta_k, \psi_{k+1}, \ldots, \psi_{k+n})$ that satisfies Eqs. (1), (3), such that the action $x_{it}^* = \theta_i(S_t)$ satisfies the Kantian rule (2), and usual transversality conditions hold.

## 4 An Application: Kant-Nash Equilibrium in a Modified Levhari-Mirman Model

In this section, we apply the concept of Kant-Nash equilibrium to the Levhari-Mirman model of fishery (Levhari and Mirman 1980). We consider a slightly more general version of the Kantian behavior rule, using the concept of generalized Kant-Nash equilibrium explained in Sect. 2. We assume that a Kantian agent would use the following test to determine her extraction level.

The test for the appropriateness of an action level $x_i^*$ that each Kantian agent must carry out consists of asking herself the following question:

*If I were to scale up or scale down of my effort level by any non-negative factor $\lambda \neq 1$, and if all other Kantian agents in the community, $j \in K - \{i\}$, were to scale up or down their effort levels by the same factor, while the Nashian agents were to scale up or down their effort levels by a factor $\mu(\lambda)$, would my utility level be (weakly) lower?*

In the definition of a Kant-Nash equilibrium that we adopted in Sect. 3, $\mu(\lambda) = 1$ identically. In this section, we allow $\mu(\lambda)$ to be different from unity. This means that Kantians consider Non-Kantians as members of the community. How should $\mu(\lambda)$ be specified? It seems sensible to suppose that $\mu(\lambda) \neq \lambda$ if and only if $\lambda \neq 1$. An operational specification would be to introduce a parameter $\tau$, such that

$$\mu(\lambda) = (\lambda - 1)\tau + 1 \text{ where } 0 \leq \tau \leq 1 \tag{4}$$

so that $\mu'(\lambda) = \tau \leq 1$. This means that if $\lambda = 1$ (neither scaling up nor down) then $\mu = 1$ too; if $\lambda > 1$, then $\mu(\lambda) \geq 1$, and $\mu(\lambda) \leq \lambda$; and if $\lambda < 1$ (scaling down), then $1 \geq \mu(\lambda) \geq \lambda$. The resulting equilibrium may be called a *generalized Kant-Nash Equilibrium*. The parameter $\tau$ may be called the Kantian's *degree of inclusiveness*.

Let $S \in [0, 1]$ be the state variable representing a natural asset at the beginning of the current period. The highest value that $S$ can take is 1. Let $S'$ be the value of $S$ at the beginning of the next period. Extraction in any period is bounded above by the stock level, i.e., $Q_t \leq S_t$. Following Levhari and Mirman (1980), we assume that

$$S' = (S - Q)^{\alpha} \text{ where } 0 < \alpha < 1.$$

We assume the function $u(.)$ is logarithmic, thus $u(x_i) = \ln x_i$. Furthermore, there is a scrap value function

$$Z(S) = \ln(\gamma S), \text{ where } \gamma \geq 0.$$

In their paper, Levhari and Mirman (1980) derived the value function for their infinite horizon game by solving finite-horizon games, and taking the limit as the horizon tends to infinity. We will adopt the same solution procedure for our game.

## 4.1 Solution for the One-Period Horizon Game

Since all Nashians are identical, and all Kantians behave identically, we will focus on the symmetric generalized Kant-Nash equilibrium. In the one-period-horizon game, each Nashian agent $j$ chooses $y_j$ to maximize

$$\ln y_j + g \ln\left(S - Q_{-j} - y_j\right) + \beta\left[\ln\gamma + \alpha\ln(S - Q_{-j} - y_j)\right]$$

where $0 < \beta < 1$ is the discount factor. Each Kantian agent $i$ is in equilibrium if and only if

$$1 = \arg\max_{\lambda}\{\ln\lambda x_i + g\ln\left[S - n\mu(\lambda)y - k\lambda x_i\right]$$
$$\beta\ln\gamma + \alpha\beta\ln\left[S - n\mu(\lambda)y - k\lambda x_i\right]\}$$

where $\mu(\lambda) = (\lambda - 1)\tau + 1$. In this section, we set the parameter $g$ to be equal to zero. In Sect. 5, this parameter will take on a positive value, to reflect preferences for amenity services.

To ensure the existence of a generalized Kant-Nash equilibrium for this specific fishery model, we make the following assumption:

**Assumption A1** $1 - \tau(m - k) > 0$.

To satisfy this assumption, we must rule out the case where $\tau = 1$ and $n \geq 1$. In other words, if there is at least one Nashian, and if Kantians are inclusive (they set $\tau = 1$ in their test), then in this specific fishery model, there does not exist an equilibrium. The intuition is as follows. If all agents are Kantians ($k = m$), then of course an equilibrium exists: it is the cooperative solution. But as soon as an agent changes her moral attitude (i.e., becoming a Nashian), she would want to increase her fish harvest, and the remaining $m - 1$ inclusive Kantians (with $\tau = 1$) would react by catching less, which would unfortunately induce the Nashian to catch more, and so on, and this process does not converge to an equilibrium.[7]

Under Assumption A1, there exists a unique generalized Kant-Nash equilibrium for the one-period-horizon game. The equilibrium extraction levels of Nashian and Kantian agents are, respectively,

$$y = \frac{S}{(m-k)(1-\tau)+(1+b)}, b \equiv \alpha\beta \tag{5}$$

$$x = \left(\frac{1-\tau(m-k)}{k}\right)\left(\frac{S}{(m-k)(1-\tau)+(1+b)}\right) \tag{6}$$

For the one-period-horizon game, the equilibrium payoff function of a representative Nashian is denoted by $V_N^{(1)}$, where the superscript indicates that there is only one period to go. Then, using (1), (5), and (6), we obtain

$$V_N^{(1)}(S) = (1+b)\ln S + \eta_N^{(1)} + \phi^{(1)} + \beta \ln \gamma \tag{7}$$

where

$$\eta_N^{(1)} = \ln\left(\frac{1}{(m-k)(1-\tau)+(1+b)}\right)$$

$$\phi^{(1)} = b\ln\left(\frac{b}{(m-k)(1-\tau)+(1+b)}\right)$$

Note that $\phi^{(1)}$ does not have a subscript because this term is the same for Nashian and Kantian players. For Kantians, the equilibrium payoff function is obtained in a similar fashion, using (3), (5), and (6):

$$V_K^{(1)}(S) = (1+b)\ln S + \eta_K^{(1)} + \phi^{(1)} + \beta \ln \gamma \tag{8}$$

where it can be shown that

$$\eta_K^{(1)} = \ln\left(\frac{(1-\tau(m-k))k^{-1}}{(m-k)(1-\tau)+(1+b)}\right)$$

We observe that Nashians achieve higher payoffs than Kantians. The difference between the payoff is

$$V_N^{(1)}(S) - V_K^{(1)}(S) = \eta_N^{(1)} - \eta_K^{(1)} = \ln\left(\frac{k}{1-\tau(m-k)}\right) > 0$$

[7] I am indebted to a reviewer for raising this pertinent issue.

## 4.2 Solution for the Two-Period-Horizon Game

Now, consider the game where all agents have two periods to go. All agents know their equilibrium payoffs of the one-period-to-go subgame: they are given by Eqs. (7) and (8). Then, given the opening stock $S$, the Nashian agent $i$ chooses the current period extraction level $y_i$ to maximize

$$R_i^{(2)} = u(y_i) + \beta V_{Ni}^{(1)}(S')$$

And Kantians will be in equilibrium if and only if

$$1 = \arg\max_{\lambda} u(\lambda x) + \beta V_{Ki}^{(1)}(S')$$

Thus, if $T = 2$, the Nashian agent's equilibrium exploitation in the first period when there are two periods to go is

$$y_N^{(2)} = \frac{S}{(m-k)(1-\tau) + (1+b) + b^2}$$

and, for Kantians, their equilibrium exploitation is only a fraction of the Nashian agent's exploitation:

$$x_K^{(2)} = \left(\frac{1-\tau(m-k)}{k}\right) y_N^{(2)}$$

The equilibrium payoff functions are as follows. For each Nashian,

$$V_N^{(2)}(S) = (1+b+b^2)\ln S + A_N^{(2)} + B^{(2)} + \beta^2 \ln \gamma$$

with

$$A_N^{(2)} = \eta_N^{(2)} + \beta\eta_N^{(1)}, \text{with} \eta_N^{(2)} \equiv \ln\left(\frac{1}{(m-k)(1-\tau) + 1 + b + b^2}\right)$$

and

$$B^{(2)} = \phi^{(2)} + \beta\phi^{(1)}$$

where

$$\phi_N^{(2)} \equiv (1+b+b^2-1)\ln\left(\frac{1+b+b^2-1}{(m-k)(1-\tau) + 1 + b + b^2}\right)$$

For each Kantian, the equilibrium payoff function is

$$V_K^{(2)}(S) = (1 + b + b^2)\ln S + A_K^{(2)} + B^{(2)} + \beta^2 \ln \gamma$$

where

$$A_K^{(2)} = \eta_K^{(2)} + \beta \eta_K^{(1)}$$

$$\eta_K^{(2)} = \ln\left(\frac{(1 - \tau(m-k))k^{-1}}{(m-k)(1-\tau) + 1 + b + b^2}\right)$$

Thus the Nashian payoff exceeds the Kantian payoff by

$$V_N^{(2)}(S) - V_K^{(2)}(S) = A_N^{(2)} - A_K^{(2)} = (1 + \beta)\ln\left(\frac{k}{1 - \tau(m-k)}\right)$$

In other words, the Kantian payoff is equal to the Nashian payoff minus $(1 + \beta)\ln[k/(1 - \tau(m-k))]$.

## 4.3 Solution for the q-Period-Horizon Game

Given the opening stock $S$, Nashian agent $i$ chooses the first period exploitation level $y_i$ to maximize

$$R_i^{(q)} = u(y_i) + \beta V_{Ni}^{(q-1)}(S')$$

And Kantians will be in equilibrium if and only if

$$1 = \arg\max_{\lambda} u(\lambda x) + \beta V_{Kj}^{(q-1)}(S')$$

For $T = q$, the Nashian agent's equilibrium first period exploitation level is

$$y_N^{(q)} = \frac{S}{(m-k)(1-\tau) + \left(\left(\sum_{s=0}^{q-1} b^s\right) + b^q\right)}$$

and the Kantian agent's exploitation in period 1 is

$$x_K^{(q)} = \left(\frac{1 - \tau(m-k)}{k}\right) y_N^{(q)}$$

The value function for Nashians is

$$V_N^{(q)}(S) = \left( \left( \sum_{s=0}^{q-1} b^s \right) + b^q \right) \ln x + A_N^{(q)} + B_N^{(q)} + \beta^q \gamma$$

with

$$A_N^{(q)} = \eta_N^{(q)} + \beta \eta_N^{(q-1)} + \beta^2 \eta_N^{(q-2)} + \dots \beta^{q-1} \eta_N^{(1)}$$

$$\eta_N^{(q)} \equiv \ln \left( \frac{1}{(m-k)(1-\tau) + \left( \sum_{s=0}^{q-1} b^s \right) + b^q} \right)$$

and

$$B^{(q)} = \phi^{(q)} + \beta \phi^{(q-1)} + \dots + \beta^{q-1} \phi^{(1)},$$

$$\phi^{(q)} \equiv \left( \left( \sum_{s=0}^{q-1} b^s \right) + b^q - 1 \right) \ln \left( \frac{\left( \sum_{s=0}^{q-1} b^s \right) + b^q - 1}{(m-k)(1-\tau) + \left( \sum_{s=0}^{q-1} b^s \right) + b^q} \right)$$

We can show the Kantian payoff is equal to the Nashian payoff minus $(1 + \beta + \beta^2 + \dots + \beta^{q-1}) \ln [k / (1 - \tau(m-k))]$

$$V_N^{(q)}(S) - V_K^{(q)}(S) = \ln \left( \frac{k}{1 - \tau(m-k)} \right)$$

Note that the difference is independent of $S$. This property is due to the logarithmic function. We conjecture that if we assume a different utility function the difference would depend on $S$. However, one would have to rely on numerical calculations, as it is probably impossible to find simple closed form solutions.

## *4.4 The Infinite-Horizon Problem*

Taking the limit as $q$ tends to infinity, we obtain the equilibrium strategies of Nashian and Kantian players for the infinite horizon problem. We find that the equilibrium strategies of the Nashians and the Kantians depend only on the current stock level, $S$, and are independent of the calendar time. For Nashians,

$$y = \frac{(1-b)S}{(m-k)(1-\tau)(1-b) + 1}$$

and for Kantians,

$$x = \left(\frac{1-\tau(m-k)}{k}\right)\frac{(1-b)S}{(m-k)(1-\tau)(1-b)+1}$$

The value function of the representative Nashian is

$$V_N(S) = \frac{1}{1-b}\ln S + \frac{1}{1-\beta}\ln\left[\frac{1-b}{(m-k)(1-\tau)(1-b)+1}\right]$$

$$+\frac{1}{1-\beta}\left(\frac{b}{1-b}\right)\ln\left(\frac{b}{(m-k)(1-\tau)(1-b)+1}\right)$$

and the value function of the representative Kantian is

$$V_K(S) = V_N(S) - \frac{1}{1-\beta}\ln\left(\frac{k}{1-\tau(m-k)}\right)$$

Along the equilibrium path,

$$S_{t+1} = \left(\frac{b}{(m-k)(1-\tau)(1-b)+1}\right)^{\alpha} S_t^{\alpha}$$

The steady state level of the stock is

$$S^* = \left(\frac{b}{(m-k)(1-\tau)(1-b)+1}\right)^{\frac{\alpha}{1-\alpha}}$$

It is easy to verify that the steady state is stable: starting at any positive $S_0$, the stock will converge to $S^*$.

Using the above analysis, we obtain the following results. (Detailed proofs are available upon request.)

**Proposition 1** *The Kant-Nash equilibrium in feedback strategies display the following properties.*

(a) *$V_N(S)$ is increasing in $\tau$ (in the Kantians' degree of inclusiveness) and in $k$ (the population share of Kantians).*
(b) *A sufficient condition for $V_K(S)$ to increase in $k$ is $\tau m - 1 \geq 0$.*
(c) *Assume that $\tau m - 1 > 0$, and that $k$ is sufficiently large such that $1 - \tau(m - k) > 0$. Then as $k$ increases from $k$ to $k + 1$ or higher values, the gap between $V_N(S) - V_K(S)$ becomes smaller.*
(d) *The steady state stock increases in $k$.*
(e) *The steady state stock increases in $\tau$ provided that Assumption A1 is satisfied.*
(f) *The pure Nash steady state stock level , i.e., when $n = m$, is smaller than the Kant-Nash steady state level if $k \geq 2$.*

## 5 Extension: The Case Where the Resource Yields Amenity Values

This section extends the model to the case where the resource has amenity values. Assume that members of the community enjoy a public good: the amenity services provided by the biomass. Assume that the amenity service level in period $t$ depends on both the stock level $S_t$ and the exploitation activities, $Q_t$

$$G_t = G(S_t, Q_t)$$

with $G_S > 0$ and $G_Q < 0$. The utility level of Kantian agent $i$ in period $t$ is

$$U(x_{it}, G_t) = u_i(x_{it}) + w_i(G_t)$$

The same assumption is made for the utility function of Nashian agents. Assume and the amenity service level is given by

$$G_t = G\left(S_t, y_{j\tau} + \sum_{h \in N-\{j\}} \psi_h(S_t) + \sum_{f \in K} \theta_f(S_t)\right)$$

We now modify the model of Levhari and Mirman (1980) to allow for the enjoyment of environmental quality (amenity services). The parameter for this enjoyment is denoted by $g \geq 0$. (In the model of Levhari and Mirman (1980), $g = 0$ identically, and there are no Kantian agents.)

The level of environmental services delivered to the agents during period $t$ is assumed to be

$$G_t = G(S_t, X_t) = S_t - Q_t$$

And we suppose that

$$U(x_i, G) = \ln x_i + g \ln G \text{ where } g > 0.$$

The equilibrium for the one-period game is similar to the one described in the preceding section, with only a minor modification, namely

$$y = \frac{S}{(m-k)(1-\tau) + (1+g+b)}, \quad b \equiv \alpha\beta$$

$$x = \left(\frac{1-\tau(m-k)}{k}\right)\left(\frac{S}{(m-k)(1-\tau) + (1+g+b)}\right)$$

$$V_N^{(1)}(S) = (1+g+b)\ln S + \eta_N^{(1)} + \phi^{(1)} + \beta \ln \gamma$$

$$\eta_N^{(1)} = \ln\left(\frac{1}{(m-k)(1-\tau)+(1+g+b)}\right)$$

$$\phi^{(1)} = (g+b)\ln\left(\frac{g+b}{(m-k)(1-\tau)+(1+g+b)}\right)$$

Similarly, for the two-period model, one makes only a few modifications, such as

$$R_i^{(2)} = U(y_i, G(S, Q_{-i}+y_i)) + \beta V_N^{(1)}(S')$$

Then, if $T = 2$, the Nashian agent's equilibrium exploitation in the first period when there are two periods to go is

$$y_N^{(2)} = \frac{S}{(m-k)(1-\tau)+(1+g)(1+b)+b^2}$$

and, for Kantians, their equilibrium exploitation is only a fraction of the Nashian agent's exploitation:

$$x_K^{(2)} = \left(\frac{1-\tau(m-k)}{k}\right) y_N^{(2)}$$

The equilibrium payoff functions are as follows. For each Nashian,

$$V_N^{(2)}(S) = ((1+g)(1+b)+b^2)\ln S + A_N^{(2)} + B^{(2)} + \beta^2 \ln\gamma$$

with

$$A_N^{(2)} = \eta_N^{(2)} + \beta\eta_N^{(1)}, \text{ with } \eta_N^{(2)} \equiv \ln\left(\frac{1}{(m-k)(1-\tau)+(1+g)(1+b)+b^2}\right)$$

and

$$B^{(2)} = \phi^{(2)} + \beta\phi^{(1)}$$

where

$$\phi_N^{(2)} \equiv ((1+g)(1+b)+b^2-1)\ln\left(\frac{(1+g)(1+b)+b^2-1}{(m-k)(1-\tau)+(1+g)(1+b)+b^2}\right)$$

For each Kantian, the equilibrium payoff function is

$$V_K^{(2)}(S) = ((1+g)(1+b)+b^2)\ln S + A_K^{(2)} + B^{(2)} + \beta^2 \ln\gamma$$

where

$$A_K^{(2)} = \eta_K^{(2)} + \beta \eta_K^{(1)}$$

$$\eta_K^{(2)} = \ln\left(\frac{(1-\tau(m-k))k^{-1}}{(m-k)(1-\tau)+(1+g)(1+b)+b^2}\right)$$

Thus the Nashian payoff exceeds the Kantian payoff by

$$V_N^{(2)}(S) - V_K^{(2)}(S) = A_N^{(2)} - A_K^{(2)} = (1+\beta)\ln\left(\frac{k}{1-\tau(m-k)}\right)$$

In other words, the Kantian payoff is equal to the Nashian payoff minus $(1+\beta)\ln[k/(1-\tau(m-k))]$.

For $T = q$, the Nashian agent's equilibrium first period exploitation level is

$$y_N^{(q)} = \frac{S}{(m-k)(1-\tau)+\left(\left(\sum_{s=0}^{q-1} b^s\right)+b^q\right)}$$

and the Kantian agent's exploitation in period 1 is

$$x_K^{(q)} = \left(\frac{1-\tau(m-k)}{k}\right) y_N^{(q)}$$

The value function for Nashians is

$$V_N^{(q)}(S) = \left((1+g)\left(\sum_{s=0}^{q-1} b^s\right) + b^q\right)\ln x + A_N^{(q)} + B_N^{(q)} + \beta^q \gamma$$

with

$$A_N^{(q)} = \eta_N^{(q)} + \beta\eta_N^{(q-1)} + \beta^2\eta_N^{(q-2)} + \ldots \beta^{q-1}\eta_N^{(1)}$$

$$\eta_N^{(q)} \equiv \ln\left(\frac{1}{(m-k)(1-\tau)+(1+g)\left(\sum_{s=0}^{q-1} b^s\right)+b^q}\right)$$

and

$$B^{(q)} = \phi^{(q)} + \beta\phi^{(q-1)} + \ldots + \beta^{q-1}\phi^{(1)},$$

$$\phi^{(q)} \equiv \left((1+g)\left(\sum_{s=0}^{q-1} b^s\right) + b^q - 1\right)\ln\left(\frac{(1+g)\left(\sum_{s=0}^{q-1} b^s\right)+b^q-1}{(m-k)(1-\tau)+(1+g)\left(\sum_{s=0}^{q-1} b^s\right)+b^q}\right)$$

The Kantian payoff is equal to the Nashian payoff minus $(1 + \beta + \beta^2 + \ldots + \beta^{q-1}) \ln [k/(1 - \tau(m - k))]$

$$V_N^{(q)}(S) - V_K^{(q)}(S) = \ln\left(\frac{k}{1 - \tau(m - k)}\right)$$

By taking the limit as $q$ tends to infinity, we can obtain the equilibrium strategies of Nashian and Kantian players for the infinite horizon problem. The equilibrium strategies depend only on $S$. For Nashians,

$$y = \frac{(1 - b)S}{(m - k)(1 - \tau)(1 - b) + (1 + g)}$$

and for Kantians,

$$x = \left(\frac{1 - \tau(m - k)}{k}\right) \frac{(1 - b)S}{(m - k)(1 - \tau)(1 - b) + (1 + g)}$$

The value function of the representative Nashian is

$$V_N(S) = (1 + g)\frac{1}{1 - b} \ln S + \frac{1}{1 - \beta} \ln\left[\frac{1 - b}{(m - k)(1 - \tau)(1 - b) + (1 + g)}\right]$$

$$+ \frac{1}{1 - \beta}\left(\frac{g + b}{1 - b}\right) \ln\left(\frac{g + b}{(m - k)(1 - \tau)(1 - b) + (1 + g)}\right)$$

and that of the representative Kantian is

$$V_K(S) = V_N(S) - \frac{1}{1 - \beta} \ln\left(\frac{k}{1 - \tau(m - k)}\right)$$

Along the equilibrium path,

$$S_{t+1} = \left(\frac{g + b}{(m - k)(1 - \tau)(1 - b) + (1 + g)}\right)^{\alpha} S_t^{\alpha}$$

The steady state level of the stock is

$$S^* = \left(\frac{g + b}{(m - k)(1 - \tau)(1 - b) + (1 + g)}\right)^{\frac{\alpha}{1-\alpha}}$$

Starting at any positive $S_0$, the stock will converge to $S^*$.

We obtain the following result:

**Proposition 2** *The Kant-Nash equilibrium in feedback strategies display all the properties stated in Proposition 1, and the following additional property: Regardless of the sign of $\tau m - 1$, if $g$ is sufficiently great, then an increase in $k$ will increase social welfare.*

## 6 Concluding Remarks

The idea that pro-socialness can help attenuate the tragedy of the commons has a long history. One can find it discussed in the works of Smith (1790), Gordon (1954), Laffont (1975), Ostrom (1990), Roemer (2010, 2015), and many others. Most of these discussions have been set in a static framework. Our contribution is twofold: First, we formalize the concept of interaction between Kantian agents and Nashian agents. Second, we apply the concept of Kant-Nash equilibrium to a dynamic game and show how it may shed light on games of common property resource exploitation when not all agents are Nashian. We have been able to show that social welfare increases with the Kantian population share: Given the total population, as the percentage share of the Kantians increases, social welfare increases as a result. It is hoped that our discussions of ethics could go some way to de-emphasize the 'homo œconomicus' conception of human behavior taught in standard economics courses. Of course, we must be aware of Arrow's caution: "One must not expect miraculous transformations in human behavior. Ethical codes, if they are viable, should be limited in scope." (Arrow 1973, p. 316).

An interesting idea for future research is the study of evolutionary dynamics toward a Kantian society.[8] According to Clément et al. (2000), the process of achieving universal justice is far from being straightforward for Kant (1795). It would necessitate the establishment of a Society of Nations, in other words a global social contract:

> Kant asserts at the same time that the future of our species is ultimately the rule of law and universal peace, and that, nevertheless, the establishment of public justice - the greatest problem for the human species, the most difficult one - can never be considered as a settled affair, and only the establishment of a "society of nations" subject to international law will allow man access to peace and the rule of law (the condition for true autonomy) and truly overcome his original savagery.[9]

[8]A small step in this direction has been taken in Long (2018), using an overlapping generation models in which parents have incentives to transmit prosocialness to their children.

[9]The above paragraph is translated from the French text: "Kant affirme à la fois que le devenir de notre espèce a pour finalité le règne de la loi et la paix universelle, et que, pourtant, l'établissement de la justice publique- le "plus grand problème pour l'espèce humaine, le plus difficile- ne peut jamais être considéré comme une affiare réglée. Seule l'établissement d'une "société des nations" soumise à une législation internationale, permetra à l'homme d'accéder à la paix et à l'ordre juridique (condition de toute véritable autonomie) et de surmonter véritablement sa sauvagerie originelle." I thank a reviewer for this reference.

# References


Arrow, K. J. (1973). Social responsibility and economic efficiency. *Public Policy, 21*, 303–317.

Breton, M., Sbragia, L., & Zaccour, G. (2010). A dynamic model of international environmental agreements. *Environmental and Resource Economics, 45*, 25–48.

Clément, E., Demonque, C., Hansen-Love, L., & Kahn, P. (2000). *La Philosophie de A à Z*. Paris: Hatier.

Gordon, S. (1954). The economic theory of a common property resource: The fishery. *Journal of Political Economy, 62*(2), 124–142.

Grafton, R. Q., Kompas, T., & Long, N. V. (2017). A brave new world? Kantian-Nashian interactions and the dynamics of global climate change mitigation. *European Economic Review, 99*, 31–42.

Hardin, G. (1968). The tragedy of the commons. *Science, 162*, 1243–1248.

Harsanyi, J. (1980). Rule utilitarianism, rights, obligations, and the theory of rational behavior. *Theory and Decision, 12*(2), 115–133.

Hill, T. E. Jr., & Zweig, A. (2002). *Groundwork for the metaphysics of morals*. Oxford: Oxford University Press.

Johnson, R., & Cureton, A. (2018). Kant's moral philosophy. In *The Stanford encyclopedia of philosophy*. Palo Alto: Stanford University Press.

Kant, I. (1785). *Grundlegung zur Metaphysik der Sitten*. Translated by Mary J. Gregor. (1998). *Groundwork of the metaphysics of morals*. Cambridge: Cambridge University Press.

Kant, I. (1795). *Zum ewigen Frieden. Ein philosophischer Entwurf*. Perpetual peace: A philosophical sketch. Downloaded from mtholyoke.edu. Accessed 20 December 2018.

Laffont, J.-J. (1975). Macroeconomic constraints, economic efficiency and ethics: An introduction to Kantian economics. *Economica, 42*, 430–437.

Levhari, D., & Mirman, L. (1980). The great fish wars: An example of using a dynamic Nash-Cournot solution. *Bell Journal of Economics, 11*, 322–334.

Long, N. V. (2016). Kant-Nash equilibrium in a quantity-setting oligopoly. In P. von Mouche & F. Quartieri (Eds.), *Equilibrium theory for Cournot oligopolies and related games: Essays in honour of Koji Okuguchi* (pp. 179–202). New York: Springer.

Long, N. V. (2017). Mixed-strategy Kant-Nash equilibrium in games of voluntary contributions to a public good. In W. Buchholz & D. Ruebbelke (Eds.), *The theory of externalities and public goods* (pp. 107–126). New York, Springer.

Long, N. V. (2018). Warm glow and the transmission of pro-socialness across generations. *Journal of Public Economic Theory* (in press). https://doi.org/10.1111/jpet.12347.

Muller, J. Z. (1993). *Adam Smith in his time and ours: Designing the decent society*. New York: The Free Press.

Ostrom, E. (1990). *Governing the commons: The evolution of institutions for collective actions*. Cambridge: Cambridge University Press.

Roemer, J. E. (2010). Kantian equilibrium. *Scandinavian Journal of Economics, 112*(1), 1–24.

Roemer, J. E. (2015). Kantian optimization: A microfoundation for cooperation. *Journal of Public Economics, 127*, 45–57.

Russell, B. (1945). *A history of western philosophy*. London: George Allen and Unwin.

Sen, A. (1977). Rational fools: A critique of the behavioral foundation of utility. *Philosophy and Public Affairs, 6*(4), 317–344.

Sethi, R., & Somanathan, E. (1996). The evolution of social norms in common property resource use. *American Economic Review, 86*(4), 766–788.

Smith, A. (1790 [2002]). *The Theory of Moral Sentiments* (revised edition, 1790). Edinburg: T. Creech and J. Bell & Co. Republished in the series Cambridge Texts in the History of Philosophy, edited by Knud Haakonssen, Cambridge University Press, 2002.

Smith, V. L. (2003). Constructivist and ecological rationality in economics. *American Economic Review, 93*(3), 456–508.

# Frugals, Militants and the Oil Market

**Etienne Billette de Villemeur and Pierre-Olivier Pineau**

**Abstract** The oil market has often been modeled as an oligopoly where the strategic players are producers. With climate change, a new sort of game appeared, where environmental militants play a significant role by opposing some projects, to contain oil production. At the same time, consumers continue to use increasing amounts of oil, independently of oil price fluctuations. Should we oppose oil projects, reduce demand or both? We investigate in this paper the double prisoner's dilemma in which individuals find themselves, with respect to oil consumption and their environmental stance towards the oil industry. We find that the collective outcome of such game is clearly better when a frugal behaviour is adopted, without being militant. The Nash equilibrium, resulting from the individual strategies, leads by contrast to the worst possible outcome: high prices, high consumption and high environmental (negative) impact. An effective environmental action should avoid opposing oil supply sources (a costly militant act) and help consumers becoming more frugal.

**Keywords** Prisoner's dilemma · Oil production · Militancy · Frugality

## 1 Introduction

A large consensus exists on the necessity to mitigate climate change. A reduction of $CO_2$ emissions is needed for this to happen, and as 65% of all greenhouse gases are related to fossil fuel and industrial processes (IPCC 2014), fossil fuel consumption

E. Billette de Villemeur
Université de Lille, Lille, France
e-mail: etienne.de-villemeur@univ-lille.fr

P.-O. Pineau (✉)
Chair in Energy Sector Management, HEC Montréal Energy Sector Management, Montréal, QC, Canada
e-mail: pierre-olivier.pineau@hec.ca



P.-O. Pineau et al. (eds.), *Games in Management Science*,
International Series in Operations Research & Management Science 280,
https://doi.org/10.1007/978-3-030-19107-8_15

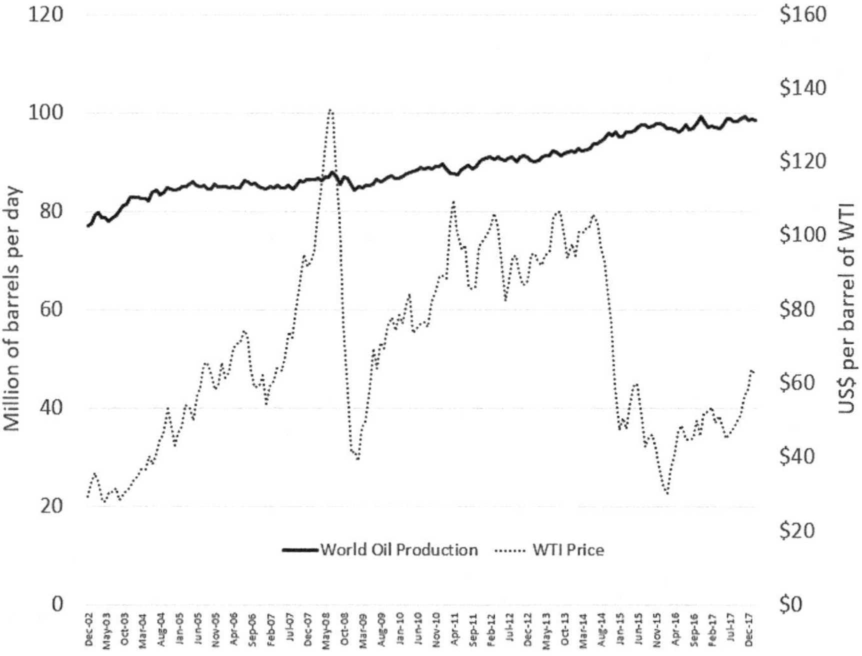


**Fig. 1** World oil production and price, 2002–2017 (EIA 2018)

has to decline and industrial processes have also to change. Emissions from the combustion of coal and oil are particularly important (IEA 2017). But while coal consumption and coal-related emissions have peaked in 2014, oil consumption continues to grow. So does oil production, as consumption and production follow each other very closely—with only short term stocks and strategic reserves creating a difference. As illustrated in Fig. 1, oil production is on an almost linearly increasing path, from about 80 millions barrels per day in 2002 to close to 100 millions barrels per day at the end of 2017 (EIA 2018). Oil prices, perhaps surprisingly, have no apparent impact on consumption and production. Indeed, despite large swings in oil prices in the 2002–2017 period, from $28 to $134 per barrel of Western Texas Intermediate (WTI), oil demand and production continued their steady growth. This translates in very inelastic measures of the price elasticity of oil and oil products demands (usually around −0.2), as many econometric studies conclude. See, for instance, (Labandeira 2017).

While high oil prices have not discouraged consumers to use oil, environmental militants have been very active to oppose oil projects. Movements such as *Keep It in the Ground* try to "revoke the social license of the fossil fuel industry" and "fight iconic battles against fossil fuel infrastructure" (350.org 2018). Their hope is that by opposing oil development, hence by limiting supply, consumption will go down, and so would emissions. Climate change would consequently be mitigated. Some fossil fuel infrastructure can indeed be abandoned by promoters after a "successful" opposition. In Canada, for instance, strong opposition to some major oil pipeline projects (Energy East and TransMountain) has pushed their promoters to renounce developing them.

Despite such opposition, however, oil consumption has not decreased. It is supply that has been affected: some oil projects are removed from the supply mix, and more expensive ones are selected. While it's impossible to directly link the cancellation of one project to the development of another, one could easily conjecture that when oil investments are not made in, for instance, Alberta (Canada), because of some strong local opposition, it will lead to some equivalent investments made in the United States, Brazil, Iraq or Libya, where oil production can grow and is growing (see, for instance, IEA (2018), for some current numbers and forecasts). In short, oil production does not decline after some oil project opposition, but marginally more expensive projects, with less opposition, are chosen. See Herfindahl (1967) and the subsequent literature on the order of extraction of an exhaustible resource. Such price increase indirectly makes renewable energy more competitive. However, so far new renewable energy sources (such as wind or solar) are added to the global production mix, especially in electricity generation, but aren't substitute to oil in the transportation sector, where most of the oil is consumed.

In many cases, when environmental militants oppose oil projects, they do not directly call for a lower oil consumption from individuals. Greenpeace International, for instance, asks its website visitors to "Join the wave of resistance against pipelines", but does not advise to use less oil products, to question friends about their vehicle choice or to adopt a frugal energy consumption level (see Greenpeace International 2018). Maybe they assume that displaying "resistance" is more self-satisfying than not, while reducing oil consumption is too individually demanding. Could it therefore be a better strategy to be an environmental militant than to adopt (and possibly promote) a frugal lifestyle? Of course, the two are separate decisions and can be done simultaneously. But given the price inelasticity of oil demand, as illustrated before, supply side strategies of environmental militants may not have the intended results.

This paper belongs to the family of papers dealing with pollution challenges within a game theoretic framework, to which Georges Zaccour has significantly contributed. See, for instance, Petrosjan and Zaccour (2003), Breton et al. (2010) or Jørgensen et al. (2010), among many others. Our paper also considers the action of the civil society (or environmental groups) in the absence of a central authority, where strategic choices can be made to the benefit, or detriment, of all, as in Ngendakuriyo and Zaccour (2017), which focuses on corruption. Contrary to these papers, however, we limit ourselves to a static context.

More specifically, this paper attempts to disentangle the different aspects related to our specific situation. Given the two sets of choices mentioned above, being an environmental militant or not and adopting a frugal level of energy consumption or not, what are the individual and collective outcomes? What are the environmental impacts of these choices, but also the price and welfare impacts?

We offer some answers to these questions, by studying the strategic situations related to the two sets of choices. In both cases, individuals face a prisoner's dilemma: they would be better off with a lower consumption level (because of the global environmental impact) and no opposition to oil projects (because of the lower prices), only if all did the same. But gratification from higher consumption and adopting a militant environmental stance creates incentives to defect.

While we make some simplifying assumptions, notably that oil demand is strictly price-inelastic, our analysis shows that welfare gains come from lower consumption levels. Militancy can be costly and benefit the oil industry in ways that may not be fully understood by oil projects opponents. However, the assumption on price-elasticity is made for the sake of clarity in the exposition, but would not change the main results if relaxed.

We present the model in the next section, the individual strategies and the market equilibrium. Then we investigate the four polar collective outcomes of the game, and compare their price, quantity (equivalent to the environmental impact) and welfare levels.

# 2 The Model

Consider a population with $N$ identical individuals endowed with a utility

$$\mathcal{U}(q, s; p, Q) = v(p, q) + b(s) - e(Q),$$

where $v(p, q)$ stands for the net utility from individual consumption $q$ at price $p$, $b(s)$ stands for the benefits from environmental stance $s$ and $e(Q)$ for the individual environmental costs, a function of total consumption $Q$.

Individual consumption $q$ can be either *average* or *frugal:* $q \in \{a; f\}$. Environmental stance is either *militant* or *not*: $s \in \{m; \emptyset\}$. Collective consumption $Q$ is determined by the interplay of supply and demand.

## 2.1 The Game in Individual Strategies

Let $N$ be the total number of players and denote by $N_f$ and $N_m$ the number of "frugals" and the number of militants. The market equilibrium depends upon the individual strategies of all players. We denote, respectively, by $p^* = p(N_f, N_m)$ and $Q^* = Q(N_f, N_m)$ the equilibrium price and quantity outcomes. As we shall see—and as expected—$Q^* = Q(N_f, N_m)$ is non-increasing in both of its arguments.

We make the following assumptions:

**Assumption 1** *In regard to their environmental impact, individuals find it individually too costly to adopt a frugal behaviour:*

$$\mathcal{U}\left(f, s, Q\left(N_f; N_m\right)\right) < \mathcal{U}\left(a, s, Q\left(N_f - 1; N_m\right)\right),$$

*for all* $N_f \in [\![1; N]\!]$ *and whatever the values of* $s \in \{m; \emptyset\}$ *and* $N_m \in [\![0; N]\!]$.

**Assumption 2** *Individuals find it individually profitable to adopt a stance of environmental militant:*

$$\mathcal{U}\left(q, m, Q\left(N_f; N_m\right)\right) > \mathcal{U}\left(q, \emptyset, Q\left(N_f; N_m - 1\right)\right),$$

*for all* $N_m \in [\![1; N]\!]$ *and whatever the values of* $q \in \{a; f\}$ *and* $N_f \in [\![0; N]\!]$.

**Assumption 3** *It would be collectively rational to adopt a frugal behaviour:*

$$\mathcal{U}\left(f, s, Q\left(N; N_m\right)\right) > \mathcal{U}\left(a, s, Q\left(0; N_m\right)\right),$$

*whatever the values of* $s \in \{m; \emptyset\}$ *and* $N_m \in [\![0; N]\!]$.

Given these assumptions it is clear that:

**Lemma 1** *The dominant individual strategies are*

$$(q; s) = (a, m).$$

## 2.2 The Market Equilibrium

As mentioned in the introduction, demand is pretty insensitive to prices. We assume that total demand $D$ thus depends only upon the number of frugals, so that

$$D\left(N_f\right) = q_a\left(N - N_f\right) + q_f N_f,$$

where $q_a > q_f$. On the supply side, price matters. Moreover, it is directly impacted by militancy. Again, for simplicity we suppose that:

$$S\left(p, N_m\right) = \sup\left\{0; \beta\left[p - \left(\underline{p} + cN_m\right)\right]\right\},$$

where $\beta$ is the positive slope of the supply curve, $\underline{p}$ is the minimum price at which production can take place with no militancy and $c$ is the individual impact of militancy on such minimum price.

We assume competitive markets. By definition, at equilibrium $D = S$ so that the equilibrium price is given by

$$\begin{aligned} p\left(N_f, N_m\right) &= \underline{p} + cN_m + \beta^{-1}\left[q_a\left(N - N_f\right) + q_f N_f\right] \\ &= \underline{p} + \beta^{-1} N q_a - \beta^{-1}\left(q_a - q_f\right) N_f + cN_m. \end{aligned} \quad (1)$$

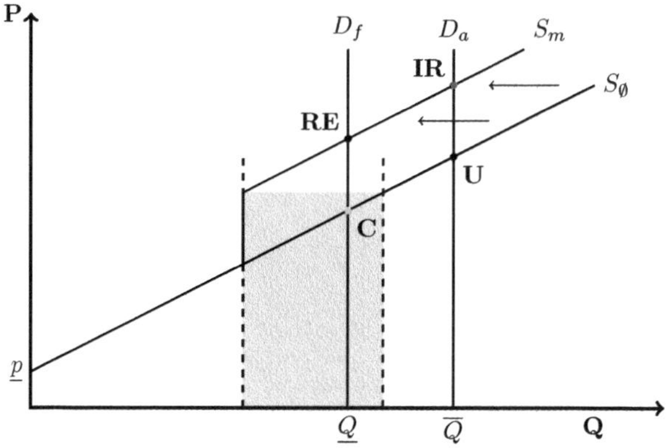


**Fig. 2** The four polar collective outcomes: individually rational (**IR**), unconcerned (**U**), cooperative (**C**) and radical environmentalist (**RE**). When militants manage for the projects associated to the grey area to be cancelled, the subsequent supply curve is shifted to the left

This says that the price increases with the number of militants, $N_m$, but decreases with the number of frugals, $N_f$. By contrast, the equilibrium quantity is a function of the number of frugals only[1]:

$$Q\left(N_f, N_m\right) = q_a N - \left(q_a - q_f\right) N_f, \qquad (2)$$
$$\equiv Q\left(N_f\right).$$

In words, the number of militants has an impact only on price (hence on consumer welfare) but *not* on equilibrium demand—hence upon the environmental impact. This is illustrated in Fig. 2, where $S_m$ is the supply curve after militant have removed some oil projects (the grey area). By shifting supply to the left, equilibrium prices rise. The demand can either be frugal ($D_f$) or average ($D_a$), depending on the number of frugals. This leads to the four "polar" outcomes: individually rational (**IR**), unconcerned (**U**), cooperative (**C**) and radical environmentalist (**RE**).

## 2.3 Discussion

### 2.3.1 Individual Cost of Militancy

Coming back to Assumption 2 upon the payoff of militancy, we assumed that

$$b(m) - b(\emptyset) > v\left(p\left(N_f; N_m - 1\right), q\right) - v\left(p\left(N_f; N_m\right), q\right).$$

Decomposing the net utility into gross utility net of spendings (that is substituting $w(q) - pq$ to $v(p, q)$) this amounts to:

[1]This is a direct consequence of the assumption on demand *in*elasticity.

$$b(m) - b(\emptyset) > \left[p\left(N_f; N_m\right) - p\left(N_f; N_m - 1\right)\right] q. \tag{3}$$

This means that the monetary costs (through the price impact) to the consumers of their militancy cannot counterweight the benefits from their environmental stance. This is fully consistent with the (negligible) price inelasticity of their individual demand.

### 2.3.2 Impact on the Oil Industry

Yet militancy increases consumer financial burden which directly profits the industry. In fact, for any $N_f$, hence for any given level of total demand, the industry revenues $R$ are an increasing function of the number of militants:

$$\begin{aligned} R\left(N_f; N_m\right) &= p\left(N_f; N_m\right) Q\left(N_f\right) \\ &= R\left(N_f; 0\right) + cN_m Q\left(N_f\right), \end{aligned}$$

from Eqs. (1) and (2). Paradoxically, therefore, militancy is beneficial to the oil industry, except of course for those producers who have been excluded from the market. In fact, by reducing total supply, militancy is akin to the action of an oil cartel. The main difference is that the production reduction is not evenly shared by all producers, but obtained by excluding some specific oil production sites (which corresponds to the grey area in Fig. 2).

### 2.3.3 Social Costs of Militancy

While the individual costs of militancy are smaller than its individual benefits (see Eq. (3)), it is also borne by everyone, through the price increase it triggers. We assume that the individual (psychological) benefits from taking a militant environmental stance are smaller than the financial costs it imposes on all consumers. Formally

**Assumption 4** *Individual (psychological) benefits from taking a militant environmental stance are smaller than the financial costs it imposes on all consumers.*

$$b(m) - b(\emptyset) < \left[p\left(N_f; N_m\right) - p\left(N_f; N_m - 1\right)\right] Q\left(N_f\right),$$

*for any $N_m \geq 1$ and any $N_f$.*

## 3 Collective Outcomes

We now consider the collective outcomes of the strategic game. There is a double prisoner's dilemma, one in each of the strategic variables $q \in \{a, f\}$ and $s \in \{m, \emptyset\}$. We identify four polar collective outcomes.

## 3.1 Four Polar Collective Outcomes

Let $\overline{Q} = q_a N$ and $\underline{Q} = q_f N$.

### 3.1.1 Individually Rational Outcome

As already mentioned in Lemma 1, it is a dominant strategy for individuals to be an average consumer and a militant. Therefore, the individually rational outcome is $(N_f; N_m) = (0, N)$ and

$$Q^{IR} = \overline{Q}, \qquad p^{IR} = \underline{p} + \beta^{-1}\overline{Q} + cN$$

so that

$$\mathcal{U}^{IR} = w(q_a) - p^{IR} q_a + b(m) - e(\overline{Q}).$$

### 3.1.2 Cooperative Outcome

If players were to cooperate, they would be frugal and abstain from militancy. Therefore, the cooperative outcome is $(N_f; N_m) = (N, 0)$ and

$$Q^C = \underline{Q} \qquad p^C = \underline{p} + \beta^{-1}\underline{Q}$$

so that

$$\mathcal{U}^C = w(q_f) - p^C q_f + b(\emptyset) - e(\underline{Q}).$$

### 3.1.3 Outcome of an Unconcerned Population

If consumers are unconcerned so that they all maintain an average consumption and do not bother to take a militant position, the collective outcome is $(N_f; N_m) = (0, 0)$ and

$$Q^U = \overline{Q} \qquad p^U = \underline{p} + \beta^{-1}\overline{Q}$$

so that

$$\mathcal{U}^U = w(q_a) - p^U q_a + b(\emptyset) - e(\overline{Q}).$$

#### 3.1.4 Outcome of a Radical Environmentalist Population

If individuals are all frugal and engaged in militancy, despite its costs, then $\left(N_f; N_m\right) = (N, N)$ and

$$Q^{RE} = \underline{Q} \qquad p^{RE} = \underline{p} + \beta^{-1}\underline{Q} + cN$$

so that

$$\mathcal{U}^{RE} = w\left(q_f\right) - p^{RE}q_f + b\left(m\right) - e\left(\underline{Q}\right).$$

### 3.2 *Discussion*

#### 3.2.1 Quantity and Price Comparisons

In terms of quantities, hence environmental impact, the comparison between the four cases is pretty straightforward:

$$Q^{RE} = Q^{C} = \underline{Q} < \overline{Q} = Q^{U} = Q^{IR}.$$

There are also simple comparisons between some prices:

$$p^{RE} < p^{IR} \quad \text{and} \quad p^{C} < p^{U};$$
$$p^{C} < p^{RE} \quad \text{and} \quad p^{U} < p^{IR};$$

Hence $p^{C} < p^{IR}$. However, the comparison $p^{RE}$ and $p^{U}$ is a priori ambiguous.

The difference between both prices depends upon the elasticity of supply and the difference between the average and frugal demands. More precisely,

$$p^{RE} - p^{U} = cN - \beta^{-1}\left(Q^{U} - Q^{RE}\right) = \left[c - \beta^{-1}\left(q_a - q_f\right)\right]N.$$

In words, the price will be higher with a population of **R**adical **E**nvironmentalists than with an **U**nconcerned population if (and only if) the sole impact of their own militancy upon the equilibrium price is sufficient to induce individuals to reduce their demand by a larger amount than that associated to shifting from average to frugal consumption. In all other cases, that is when

$$\beta c < q_a - q_f, \tag{4}$$

we have $p^{RE} < p^{U}$. It is thus fair to assume that

$$p^{C} < p^{RE} \leq p^{U} < p^{IR}.$$

### 3.2.2 Welfare Comparisons

In terms of welfare, the pairwise comparison of the four polar outcomes is less straightforward. We have

$$\begin{aligned}\mathcal{U}^{IR}-\mathcal{U}^{RE}&=\left[w\left(q_a\right)-p^{IR}q_a+b\left(m\right)-e\left(\overline{Q}\right)\right]-\left[w\left(q_f\right)-p^{RE}q_f+b\left(m\right)-e\left(\underline{Q}\right)\right]\\&=\left[w\left(q_a\right)-p^{IR}q_a\right]-\left[w\left(q_f\right)-p^{RE}q_f\right]-\left[e\left(\overline{Q}\right)-e\left(\underline{Q}\right)\right].\end{aligned}$$

By Assumption 3 upon the collective rationality of frugal behaviour

$$v\left(p\left(0;N_m\right),a\right)-v\left(p\left(N;N_m\right),f\right)<e\left[Q\left(0;N_m\right)\right]-e\left[Q\left(N;N_m\right)\right]=e\left(\overline{Q}\right)-e\left(\underline{Q}\right),$$

so that, substituting $w\left(q\right)-pq$ to $v\left(p,q\right)$we have:

$$\left[w\left(q_a\right)-p\left(0;N_m\right)q_a\right]-\left[w\left(q_f\right)-p\left(N;N_m\right)q_f\right]<e\left(\overline{Q}\right)-e\left(\underline{Q}\right),$$

for any $N_m$. Let $N_m=N$. We have $p\left(0;N\right)=p^{IR}$ and $p\left(N;N\right)=p^{RE}$ so that we can conclude:

$$\mathcal{U}^{IR}-\mathcal{U}^{RE}<0.$$

Moreover, Assumption 4 on the social cost of militancy says that

$$b\left(m\right)-b\left(\emptyset\right)<\left[p\left(N_f;N_m\right)-p\left(N_f;N_m-1\right)\right]Q\left(N_f\right)=cQ\left(N_f\right)$$

which implies that $b\left(m\right)-b\left(\emptyset\right)<\left[p\left(N_f;N\right)-p\left(N_f;0\right)\right]Q\left(N_f\right)$. What is of more interest is that, for $N_f=0$ and $N_f=N$ it also states

$$\begin{aligned}b\left(m\right)-b\left(\emptyset\right)&<\left[p\left(0;N_m\right)-p\left(0;N_m-1\right)\right]q_aN=cNq_a,\\b\left(m\right)-b\left(\emptyset\right)&<\left[p\left(N;N_m\right)-p\left(N;N_m-1\right)\right]q_fN=cNq_f.\end{aligned}$$

As a consequence, we also have

$$\begin{aligned}\mathcal{U}^{IR}-\mathcal{U}^{U}&=\left[w\left(q_a\right)-p^{IR}q_a+b\left(m\right)-e\left(\overline{Q}\right)\right]-\left[w\left(q_a\right)-p^{U}q_a+b\left(\emptyset\right)-e\left(\overline{Q}\right)\right]\\&=\left[b\left(m\right)-b\left(\emptyset\right)\right]-\left(p^{IR}-p^{U}\right)q_a\\&=\left[b\left(m\right)-b\left(\emptyset\right)\right]-cNq_a\\&<0,\end{aligned}$$

from Assumption 2.

Similarly, we have

$$\begin{aligned}\mathscr{U}^{RE}-\mathscr{U}^{C} &= \left[w\left(q_f\right)-p^{RE}q_f+b\left(m\right)-e\left(\underline{Q}\right)\right]-\left[w\left(q_f\right)-p^{C}q_f+b\left(\emptyset\right)-e\left(\underline{Q}\right)\right]\\ &= \left[b\left(m\right)-b\left(\emptyset\right)\right]-\left(p^{RE}-p^{C}\right)q_f\\ &= \left[b\left(m\right)-b\left(\emptyset\right)\right]-cNq_f\\ &< 0,\end{aligned}$$

again from Assumption 2.

We now compare $\mathscr{U}^{U}$ to both $\mathscr{U}^{RE}$ and $\mathscr{U}^{C}$.

We have

$$\begin{aligned}\mathscr{U}^{U}-\mathscr{U}^{C} &= \left[w\left(q_a\right)-p^{U}q_a+b\left(\emptyset\right)-e\left(\overline{Q}\right)\right]-\left[w\left(q_f\right)-p^{C}q_f+b\left(\emptyset\right)-e\left(\underline{Q}\right)\right]\\ &= \left[w\left(q_a\right)-p^{U}q_a\right]-\left[w\left(q_f\right)-p^{C}q_f\right]-\left[e\left(\overline{Q}\right)-e\left(\underline{Q}\right)\right].\end{aligned}$$

Assumption 3 upon the collective rationality of frugal behaviour implies that

$$e\left(\overline{Q}\right)-e\left(\underline{Q}\right)>\left[w\left(q_a\right)-p^{U}q_a\right]-\left[w\left(q_f\right)-p^{C}q_f\right],$$

so that we obtain

$$\mathscr{U}^{U}-\mathscr{U}^{C}<0.$$

Finally

$$\begin{aligned}\mathscr{U}^{U}-\mathscr{U}^{RE} &= \left[w\left(q_a\right)-p^{U}q_a+b\left(\emptyset\right)-e\left(\overline{Q}\right)\right]-\left[w\left(q_f\right)-p^{RE}q_f+b\left(m\right)-e\left(\underline{Q}\right)\right]\\ &= \left[w\left(q_a\right)-p^{U}q_a\right]-\left[w\left(q_f\right)-p^{RE}q_f\right]-\left[b\left(m\right)-b\left(\emptyset\right)\right]-\left[e\left(\overline{Q}\right)-e\left(\underline{Q}\right)\right].\end{aligned}$$

Observe that $p^{RE}=\underline{p}+\beta^{-1}\underline{Q}+cN=p^{C}+cN$. It follows that

$$\begin{aligned}\mathscr{U}^{U}-\mathscr{U}^{RE} = &\left\{cNq_f-\left[b\left(m\right)-b\left(\emptyset\right)\right]\right\}\\ &-\left\{\left[e\left(\overline{Q}\right)-e\left(\underline{Q}\right)\right]-\left[\left(w\left(q_a\right)-p^{U}q_a\right)-w\left(q_f\right)-p^{C}q_f\right]\right\}\end{aligned}$$

where from Assumption 3 upon the collective rationality of frugal behaviour and from Assumption 4 upon the social cost of militancy,

$$\begin{aligned}e\left(\overline{Q}\right)-e\left(\underline{Q}\right) &> \left[w\left(q_a\right)-p^{U}q_a\right]-\left[w\left(q_f\right)-p^{C}q_f\right],\\ cNq_f &> b\left(m\right)-b\left(\emptyset\right)\end{aligned}$$

so that both terms are positive and the sign of $\mathcal{U}^U - \mathcal{U}^{RE}$ is indeterminate. It depends upon the relative magnitude of the costs of militancy and the environmental costs. If the latter dominates, $\mathcal{U}^{RE} > \mathcal{U}^U$, the converse otherwise.

To summarize

$$\mathcal{U}^{IR} < \mathcal{U}^{RE} < \mathcal{U}^C,$$

$$\mathcal{U}^{IR} < \mathcal{U}^U < \mathcal{U}^C.$$

## 4 Conclusion

The double prisoner's dilemma leads, unsurprisingly, to the worst welfare outcome. Demand policies, targeting individual behaviours, or simply individual action to reduce demand, are more effective than supply strategies to improve environmental outcomes. Supply strategies are not only ineffective but are beneficial to the industry by raising its revenues. This is definitely something most environmental militants do not intend.

This paper is a first step toward the analysis of the effects of militancy on oil markets. Further developments could include the assessment of the impact of some elasticity in oil demand on the these results and the study of distributional effects of oil price increases induced by militancy. Indeed, many lower income oil consumers already spend a higher percentage of their income on energy, despite using less of it, than higher income ones. They bear a greater cost when oil becomes more expensive and could be collateral victims of environmental militancy. On the other hand, if higher income consumers became more frugal, it would provide both financial and environmental relief to everyone—but especially for the poorest, who are often, also, the most exposed to environmental problems.

**Acknowledgements** We want to acknowledge the useful comments of two referees.

## References

350.org. (2018). About 350. New York: 350.org. Retrieved July 18, 2018, from https://350.org/about/

Breton, B., Sbragia, L., & Zaccour, G. (2010). A dynamic model for international environmental agreements. *Environmental and Resource Economics, 45*(1), 25–48.

EIA. (2018). *International data*. Washington, DC: Energy Information Administration.

Greenpeace International (2018) Act. Amsterdam: Greenpeace International. Retrieved July 18, 2018, from https://www.greenpeace.org/international/act/

Herfindahl, O. C. (1967). Depletion and economic theory. In M. M. Gaffney (Ed.), *Extractive resources and taxation* (pp. 63–69). Madison, WI: Wisconsin University Press.

IEA. (2017). *Key world energy statistics*. Paris: International Energy Agency.

IEA. (2018). *Oil market report 2018*. Paris: International Energy Agency.

IPCC. (2014). Climate change 2014: Mitigation of climate change. In O. Edenhofer, R. Pichs-Madruga, Y. Sokona, E. Farahani, S. Kadner, K. Seyboth, et al. (Eds.), *Contribution of working group III to the fifth assessment report of the intergovernmental panel on climate change*. Cambridge: Cambridge University Press.

Jørgensen, S., Martín-Herrán, G., & Zaccour, G. (2010). Dynamic games in the economics and management of pollution. *Environmental Modeling & Assessment, 15*(6), 433–467.

Labandeira, X., Lageaga, J. M., & Lopez-Otero, X. (2017). A meta-analysis on the price elasticity of energy demand. *Energy Policy, 102*, 549–68.

Ngendakuriyo, F., & Zaccour, G. (2017). Should civil society organizations cooperate or compete in fighting a corrupt government? *Mathematical Social Sciences, 85*, 30–36.

Petrosjan, L., & Zaccour, G. (2003). Time-consistent Shapley value allocation of pollution cost reduction. *Journal of Economic Dynamics and Control, 27*(3), 381–398.

# Compliance with Social Norms as an Evolutionary Stable Equilibrium

**Francisco Cabo, Ana García-González, and Mercedes Molpeceres-Abella**

**Abstract** This paper analyzes the compliance with social norms optimally established by a benevolent central planner. Since compliance is costly, agents have an incentive to free-ride on others, in a public good game. We distinguish two types of agents: standard pro-self agents (Sanchos) whose payoffs are defined by a prisoner's dilemma game dominated by the non-compliance strategy, and pro-social Quixotes, who still have an incentive to free-ride, although prefer compliance over mutual defection (as in a snowdrift game). Compliance is analyzed in a two-population evolutionary game considering an imitative revision protocol. Individuals from one population play against and imitate agents within their own but also the other population. Inter-population interaction and imitation allow us to investigate under which circumstances some Sanchos might imitate compliant Quixotes, so escaping the non-compliance equilibrium characteristic of an isolated population of Sanchos. Correspondingly, we analyze the conditions under which the interaction with the population of selfish Sanchos increases or decreases the compliance rate among altruistic Quixotes.

**Keywords** Two-population evolutionary game · Heterogeneous preferences · Prisoner's dilemma game · Snowdrift game · Social norms

## 1 Introduction

Compliance with social norms can be analyzed as a collective action problem. Although collective actions may be jointly beneficial, a classical result, as stated by

F. Cabo (✉) · A. García-González
IMUVa, Universidad de Valladolid, Valladolid, Spain
e-mail: pcabo@eco.uva.es; anagar@eco.uva.es

M. Molpeceres-Abella
Universidad de Valladolid, Valladolid, Spain
e-mail: abella@eco.uva.es



P.-O. Pineau et al. (eds.), *Games in Management Science*,
International Series in Operations Research & Management Science 280,
https://doi.org/10.1007/978-3-030-19107-8_16

Olson (1971), is that no self-interested person would contribute in the production of a public good (except in the case of a very small group or under a mechanism like coercion). This zero contribution thesis corresponds to the Nash equilibrium in a prisoner's dilemma game. For the public good game (the generalization of the prisoner's dilemma to a large number of players), defection is the dominant strategy. Thus, classical game theory and also evolutionary game theory predict zero contribution (see, for example, Miller and Andreoni 1991). Nevertheless, experimental economics seems to contradict this zero contribution hypothesis. For example, Andreoni (1988) and later Ostrom (2000) revise the literature on public goods experiments, finding positive levels of contribution in one-shot and repeated games. Many examples of collective actions also exist in everyday life: voting, not cheating on taxes, contributions to voluntary associations, or recycling.

The literature provides alternative explanations to reconcile the standard theory with the evidence of individual behavior. Andreoni (1988) suggests that individual behavior, at least in the laboratory, is not exclusively oriented by self-interest but also by factors like altruism, social norms, or bounded rationality. Here, we focus on the idea that individuals, or at least some individuals, can derive enjoyment from collaboration independently of how much their actions benefit others. This effect is called intrinsic satisfaction in de Young (1996), or "warm-glow" in the more standard terminology, like in Andreoni (1990). As long as social norms are perceived as ideal forms of behavior, compliance provides a warm-glow to each particular compliant agent, which is a supplementary benefit to the public good.[1] Alternatively, individuals get dis-utility from doing bad. Thus, as long as not following social norms is perceived as doing bad, defection involves a cold-prickle, following the terminology in Andreoni (1995), who highlighted the asymmetry between doing good versus not doing bad.

An economy populated by different types of individuals, some more inclined to collaborate than others was proposed by Ostrom (2000), from an evolutionary perspective. The existence of "norm-using" individuals, unlike the standard rational egoists, helps bring the theory closer to the findings in the laboratory.[2] More recently, Grafton et al. (2017) consider the society as a mixture of standard "Nashian" individuals (who maximize their utility taking as given the strategies of others) and "Kantian" individuals (who do not do anything that they would

[1]Other factors not analyzed here are inequity aversion, reciprocal altruism, sense of identity or preference for efficiency (see Alger and Weibull 2013 and the references therein). The temporal dimension can also make compliance more rewarding. For pollution problems, current abatement activities can be beneficial when the future cleaner environment is taken into account. An interesting example is Breton et al. (2010), who propose a dynamic game for international environmental agreements that takes into account the pollution stock dynamics, as well as the membership dynamics, defined by an evolutionary process.

[2]The mixture of rational egoists and norm-users results from an indirect evolutionary approach, proposed by Güth and Yaari (1992) and Güth (1995). Preferences are also considered as strategies subject to evolutionary selection in Alger and Weibull (2013), for whom natural selection has built preferences as a convex combination between selfishness and morality.

not like if everyone behaves in the same way). We consider a society composed of two different types of agents: rational egoists or pro-self individuals, which we call Sanchos; and norm-using or pro-social individuals, called Quixotes. Both follow a "Nashian" behavior, and their strategic behavior changes according to the evolutionary dynamics.

The preferences of rational egoist Sanchos are captured by a payoff matrix of the prisoner's dilemma type with a strong incentive to free-ride (the most widely used paradigm to study human cooperation). In contrast, pro-social[3] Quixotes experience a warm-glow from compliance, together with a cold-prickle from defection. The free-riding incentive still persists when the opponent complies, however, if the opponent defects, compliance is their best strategy. Thus, for pro-social Quixotes the compliance dilemma is relaxed, described as a snowdrift game (also known as hawk-dove or chicken game), as proposed, for example, in Doebeli et al. (2004). Our setting follows Cabo and García-González (2018) who analyze a two-population evolutionary game in between a symmetric and an asymmetric game. Individuals belonging to one population play against and imitate individuals within their own and also the other population. The evolutionary dynamics converges towards a mixed strategy equilibrium at which some Sanchos might find it attractive to imitate the compliant behavior of Quixotes, though it is not in their best interest. This answers the question which typically intrigues a reader of *"The Ingenious Gentleman Don Quixote of La Mancha"*[4]: What led the pragmatic and rational Sancho to follow the "madness" of the morally oriented Don Quixote? Or put differently: under which circumstances does the existence of a population of pro-social individuals induce standard pro-self individuals to comply with the social rules when it is in their best interest not to do so? A positive answer requires a much higher reward to compliance for Quixotes than for Sanchos.

We extend Cabo and García-González (2018) by relaxing the assumption that the global population is equally divided between Sanchos and Quixotes. We analyze how compliance decisions in either population are affected by their relative sizes. Considering a pairwise imitation protocol, the relative size of each population determines the likelihood of being paired with an individual within this particular population. At the same time, it directly influences the share of compliance in the global population and, in consequence, the incentives to free-ride when others comply or to comply when others do not.

The analysis shows that the likelihood that some Sanchos imitate the compliant behavior of Quixotes increases with the ratio of Sanchos in the overall population.[5] Conversely, when the ratio of Sanchos is very small, they would certainly free-

[3]In the same way as Don Quixote sought to behave in accordance with his reading of chivalry, pro-social individuals perceive social norms as ideal forms of behavior.

[4]Originally titled: El ingenioso hidalgo don Quijote de la Mancha, by Miguel de Cervantes Saavedra (1605 Part I and 1615 Part II).

[5]Similar result is obtained in Bontems and Rotillon (2000), for pollution compliance in a population divided between honest polluters (always comply) and opportunistic individuals.

ride on compliant Quixotes, unless these latter get a very large warm-glow from compliance. This warm-glow makes it more likely compliance among Sanchos, and it also increases the actual compliance rates for both agents, Sanchos and Quixotes.

A larger ratio of Sanchos undoubtedly reduces the share of compliance in the global population. However, its effect on the share of compliance within the population of Quixotes is twofold. A snowdrift effect induces Quixotes to increase compliance because the compliance rate in the global population is smaller. Conversely, an imitation effect leads Quixotes to reduce compliance following the non-compliant behavior of Sanchos. The positive effect becomes more important the greater the warm-glow, making more likely that a higher ratio of Sanchos induces higher compliance among Quixotes. On the other hand, the ratio of Sanchos first raises compliance within this population although, with less and less Quixotes, Sanchos hardly meet and imitate Quixotes and compliance rates decrease to zero.

The rest of the paper is organized as follows. Section 2 presents the preferences of Sanchos and Quixotes. The two-population evolutionary game and the associated dynamics are introduced in Sect. 3. The two types of equilibria with and without compliance among Sanchos are computed in Sect. 4. Section 5 studies how compliance in each population is affected by the discrepancy between Sanchos and Quixotes, and by the relative sizes of these populations. It also presents some real-life examples. Section 6 concludes.

## 2 Two Distinct Populations

This section explains the distinct behavior of pro-self Sanchos and norm-using Quixotes when they face a collective action problem like compliance with social norms. We assume that if the social norms are followed by all individuals, then they all will be better off.[6] The collective action problem is perceived as: a prisoner's dilemma game for Sanchos, and a snowdrift game for Quixotes.

The population of Sanchos represents the standard rational agents, whose payoff matrix collects the canonical representation of a collective action problem:

| | C | D |
|---|---|---|
| C | $(1, 1)$ | $(1-d, 1+c-d)$ |
| D | $(1+c-d, 1-d)$ | $(0, 0)$ |

Net gains when both players comply are normalized to 1. Parameter $c$ represents the effort or the cost of compliance, and $d$ is the damage caused by a non-compliant player. Thus, the gap $c - d > 0$ determines the incentive to free-ride on the other,

[6]Norms are optimally chosen by a benevolent central planner, or they are the result of a social learning process.

which induces the agent to disobey as long as his opponent complies. Moreover, in a prisoner's dilemma game a player prefers defection also when the opponent defects. This incentive is measured by $1 - d < 0$, which is less pronounced than the free-riding incentive if the opponent complies:

$$\sigma \equiv c - d - (d - 1) > 0, \quad \text{with} \quad 1 < d < c. \tag{1}$$

Notice that the restriction $1 < d < c$ ensures that defection is the dominant strategy among Sanchos. The restriction $\sigma > 0$ is that the sum of the off-diagonal payoffs is greater than the sum of the diagonal payoffs. Thus, in the population of Sanchos the player's gains from defection, as opposed to compliance, is greater when the opponent complies, i.e. $c - d > d - 1$. Under this prisoner's dilemma specification, the unique Nash equilibrium is the standard situation of mutual defection.

The population of Quixotes attains a greater reward for compliance and a lower reward for defection than the population of Sanchos. They attach a warm-glow to compliance, $wg \geq 0$, and a cold-prickle to defection, $cp \geq 0$. Thus, their payoffs can be computed as:

| | C | D |
|---|---|---|
| C | $(1 + wg, 1 + wg)$ | $(1 - d + wg, 1 + c - d - cp)$ |
| D | $(1 + c - d - cp, 1 - d + wg)$ | $(-cp, -cp)$ |

The warm-glow associated with compliance and the cold-prickle associated with defection makes compliance more attractive and defection less so. The social dilemma for the population of Quixotes is described as a snowdrift (Hawk-Dove or chicken) game. This implies that the joint effect of the warm-glow and the cold-prickle is not enough to counterbalance the free-riding incentive when the opponent complies: $wg + cp < c - d$. In contrast, this joint effect is strong enough to induce compliance when the opponent disobeys $wg + cp > d - 1$. In what follows, this joint effect will be denoted as $\varepsilon = wg + cp$. In consequence, the warm-glow from compliance and the cold-prickle from defection can be expressed as: $wg = \alpha\varepsilon$, $cp = (1 - \alpha)\varepsilon$, with $\alpha \in [0, 1]$. While $\varepsilon$ is a quantitative measure of the distance between Sanchos and Quixotes, parameter $\alpha$ can be interpreted as a qualitative measure of this distance that measures the relative importance of the warm-glow over the cold-prickle. Note that $wg/cp = \alpha/(1 - \alpha)$, and it moves between 0 and $\infty$ as $\alpha$ moves between 0 and 1. Given this notation, the snowdrift structure of this matrix requires conditions in (1) together with:

$$d - 1 < \varepsilon < c - d. \tag{2}$$

For such a game, mutual defection is no longer a Nash equilibrium. It is characterized by a Nash equilibrium in mixed strategies[7]:

$$(\Delta, 1-\Delta) = \left( \frac{\varepsilon - (d-1)}{\sigma}, \frac{c-d-\varepsilon}{\sigma} \right) \in (0,1) \times (0,1). \tag{3}$$

The expression $\Delta$ represents a relative measure of the incentive to comply when the opponent disobeys. Correspondingly, $1 - \Delta$ provides the relative measure of the free-riding incentive when the opponent complies. These two values are positive under condition (2) and the assumption $\sigma > 0$. The greater the absolute distance which separates Sanchos and Quixotes, $\varepsilon$, the greater the incentive to comply if the opponent disobeys, and the lower the free-riding incentive if the opponent complies (see Fig. 1). However, $\Delta$ is independent of $\alpha$, i.e. of whether Quixotes attach a large warm-glow to compliance or a strong cold-prickle to non-compliance.

In addition to the definition of $\Delta$, we introduce two new expressions:

$$\Delta_{wg} = \frac{\alpha\varepsilon - (d-1)}{\sigma}, \qquad \Delta_{cp} = \frac{(1-\alpha)\varepsilon - (d-1)}{\sigma}. \tag{4}$$

These expressions can be positive or negative. A positive $\Delta_{wg}$ implies that the warm-glow from compliance for Quixotes, $\alpha\varepsilon$, is by itself a strong enough incentive to induce compliance when the opponent defeats. Likewise, a positive $\Delta_{cp} > 0$ implies that the Quixote's preference for compliance when the opponent defeats could be based only on the cold-prickle she associates to defection (with no need of warm-glow from compliance).

In what follows, unless said otherwise, the illustrative examples will consider the following parameters' values:

$$c = 1.6, \; d = 1.1, \; \varepsilon = 0.4. \tag{5}$$

For these parameters it holds that $\sigma = 0.4 > 0$, $\Delta = 0.75$, $d-1 = 0.1 < c-d = 0.5$, satisfying conditions (1) and (2). Moreover, $\Delta_{wg} = \alpha - 1/4$ and $\Delta_{cp} = 3/4 - \alpha$.

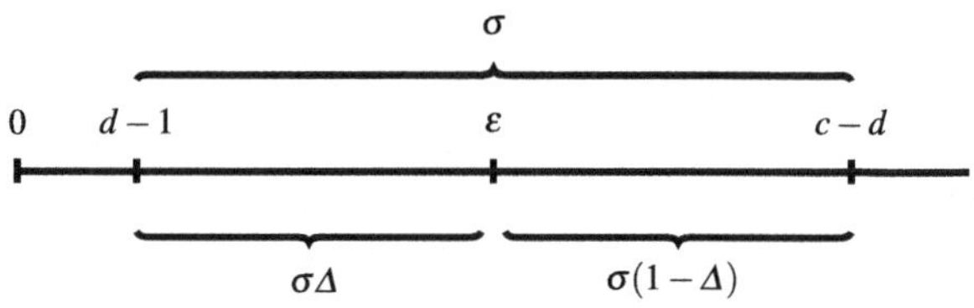


**Fig. 1** Incentives to comply/free-ride

[7]As well as two asymmetric pure Nash equilibria, (1, 0), (0, 1).

## 3 Interaction and Imitation Between and Within Populations

The first specificity of the two-population game analyzed here is that individuals in one population play against individuals of their own kind but also of the other population. We assume a unit mass population divided between Sanchos and Quixotes. The share of Sanchos in the overall population is represented by constant $s \in (0, 1)$, and hence $1-s$ represents the ratio of Quixotes. Both types of players have the same set of two strategies: compliance and defection, $\{C, D\}$. Importantly, while they can switch between strategies, their preferences do not vary, and hence, they remain within their population. Thus, the set of social states in this two-population game can be written as $X = \{\mathbf{x} = (ps, (1-p)s, q(1-s), (1-q)(1-s)) : p, q \in [0, 1]\}$, with $p$ (resp. $q$) the ratio of agents in the population of Sanchos (resp. Quixotes) who comply.

For each social state, $\mathbf{x}$, the payoff function for the two-population game can be computed as $F\mathbf{x}'$ with

$$F = \begin{pmatrix} 1 & 1-d & 1 & 1-d \\ 1+c-d & 0 & 1+c-d & 0 \\ 1+\alpha\varepsilon & 1-d+\alpha\varepsilon & 1+\alpha\varepsilon & 1-d+\alpha\varepsilon \\ 1+c-d-(1-\alpha)\varepsilon & -(1-\alpha)\varepsilon & 1+c-d-(1-\alpha)\varepsilon & -(1-\alpha)\varepsilon \end{pmatrix}.$$

For simplicity we will refer to the total population which complies (adding up Sanchos and Quixotes), denoted by $y = ps + q(1-s)$. Then the payoffs $F\mathbf{x}'$ for the two populations can be rewritten as a function of $y$[8]:

$$\pi(y) = \begin{pmatrix} \pi_C^S(y) \\ \pi_D^S(y) \\ \pi_C^Q(y) \\ \pi_D^Q(y) \end{pmatrix} = \begin{pmatrix} 1-d+dy \\ y(1+c-d) \\ 1-d+dy+\alpha\varepsilon \\ y(1+c-d)-(1-\alpha)\varepsilon \end{pmatrix}. \tag{6}$$

From (6) it is clear that the expected payoff for Quixotes surpasses the expected payoff for Sanchos in the warm-glow, $\alpha\varepsilon$, while defection is less rewarding in the cold-prickle, $(1-\alpha)\varepsilon$.

$$\pi_C^Q(y) = \pi_C^S(y) + \alpha\varepsilon, \quad \pi_D^Q(y) = \pi_D^S(y) - (1-\alpha)\varepsilon.$$

As $\alpha$ approaches 1, the differences between Sanchos and Quixotes particularize on the warm-glow from compliance, and the payoff to defection converges for the two populations. Conversely, when $\alpha$ tends to 0, the gap between populations

[8]Superscripts S and Q refer to Sanchos and Quixotes, respectively. Subscripts C and D to compliance and defection.

particularizes in the cold-prickle from non-compliance, while compliance is equally valued by the two types of agents.

Comparing the payoffs between strategies for the same population it follows:

$$\pi^{S}_{D}(y) - \pi^{S}_{C}(y) = (d-1) + \sigma y > 0, \qquad \pi^{Q}_{D}(y) - \pi^{Q}_{C}(y) = \sigma(y - \Delta).$$

Within the population of Sanchos, defection always dominates compliance regardless of $y$. As for Quixotes, whether they prefer compliance or defection depends on how the share of compliance in the global population compares to $\Delta$. As shown in expression (3), the expression $\Delta$ (which represents the NE or the ESS if only a single population of Quixotes existed) takes values within the interval (0, 1), due to the snowdrift structure of the payoff matrix for Quixotes. These later would prefer to comply when few individuals comply, $y < \Delta$, and would prefer to defect in the opposite case.

A population game defined exclusively for the population of Sanchos would be characterized by the convergence towards an evolutionary stable strategy of zero compliance. Similarly, for a population of Quixotes, the compliance rate at the equilibrium would be $\Delta$. Our main interest is to analyze the equilibrium for the two-population game proposed. With that aim, the imitation mechanism must be specified. And it is this mechanism which constitutes the second main feature of the proposed two-population game.

Considering an imitative revision protocol, the temporal evolution of the compliance rate in each population is determined by the share of non-compliance times the probability of switching to compliance, minus the share of compliance times the probability of switching to defection:

$$\dot{p} = (1-p)\rho^{S}_{DC} - p\rho^{S}_{CD}, \tag{7}$$

$$\dot{q} = (1-q)\rho^{Q}_{DC} - q\rho^{Q}_{CD}, \tag{8}$$

where $\rho^{h}_{ij}$ is the probability that an individual in population $h \in \{S, Q\}$ playing strategy $i \in \{C, D\}$ switches to the alternative strategy $j \in \{C, D\}$, $j \neq i$. This probability is determined by the likelihood that a revising agent is paired with an individual playing the alternative strategy, times the conditional imitation rate, $r^{h}_{ij}$. In the standard formulation of multi-population games (see, for example, Sandholm 2010), when an individual in population $h$ receives a revision opportunity, she can only be paired with other individuals within her own population. Hence the likelihood of meeting someone playing the alternative strategy is given by the ratio of individuals in population $h$ playing this alternative strategy $j$. Thus, for Sanchos, the probability of switching from compliance to defection would read $\rho^{S}_{CD} = (1-p)r^{S}_{CD}$, and the mirror probability of switching from defection to compliance, $\rho^{S}_{DC} = pr^{S}_{DC}$ (similarly for Quixotes, changing $p$ by $q$, and $r^{S}_{ij}$ by $r^{Q}_{ij}$). Our proposal adds the possibility that she could also imitate the behavior of the agents belonging to the alternative population. Thus the probabilities of switching her strategy depend on the likelihood of being paired with someone in her own and also in the other population, who plays the alternative strategy. Hence, for Sanchos and Quixotes, this probabilities read:

$$\rho^{S}_{CD} = (1-p)sr^{SS}_{CD} + (1-q)(1-s)r^{SQ}_{CD}, \quad \rho^{S}_{DC} = psr^{SS}_{DC} + q(1-s)r^{SQ}_{DC}, \tag{9}$$

$$\rho^{Q}_{CD} = (1-p)sr^{QS}_{CD} + (1-q)(1-s)r^{QQ}_{CD}, \quad \rho^{Q}_{DC} = psr^{QS}_{DC} + q(1-s)r^{QQ}_{DC}, \tag{10}$$

where $r^{hk}_{ij}$, with $h, k, \in \{S, Q\}$, and $i, j \in \{C, D\}$, represents the conditional imitation rate of an agent in population $h$ playing strategy $i$ who is paired with an agent in population $k$ playing strategy $j \neq i$.

Assuming a revision protocol governed by pairwise imitation, the conditional imitation rate is proportional to the gap between the payoffs of the randomly chosen opponent and the revising player. Thus, the conditional imitation rate of an $i$-player in population $h$ who meets a $j$-player in population $k$ reads:

$$r^{hk}_{ij} \equiv r^{hk}_{ij}(y) = [\pi^{k}_{j}(y) - \pi^{h}_{i}(y)]_{+}, \tag{11}$$

where $[z]_{+} = z$ if $z > 0$ and 0 otherwise. It is important to notice that imitation is only driven by the gap between payoffs, and individuals in one population will equally imitate agents from their same or the other population.[9]

A compliance individual, who can be a Sancho or a Quixote, can be paired with non-compliant Sanchos or non-compliant Quixotes. This gives a matrix of payoffs comparisons describing the conditional imitation rates from compliance to defection, $G_{CD}(y) = \left(r^{hk}_{CD}(y)\right)_{2\times 2}$, with $h, k \in \{S, Q\}$. Likewise, the matrix $G_{DC}(y) = \left(r^{hk}_{DC}(y)\right)_{2\times 2}$ collects the conditional imitation rates from defection to compliance.[10]

$$G_{CD}(y) = \begin{pmatrix} [\pi^{S}_{D} - \pi^{S}_{C}]_{+} & [\pi^{Q}_{D} - \pi^{S}_{C}]_{+} \\ [\pi^{S}_{D} - \pi^{Q}_{C}]_{+} & [\pi^{Q}_{D} - \pi^{Q}_{C}]_{+} \end{pmatrix} = \sigma \begin{pmatrix} y + \frac{d-1}{\sigma} & \left[y - \Delta_{cp}\right]_{+} \\ \left[y - \Delta_{wg}\right]_{+} & [y - \Delta]_{+} \end{pmatrix}. \tag{12}$$

$$G_{DC}(y) = \begin{pmatrix} [\pi^{S}_{C} - \pi^{S}_{D}]_{+} & [\pi^{Q}_{C} - \pi^{S}_{D}]_{+} \\ [\pi^{S}_{C} - \pi^{Q}_{D}]_{+} & [\pi^{Q}_{C} - \pi^{Q}_{D}]_{+} \end{pmatrix} = \sigma \begin{pmatrix} 0 & \left[\Delta_{wg} - y\right]_{+} \\ \left[\Delta_{cp} - y\right]_{+} & [\Delta - y]_{+} \end{pmatrix}. \tag{13}$$

These matrices show the conditional imitation rates of a row individual belonging to a given population that meets a column individual from her or the other population playing the alternative strategy.

[9]One might introduce asymmetries, assuming that individuals are more willing to imitate their own kind rather than individuals belonging to the other population. We restrict to the symmetric case for simplicity of the exposition.

[10]The $y$ argument in the payoffs functions is removed when no confusion can arise.

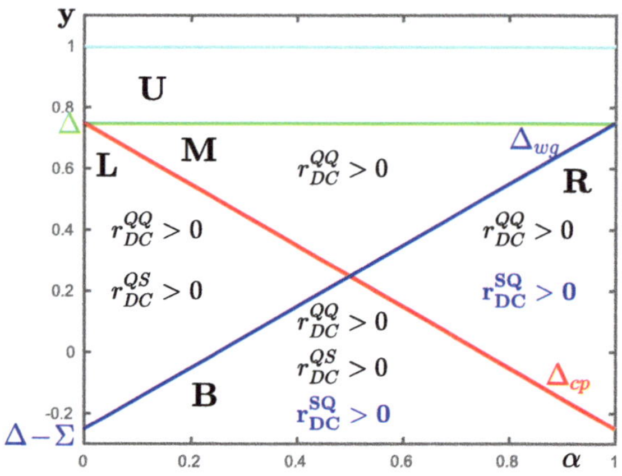


**Fig. 2** Regions

These matrices help distinguish five regions in the $\alpha - y$ plane (see Fig. 2). The payoffs comparison is dependent on the share of the total population that complies, $y$, and on whether Quixotes attain great warm-glow from compliance ($\alpha$ close to one), or strong cold-prickle from rules transgression ($\alpha$ close to zero). The warm-glow is relatively more important than the cold-prickle if $\alpha > 1/2$, and vice versa.

From the prisoner's dilemma structure of the game for Sanchos, compliant Sanchos will always imitate non-compliant Sanchos and never the reverse: $r^{\mathrm{SS}}_{\mathrm{CD}} > 0$ and $r^{\mathrm{SS}}_{\mathrm{DC}} = 0$. As for the other matching pairs, Fig. 2 displays five different regions, and highlights in each region when a non-compliant agent does worse than (and hence imitate), a compliant individual, either belonging to her own or the alternative population.[11] From the snowdrift structure of the game for Quixotes, when the share of compliance in the global population is small, $y < \Delta$, non-compliance is highly detrimental for Quixotes who will be willing to imitate compliant Quixotes, $r^{\mathrm{QQ}}_{\mathrm{DC}} > 0$. Conversely in region **U**, $y > \Delta$ and the opposite occurs: $r^{\mathrm{QQ}}_{\mathrm{CD}} > 0$.

If the warm-glow from compliance for Quixotes is strong ($\Delta_{wg} > 0$), and if $y < \Delta_{wg}$, non-compliant Sanchos might find attractive to imitate compliant Quixotes, $r^{\mathrm{SQ}}_{\mathrm{DC}} > 0$ (in regions **R** and **B** ). However, if $y > \Delta_{wg}$, then compliant Quixotes would imitate non-compliant Sanchos, $r^{\mathrm{QS}}_{\mathrm{CD}} > 0$. On the other hand, if the cold-prickle from defection is strong, ($\Delta_{cp} > 0$) and compliance is small $y < \Delta_{cp}$, then non-compliant Quixotes would be inclined to imitate compliant Sanchos, $r^{\mathrm{QS}}_{\mathrm{DC}} > 0$. Conversely, for $y > \Delta_{cp}$, compliant Sanchos would imitate non-compliant Quixotes, $r^{\mathrm{SQ}}_{\mathrm{CD}} > 0$.

[11]In the cases for which nothing is said, non-compliance provides a higher payoff.

# 4 Two Asymptotically Stable Equilibria

The evolutionary dynamics is characterized by the mean dynamics in (7)–(8), the probabilities of switching strategies in (9)–(10), and the conditional imitation rates defined in (12)–(13). For this evolutionary dynamics in variables $(p, q)$, presented in (26)–(27) in the Appendix, next proposition characterizes the different possible equilibria.

**Proposition 1** *The evolutionary dynamics in (26)–(27) presents two unstable equilibria, characterized by either full compliance,* $(p^*, q^*) = (1, 1)$, *or zero compliance* $(p^*, q^*) = (0, 0)$. *Moreover, under condition (2) there also exists a unique asymptotically stable fixed point of the evolutionary dynamics. This equilibrium can be of two types*[12]*:*

1. *Scenario* $Q$ *can be characterized by an upper bound on the warm-glow from compliance:*

$$wg \equiv \alpha\varepsilon \leq s(d-1) + (1-s)(c-d) \equiv \widehat{\alpha\varepsilon}, \tag{14}$$

*or equivalently, in terms of the share of Sanchos in the overall population:*

$$s \leq 1 - \frac{\alpha\varepsilon - (d-1)}{\sigma} \equiv 1 - \Delta_{wg} \equiv \hat{s}. \tag{15}$$

*Under this condition, the equilibrium lies within regions* **L** *or* **M***, with*

$$p^*_Q = 0, \qquad q^*_Q = \frac{1 + \Delta - \sqrt{(1-\Delta)^2 + 4(1-\alpha)s\varepsilon/\sigma}}{2(1-s)} \in (0, 1). \tag{16}$$

2. *The scenario* $SQ$ *is characterized by the opposite condition:*

$$wg \equiv \alpha\varepsilon > s(d-1) + (1-s)(c-d) \equiv \widehat{\alpha\varepsilon}. \tag{17}$$

*or equivalently,*

$$s > 1 - \Delta_{wg} \equiv \hat{s}. \tag{18}$$

*The equilibrium lies within regions* **R** *or* **B***, with:*

$$p^*_{SQ} = \frac{-2\sigma(1-s) - (d-1) + \sqrt{[d-1]^2 + 4\sigma\alpha\varepsilon(1-s)}}{2\sigma s} \in (0, 1), \quad q^*_{SQ} = 1. \tag{19}$$

[12]Subscript $Q$ highlights that only Quixotes comply, while subscript SQ indicates that both Quixotes and Sanchos comply.

*Proof* See Appendix.

Under condition (14), $(0, q_Q^*)$ is the unique asymptotically stable equilibrium, while under condition (17), the unique asymptotically stable equilibrium is given by $(p_{SQ}^*, 1)$. The evolution of the share of compliance for Sanchos and Quixotes towards either of these two equilibria is depicted in Figs. 3 and 4, for the parameters' values in (5). Additionally, Fig. 3 considers ($s = 0.5,\ \alpha = 0.6$) a lower ratio of Sanchos and a smaller warm-glow than Fig. 4 ($s = 0.8,\ \alpha = 0.7$).

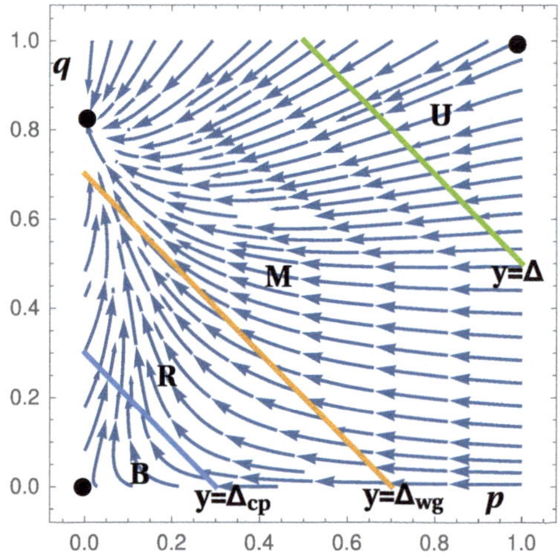


**Fig. 3** $(p, q)$ for $s = 0.5,\ \alpha = 0.6$

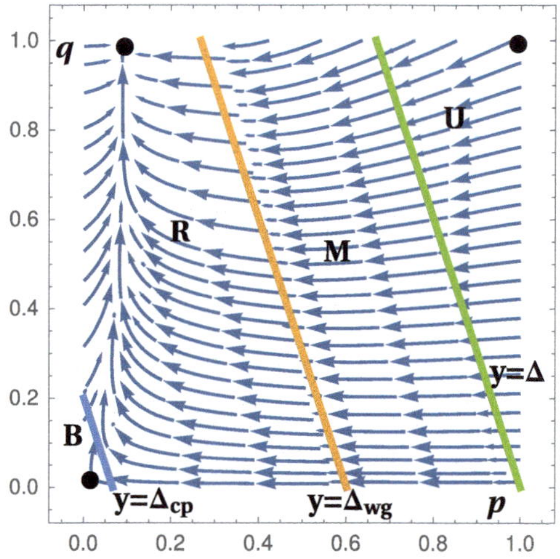


**Fig. 4** $(p, q)$ for $s = 0.8,\ \alpha = 0.7$

Under condition (17), compliant Quixotes do better than non-compliant Quixotes, and more importantly, also better than non-compliant Sanchos. And this remains true no matter how large the ratio of compliant Quixotes. Therefore, all Quixotes would end up complying. When non-compliant Sanchos who revise their strategy are paired to compliant Quixotes, some switch to compliance, which leaves a positive rate of compliance among formerly disobedient Sanchos, $p_{SQ} > 0$. The equilibrium in regions **B** or **R** is then characterized by a unitary compliance rate for Quixotes and a positive compliance rate for Sanchos (see Fig. 4). Conversely, under condition (14), there is a threshold above which non-compliant Sanchos do better. Hence, as the number of compliant Quixotes increases, non-compliance becomes more attractive and ends up being the dominant strategy in the population of Sanchos.

**Corollary 1** *The threshold $\widehat{\alpha\varepsilon} \in (d-1, c-d)$ tends to $c-d$ when $s$ tends to 0, and to $d-1$ when $s$ tends to 1. Therefore, if $\alpha\varepsilon < d-1 < \widehat{\alpha\varepsilon}$, the equilibrium in scenario $Q$ is the only feasible equilibrium regardless of the value of $s$.*

As Fig. 5 shows, scenario $SQ$ involving compliance among Sanchos requires Quixotes who attach a sufficiently large warm-glow to compliance, $wg > \widehat{\alpha\varepsilon}$. The greater the ratio of Sanchos in the overall population, $s$, the lower will be (as we will see later on) the share of compliance in the total population, $y$, and hence the stronger the reward that Quixotes obtain from compliance. Therefore, when paired with them, non-compliant Sanchos will imitate their compliant behavior. Conversely, if the ratio of Sanchos in the global population is very small, $s \to 0$, then the threshold $\widehat{\alpha\varepsilon}$ rises towards $c-d$. But since $\alpha\varepsilon \leq \varepsilon < c-d$ condition $wg > \widehat{\alpha\varepsilon}$ becomes highly demanding.

According to this corollary, a solution $SQ$, with a positive compliance rate among Sanchos, is never feasible, if the warm-glow associated to compliance is not enough, by itself, to induce compliance among Quixotes ($\alpha\varepsilon < d-1$, i.e. $\Delta_{wg} < 0$). That is, if the willingness to comply for Quixotes when many other do not require a strong cold-prickle from defection, then the payoff to compliant Quixotes is never large enough to induce imitation to non-compliant Sanchos. Therefore, Sanchos stick to their non-compliant dominant strategy. This is true regardless of how large the quantitative difference between Sanchos and Quixotes might be.

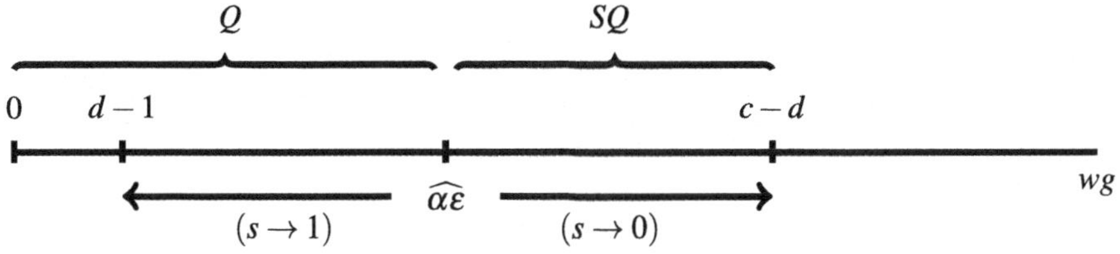


**Fig. 5** Equilibrium type and warm-glow

**Corollary 2** *A positive share of compliance among Quixotes requires:*

$$\varepsilon > \underline{\varepsilon}(s) \equiv \frac{d-1}{1-s(1-\alpha)}, \quad \textit{with} \quad \underline{\varepsilon}(s) \in \left(d-1, \frac{d-1}{\alpha}\right), \ \underline{\varepsilon}'(s) \leq 0. \tag{20}$$

Conditions in (2) are not enough to guarantee compliance in the two-population game. If the share of Sanchos in the overall population, $s$, tends to 0, then $\underline{\varepsilon}$ tends to $d-1$ and the condition (20) coincides with the lower bound in (2). However, if $s \to 1$ condition (20) converges to the condition of a positive $\Delta_{wg}$, i.e. the warm-glow from compliance must be enough to induce compliance among Quixotes even with no cold-prickle from defection. Thus, $\Delta_{wg} > 0$ is a sufficient condition for $q_Q^* > 0$, regardless of the value of $s$.

The following proposition presents the main features of the two equilibria described in Proposition 1 and depicted in Figs. 3 and 4.

**Proposition 2** *The equilibrium under scenario $Q$ is characterized by $y_Q^* > \Delta_{wg}$, $p_Q^* = 0$ and $q_Q^*$ given in (16) satisfying:*

$$q_Q^* \gtreqless \Delta \Leftrightarrow wg \equiv \alpha\varepsilon \gtreqless d - 1 + \sigma\Delta^2(1-s), \tag{21}$$

*or equivalently*

$$q_Q^* \gtreqless \Delta \Leftrightarrow s \gtreqless 1 - \frac{\Delta_{wg}}{\Delta^2}. \tag{22}$$

*The equilibrium under scenario $SQ$ is characterized by $y_{SQ}^* < \Delta_{wg}$, $q_{SQ}^* = 1$ and $p_{SQ}^* \in (0, 1)$ given in (19).*

*Proof* See Appendix.

To interpret this proposition in terms of the warm-glow, notice first that $d - 1 \leq d - 1 + \sigma\Delta^2(1-s) \leq \widehat{\alpha\varepsilon}$. As represented in Fig. 6, if the warm-glow satisfies $wg \in (d-1, d-1+\sigma\Delta^2(1-s))$, only Quixotes comply, and the share of compliance in this population remains below its equilibrium value in a single population of Quixotes, $\Delta$. A larger warm-glow, $wg \in (d-1+\sigma\Delta^2(1-s), \widehat{\alpha\varepsilon})$, would push their compliance rate above $\Delta$, but Sanchos still refuse to comply. They will only

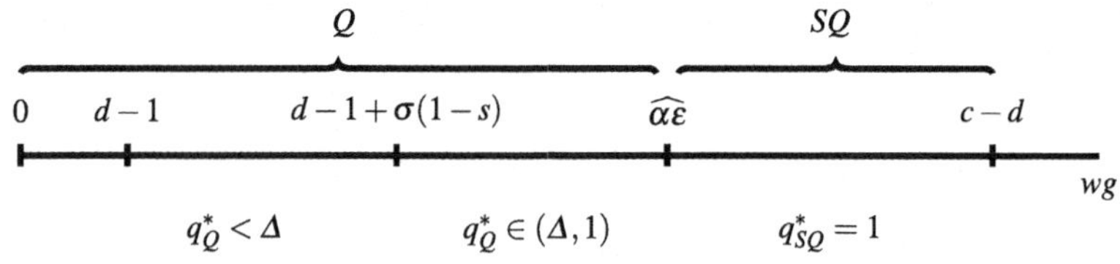


**Fig. 6** Share of compliance among Quixotes and warm-glow

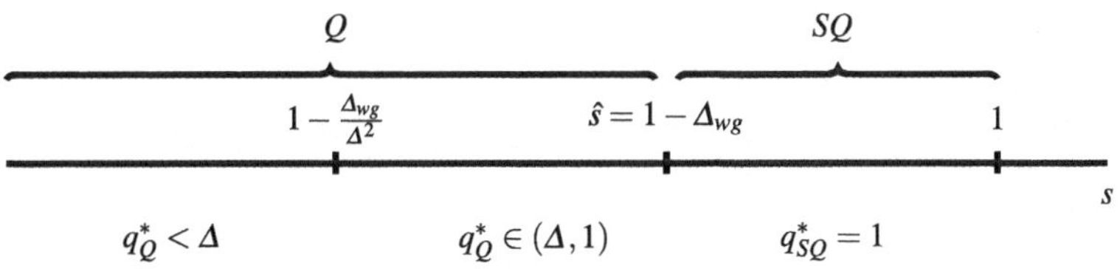


**Fig. 7** Share of compliance among Quixotes and $s$

start to imitate Quixotes if the warm-glow that these latter attach to compliance is sufficiently strong, satisfying $wg > \widehat{\alpha\varepsilon}$.

Proposition 2 can also be read in terms of the ratio of Sanchos in the overall population, $s$, as shown in Fig. 7. Assuming $\Delta_{wg} > 0$ (otherwise, scenario $SQ$ is never feasible), then, since $\Delta \in (0, 1)$, it is easy to see that:

$$1 - \frac{\Delta_{wg}}{\Delta^2} < 1 - \Delta_{wg}(\equiv \hat{s}) < 1.$$

If the ratio of Sanchos in the overall population is small, $s < \hat{s}$, only Quixotes comply. They will comply above the equilibrium in the single population game, $q_{\mathrm{Q}}^* > \Delta$, if $s \in (1-\Delta_{wg}/\Delta^2, \hat{s})$, and they will comply below $\Delta$ if the ratio of Sancho is smaller $s < 1-\Delta_{wg}/\Delta^2$. So, when the ratio of Quixotes is very large, they will be more strongly inclined to imitate the non-compliant behavior of Sanchos, showing a small share of compliance.

To better understand this dual possibility of over- and under-compliance, let us start with a single population of Quixotes at the equilibrium $q^* = \Delta$. The incorporation of some Sanchos within this population will have a twofold effect on the compliance decision of Quixotes. Firstly, a positive snowdrift effect is associated with the snowdrift nature of the game for Quixotes. Since we are in an equilibrium with zero compliance among Sanchos, the share of compliance in the global population immediately diminishes. Because each Quixote prefers compliance over universal disobedience, then a lower compliance rate increases the Quixote's incentive to comply. Secondly, a negative imitation effect is linked to the imitative revision protocol. With no Sanchos, compliant Quixotes only compare their payoff against the lower payoff of non-compliant Quixotes. However, with the entrance of some Sanchos who do not comply in this $Q$ scenario, compliant Quixotes can now be paired with non-compliant Sanchos who enjoy a larger payoff (in regions **L** and **M** in Fig. 2). This negative imitation effect, given by $q(1-p)s[y-\Delta_{wg}]_+$, is stronger the greater the ratio of Sanchos, $s$, and it is weaker the stronger is the warm-glow from compliance.

In scenario $Q$, the share of compliance in the global population, $y_{\mathrm{Q}}^*$, is at least as large as $\Delta_{wg}$ (according to Proposition 2, and also shown in Fig. 3). This ratio grows with the $wg$. However, even when Quixotes comply above $\Delta$, their over-compliance is not enough to counterbalance the zero compliance among Sanchos.

The compliance rate in the global population is always lower than the compliance rate in a world without Sanchos[13]: $y_Q^* \leq \Delta$.

In scenario $SQ$, the warm-glow surpasses $\widehat{\alpha\varepsilon}$, satisfying condition (17), then all Quixotes comply, and some Sanchos also imitate their compliant behavior. Now, although $\Delta_{wg}$ is larger, it serves as an upper bound for the share of compliance in the global population, $y_{SQ}^* \leq \Delta_{wg} \leq \Delta$. Then, in this scenario, the share of compliance in the global population is again lower than its value in the case of a single population of Quixotes, $y_{SQ}^* \leq \Delta$.

Propositions 1 and 2 suggest that large shares of compliance are associated with a strong warm-glow and a large ratio of Sanchos in the overall population. The following subsection analyzes whether this is a monotonous result. Does a higher warm-glow/a larger ratio of Sanchos increase compliance among Quixote and among Sanchos monotonously?

## 5 Differences Between Populations and Compliance Shares

This section analyzes how the share of compliance in each populations and the share of compliance in the global population are affected by the degree of dissimilarity between them. In particular we focus, on the one hand, on the gap that separates payoffs between the two populations, the warm-glow from compliance and the cold-prickle from defection experienced by Quixotes. On the other hand, we analyze how compliance is affected by the relative size of each population.

### 5.1 *The Distance Between Sanchos and Quixotes,* $(\boldsymbol{\varepsilon}, \boldsymbol{\alpha})$

From condition (17) it follows that the feasibility of an equilibrium with positive compliance among Sanchos increases with the absolute distance between Sanchos and Quixotes, $\varepsilon$, and particularly, with the warm-glow that Quixotes associate to compliance, rather than the cold-prickle to defection, measured by $\alpha$. Furthermore, as stated in the proposition below, the compliance rates for both types of agents and in both types of equilibria also increase.

**Proposition 3** *Under scenario Q,* $p_Q^* = 0$ *and*

$$\frac{\partial q_Q^*}{\partial \alpha}, \frac{\partial q_Q^*}{\partial \varepsilon} > 0 \Rightarrow \left(\frac{\partial y_Q^*}{\partial \alpha} = (1-s)\frac{\partial q_Q^*}{\partial \alpha} > 0, \frac{\partial y_Q^*}{\partial \varepsilon} = (1-s)\frac{\partial q_Q^*}{\partial \varepsilon} > 0\right), \quad \forall s \in (0, 1).$$

[13]This can be immediately proved from (16), provided that $y_Q^* = q_Q^*(1 - s)$. Graphically, as shown in Fig. 2, an equilibrium in region **L** or **M** always lies below the $y_Q^* = \Delta$ line.

*Under scenario SQ, $q^*_{SQ} = 1$ and*

$$\frac{\partial p^*_{SQ}}{\partial \alpha}, \frac{\partial p^*_{SQ}}{\partial \varepsilon} > 0 \Rightarrow \left( \frac{\partial y^*_{SQ}}{\partial \alpha} = s\frac{\partial p^*_{SQ}}{\partial \alpha} > 0, \frac{\partial y^*_{SQ}}{\partial \varepsilon} = s\frac{\partial p^*_{SQ}}{\partial \varepsilon} > 0 \right), \quad \forall s \in (0, 1).$$

*Proof* See Appendix.

At the equilibrium, the compliance rates for Sanchos (when they comply) and for Quixotes (when only they comply) increase with the discrepancy between the payoffs obtained by Sanchos and Quixotes, as measured by $\varepsilon$, and in particular, by the warm-glow from compliance for Quixotes, $\varepsilon\alpha$. Consequently, also the share of compliance in the global population rises with this discrepancy in payoffs between Sanchos and Quixotes.

Figure 8 presents the level curves for the compliance rate in the global population in the $\varepsilon - \alpha$ plane for two different values $s = 0.25$ and $s = 0.8$. The level curve $y^* = 1 - s$ is the frontier delimiting the two scenarios and represents the case in which all Quixotes comply, $q = 1$ but still no Sancho imitates this behavior, $p = 0$. In the gray region above this dotted line, the figure plots the level curves when all Quixotes comply together with some Sanchos, $y^*_{SQ} \geq 1 - s$; and below this dotted line, the level curves when only Quixotes comply, $y^*_{Q} \in [0, 1 - s]$. The region below the curve $y^* = 0$ represents $(\varepsilon, \alpha)$ combinations with null compliance among Sanchos and Quixotes.

This figure also highlights the result in Proposition 1, according to which the existence of an equilibrium with a positive compliance rate among Sanchos requires a sufficiently large gap between the payoffs attained by Sanchos and Quixotes. To reach this type of equilibrium, the less dissimilar the payoffs are ($\varepsilon$ small), the more strongly the Quixotes must value compliance ($\alpha$ large), and vice versa. If Quixotes are not too distinct from Sanchos, the latter would not imitate the former and only Quixotes would comply. In fact, if Quixotes attain roughly the same (but slightly higher) satisfaction from compliance, then it is the Quixotes who imitate Sanchos, ending up with a low, or even a zero compliance rate in the white region below the $y^* = 0$ curve.

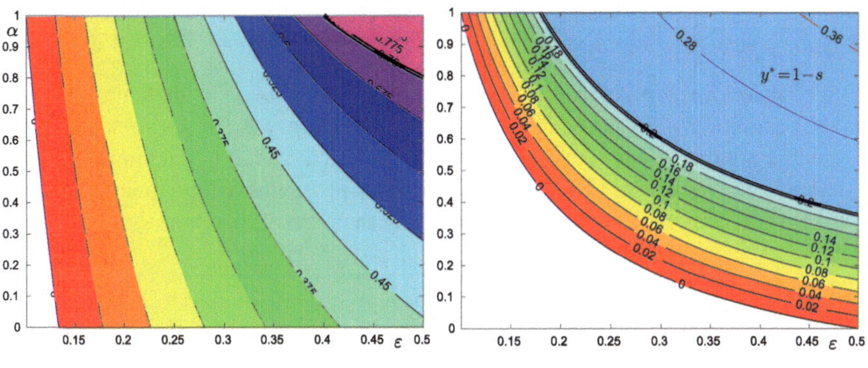


**Fig. 8** $y^*$ level curves for $s = 0.25$ (left); and $s = 0.8$ (right)

By comparing the two graphs in Fig. 8, we observe three expected results. Firstly, the share of compliance in the global population decreases with the percentage of Sanchos in the global population. Any point in the $\varepsilon - \alpha$ plane is characterized by a greater share of compliance in Fig. 8 (left) (with a lower ratio of Sanchos in the overall population). Secondly, the warm-glow required to induce compliance among Sanchos decreases with the ratio of Sanchos in the overall population. The gray area above the curve $y^* = 1 - s$ is greater for $s = 0.8$ (right) than for $s = 0.25$ (left). Finally, a greater ratio of Sanchos also makes it easier for some Quixotes to imitate the non-compliant behavior of Sanchos. Thus, a situation with null compliance in the overall population becomes more likely, represented by a greater white region below the level curve $y^* = 0$. The effect of the ratio of Sanchos in the overall population is more deeply analyzed in the following subsection.

## 5.2 The Relative Size of Each Population

This section studies how the compliance rates in each population, as well as in the global population, are affected by the size of the population of Sanchos.[14] It is worth recalling that we consider the size of the two populations as constant. Individuals may change their strategies, but their preferences remain unchanged, i.e. they remain within their population. We analyze how differences in populations' sizes affect, first the share of compliance in the global population, and second the shares of compliance in each population.

**Proposition 4** *The share of compliance in the global population decreases with the ratio of Sanchos in the overall population. This is true under scenario $Q$ and scenario $SQ$:*

$$\frac{\partial y_Q^*}{\partial s} < 0, \qquad \frac{\partial y_{SQ}^*}{\partial s} < 0. \tag{23}$$

*Proof* Under scenario $Q$, the share of compliance in the global population is $y_{\mathrm{Q}}^* = q_{\mathrm{Q}}^*(1-s)+0s$, with $q_{\mathrm{Q}}^*$ given in (16). The derivative with respect to $s$ is unequivocally negative. Likewise, for scenario $SQ$, $y_{\mathrm{Q}}^* = 1(1-s) + p_{\mathrm{SQ}}^* s$, with $p_{\mathrm{SQ}}^*$ given in (19). Again its derivative with respect to $s$ is unequivocally negative.

The share of compliance in the global population decreases with the relative size of the population of Sanchos. The reasoning is straightforward under scenario $Q$: the larger the number of non-compliant individuals in the overall population, the lower the share of compliance. Under scenario $SQ$, some Sanchos comply; meanwhile, all Quixotes do comply. Hence, again, the greater the ratio of Sanchos, the lower must

[14]Because we have normalized the total population to 1, the ratio of Quixotes is just the complementary of the ratio of Sanchos.

be the share of compliance in the global population. Although the ratio of Sanchos in the overall population has an undeniable discouraging effect in the share of compliance in the global population, its effect in each population is not so clear-cut.

**Proposition 5** *The share of compliance within the population of Quixotes under both scenarios can be expressed as a function of the ratio of Sanchos, s:*

$$q^*(s) = \begin{cases} \dfrac{1+\Delta-\sqrt{(1-\Delta)^2+4(1-\alpha)s\varepsilon/\sigma}}{2(1-s)} & \textit{if } s \leq \hat{s}\ (wg \leq \widehat{\alpha\varepsilon}), \\ 1 & \textit{if } s > \hat{s}\ (wg > \widehat{\alpha\varepsilon}). \end{cases}$$

*This function satisfies $q^*_Q(0) = \Delta$, $q^*_Q(\hat{s}) = 1$, and it reaches its minimum at:*

$$\underline{s}_q = 1 + \sigma \frac{2\Delta_{wg} - (1+\Delta)\sqrt{\Delta_{wg}}}{(1-\alpha)\varepsilon}. \tag{24}$$

*Thus, three situation are feasible:*

*1. $\alpha\varepsilon > d-1$, i.e. $\Delta_{wg} > 0$ and, denoting by $\tilde{\alpha} = 1 - \Delta(1-\Delta)\sigma/\varepsilon$:*

*(a) $\alpha < \tilde{\alpha}$. Then $\underline{s}_q \in (0,1)$ and $(q^*)'(s) < 0$ for $s \in (0, \underline{s}_q)$ and $(q^*)'(s) > 0$ for $s \in (\underline{s}_q, \hat{s})$.*

*(b) $\alpha \geq \tilde{\alpha}$. Then $\underline{s}_q \leq 0$ and $(q^*)'(s) > 0$ for any $s \in (0, \hat{s})$.*

*2. $\alpha\varepsilon < d-1$, i.e. $\Delta_{wg} < 0$. Then $\underline{s}_q$ is a complex number and $(q^*)'(s) < 0$ for all $s \in (0,1)$.*

*Proof* See Appendix.

As already mentioned, the relative size of the Sanchos' population has a twofold effect on the compliance decision of Quixotes: the snowdrift effect, which induces Quixotes to increase compliance in a world with a lower compliance rate, due to the enlargement of the portion of Sanchos; and the imitation effect, which induces Quixotes to imitate the non-compliant behavior of Sanchos.

If the warm-glow is enough to induce compliance when others disobey, $wg > d-1$, then two situations are possible. For a relatively small $\alpha \in ((d-1)/\varepsilon, \tilde{\alpha})$, Fig. 9 shows that for an initially very small ratio of Sanchos the imitation effect is stronger. An increment in this population' size would reduce compliance among Quixotes who tend to imitate the non-compliant behavior of Sanchos. As the ratio of Sanchos rises, the share of compliance decreases (because there are more Sanchos who do not comply, and because less Quixotes comply). In consequence, the snowdrift effect becomes stronger than the imitation effect and compliance start rising among Quixotes. Incidentally, when the ratio of Sanchos reaches $\hat{s}$, all Quixotes comply. Conversely, if the warm-glow is relatively strong wrt the cold-prickle, $\alpha \geq \tilde{\alpha}$, then the snowdrift effect is stronger than the imitation effect even when the size of the Sanchos' population is very small. The share of compliance among Quixotes increases monotonously with the number of Sanchos in the overall population, up until every Quixote complies (see Fig. 10).

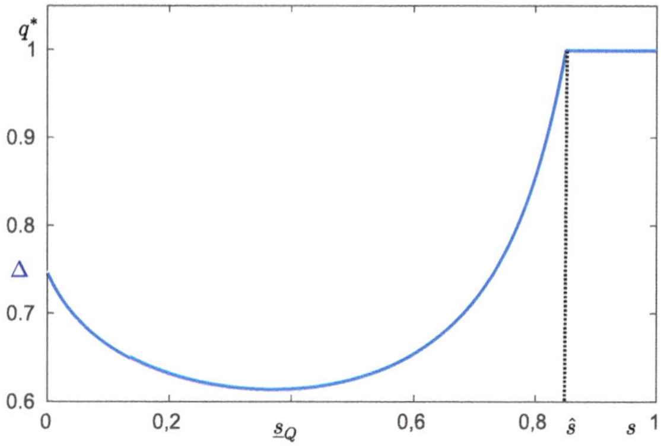


**Fig. 9** $q^*(s)$ for $\alpha = 0.4 < \tilde{\alpha} = 0.81$

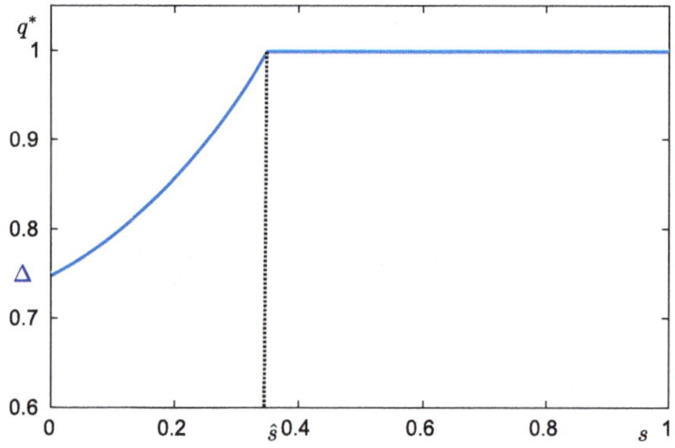


**Fig. 10** $q^*(s)$ for $\alpha = 0.9 > \tilde{\alpha} = 0.81$

By contrast, if the warm-glow is not enough to guarantee compliance when others defect, then the share of compliance in the population of Quixotes decreases monotonously with the ratio of Sanchos. The imitation effect is stronger regardless of each population's size. More and more Quixotes imitate the non-compliant behavior of a larger population of non-compliant Sanchos, as shown in Fig. 11.

The following proposition analyzes how the share of compliance in the population of Sanchos is affected by the relative size of this population.

**Proposition 6** *The share of compliance within the population of Sanchos in both scenarios can be defined as a function of the ratio of Sanchos:*

$$p^*(s)= \begin{cases} 0 & \text{if } s \leq \hat{s} \ (wg \leq \widehat{\alpha\varepsilon}), \\ \dfrac{-2\sigma(1-s)-(d-1)+\sqrt{[d-1]^2+4\sigma\alpha\varepsilon(1-s)}}{2\sigma s} & \text{if } s > \hat{s} \ (wg > \widehat{\alpha\varepsilon}). \end{cases}$$

*This function satisfies* $p^*(\hat{s}) = 0$, *and it reaches its maximum at:*

$$\bar{s}_p = \frac{-2\sigma(1-\Delta_{wg}) + (c-d+\sigma)\sqrt{1-\Delta_{wg}}}{\alpha\varepsilon}. \tag{25}$$

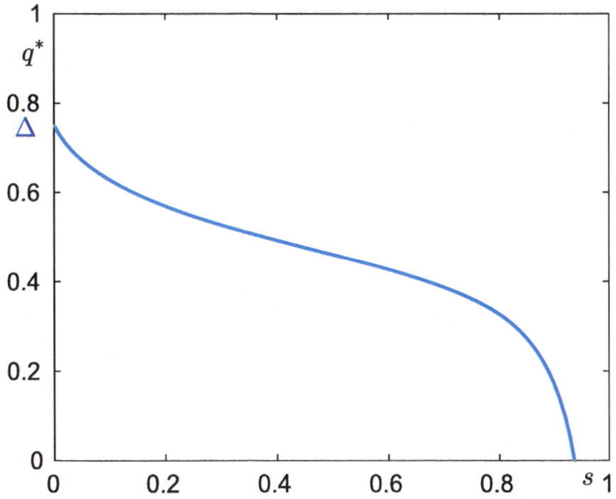


**Fig. 11** $q^*(s)$ for $\alpha = 0.2$, with $\Delta_{wg} < 0$

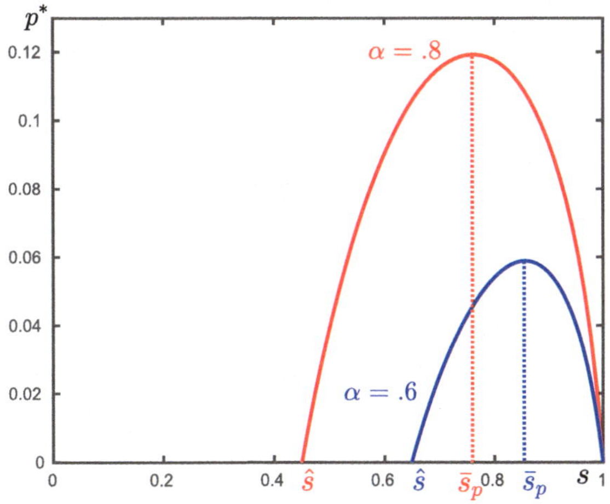


**Fig. 12** $p^*(s)$ for different $\alpha$

*Proof* Considering $p^*_{SQ}$ in (19) as a function of $s$, with the help of Mathematica, we compute the values at which $(p^*_{SQ})'(s) = 0$. This equation has a negative root and a positive root given by $\bar{s}_p$ in (25). From (19), we know that $p^*_{SQ}(\hat{s}) = p^*_{SQ}(1) = 0$, and $p^*_{SQ}(s)$ is continuous and strictly positive within the interval $(\hat{s}, 1)$. Moreover there is a unique positive value $\bar{s}_p$ satisfying $(p^*_{SQ})'(\bar{s}_p) = 0$. In consequence, this value $\bar{s}_p$ must lie within the interval $\in (\hat{s}, 1)$ and $p^*_{SQ}(s)$ must reach a maximum at this point.

As shown in Fig. 12, the relative size of the Sanchos' population has a non-monotonous effect on the compliance rate of these individuals. Again two forces are at stake here. As the number of Sanchos in the overall population grows, and the share of compliance in the global population decreases, also increases the reward to compliant Quixotes. In consequence more Sanchos are boosted to imitate their compliant behavior. This is the prevailing force up until $\bar{s}_p$. From this value on, a second and negative force dominates. As the number of Quixotes decays, although all of them comply, it becomes less and less likely for Sanchos to be paired with and to imitate their compliant behavior. Sanchos are more and more often paired to other Sanchos and the non-compliant dominant strategy in this population starts growing with $p^*_{SQ}(s)$ tending to zero as $s$ goes to 1.

**Corollary 3** *The value $\bar{s}_p$ decreases with the absolute distance which separates Sanchos from Quixotes and with the relative importance of the warm-glow with respect to the cold-prickle:*

$$\frac{\partial \bar{s}_p}{\partial \alpha} < 0, \quad \frac{\partial \bar{s}_p}{\partial \varepsilon} < 0,$$

*for all $\alpha \in (0, 1)$ and $\varepsilon$ satisfying condition (2).*

*The compliance rates increase with the absolute distance which separates Sanchos from Quixotes and with the relative importance of the warm-glow with respect to the cold-prickle:*

$$\frac{\partial p^*_{SQ}(\bar{s}_p)}{\partial \alpha} > 0, \quad \frac{\partial p^*_{SQ}(\bar{s}_p)}{\partial \varepsilon} > 0.$$

*In particular, the maximum is higher.*

*Proof* See Appendix.

According to this corollary, the more dissimilar the two populations, or the more biased towards a greater warm-glow from compliance rather than a cold-prickle from defection, the wider is the interval with positive compliance among Sanchos. Moreover, the compliance rate among Sanchos, $p^*_{SQ}(s)$, reaches its maximum for a lower ratio of Sanchos in the overall population. And this maximum is characterized by a larger compliance rate, as shown in Figs. 12 and 13.

## 5.3 Real-Life Examples

In our everyday life decisions, we continuously compare our gains with those of others implementing different options. We clearly do so when we take purchasing decisions as: whether to try a new restaurant instead of our favorite one, where to go on vacations, or whether to follow the fashion in order to look more attractive. Likewise, we also observe other people satisfaction in activities like paying taxes, contributing to voluntary associations, or recycling. We collect information from our friends and acquaintances and, more and more, we share experiences with unknown people through the use of the social networks and Internet. And we believe that it is seldom the case that we share their same preferences. Indeed although (for simplicity) we are assuming only two type of individuals, one could think that there are not just but many. Indeed, going to the extreme, each individual could have

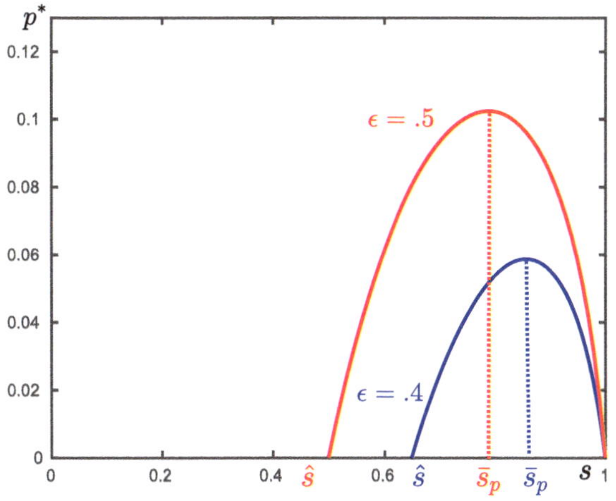


**Fig. 13** $p^*(s)$ for different $\varepsilon$

his specific payoff matrix, which could be very similar or far dissimilar to his new neighbor.

An example of how we imitate others and obtain different reward are marketing campaigns. The fancy car or the pair of jeans that we buy will rarely lead us to the success that actors pretend to enjoy in the commercials. More connected to social norms, an example could be the fund-raising television gala for charitable causes. We observe on TV how rewarding is donating for the celebrities, who act as role models or influencers. This induces some people, who would have not done otherwise, to collaborate. Marketing campaigns are often undertaken by the fiscal service on tax collection, or by the traffic authorities on road safety.

Examples of individuals having different preferences can be more clear when confronting people from clearly different cultures. An extreme example could be the missionaries, who are relatively few, whose preferences are rather different from the locals', and find highly rewarding to comply with their moral precepts.[15] Thus, as Proposition 1 predicts, it is very likely that some of the locals embrace the missionaries' religion.

To highlight the main findings of the paper, consider individuals who have grown up in a pleasant neighborhood with a well-established culture of following social norms, either because neighbors really obtain satisfaction from compliance, or because they dislike or fear the penalty associated with defection. Assume that some of them move to a dirty and noisy neighborhood where people do not find it attractive to follow social rules. Then, some of the primitive neighbors might imitate the compliant behavior of the newcomers. This is more likely the more different the preferences of the new and the primitive neighbors are, and will only occur if these latter observe that the newcomers enjoy compliance, rather than dislike defection.

[15] Please note that we do not mean any superiority of the missionaries' morality over other religions.

The compliance decisions of newcomers are influenced by two effects. Because of the deteriorated living environment, the snowdrift effect induces them to comply. Conversely, the inappropriate behavior of the primitive neighbors discourages compliance. These two effects are stronger the smaller is the share of newcomers. As stated in Proposition 5 and Figs. 9, 10, and 11, if the new neighbors dislike defection, rather than enjoy compliance, the imitation effect is stronger and they will reduce their effort, the smaller their relative size. Conversely, if they attain a strong satisfaction from compliance, the snowdrift effect prevails and the lower their numbers, the more they will comply.

Conventional neighbors also value a clean and quiet environment, but they prefer others to pay the cost. As Proposition 6 and Figs. 12 and 13 state, if there are many newcomers, conventional neighbors will free-ride on them and enjoy a better neighborhood. Conversely, if the number of newcomers is sufficiently small, the satisfaction of these latter is so large that some of the primitive neighbors will imitate them, aiming at a higher welfare. However, with even less newcomers, primitive neighbors seldom meet them, and therefore very few will imitate their compliant behavior.

Both newcomers and conventional neighbors will comply higher the more the newcomers' preferences differ from the primitive neighbors', and specially the greater their inner satisfaction from compliance.

## 6 Conclusions

The paper analyzes the compliance with social norms as a social dilemma involving two types of individuals. The dilemma for pro-self Sanchos is described by a prisoner's dilemma game. On the other hand, pro-social Quixotes still have an incentive to free-ride when others comply, but are willing to pay the cost of compliance when others deviate. For them, the social dilemma is described as a snowdrift game.

We have analyzed the compliance decisions of pro-self and pro-social norm-using individuals, when their populations are not isolated from each other. Individuals in one population play against and imitate individuals from their own and from the other population. A two-population evolutionary game is defined involving Sanchos, with a payoffs matrix characteristic of a prisoner's dilemma game, and Quixotes, who obtain warm-glow from compliance and cold-prickle from defection, and whose payoffs matrix is characteristic from a snowdrift game.

Evolutionary dynamics is defined considering an imitative revision protocol, in particular, pairwise imitation. The imitative dynamics admits a unique asymptotically stable equilibrium. The nature of this stable equilibrium depends on the characteristics which describe the populations of Sanchos and Quixotes, as well as on the size of these populations. If the warm-glow that Quixotes attach to compliance is not too large, or equivalently if the ratio of Sanchos in the global population is low, then this equilibrium is characterized by zero compliance among

Sanchos and a positive compliance rate (but not complete compliance) for Quixotes. Alternatively, if the warm-glow for Quixotes is high, then all Quixotes comply together with some Sanchos. What determines this latter equilibrium is a high degree of dissimilarity between Sanchos and Quixotes and particularly, a strong warm-glow from compliance rather than a strong cold-prickle from defection. Moreover, the size of the population of Sanchos also increases the likelihood that some Sanchos imitate the compliant behavior of Quixotes.

In the first equilibrium type, where only Quixotes comply, they can comply above or below their compliance in the case of a single population of Quixotes. Thus, at the equilibrium, they could over-comply to compensate the disobedient behavior of Sanchos, or imitate them and under-comply. Over-compliance would require a strong warm-glow from compliance or a large ratio of Sanchos in the overall population.

The absolute distance that separates Quixotes from Sanchos, or the relative size of the warm-glow over the cold-prickle, facilitates an equilibrium with a positive compliance rate among Sanchos. Moreover, the wider this gap, the higher the compliance rate in each population, and consequently in the global population. Conversely, the compliance rate in the global population decreases with the ratio of (reluctant-to-comply) Sanchos in the overall population. However, although compliance decreases globally with the percentage of Sanchos, its effect over the share of compliance in each population is diverse. In the equilibrium where only Quixotes comply, the ratio of Sanchos in the overall population may increase compliance if the warm-glow is large, or reduce compliance if it is small. With a moderate warm-glow, the share of compliance for Quixotes is a u-shaped function of the ratio of Sanchos in the overall population. In the equilibrium with full compliance among Quixotes and partial compliance among Sanchos, a larger percentage of Sanchos increases their compliance rate initially. However, as this ratio becomes close to one the non-compliant strategy becomes dominant and the share of compliance among Sanchos decreases to zero.

**Acknowledgements** This study was funded by the Spanish Government (research projects ECO2014-52343-P and ECO2017-82227-P) and by the Regional Government of Castilla y León (research projects VA024P17 and VA105G18), co-financed by FEDER funds.

## Appendix: Proof of Propositions

*Proof of Proposition 1* The system dynamics reads:

$$\frac{\dot{p}}{\sigma} = (1-p)q(1-s)[\Delta_{wg} - y]_+ \\ -p\left\{(1-p)s\left(y + \frac{d-1}{\sigma}\right) + (1-q)(1-s)[y - \Delta_{cp}]_+\right\}, \tag{26}$$

$$\frac{\dot{q}}{\sigma} = (1-q)\{q(1-s)[\Delta - y]_+ + ps[\Delta_{cp} - y]_+\}$$
$$-q\{(1-q)(1-s)[y-\Delta]_+ + (1-p)s[y-\Delta_{wg}]_+\}. \quad (27)$$

From this system, the evolution of the compliance rates among Sanchos, $p$, and Quixotes, $q$, is analyzed separately for each of the five regions in the $\alpha - y$ plane, resumed in Fig. 2. We can distinguish two situations depending on whether $\alpha < 1/2$ ($\Delta_{wg} < \Delta_{cp}$), or $\alpha > 1/2$ ($\Delta_{wg} > \Delta_{cp}$). The possible equilibria in each region and their stability are also studied.

**U**: $y > \Delta$. The dynamics reads:

$$\dot{p} = -p\{(1-y)(y-\Delta)\sigma + [(1-p)s + (1-q)(1-s)\alpha]\varepsilon]\} \leq 0, \quad (28)$$
$$\dot{q} = -q\{(1-y)(y-\Delta)\sigma + (1-p)s(1-\alpha)\varepsilon]\} \leq 0. \quad (29)$$

$\dot{p} < 0$, except if $p = 0$ or $p = q = 1$, when $\dot{p} = 0$. Similarly, $\dot{q} < 0$, except for $q = 0$ or $p = q = 1$, when $\dot{q} = 0$. The point $(0, 0) \notin \mathbf{U}$, while $(1, 1) \in \mathbf{U}$. Thus $(1, 1)$ is the only equilibrium in this region and it is unstable.

**M**: $\max\{\Delta_{wg}, \Delta_{cp}\} < y \leq \Delta$. The dynamics in this region reads:

$$\dot{p} = -p\{(1-p)s[(y-\Delta)\sigma + \varepsilon] + (1-q)(1-s)(y-\Delta_{cp})\sigma\} \leq 0,$$
$$\dot{q} = q\{(1-y)(\Delta - y)\sigma - (1-p)s(1-\alpha)\varepsilon\}.$$

As in region **U**, $\dot{p} < 0$, except if $p = 0$ or $p = q = 1$, when $\dot{p} = 0$, but $(1, 1) \notin \mathbf{M}$. Furthermore:

$$\dot{q} \gtrless 0 \Leftrightarrow (1-y)(\Delta - y)\sigma \gtrless (1-\alpha)(1-p)s\varepsilon.$$

From this dynamics, the only possible equilibrium in this region must satisfy $p = 0$. In this situation $\dot{q} = 0$ under equation:

$$[1 - q(1-s)][\Delta - q(1-s)] = (1-\alpha)s\frac{\varepsilon}{\sigma}. \quad (30)$$

Or equivalently, for $p = 0$, $\dot{q}/q$ is given by the second order polynomial in $q$:

$$q^2(1-s)^2 - q(1-s)(1+\Delta) + \Delta - (1-\alpha)s\frac{\varepsilon}{\sigma},$$

which has one stable and one unstable root. The stable root is given by (16).

**L**: $\Delta_{wg} < y \leq \Delta_{cp} \leq \Delta$ ($\alpha < 1/2$). The dynamics in this region reads:

$$\dot{p} = -p(1-p)s[y + d - 1]\sigma \leq 0,$$
$$\dot{q} = \{(1-q)y + q(1-p)s\}(\Delta - y)\sigma - s\varepsilon[\alpha(p-q) + (1-p)q)].$$

Still in this region $\dot{p} < 0$, except if $p = 0$ or $p = 1$, when $\dot{p} = 0$. For Quixotes:

$$\dot{q} \gtrless 0 \Leftrightarrow \{(1-q)y + q(1-p)s\}(\Delta - y)\sigma \gtrless s\varepsilon[\alpha(p-q) + (1-p)q)].$$

The equilibrium in this region also requires $p = 0$. Plugging this into the dynamics of $\dot{q}$ and equating to zero leads again to Eq. (30).

The equilibrium $(0, q^*_{\mathrm{Q}})$ may lie within this region **M** or region **L**. It will be located in region **M** if $\max\left\{\Delta_{wg}, \Delta_{cp}\right\} < q^*_{\mathrm{Q}}(1-s) \leq \Delta$, and it will lie in region **L** if $\Delta_{wg} < q^*_{\mathrm{Q}}(1-s) \leq \Delta_{cp} \leq \Delta$.

**R**: $\Delta_{cp} < y \leq \Delta_{wg} \leq \Delta$ $(\alpha > 1/2)$

$$\dot{p} = \{p(1-y) + (1-s)q(1-p)\}(\Delta - y)\sigma - \varepsilon\{(1-p)y + (1-s)\alpha(p-q)\},$$
$$\dot{q} = (1-q)q(1-s)(\Delta - y)\sigma \geq 0.$$

In this region $\dot{q} \geq 0$, and $\dot{q} = 0$ if $q = 0$ (which does not belong to **R**) or $q = 1$. For Sanchos, $\dot{p} \gtrless 0$ if and only if:

$$\left[y + (1-s)p\frac{1-q}{1-p}\right](\Delta - y)\sigma \gtrless \varepsilon\left[y + (1-s)\alpha\frac{p-q}{1-p}\right],$$

In the limiting case of a very small ratio of compliant Sanchos, $\dot{p}$ can be approximated by:

$$\dot{p}|_{p=0} = (1-s)q\sigma\left[\Delta_{wg} - y\right],$$

which in this region is positive, increasing in $\alpha$ and decreasing in $s$.

The equilibrium in this region requires $q = 1$, and hence $y = ps + 1 - s$. Plugging this into the dynamics $\dot{p}$, it follows that $p$ must be either equal to 1 (but $p = q = 1$ does not belong to **R**) or satisfy equation:

$$[ps + (1-s)](\Delta - ps - (1-s))\sigma = \varepsilon[ps + (1-s)(1-\alpha)]. \tag{31}$$

Equivalently, when $q = 1$, the expression $\dot{p}/((1-p)\sigma)$ is given by the second order polynomial in $p$:

$$-p^2s^2 - ps\left[2(1-s) + d - 1\right] + (1-s)\left[\Delta_{wg} - (1-s)\right],$$

which has one stable and one unstable root. The stable root is given by (19).

**B**: $y \leq \min\left\{\Delta_{wg}, \Delta_{cp}\right\} \leq \Delta$. The dynamics reads:

$$\dot{p} = (1-p)\{(\Delta - y)y\sigma - \varepsilon[y - q(1-s)\alpha)]\},$$
$$\dot{q} = (1-q)\left\{ps\sigma(\Delta_{cp} - y) + q(1-s)(\Delta - y)\sigma\right\} \geq 0.$$

Since $y < \Delta_{cp}$ then $\dot{q} \geq 0$ except if $q = 1$ or $p = q = 0$, when $\dot{q} = 0$. For Sanchos:

$$\dot{p} \gtrless 0 \Leftrightarrow (\Delta - y)y\sigma \gtrless \varepsilon[y - q(1-s)\alpha].$$

This system has three equilibria. Two unstable equilibria $(p^*, q^*) = (0, 0)$, $(p^*, q^*) = (1, 1)$ (which does not belong to this region), and a stable equilibrium with $q = 1$ and $p$ given by Eq. (31).

The equilibrium $\left(p^*_{\text{SQ}}, 1\right)$ may belong to region **R** or region **B**. It will lie in region **R** if $\Delta_{cp} < p^*_{\text{SQ}}s + (1-s) \leq \Delta_{wg} \leq \Delta$, and it will lie in region **B** if $p^*_{\text{SQ}}s + (1-s) \leq \min\{\Delta_{cp}, \Delta_{wg}\} \leq \Delta$.

It is easy to see that $q^*_{\text{Q}} > 0$ for all $s \in (0, 1)$. Moreover, under condition $s < 1 - \Delta_{wg} \equiv \hat{s}$, it can be seen that $q^*_{\text{Q}} < 1$.

It is also immediate to see that condition $s < 1 - \Delta_{wg}$ implies $p^*_{\text{SQ}} > 0$. Likewise, $p^*_{\text{SQ}} < 1$ under condition $-\alpha\varepsilon s < c - d - \alpha\varepsilon$. But this inequality is always true under condition (2).

The proof of the asymptotic stability can be found in Cabo and García (2018).

*Proof of Proposition 2* Conditions (21) and (22) are straightforward from the definition of $q^*_{\text{Q}}$ in (16).

Under scenario $Q$, condition $y^*_{\text{Q}} > \Delta_{wg}$ is equivalent to $q^*_{\text{Q}}(1-s) > \Delta_{wg}$, or:

$$\frac{1 + \Delta - \sqrt{(1-\Delta)^2 + 4(1-\alpha)s\varepsilon/\sigma}}{2} > \Delta - (1-\alpha)\frac{\varepsilon}{\sigma},$$

or equivalently:

$$1 - \Delta + 2(1-\alpha)\frac{\varepsilon}{\sigma} > \sqrt{(1-\Delta)^2 + 4(1-\alpha)s\varepsilon/\sigma}.$$

The LHS is positive, hence, if we square both sides, after some algebra, we end up with condition:

$$s < 1 - \Delta_{wg},$$

which is equivalent to condition (14) that characterizes scenario $Q$.

Likewise, under scenario $SQ$, condition $y^*_{\text{SQ}} < \Delta_{wg}$ is equivalent to $p^*_{\text{SQ}}s + (1 - s) < \Delta_{wg}$. Following a similar analysis, we end up with condition $s > 1 - \Delta_{wg}$, which is equivalent to condition (17) that characterizes scenario $SQ$.

*Proof of Proposition 3* In the scenario with $p^*_{\text{Q}} = 0$ and $q^*_{\text{Q}} > 0$,

$$\frac{\partial q^*_{\text{Q}}}{\partial \alpha} = \frac{s\varepsilon}{(1-s)\sigma\sqrt{(1-\Delta)^2 + 4(1-\alpha)s\varepsilon/\sigma}} > 0,$$

and

$$\frac{\partial q^*_{\mathrm{Q}}}{\partial \varepsilon} = \frac{\sqrt{(1-\Delta)^2 + 4(1-\alpha)s\varepsilon/\sigma} + 1 - \Delta - 2(1-\alpha)s}{2\sigma(1-s)\sqrt{(1-\Delta)^2 + 4(1-\alpha)s\varepsilon/\sigma}}.$$

This latter derivative is positive if and only if:

$$\sqrt{(1-\Delta)^2 + 4(1-\alpha)s\varepsilon/\sigma} > 2(1-\alpha)s - (1-\Delta).$$

If the RHS of this inequality was negative, $\partial q^*_{\mathrm{Q}}/\partial\varepsilon > 0$. Conversely, if the RHS was positive, raising both sides to the square, and rearranging terms,

$$s\alpha > s - \left[1 - \Delta + \frac{\varepsilon}{\sigma}\right].$$

But according to (15), an equilibrium with no compliance among Sanchos requires $s < 1 - \Delta + (1-\alpha)\varepsilon/\sigma < 1 - \Delta + \varepsilon/\sigma$. Therefore, the RHS in inequation above is negative and hence, it always holds, which proves $\partial q^*_{\mathrm{Q}}/\partial\varepsilon > 0$.

In the scenario with $q^*_{\mathrm{SQ}} = 1$ and $p^*_{\mathrm{SQ}} > 0$,

$$\frac{\partial p^*_{\mathrm{SQ}}}{\partial \alpha} = \frac{(1-s)\varepsilon}{s\sqrt{(\varepsilon - \sigma\Delta)^2 + 4\sigma\alpha\varepsilon(1-s)}} > 0,$$

$$\frac{\partial p^*_{\mathrm{SQ}}}{\partial \varepsilon} = \frac{(1-s)\alpha}{s\sqrt{(\varepsilon - \sigma\Delta)^2 + 4\sigma\alpha\varepsilon(1-s)}} > 0.$$

The marginal effect of $\alpha$ and $\varepsilon$ in the global compliance rates immediately follows, provided that $y^*_{\mathrm{Q}} = q^*_{\mathrm{Q}}(1-s)$ and $y^*_{\mathrm{SQ}} = p^*_{\mathrm{SQ}}s + (1-s)$.

*Proof of Proposition 5* Considering $q^*_{\mathrm{Q}}$ in (16) as a function of $s$, one can prove that $(q^*_{\mathrm{Q}})'(s) \gtreqless 0$ if and only if:

$$(1-\alpha)^2(1-s)^2\left(\frac{\varepsilon}{\sigma}\right)^2 - \Delta_{wg}\left(1 - \Delta^2 + 4(1-\alpha)\frac{\varepsilon}{\sigma}\right) \lesseqgtr 0. \tag{32}$$

It is ease to see that $\alpha\varepsilon < d - 1$ implies that the expression in (32) is positive, and therefore, $(q^*_{\mathrm{Q}})'(s) < 0$, regardless of the value of $s$.

For the more general case of $\alpha\varepsilon > d - 1$, the expression in (32) vanishes for two values of $s$, one greater than one and the other given by $\underline{s}_q$ in (24). From Proposition 1 we know that $q^*_{\mathrm{Q}} < 1$ for all $s \in (0, \hat{s})$ and $q^*_{\mathrm{Q}}(\hat{s}) = 1$.

- If $\underline{s}_q \in (0, \hat{s})$, then, being $q^*_{\mathrm{Q}}(s)$ a continuous function, it must hold true that $(q^*_{\mathrm{Q}})'(s) < 0$ for $s \in (0, \underline{s}_q)$ and $(q^*_{\mathrm{Q}})'(s) > 0$ for $s \in (\underline{s}_q, \hat{s})$. Therefore, if $\underline{s}_q \in (0, \hat{s})$, $q^*_{\mathrm{Q}}(s)$ would start at $q^*_{\mathrm{Q}}(0) = \Delta$, decrease to reach its minimum at $q^*_{\mathrm{Q}}(\underline{s}_q)$, and increase from that point till $q^*_{\mathrm{Q}}(\hat{s}) = 1$.

- It is nonetheless possible that $\underline{s}_q < 0$ and in that case, $q_{\mathrm{Q}}^*(s)$ would monotonously increase from $q_{\mathrm{Q}}^*(0) = \Delta$, for all $s \in (0, \hat{s})$, again to $q(\hat{s}) = 1$.

We can compute, with the help of Mathematica, the unique value of $\alpha$ at which $\underline{s}_q = 0$, given by $\tilde{\alpha} = 1 - \Delta(1 - \Delta)\sigma/\varepsilon$. Moreover, after some algebra, it can be proven that $\partial \underline{s}_q / \partial \alpha < 0$ if and only if:

$$(1 + \Delta)^2 (1 - \alpha)^2 \left(\frac{\varepsilon}{\sigma}\right)^2 + 4(1 - \Delta)^2 \Delta_{wg} \Delta > 0.$$

And this is true whenever $\alpha\varepsilon > d - 1$ and hence $\Delta_{wg} > 0$. In consequence, if $\alpha > \tilde{\alpha}$, $\underline{s}_q < 0$ and $q_{\mathrm{Q}}^*(s)$ increases for all $s \in (0, \hat{s})$, while for $\alpha < \tilde{\alpha}$, $\underline{s}_q > 0$ and $q_{\mathrm{Q}}^*(s)$ shows a u-shape within $(0, \hat{s})$.

*Proof of Corollary 3* The derivatives $\partial \bar{s}_p / \partial \alpha$ and $\partial \bar{s}_p / \partial \varepsilon$ are both negative under the same condition:

$$[2(c - d) - \alpha\varepsilon](c - d + \sigma) > 4(c - d)\sigma \sqrt{(1 - \Delta_{wg})}.$$

After some tedious algebra, this condition can be transformed to:

$$(1 - \Delta_{wg})[(c - d - \sigma)^2 + 2(c - d)] + 4\alpha\varepsilon(c - d) > 0,$$

which clearly holds.

Proving that $\partial \hat{s} / \partial wg < 0$ is straightforward.

Finally, as stated in Proposition 3, $\partial p_{\mathrm{SQ}}^* / \partial \alpha > 0$ and $\partial p_{\mathrm{SQ}}^* / \partial \varepsilon >$. Therefore, also the maximum is reached at a higher value.

## References


Alger, I., & Weibull, J. W. (2013). Homo Moralis-preference evolution under incomplete information and assortative matching. *Econometrica, 81*, 2269–2302. https://doi.org/10.3982/ECTA10637

Andreoni, J. (1988). Why free ride? *Journal of Public Economics, 37*(3), 291–304. https://doi.org/10.1016/0047-2727(88)90043-6

Andreoni, J. (1990). Impure Altruism and donations to public goods: A theory of warm-glow giving. *Journal of Econometrics, 100*(401), 464–477. https://doi.org/10.2307/2234133

Andreoni, J. (1995). Warm-glow versus cold-prickle: The effects of positive and negative framing on cooperation in experiments. *Quarterly Journal of Economics, 10*(1), 1–21.

Bontems, P., & Rotillon, G. (2000). Honesty in environmental compliance games. *European Journal of Law and Economics, 10*(1), 31–41. https://doi.org/10.1023/A:1018786721348

Breton, M., Sbragia, L., & Zaccour, G. (2010). A dynamic model for international environmental agreements. *Environmental and Resource Economics, 45*, 25–48. https://doi.org/10.1007/s10640-009-9304-6

Cabo, F., & García-González, A. (2018). Inter-population interaction in a world of Quixotes and Sanchos. *Journal of Evolutionary Economics*. Online-first.

de Young, R. (1996). Some psychological aspects of reduced consumption behavior: The role of intrinsic satisfaction and competence motivation. *Environment and Behavior, 28*(3), 358–409. https://doi.org/10.1177/0013916596283005

Doebeli, M., Hauert, C., & Killingback, T. (2004). The evolutionary origin of cooperators and defectors. *Science, 5697*, 859–862. https://doi.org/10.1126/science.1101456

Grafton, R. Q., Kompas, T., & Long, N. V. (2017). A brave new world? Kantian–Nashian interaction and the dynamics of global climate change mitigation. *European Economic Review, 99*, 31–42. https://doi.org/10.1016/j.euroecorev.2017.04.002

Güth, W. (1995). An evolutionary approach to explaining cooperative behavior by reciprocal incentives. *International Journal of Games Theory, 24*(4), 323–344. https://doi.org/10.1007/BF01243036

Güth, W., & Yaari, M. (1992). Explaining reciprocal behavior in simple strategic games: An evolutionary approach. In U. Witt (Ed.), *Explaining process and change: Approaches to evolutionary economics* (pp. 23–34). Ann Arbor, MI: University of Michigan Press.

Miller, J. H., & Andreoni, J. (1991). Can evolutionary dynamics explain free riding in experiments? *Economics Letters, 36*(1), 9–15. https://doi.org/10.1016/0165-1765(91)90047-O

Olson, M. (1971). *The logic of collective action: Public goods and the theory of groups.* Cambridge: Harvard University Press.

Ostrom, E. (2000). Collective action and the evolution of social norms. *The Journal of Economic Perspectives, 14*(3), 137–158. https://doi.org/10.1257/jep.14.3.137

Sandholm, W. H. (2010). *Population games and evolutionary dynamics*. Cambridge, MA: MIT Press

# Building Efficient Institutions: A Two-Stage Differential Game

**Fabien Ngendakuriyo and Puduru Viswanada Reddy**

**Abstract** We consider a two-stage dynamic game with a corrupt government and civil society as its players. We characterize open-loop Nash equilibria and an interior switching time from a regime with high government corruption which persists in the first stage (bad regime) to a free-corruption regime and greater institutional quality (good regime, second stage). We found that an increase of optimism (pessimism) in the society will lead the civil society to invest less (more) efforts to fight corruption whereas a corrupt government will invest more (less) efforts in repression policy. Overall, the numerical results show that the higher the efficiency of the civil monitoring effort, the efficiency of institutions and the lower the discount rate; the higher the inertia which will lead the economy to a much earlier switch to good regime with low corruption as the jump occurs early.

**Keywords** Corruption · Differential games · Regime switching

## 1 Introduction

Several studies in economics literature demonstrated that institutions are key factor determining growth and economic development (see, e.g., Acemoglu et al. 2005; De 2010; Bidner and François 2010). In a more comprehensive way, Lloyd and Lee (2018) reviewed the recent (post-2000) literature on the institutional economics analysis and pointed out that several theoretical and empirical studies reminded us that institutions matter and are a significant determinant of the long-run growth/prosperity performance of economies. In contrast, the factors leading to

F. Ngendakuriyo (✉)
Ministry of Finance Québec, Montréal, Québec, Canada
e-mail: fabien.ngendakuriyo@finances.gouv.qc.ca

P. Viswanada Reddy
Department of Electrical Engineering, Indian Institute of Technology Madras, Chennai, India
e-mail: vishwa@ee.iitm.ac.in

 
P.-O. Pineau et al. (eds.), *Games in Management Science*,
International Series in Operations Research & Management Science 280,
https://doi.org/10.1007/978-3-030-19107-8_17

failure and poor institutions such as corruption are detrimental to economic development and ultimately rise rent-seeking activities in the economy. It is in this context that Ngendakuriyo (2013) and Ngendakuriyo and Zaccour (2013, 2017) analyzed how to affect institutional change in the context where a corrupt government faces an active civil society engaged in the fight against the institutionalized-corruption. These papers analyzed dynamic interaction between citizens and government where the civil society monitoring efforts strive to improve the quality of institutions whereas the (corrupt) government's repression has the opposite effect.

In particular, following Hirschman (1970) and Dowding et al. (2000), Ngendakuriyo (2013) characterized a permanent interaction between an active civil society and a corrupt government, where civil society may protest or not against the government abuse and the government may retaliate or not any protest from the citizens. Two set of strategies were differentiated: $S_1$ for civil society, with $S_1 = \{Voice, Loyalty\}$ and $S_2$ for Government, with $S_2 = \{Retaliate, Not\ Retaliate\}$. Therefore, two cases were solved, namely a one-agent differential game (Voice, Not Retaliate) and a two-agent open-loop differential game (Voice, Retaliate).

Furthermore, Ngendakuriyo and Zaccour (2013) extended the two-agent differential game analysis (Voice, Retaliate) by characterizing the subgame-perfect feedback equilibria and assessing the circumstances under which the players are better off precommitting to a course of action (i.e., playing open-loop strategies) than adopting state-dependent strategies (i.e., feedback or Markov-perfect strategies). In addition, Ngendakuriyo and Zaccour (2017) extended the latter paper by introducing citizens heterogeneity and sequential strategic interaction through a leader–follower game, with the government acting as a leader while civil society organizations are followers who may compete or coordinate their strategies.

These situations raise the question of how to ensure that a "best/optimum" institutional level (free-corruption regime) is attained which will augment the amount of labor devoted to the productive activities. Yet Argandona (1991) argued that in the short term the institutional change may be efficient or not and that the social evolution will bring an efficiency optimum in a very long time span when experience can accumulate (and the opposition to the innovation may be defeated) and low transaction costs associated with new (best) institution may prevail. In addition, besides the policy or authority-induced institutional changes which can not guarantee efficiency, the author highlighted that changes in values and attitudes are also a causal factor of institutional change. Self-explanatory examples provided are the growing popular attitude against slavery where many voices had protested against it over centuries and the changing role of women in our society which is also the result of a change in ideas and values. Obviously, the inertia in the society tends to maintain the status-quo. In that sense, when the institutional quality is high the inertia pushes it up and the initial prevailing institutions will improve. Conversely, when the institutional quality tends to flow down the inertia pushes it down and the initial prevailing institutions will further deteriorate.

Likewise, it is of great interest to address the following question in the context of our institutional game: Under which circumstances a society with high prevalence of endogenous corruption may eradicate the corruption and switch to a society with free or very low corruption in the process of building efficient institutions? For example, the corruption perception index (CPI) of transparency international ranks countries and territories based on how corrupt their public sector is perceived to be on a scale of 0 to 100, where 0 means that a country is perceived as highly corrupt and 100 means it is perceived as very clean.[1] The scores and the ranks may change in both directions, i.e., a country may lose or gain in terms of scores or ranks over time. Intuitively, the countries with CPI scores above 50 may have more chances to improve in corruption eradication while it may require tremendous efforts to reduce the corruption in the countries with CPI scores below 50. It would be thus interesting to assess under which circumstances a country can leave the group of the most corrupt countries to be classified among the least corrupt countries as it is well known that good governance and institutions are the key engine for growth and economic development.

To answer the above research question, we extend the two-agent open-loop differential game (Voice, Retaliate) characterized in Ngendakuriyo and Zaccour (2013) by introducing the effects of inertia in the society which induce positive (negative) feedback depending on the social perception of the prevailing institutional quality. Throughout the paper, the government's retaliation is considered as government pressure to civil society through fighting back against the citizens' initiatives in fight against government corruption. This will allow us to characterize an equilibrium switching time where the equilibrium moves from a regime with corruption to a free-corruption regime. Obviously, characterizing strategies which exhibit multiple steady states reflects the Skiba phenomenon (see Skiba 1978; Dawid et al. 2017), which is largely inherent to most dynamic optimization problems displaying coexisting multiple stable steady states. Here the idea implies a two-stage analysis where government corruption persists in the first stage (bad regime) while questioning the level of efforts to be invested by the society to jump in the second stage (good regime) with low corruption and greater institutional quality.

The rest of the paper is organized as follows. Section 2 introduces the model and Sect. 3 discusses the equilibria. Section 4 presents the numerical results while Sect. 5 briefly concludes.

## 2 The Economy

We slightly modify the model in Ngendakuriyo and Zaccour (2013) by introducing the effects of optimism/pessimism in the society. We therefore consider that the economy is populated by a continuum of identical consumers who inelastically

[1] Since 1995, the Non-Governmental Organization Transparency International (henceforth TI) provides an index of perception of corruption for a number of countries across the world.

supply labor to produce the output according to the following additive production function for tractability purposes:

$$Y(t) = \alpha q(t) + \theta(t) F(L(t)),$$

where $L(t)$ is the amount of labor, $q(t)$ is the institutional quality, and $\theta(t)$ is the total factor productivity (TFP) at time $t$, with $t \in [0, \infty)$ and $\alpha$ is a positive parameter measuring the external impact of the institutional quality in the economy.

We assume that the evolution of the institutions can be well approximated by the following linear-differential equation:

$$\dot{q}(t) = \begin{cases} bw(t) - \beta x(t) - a, & q(t) < q_{th} \\ bw(t) - \beta x(t), & q(t) = q_{th} \\ bw(t) - \beta x(t) + a, & q(t) > q_{th} \end{cases}, \quad q(0) = q_0 \geq 0 \tag{1}$$

where $w(t)$ is the civil monitoring effort, $x(t)$ is the government pressure, and $b$ and $\beta$ are positive parameters. The parameter $b$ captures the efficiency of the civil society's monitoring while $\beta$ captures the impact of government pressure on the institutional change and the term $a$ captures the degree of inertia in the society with $a \geq 0$. If the perception of the institutional quality is too damaged, unless a strong social activism, it will deteriorate even worse. More precisely, we assume that if institutional quality improves beyond a certain threshold $q_{th}$, then optimism in the society sets in and the inertia will push it up that further improves institutional quality. Likewise, if institutional quality deteriorates below the threshold $q_{th}$, then inertia will push it down and the institutional quality will continue decreasing.

The output produced is shared between a corrupt agent (Government) and a non-corrupt one (Consumer). The corrupt agent takes a share $\phi$ of the public goods production, where $\phi \in (0, 1)$ and the non-corrupt agent the share $(1 - \phi)$. In our framework, we assume that corruption depends positively on the government pressure such that $\phi = \phi(x)$ with $\phi'(x) > 0$.

The consumers' participation in the civil society reduces the amount of labor devoted to the production sector. Assuming that the time available to each consumer is normalized to one, then the time-allocation constraint for consumers is $L(t) + w(t) = 1$. Consequently, the level of consumption at time $t$ is $C(t) = (1 - \phi(x(t)))Y(t)$ and the production function becomes

$$Y(t) = \alpha q(t) + \theta F(1 - w(t)).$$

To keep the model as simple as possible, without however much loss in qualitative insight, we retain an additive specification of the production function with an AK form for the second term with constant TFP, $Y(t) = \alpha q(t) + \theta L(t)$; a quadratic cost function of civil monitoring effort $f(w(t)) = \dfrac{w^2(t)}{2}$ and a quadratic cost function of implementing punishment mechanisms $g(x(t)) = \frac{x^2(t)}{2}$. Furthermore, linear utility functions for players, corresponding to their shares in production, that is,

$$U_G(t) = \phi(x)[\alpha q(t) + \theta(t)(1 - w(t))],$$
$$U_C(t) = (1 - \phi(x))[\alpha q(t) + \theta(t)(1 - w(t))],$$

where $G$ stands for $G$overnment and $C$ for $C$onsumer. Furthermore, we take a linear corruption technology $\phi(x) = \kappa x$. Denoting by $\rho$ the common discount rate, the optimization problems of Government and a representative Consumer are as follows:

$$V_G = \max_{x_t} \int_0^\infty e^{-\rho t} [U_G(t) - g(x(t))] dt,$$
$$V_C = \max_{w_t} \int_0^\infty e^{-\rho t} [U_C(t) - f(w(t))] dt,$$

subject to the dynamics in (1).

Omitting from now on the time argument when no ambiguity may arise and substituting for $U_G, g(x), U_C$, and $f(w)$, the two players' optimization problems become

$$V_G = \max_{x_t} \int_0^\infty e^{-\rho t} \left[\kappa x(\alpha q + \theta(1 - w)) - \frac{x^2}{2}\right] dt, \tag{2}$$

$$V_C = \max_{w_t} \int_0^\infty e^{-\rho t} \left[(1 - \kappa x)(\alpha q + \theta(1 - w)) - \frac{w^2}{2}\right] dt. \tag{3}$$

subject to the dynamics in (1).

To wrap up, by (1) and (2)–(3) we have defined a two-player two-stage differential game with two control variables $w(t), x(t)$ and one state variable $q(t)$, with $w \in [0, 1]$ and $0 \leqslant x < \frac{1}{\kappa}$.

## 3 Equilibria

We characterize and compare the open-loop Nash equilibria based on hybrid maximum principle in the two regime, namely regime 1 where $q(t) < q_{th}$ and regime 2 where $q(t) > q_{th}$. We also characterize the equilibrium switching time where the equilibrium moves from regime 1 to regime 2. Although an open-loop equilibrium is not sub-game perfect, a Markov-perfect solution would make the problem intractable.

Let $(w^*, x^*)$ represent the open-loop Nash equilibrium. The open-loop Nash equilibria are obtained by solving simultaneously the following two hybrid optimal control problems[2]:

$$(P_C) \quad V_C(q(0), w^*, x^*) \geq V_C(q(0), w, x^*), \ \forall w \in [0,\ 1]$$

$$\text{subject to } \dot{q}(t) = \begin{cases} (bw(t) - \beta x^*(t)) - a, & q(t) < q_{th} \\ bw(t) - \beta x(t), & q(t) = q_{th} \\ (bw(t) - \beta x^*(t)) + a, & q(t) > q_{th} \end{cases}, \quad q(0) = q_0 \geq 0,$$

$$(P_G) \quad V_G(q(0), w^*, x^*) \geq V_G(q(0), w^*, x), \ \forall x \in [0,\ 1/\kappa]$$

$$\text{subject to } \dot{q}(t) = \begin{cases} (bw^*(t) - \beta x(t)) - a, & q(t) < q_{th} \\ bw(t) - \beta x(t), & q(t) = q_{th} \\ (bw^*(t) - \beta x(t)) + a, & q(t) > q_{th} \end{cases}, \quad q(0) = q_0 \geq 0.$$

In the interior of each regime the necessary conditions follow from the regular maximum principle and at the switching instant, the necessary conditions are patched using some consistency conditions. As is usual in an infinite-horizon setting, we seek stationary strategies and focus the analysis on steady-state values.

**Proposition 1** *Assuming an interior solution in regime $i$, the unique stable steady-state open-loop Nash equilibrium is given by*

$$x_i^{ss} = \frac{b(b\alpha - \theta\rho) + a_i\rho}{b\kappa(b\alpha - \theta\rho) + \beta\rho},$$

$$w_i^{ss} = \frac{(b\alpha - \theta\rho)(\beta - a_i\kappa)}{b\kappa(b\alpha - \theta\rho) + \beta\rho}$$

*and the institutional quality by*

$$q_i^{ss} = \frac{b\kappa^2\rho\theta^2 - \alpha b\beta\kappa\theta + \alpha\beta^2 - b\rho\theta - \beta\kappa\rho\theta - \beta\kappa\rho\theta^2 + a_i\kappa^2\rho\theta^2 + a_i\rho}{\alpha\kappa\rho\,(\beta - b\kappa\theta)} + \frac{(\beta - a_i\kappa)\left(b^2 + 2\kappa\theta b\beta - \beta^2\right)}{\kappa\,(\beta - b\kappa\theta)\,(b\kappa(b\alpha - \theta\rho) + \beta\rho)}$$

*where $a_1 = -a$ and $a_2 = a$. The above steady-state values require that $0 < w_i^{ss} < 1, 0 < \kappa x_i^{ss} < 1$ and $q_i^{ss} > 0$, for $i = 1, 2$, and $q_1^{ss} < q_2^{ss}$.*

---

[2]See Tomiyama (1985), Makris (2001), Shaikh and Caines (2007), Seierstad and Stabrun (2010), and Long et al. (2017) for a study of multi-stage or hybrid optimal control problems.

*Proof* See the Appendix. ■

It can be shown that equilibrium steady-state values are positive if $b\alpha - \theta\rho > 0$, $b(b\alpha - \theta\rho) > a\rho$ and $\beta > a\kappa$ implying that $a \in [0, \min(\frac{\beta}{\kappa}, \frac{b(b\alpha-\theta\rho)}{\rho})]$. Furthermore, the condition $w_i^{ss} < 1$ is fulfilled for the values of parameter a such that $-A < a < A$, with $A = \left[b + \frac{\beta\rho}{\kappa(b\alpha-\theta\rho)} - \frac{\beta}{\kappa}\right]$. Obviously one needs $A > 0$ to guarantee that this condition holds for $a > 0$

Observe that the impact of the parameter $a$ on the control policies is as follows:

$$\frac{\partial x_1^{ss}}{\partial a} < 0, \quad \frac{\partial w_1^{ss}}{\partial a} > 0, \quad \frac{\partial x_2^{ss}}{\partial a} > 0, \quad \frac{\partial w_2^{ss}}{\partial a} < 0$$

and hence, implying that increasing pessimism in the society lead the civil society to invest more efforts to fight corruption whereas a corrupt government will invest less effort as the prevailing worse environment is favorable to rent seeking and predatory activities. Likewise, increasing optimism in the society lead the civil society to invest less efforts to fight corruption and thus more time is allocated to productive activities whereas a corrupt government will invest more effort as the prevailing environment is detrimental to rent seeking and predatory activities.

In addition, it can be shown that the impact of the parameter $a$ on the equilibrium points of the institutional quality at steady state is given by

$$\frac{dq_2^{ss}}{da} = -\frac{dq_1^{ss}}{da} = \frac{\rho\kappa^2\theta^2 - \alpha b\kappa^2\theta + \alpha\beta\kappa + \rho}{\alpha\kappa(\alpha\kappa b^2 - \kappa\rho\theta b + \beta\rho)}. \tag{4}$$

implying that increasing values of the parameter $a$, the equilibrium points move further away. Moreover, the parameters are chosen to satisfy $\rho\kappa^2\theta^2 - \alpha b\kappa^2\theta + \alpha\beta\kappa + \rho > 0$ to ensure that these steady-state values satisfy the separability condition $q_1^{ss} < q_2^{ss}$ meaning that more efficient institutions prevail in regime 2.

**Proposition 2** *Assume the equilibrium state trajectory $q^*(.)$ undergoes a transition from regime i to regime j at time $\tau$, with $i, j \in \{1, 2\}$ and $i \neq j$, that is $q^*(\tau) = q_{th}$ with $\tau \in [0, \infty]$. From the necessary conditions for optimality associated with the hybrid optimal control problems* (2) *and* (3)*, there exists an interior equilibrium switching time $\tau$ for which the equilibrium state trajectory enters regime j from regime i and ensures that the following Hamiltonian continuity conditions hold true:*

$$H_C^i(q^*(\tau), w^*(\tau), x^*(\tau), \lambda_C^i(\tau)) = H_C^j(q^*(\tau), w^*(\tau), x^*(\tau), \lambda_C^j(\tau)), \tag{5}$$

$$H_G^i(q^*(\tau), w^*(\tau), x^*(\tau), \lambda_G^i(\tau)) = H_G^j(q^*(\tau), w^*(\tau), x^*(\tau), \lambda_G^j(\tau)). \tag{6}$$

*Proof* See the Appendix. ■

It can be shown that under these matching conditions, an interior switching time, i.e., $\tau > 0$, exists and the jumps $\lambda_C^j(\tau) - \lambda_C^i(\tau)$ and $\lambda_G^j(\tau) - \lambda_G^i(\tau)$ can be calculated when the equilibrium state trajectory undergoes a transition from regime $i$ to regime $j$ with $i, j \in \{1, 2\}$ and $i \neq j$.

At steady state, the Jacobian matrix associated with the dynamical system in both the regimes (9) and (11) is given by:

$$\begin{bmatrix} -\frac{\alpha\kappa(\beta - b\kappa\theta)}{\kappa^2\theta^2+1} & \frac{b(b+\beta\kappa\theta)}{\kappa^2\theta^2+1} & \frac{\beta(\beta-b\kappa\theta)}{\kappa^2\theta^2+1} \\ \frac{\alpha^2\kappa^2}{\kappa^2\theta^2+1} & \rho - \frac{\alpha b\kappa^2\theta}{\kappa^2\theta^2+1} & -\frac{\alpha\beta\kappa}{\kappa^2\theta^2+1} \\ -\frac{\alpha^2\kappa^2}{\kappa^2\theta^2+1} & \frac{\alpha b\kappa^2\theta}{\kappa^2\theta^2+1} & \rho + \frac{\alpha\beta\kappa}{\kappa^2\theta^2+1} \end{bmatrix}.$$

The eigenvalues of the above Jacobian matrix are

$$\zeta_1 = \rho,$$

$$\zeta_{2,3} = \frac{\rho}{2} \pm \frac{\sqrt{4b\alpha\kappa^2(b\alpha - \rho\theta) + 4\beta\alpha\kappa\rho + \kappa^2\rho^2\theta^2 + \rho^2}}{2(\kappa^2\theta^2 + 1)}.$$

This implies that the equilibrium $(x_i^{ss}, w_i^{ss})$, $i = 1, 2$ is a saddle point in the space $(q, \lambda_C, \lambda_G)$ since one eigenvalue among the three is negative and the two others are positive. From the center manifold theorem, there exists a one dimensional stable manifold in both the regimes. We denote $W_1^s$ and $W_2^s$ as the stable manifolds associated with the equilibrium points in regimes 1 and 2, respectively. Next, the equilibrium co-states can be parametrized uniquely, along the stable manifolds, as functions of the equilibrium state $q^*(t)$ as $\lambda_C^i(t) = P_C^i(q^*(t))$ and $\lambda_G^i(t) = P_G^i(q^*(t))$ for $i = 1, 2$, so that $W_i^s := \{(q^*(t), P_C^i(q^*(t)), P_G^i(q^*(t))),\ t \in [0, \infty)\}$ for $i = 1, 2$. The equilibrium state trajectory then satisfies

$$\dot{q}^*(t) = \begin{cases} f(q^*(t), P_C^1(q^*(t)), P_G^1(q^*(t))) - a & q^*(t) < q_{th} \\ f(q^*(t), P_C^2(q^*(t)), P_G^2(q^*(t))) & q^*(t) = q_{th} \\ f(q^*(t), P_C^2(q^*(t)), P_G^2(q^*(t))) + a & q^*(t) > q_{th}, \end{cases} \tag{7}$$

where the function $f(.)$ is obtained by substituting for $x$ and $w$ in (1) using (8). Starting at any initial quality $q_0$ we denote $V_C^i(q_0)$ and $V_G^i(q_0)$ to be the objectives (5) and (6) evaluated along the stable manifold $W_s^i$ corresponding to regime $i$

$$V_C^i(q_0) = \int_0^\infty e^{-\rho t}\left[\left(\alpha q^*(t) + \theta(1 - w^*(t))\right)(1 - \kappa x^*(t)) - \frac{w^{*2}(t)}{2}\right] dt,$$

$$V_G^i(q_0) = \int_0^\infty e^{-\rho t}\left[\left(\alpha q^*(t) + \theta(1 - w^*(t))\right)(\kappa x^*(t)) - \frac{x^{*2}(t)}{2}\right] dt.$$

Notice, here the equilibrium objectives may not exist for all the values of $q_0$, that is the equilibrium objectives may not be global functions of $q_0$. This situation arises because the stable manifold $W_j^s$ corresponding to regime $j$ fails to have a suitable extension in regime $i$ and vice versa. These local functions capture the optimal behavior of both players with the constraint that the equilibrium state trajectory $q^*(.)$ converges to a locally stable steady state.

Since there exist multiple steady states we are interested in which steady state will be attained using the Nash equilibrium strategies. More pertinently, we are interested in locating the initial conditions where these local equilibrium objective functions intersect. It is clear that at these points[3] the both players receive equal values from the trajectories converging to different steady states. These initial conditions are defined formally as follows:

**Definition 3 (Indifference Point)** Let $q_I \in [q_{\min}, q_{\max}]$ be such that there are two equilibrium policies such that the associated state trajectories $q_a(.)$ and $q_b(.)$ satisfy $q_a(0) = q_b(0) = q_I$ and $q_a(t) \neq q_b(t)$ for some $t \in [0, \infty)$. Furthermore, if the equilibrium values satisfy the conditions $V_C^1(q_I) = V_C^2(q_I)$ and $V_G^1(q_I) = V_G^2(q_I)$, then $q_I$ is called an indifference point.

Clearly, starting at these points the players are *indifferent* in choosing between the equilibrium policies which are obtained along the stable manifolds $W_s^1$ and $W_s^2$ for regime 1 and 2, respectively.

**Definition 4 (Threshold Point)** Let $\hat{q} \in [q_{min}, q_{max}]$ be such that in the neighborhood of $\hat{q}$ there exist two equilibrium state trajectories $q_a(.)$ and $q_b(.)$ such that $\lim_{t\to} q_a(t) \neq \lim_{t\to} q_b(t)$, that is, the trajectories converge to different limit sets.

As it is not possible neither to analytically solve for the interior switching time $\tau$ nor to compare analytical expressions of the equilibrium steady-state values obtained in the two regimes, we turn to numerical analysis.

## 4 Numerical Illustration

The model has eight parameters, namely $b$, $\beta$, $\alpha$, $\kappa$, $\rho$, $\theta$, $a$, and $q_{th}$. Taking into account non-negativity constraints on control and state variables and production, as well as the stability condition of the steady state, we fix the base-case parameter values as follows: $b = \beta = 0.45$, $\alpha = 0.3$, $\kappa = 0.2$, $\rho = 0.08$, $\theta = 1$. The parameters $a$ and $q_{th}$ are chosen so that there exist two saddles on either side of the discontinuity line $q = q_{th}$. This condition is met by setting $q_{th} = \frac{1}{2}(q_1^{ss} + q_2^{ss})$. When $a = 0$ then $q_1^{ss} = q_2^{ss}$, and saddle points coincide. For the base-case scenario, consider $a = 0.15$.

[3]In the optimal control context, such points are called Skiba points or DNSS points (see Skiba 1978; Sethi 1977; Dechert and Nishimura 1983).

**Table 1** Equilibrium results for the benchmark case

| Regime | $w^{ss}$ | $x^{ss}$ | $q^{ss}$ | $Y^{SS}$ | $V_C$ | $V_G$ |
|---|---|---|---|---|---|---|
| Regime 1 (bad regime) | 0.6447 | 0.3114 | 5.7563 | 1.7900 | 21.7839 | 1.0126 |
| Regime 2 (good regime) | 0.5641 | 0.8974 | 18.5524 | 5.6607 | 59.5877 | 8.4386 |

## 4.1 Comparison

Now, we compare the two equilibrium steady-state values. For illustration purposes, we provide in Table 1 the results in the benchmark case.

These results show that the regime 2 of low or free-corruption regime yields a much higher institutional quality and output compared to regime 1 of high government corruption. Furthermore, the civil society and the government obtain higher payoffs in regime 2 than in regime 1. The rapacity of the government is also higher in the good regime meaning that inertia allows the institutional quality to grow faster and the government takes advantage by imposing a higher pressure to steal a higher share of output, and still the institutional quality does not decay. In addition, the civil society invests more monitoring efforts in regime 1 of high government corruption and less efforts once the society is in the regime 2 of greater institutional quality. In contrast, a corrupt government invests more efforts in its repression policy in the latter regime as the environment is not favorable to rent-seeking activities compared to the former. In all our numerical simulations, we observe the same phenomena as in this benchmark case, and therefore, state the following claim:

*Claim 1* The equilibrium steady-state values compare as follows:

$$w_2^{ss} < w_1^{ss}, x_2^{ss} > x_1^{ss}, q_2^{ss} > q_1^{ss},$$
$$Y_2^{ss} > Y_1^{ss}, V_C^2 > V_C^1, V_G^2 > V_G^1.$$

## 4.2 Sensitivity Analysis

Keeping $b, \beta, \alpha, \kappa, \rho, \theta$ at their base-case values, the sensitivity analysis obviously focuses on the parameters $a$ and $q_{th}$. From (4) the equilibrium points move further away from $q_{th}$ for increasing values of $a$.

Figure 1 illustrates the equilibrium payoffs when the parameter $a$ is set to its base-case value, i.e., $a = 0.15$. Notice that both the payoffs $V_C^1$ and $V_C^2$ are local as they are not defined for the entire state space. Here, the stable manifold corresponding to saddle point in regime 2 could not be extended below $\hat{q}$, and as a result $V_C^2$ is not defined for $q(0) < \hat{q}$. Again, the stable manifold corresponding to regime 1 could not be extended in regime 2. So, $V_C^1$ is defined only for $q(0) \in [0, q_{th}]$. Furthermore, it was observed that for all the initial conditions $q(0) \in [\hat{q}, q_{th}]$ $V_C^2 > V_C^1$ and

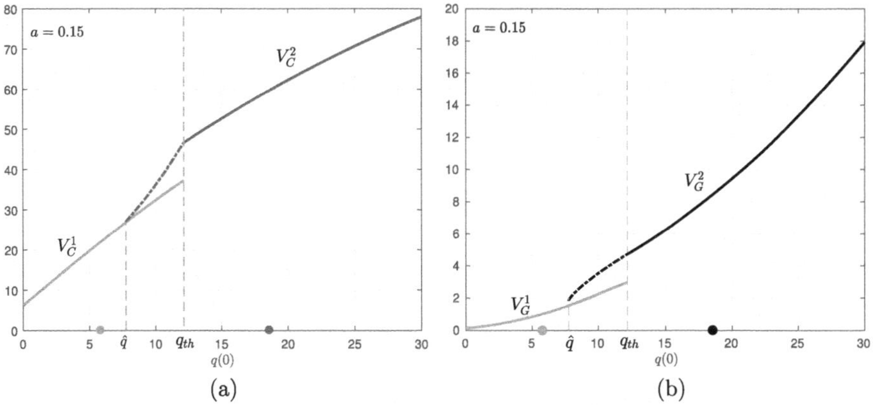


**Fig. 1** The dark bold curve represents the (local) function $V_C^2(V_G^2)$ in regime 2 and the gray bold curve represents the (local) function $V_C^1(V_G^1)$ in regime 1. The dot-dashed dark bold curve represents the local function $V_C^2(V_G^2)$. Here, the stable manifold $W_s^2$ could be extended to regime 1 for values of $q(0) \in [\hat{q},\ q_{th}]$. The equilibrium points in regime 1 and regime 2 are illustrated by the gray and dark dots, respectively, on the $q$-axis. The (local) functions $V_C^2(V_G^2)$ is continuous and has a kink at the discontinuity line $q_{th}$. (**a**) $V_C^i$ represents the equilibrium payoff associated with reaching the equilibrium in regime $i$. (**b**) $V_G^i$ represents the equilibrium payoff associated with reaching the equilibrium in regime $i$

$V_G^2 > V_G^1$. For these initial conditions, players are better off choosing the Nash equilibrium strategy which results in the state variable converging to equilibrium in regime 2. Whereas for all the initial conditions $q(0) \in [0, \hat{q})$ the players are better off choosing the Nash equilibrium strategies which results in the state variable converging to the equilibrium in regime 1. Implicitly, the initial condition $\hat{q}$ has a threshold property as defined in the previous section. Furthermore, we notice that the functions $V_C^2$ and $V_G^2$ are continuous but kinked at the discontinuity line $q_{th}$, which is a result of Hamiltonian continuity property also elaborated in the previous section.

Intuitively, the above results suggest that as building efficient institutions is a process, as long as the institutional quality is lower than a certain level far from the threshold (here, $q(0) \in [0, \hat{q})$), the institutional inertia will prevail and the economy will stick to bad regime with high corruption and social pessimism. In the contrary, if the institutional quality is initial higher and exceeds a certain level approaching the threshold (i.e., $q(0) \in [\hat{q}\ q_{th}]$), institutional changes can accumulate towards the threshold and the economy will switch to good regime with low/free corruption and high social optimism.

Figure 2 illustrates the equilibrium payoffs when the parameter $a$ is increased to 0.25. We notice that the equilibrium points in the regimes exist and move further apart. Again, there exists a threshold point $\hat{q}$ and the qualitative behavior of the Nash equilibrium strategies is similar to the case with $a = 0.15$.

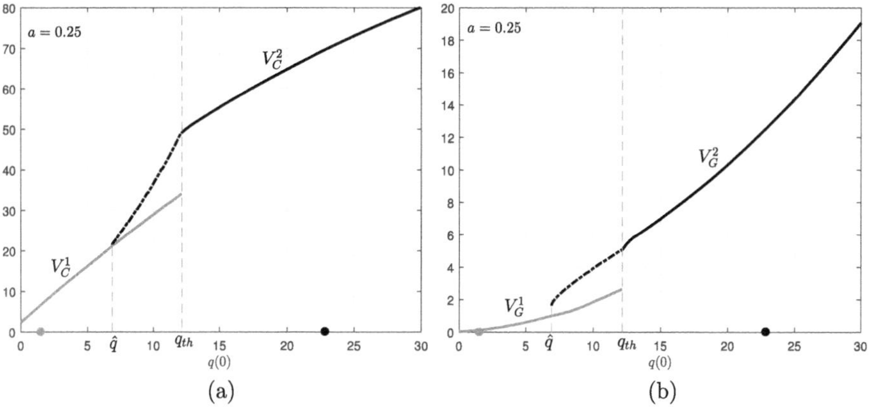


**Fig. 2** With increasing values of the parameter $a$, the equilibrium points move further away. The qualitative behavior of the value functions is similar to the case with $a = 0.15$. (**a**) $V_C^i$ represents the equilibrium payoff associated with reaching the equilibrium in regime $i$. (**b**) $V_G^i$ represents the equilibrium payoff associated with reaching the equilibrium in regime $i$

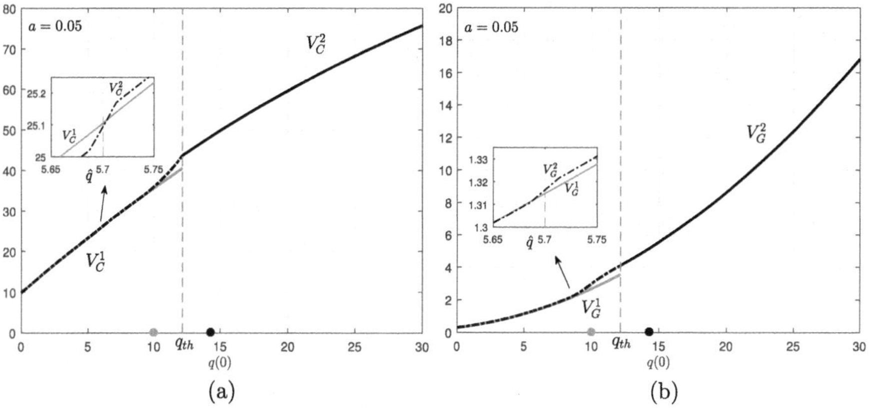


**Fig. 3** With decreasing values of the parameter $a$, the equilibrium points move closer. The function $V_C^2$ is global as the stable manifold $W_s^2$ could be extended to regime 1 for the values $q(0) \in [0,\ q_{th}]$. Here, $\hat{q}$ is both an indifference point and a threshold point. (**a**) $V_C^i$ represents the equilibrium payoff associated with reaching the equilibrium in regime $i$. (**b**) $V_G^i$ represents the equilibrium payoff associated with reaching the equilibrium in regime $i$

Figure 3 illustrates the equilibrium payoffs when the parameter $a$ is decreased to 0.05. We notice that the stable manifold in regime 2 could be extended in regime 1 for all the values $q \in [0\ q_{th}]$, and as a result $V_C^2$ is defined globally, whereas the stable manifold $W_S^1$ could not be extended in regime 2. So, $V_C^1$ is locally defined for values $q(0) \in [0,\ q_{th}]$. Furthermore, we observe that at $q > 5.68$, the value functions $V_C^2 = V_C^1$ and $V_G^2 > V_G^1$, so the players are better off choosing the Nash equilibrium strategies which result in the state variable reaching equilibrium in regime 2. For initial conditions $q = \hat{q} = 5.68$ we notice that $V_C^2 = V_C^1$ and $V_G^2 = V_G^1$. This implies that players are indifferent to choosing Nash equilibrium strategies which converge to the equilibrium points in both the regimes. Finally, for the initial conditions $q(0) < \hat{q}$ we notice that $V_C^2 < V_C^1$ and $V_G^2 \approx V_G^1$, implying that players are better off choosing Nash equilibrium strategies which result in the state variable converging to the equilibrium in regime 1. So, following Definitions 3 and 4 the initial condition $q(0) = 5.68$ is both an indifference and a threshold point.

## 4.3 Switching Time

Now, we study the effect of variation of the parameter $a$ on the switching time of the equilibrium strategies. Figures 4 and 5 illustrate the Nash equilibrium strategies of players which result in the state variable converging to the equilibrium point in regime 2. Here, the initial condition set as $q(0) = \hat{q} + \epsilon,\ \epsilon > 0$. The dark bold line represents the strategy $x^*(t)$ and the gray bold line represents the strategy $w^*(t)$.

We notice that the equilibrium strategies experience a jump at the discontinuity line $q(t) = q_{th}$. Furthermore, for higher values of $a$ the jump occurs early. Following the non-negativity conditions of the equilibrium steady-state values, it can be shown that the higher the efficiency of the civil monitoring effort $b$, the efficiency of the institutions $\alpha$ and the lower the discount rate $\rho$; the higher the inertia in the society and the economy will switch to the good regime much earlier.

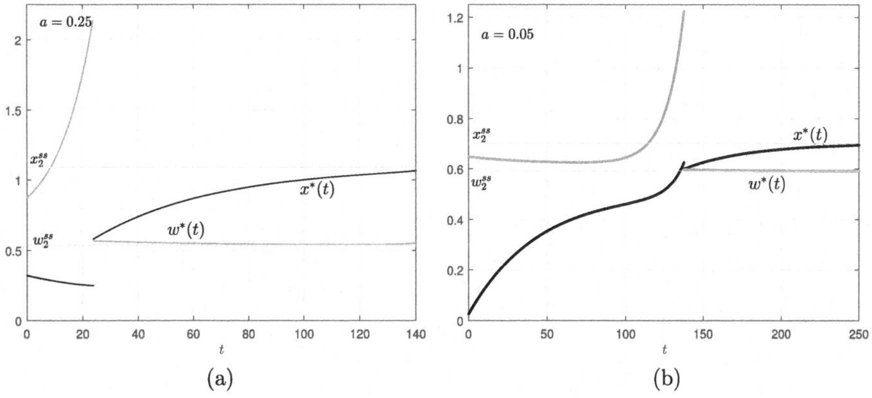


**Fig. 4** The Nash equilibrium strategies which result in state trajectories converging to the equilibrium in regime 2. The equilibrium strategies experience a jump at the discontinuity line $q(t) = q_{th}$. For higher values of $a$ the jump occurs early. (**a**) Nash equilibrium strategies with the initial condition $q_0 = \hat{q} + \epsilon = 6.9022,\ \epsilon > 0$ with $a = 0.25$. (**b**) Nash equilibrium strategies with the initial condition $q_0 = \hat{q} + \epsilon = 5.7,\ \epsilon > 0$ with $a = 0.05$

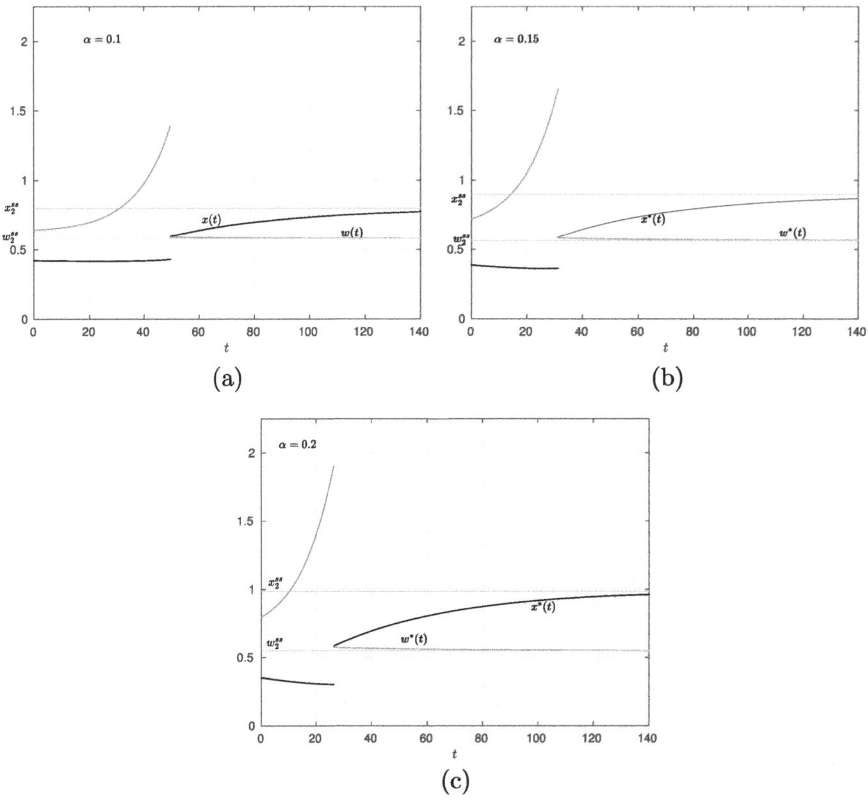


**Fig. 5** Variation of switching time with increasing values of the parameter $a$. For higher values of $a$ the jump occurs early. (**a**) $a = 0.1$. (**b**) $a = 0.15$. (**c**) $a = 0.2$

# 5 Conclusion

The paper builds upon the quasi-consensus that good governance and institutions matter as key factor determining growth and economic development, whereas factors leading to failure and poor institutions such as corruption are detrimental to economic development and ultimately increase rent-seeking activities in the economy.

Recognizing the previous work of Ngendakuriyo and Zaccour (2013) that investigates the role of citizens in improving the institutional quality through the fight against corruption, our research question was as follows: Under which circumstances a society with high prevalence of endogenous corruption may eradicate the corruption and switch to a society with free or very low corruption in her process of building efficient institutions? To answer this question, we introduced the effects of social inertia as a major parameter of institutional change and characterized open-loop Nash equilibria for both the regimes with high and low corruption. Furthermore, we characterized an optimal switching time where the equilibrium moves from a regime with high corruption to a free/low corruption regime. The open-loop strategies characterized reflected the Skiba phenomena as they exhibit jumps in the state variable at steady state.

Our numerical results show that an increase of optimism (pessimism) in the society will lead the civil society to invest less (more) efforts to fight corruption. Similarly, an increase of optimism (pessimism) in the society will lead a corrupt government to invest more (less) efforts in repression and rent-seeking activities. Overall, the numerical results show that the higher the efficiency of the civil monitoring effort and the efficiency of institutions, the higher the inertia which lead to a much earlier switch to good regime of low or free corruption as the jump occurs early.

A subsequent research avenue is of interest, namely the analysis of a game where cooperation between government and civil society is possible. Here the idea would be to set up a cooperative framework that is Pareto-improving and which prevents the government to cheat and deviate from the cooperation platform.

## Appendix

*Proof of Proposition 1 Regime 1* The current value Hamiltonian associated with problems $P_C$ and $P_G$ in regime 1 is given by:

$$H_C^1(q(t), w(t), x^*(t), \lambda_C^1(t)) = \lambda_C^1(t)\left(\left(bw(t) - \beta x^*(t)\right) - a\right) + \left[(\alpha q(t) + \theta(1 - w(t)))(1 - \kappa x^*(t)) - \frac{w^2(t)}{2}\right]$$

$$H_G^1(q(t), w^*(t), x(t), \lambda_G^1(t)) = \lambda_G^1(t)\left(\left(bw^*(t) - \beta x(t)\right) - a\right) + \left[\left(\alpha q(t) + \theta(1 - w^*(t))\right)(\kappa x(t)) - \frac{x^2(t)}{2}\right]$$

The first order conditions (assuming interior solutions) for the open-loop Nash equilibrium are given by:

$$\left.\begin{aligned}
(H_C^1)_w &= b\lambda_C^1 - \theta(1-\kappa x^*) - w^* = 0,\\
(H_C^1)_q &= \rho\lambda_C^1 - \dot{\lambda}_C^1 = \alpha(1-\kappa x^*)\\
(H_C^1)_{\lambda_C^1} &= (H_G^1)_{\lambda_G^1} = \dot{q}^* = bw^* - \beta x^* - a\\
(H_G^1)_x &= -x^* - \beta\lambda_G^1 + \kappa(\alpha q^* + \theta(1-w^*)) = 0\\
(H_G^1)_q &= \rho\lambda_C^1 - \dot{\lambda}_C^1 = \alpha\kappa x^*
\end{aligned}\right\} \tag{8}$$

The open-loop equilibrium dynamics, in $(q, \lambda_C^1, \lambda_G^1)$ coordinates, are given by:

$$\begin{bmatrix}\dot{q}^*\\ \dot{\lambda}_C^1\\ \dot{\lambda}_G^1\end{bmatrix} = \begin{bmatrix} -\frac{\alpha\kappa(\beta-b\kappa\theta)}{\kappa^2\theta^2+1} & \frac{b(b+\beta\kappa\theta)}{\kappa^2\theta^2+1} & \frac{\beta(\beta-b\kappa\theta)}{\kappa^2\theta^2+1}\\ \frac{\alpha^2\kappa^2}{\kappa^2\theta^2+1} & \rho-\frac{\alpha b\kappa^2\theta}{\kappa^2\theta^2+1} & -\frac{\alpha\beta\kappa}{\kappa^2\theta^2+1}\\ -\frac{\alpha^2\kappa^2}{\kappa^2\theta^2+1} & \frac{\alpha b\kappa^2\theta}{\kappa^2\theta^2+1} & \rho+\frac{\alpha\beta\kappa}{\kappa^2\theta^2+1}\end{bmatrix}\begin{bmatrix}q^*\\ \lambda_C^1\\ \lambda_G^1\end{bmatrix} + \begin{bmatrix} b-a-\frac{((b+\beta\kappa\theta)(\theta+1))}{(\kappa^2\theta^2+1)}\\ \frac{(\alpha(\theta\kappa^2-1))}{(\kappa^2\theta^2+1)}\\ -\frac{(\alpha\kappa^2\theta(\theta+1))}{(\kappa^2\theta^2+1)}\end{bmatrix} \tag{9}$$

The steady-state value for the control policies and the state variable in regime 1 are

$$x_1^{ss} = \frac{b(b\alpha-\theta\rho)-a\rho}{b\kappa(b\alpha-\theta\rho)+\beta\rho},\quad w_1^{ss} = \frac{(b\alpha-\theta\rho)(\beta+a\kappa)}{b\kappa(b\alpha-\theta\rho)+\beta\rho},\text{ and}$$

$$q_1^{ss} = \frac{b\kappa^2\rho\theta^2 - \alpha b\beta\kappa\theta + \alpha\beta^2 - b\rho\theta - \beta\kappa\rho\theta - \beta\kappa\rho\theta^2 - a\kappa^2\rho\theta^2 - a\rho}{\alpha\kappa\rho(\beta-b\kappa\theta)} + \frac{(\beta+a\kappa)\left(b^2+2\kappa\theta b\beta-\beta^2\right)}{\kappa(\beta-b\kappa\theta)(b\kappa(b\alpha-\theta\rho)+\beta\rho)}.$$

*Regime 2* The current value Hamiltonian associated with problems $P_C$ and $P_G$ in regime 2 is given by:

$$H_C^2(q(t), w(t), x^*(t), \lambda_C^2(t)) = \lambda_C^2(t)\left(\left(bw(t)-\beta x^*(t)\right)+a\right) + \left[\left(\alpha q(t)+\theta(1-w(t))\right)(1-\kappa x^*(t)) - \frac{w^2(t)}{2}\right]$$

$$H_G^2(q(t), w^*(t), x(t), \lambda_G^2(t)) = \lambda_G^2(t)\left(\left(bw^*(t)-\beta x(t)\right)+a\right) + \left[\left(\alpha q(t)+\theta(1-w^*(t))\right)(\kappa x(t)) - \frac{x^2(t)}{2}\right]$$

The first order conditions (assuming interior solutions) for the open-loop Nash equilibrium are given by:

$$\left.\begin{aligned}
(H_C^2)_w &= b\lambda_C^1 - \theta(1-\kappa x^*) - w^* = 0,\\
(H_C^2)_q &= \rho\lambda_C^2 - \dot{\lambda}_C^2 = \alpha(1-\kappa x^*)\\
(H_C^2)_{\lambda_C^2} &= (H_G^2)_{\lambda_G^2} = \dot{q}^* = bw^* - \beta x^* + a\\
(H_G^2)_x &= -x^* - \beta\lambda_G^2 + \kappa(\alpha q^* + \theta(1-w^*)) = 0\\
(H_G^2)_q &= \rho\lambda_C^2 - \dot{\lambda}_C^2 = \alpha\kappa x^*
\end{aligned}\right\} \tag{10}$$

The open-loop equilibrium dynamics, in $(q, \lambda_C^2, \lambda_G^2)$ coordinates, are given by:

$$\begin{bmatrix} \dot{q}^* \\ \dot{\lambda}_C^2 \\ \dot{\lambda}_G^2 \end{bmatrix} = \begin{bmatrix} -\frac{\alpha\kappa(\beta-b\kappa\theta)}{\kappa^2\theta^2+1} & \frac{b(b+\beta\kappa\theta)}{\kappa^2\theta^2+1} & \frac{\beta(\beta-b\kappa\theta)}{\kappa^2\theta^2+1} \\ \frac{\alpha^2\kappa^2}{\kappa^2\theta^2+1} & \rho-\frac{\alpha b\kappa^2\theta}{\kappa^2\theta^2+1} & -\frac{\alpha\beta\kappa}{\kappa^2\theta^2+1} \\ -\frac{\alpha^2\kappa^2}{\kappa^2\theta^2+1} & \frac{\alpha b\kappa^2\theta}{\kappa^2\theta^2+1} & \rho+\frac{\alpha\beta\kappa}{\kappa^2\theta^2+1} \end{bmatrix} \begin{bmatrix} q^* \\ \lambda_C^2 \\ \lambda_G^2 \end{bmatrix} + \begin{bmatrix} b+a-\frac{(b+\beta\kappa\theta)(\theta+1)}{\kappa^2\theta^2+1} \\ \frac{\alpha(\theta\kappa^2-1)}{\kappa^2\theta^2+1} \\ -\frac{\alpha\kappa^2\theta(\theta+1)}{\kappa^2\theta^2+1} \end{bmatrix} \tag{11}$$

The steady-state value for the control policies and state variable in regime 2 are

$$x_2^{ss} = \frac{b(b\alpha-\theta\rho)+a\rho}{b\kappa(b\alpha-\theta\rho)+\beta\rho}, \quad w_2^{ss} = \frac{(b\alpha-\theta\rho)(\beta-a\kappa)}{b\kappa(b\alpha-\theta\rho)+\beta\rho}, \text{ and}$$

$$\begin{aligned} q_2^{ss} &= \frac{b\kappa^2\rho\theta^2 - \alpha b\beta\kappa\theta + \alpha\beta^2 - b\rho\theta - \beta\kappa\rho\theta - \beta\kappa\rho\theta^2 + a\kappa^2\rho\theta^2 + a\rho}{\alpha\kappa\rho\,(\beta-b\kappa\theta)} \\ &\quad + \frac{(\beta-a\kappa)\left(b^2+2\kappa\theta b\beta-\beta^2\right)}{\kappa\,(\beta-b\kappa\theta)\,(b\kappa(b\alpha-\theta\rho)+\beta\rho)}. \end{aligned}$$

■

*Proof of Proposition 2* Using the (8) and (10) in (5) and (6) we obtain the following equations:

$$\begin{aligned} &A_c\lambda_C^{j\,2}(\tau) + B_c\lambda_C^j(\tau)\lambda_G^j(\tau) + C_c\lambda_G^{j\,2}(\tau) + (D_c+a)\lambda_C^j(\tau) + E_c\lambda_G^j(\tau) + F_c \\ &= A_c\lambda_C^{i\,2}(\tau) + B_c\lambda_C^i(\tau)\lambda_G^i(\tau) + C_c\lambda_G^{i\,2}(\tau) + (D_c-a)\lambda_C^i(\tau) + E_c\lambda_G^i(\tau) + F_c, \end{aligned} \tag{12}$$

$$\begin{aligned} &A_g\lambda_C^{j\,2}(\tau) + B_g\lambda_C^j(\tau)\lambda_G^j(\tau) + C_g\lambda_G^{j\,2}(\tau) + D_g\lambda_C^j(\tau) + (E_g+a)\lambda_G^j(\tau) + F_g \\ &= A_g\lambda_C^{i\,2}(\tau) + B_g\lambda_C^i(\tau)\lambda_G^i(\tau) + C_g\lambda_G^{i\,2}(\tau) + D_g\lambda_C^i(\tau) + (E_g-a)\lambda_G^i(\tau) + F_g. \end{aligned} \tag{13}$$

where

$A_c = \frac{(b(2\beta\kappa^3\theta^3+2\beta\kappa\theta+b))}{2(\kappa^2 t^2+1)^2}$, $B_c = \frac{(\beta(\beta\kappa^2\theta^2-b\kappa\theta+\beta))}{(\kappa^2\theta^2+1)^2}$, $C_c = \frac{(\beta^2\kappa^2\theta^2)}{2(\kappa^2\theta^2+1)^2}$,

$D_c = \frac{b(\theta+\alpha q_{th})}{\theta} - \frac{b(\theta^2+\theta+\alpha q_{th})}{\theta(\kappa^2\theta^2+1)^2} - \frac{\beta\kappa(\theta^2+\theta+\alpha q_{th})}{(\kappa^2\theta^2+1)}$, $E_c = \frac{(\beta\kappa(\theta^2+\theta+\alpha q_{th}))}{(\kappa^2\theta^2+1)^2}$, $F_c = \frac{(\theta^2+\theta+\alpha q_{th})^2}{2\theta^2(\kappa^2\theta^2+1)^2} - \frac{(\theta+\alpha q_{th})^2}{(2\theta^2)}$,

$A_g = \frac{(b^2\kappa^2\theta^2)}{2(\kappa^2\theta^2+1)^2}$, $B_g = \frac{(b(b\kappa^2\theta^2+\beta\kappa\theta+b))}{(\kappa^2\theta^2+1)^2}$, $C_g = -\frac{(\beta(2b\kappa^3\theta^3+2b\kappa\theta-\beta))}{2(\kappa^2 t^2+1)^2}$, $D_g = -\frac{(\beta\kappa^2\theta(\theta^2+\theta+\alpha q_{th}))}{(\kappa^2\theta^2+1)^2}$,

$E_g = \frac{b(\theta+\alpha q_{th})}{\theta} - \frac{b(\theta^2+\theta+\alpha q_{th})}{\theta(\kappa^2\theta^2+1)} - \frac{\beta\kappa(\theta^2+\theta+\alpha q_{th})}{(\kappa^2\theta^2+1)^2}$ and $F_g = \frac{\kappa^2(\theta^2+\theta+\alpha q_{th})^2}{2(\kappa^2\theta^2+1)^2}$.

From (12) and (13), the switching time $\tau$ as well as the jumps $\lambda_C^j(\tau) - \lambda_C^i(\tau)$ and $\lambda_G^j(\tau) - \lambda_G^i(\tau)$ can be calculated when the equilibrium state trajectory undergoes a transition from regime $i$ to regime $j$. ■

# References


Acemoglu, D., Johnson, S., & Robinson, J. (2005). Institutions as fundamental cause of long-run growth. In P. Aghion & S. N. Durlauf (Eds.), *Handbook of economic growth* (Vol. 1A, pp. 386–472). Amsterdam: Elsevier.

Argandona, A. (1991). *Values, Institutions and Ethics*. Business School-University of Navarra, Working Paper, No. 215.

Bidner, C., & François, P. (2010). Cultivating trust: Norms, institutions and the implications of scale. *Economic Journal, 121*(55), 1097–1129.

Dawid, H., Keoula, M. Y., & Kort, P. M. (2017). Numerical analysis of Markov-perfect equilibria with multiple stable steady states: A duopoly application with innovative firms. *Dynamic Games Applications, 7*, 555–577.

De, P. (2010). *Governance, Institutions and Regional Infrastructure in Asia*. ADBI Working Paper Series, No. 183.

Dechert, W. D., & Nishimura, K. (1983). A complete characterization of optimal growth paths in an aggregated model with a non-concave production function. *Journal of Economic Theory, 31*(2), 332–354.

Dowding, K., John, P., Mergoudis, T., & Van Vugt, M. (2000). Exit, voice and loyalty: Analytic and empirical developments. *European Journal of Political Research, 37*, 469–495.

Hirschman, A. O. (1970). *Exit, voice and loyalty: Responses to decline in firms, organizations and states*. Cambridge: Harvard University Press.

Lloyd, P., & Lee, C. (2018). A review of the recent literature on the institutional economics analysis of the long-run performance of Nations. *Journal of Economic Surveys, 32*(1), 1–22.

Long, N. V., Prieur, F., Puzon. K., & Tidball. M. (2017). Piecewise closed-loop Nash equilibria in differential games with regime switching strategies. *Journal of Economic Dynamics and Control, 76*, 264–286.

Makris, M. (2001). Necessary conditions for infinite-horizon discounted two-stage optimal control problems. *Journal of Economic Dynamics and Control, 25*(12), 1935–1950.

Ngendakuriyo, F. (2013). Institutional quality and growth. *Journal of Public Economic Theory, 15*(1), 157–184.

Ngendakuriyo, F., & Zaccour, G. (2013). Fighting corruption: To precommit or not? *Economics Letters, 120*, 149–154.

Ngendakuriyo, F., & Zaccour, G. (2017). Should civil society organizations cooperate or compete in fighting a corrupt government? *Mathematical Social Sciences, 85*, 30–36.
Seierstad, A., & Stabrun, S. D. (2010). Control problems with surfaces triggering jumps in the state variables. *Optimal Control Applications and Methods, 31*(2), 117–136.
Sethi, S. P. (1977). Optimal advertising policy with the contagion model. *Journal of Optimization Theory and Applications, 29*(4), 975–627.
Shaikh, M. S., & Caines, P. E. (2007). On the hybrid optimal control problem: Theory and algorithms. *IEEE Transactions on Automatic Control, 52*(9), 1587–1603.
Skiba, A. K. (1978). Optimal growth with a convex-concave production function. *Econometrica, 46*, 527–539.
Tomiyama, K. (1985). Two-stage optimal control problems and optimality conditions. *Journal of Economic Dynamics and Control, 9*(3), 317–337.

# Game Theory and Cyber Defense

**Abderrahmane Sokri**

**Abstract** The extensive use of information technology systems in military sector has changed the face of the battlefield and the nature of war. A growing body of literature argues that the game-theoretic reasoning is well-suited to many problems in cyber defense. A game between a defender and an attacker trying to gain access to computers remotely is a typical strategic interaction in this domain. This chapter discusses how game theory can be applied in cyberspace. It offers a comprehensive review of literature on the application of game theory in this area. It proposes and illustrates a new game formulation combining game theory and other techniques. The chapter highlights the recognized challenges associated with the applicability of game theory in the cyber world. It discusses how the game-theoretic formalism can be adapted to obtain sound solutions in a reasonable time.

**Keywords** Game theory · Cyber defense · Cyberattack · Cybersecurity · Common knowledge

> [The] cyber threat is one of the most serious economic and national security challenges we face as a nation. —President Barack Obama, 29 May 2009

## 1 Introduction

Revolutionary advancement in information and communication technologies (ICT) has brought many changes to the nature of war. Cyberspace has become both a crucial enabler and a critical vulnerability for military forces. It has become the new battlefield, on par with air, land, and maritime, but with its own lot of complex and

A. Sokri (✉)
Government of Canada, Department of National Defence, Ottawa, ON, Canada
e-mail: Abderrahmane.Sokri@drdc-rddc.gc.ca



P.-O. Pineau et al. (eds.), *Games in Management Science*,
International Series in Operations Research & Management Science 280,
https://doi.org/10.1007/978-3-030-19107-8_18

challenging problems. The cyber weapons could be social engineering, upgraded viruses, Trojan horses, worms, flooding denial-of-service (DoS), distributed denial-of-service (DDoS) or botnets, and advanced persistent threat (APTs) (Bernier et al. 2012; Aslanoglu and Tekir 2012).

In a social engineering attack, an attacker pieces together enough information to infiltrate an organization's network. The attacker can, for example, claim to be a new employee, repair person, or researcher and ask questions to different sources about an organization or its computer systems. A virus is a computer program designed to deliberately damage files or spread to other computers. A Trojan horse is a computer program with a good purpose that hides a damaging program that performs a malicious action. A worm is a virus that can spread from a computer to another without human interaction. It takes up memory, exhausts network bandwidth, and causes a computer to stop responding. It can also allow attackers to gain access to computers remotely. Most of these threats are included as attachments or links contained in email messages.

A DoS attack occurs when an attacker prevents legitimate users from accessing information or services such as email and online banking accounts. In this attack, an attacker overloads a network or server with information or requests. In a DDoS attack, an attacker takes advantage of security weaknesses to control multiple computers. These computers are used afterward to launch a DoS attack (McDowell 2009). These attacks can cause public or private institutions to lose important data, money, or their reputations (Liang and Xiao 2013). APTs use sophisticated techniques to monitor and extract sensitive data from a specific target over a long period of time while remaining undetected.

These cyber weapons are shaped based on the knowledge of target's vulnerabilities. The National Institute of Standards and Technology (NIST) defines vulnerability as a weakness in system security procedures, design, internal controls, or implementation that could be exploited by a threat source (NIST 2002). A vulnerability is exploitable when an attacker has the knowledge about it and the skills to exploit it.

Vulnerabilities are characterized by their dynamic nature. When a vulnerability is detected by the defender, the attacker's weapon exploiting it becomes useless and the target's defense becomes upgraded. This refers to the two paradoxes of cyber weapons. The first paradox states that cyber weapons are subject to time decay. The second paradox states that cyber weapons usage may shortly enhance the target's defense (Podins and Czosseck 2012).

Without being directly lethal, cyberattacks can cause loss of data confidentiality (e.g., unauthorized disclosure of information), integrity (e.g., unauthorized modification of information), or availability (e.g., disruption of access) (Bowen et al. 2006). It can also cause damage or destruction of equipment (Ziolkowski 2010; Podins and Czosseck 2012). The extent and severity of cyberattacks vary from local (loss of email confidentiality) to nation-wide (Ottis 2008). But without exploitable vulnerabilities, cyberattacks would be limited to DoS, DDOS, and social engineering attacks (Moore et al. 2010; Podins and Czosseck 2012).

In 2007, Estonia was the subject of the first massive nation-wide cyberattack in the world. A campaign of cyberattacks was conducted during 3 weeks against government websites, banks, critical national infrastructure, newspapers, and broadcasters. Attacks included massive DDoS, phishing, email spam, and website defacing (Aslanoglu and Tekir 2012; van Vuuren et al. 2012; Podins and Czosseck 2012).

In 2009, an APT exploited a previously unknown vulnerability in Internet Explorer to compromise systems at Google, Adobe, and more than 30 large companies. The main objective was to steal intellectual property from these security and defense contractor companies (Aslanoglu and Tekir 2012).

In 2010, the Stuxnet worm against the Iranian nuclear program was considered as the real start of cyber warfare (Adams et al. 2012). This unprecedented and highly sophisticated attack infected more than 30,000 computers in Iran. The virus continued to spread via Internet and infect about the same number of computers in other countries including the USA, the UK, China, and Germany.

This attack has changed the face of the battlefield and has broken down a common belief stating that control systems are protected, if (1) nothing on computers connects to the Internet, (2) new memory sticks are used for data exchange, and (3) viruses are detectable by the unusual behavior of computers (Miyachi et al. 2011; Aslanoglu and Tekir 2012; Podins and Czosseck 2012).

In 2013, Target Corporation came under an APT resulting in an unauthorized access to credit card numbers and personal information of 40 million customers (Acquaviva 2017). Since then, there has been a growing discussion about the best ways to protect potential target areas against offensive cyberattacks (Bier et al. 2009). To overcome these problems, a variety of protective and reactive measures have been employed. As shown in Table 1, traditional network security techniques include (1) tamperproof techniques, (2) cryptography, (3) detection and prevention techniques, (4) honeypots, and (5) technical attribution.

Although these techniques are crucial mechanisms for cybersecurity, they are not a panacea (Roy et al. 2010). They may be sufficient against casual attackers using well-known techniques, but the complex cybersecurity problem is still far from being completely solved. There is a continuous race between attackers and security specialists. When a smart security solution is proposed a smarter way to circumvent, it is found. There will be an ongoing and challenging need to design tools that protect our systems and networks against sophisticated and well-organized adversaries (Roy et al. 2010).

Many researchers including Roy et al. (2010), Zakrzewska and Ferragut (2011), Kiekintveld et al. (2015), and Tambe (2011) have argued that the game-theoretic reasoning is well-suited to many problems in network security and cyber warfare. This mathematical approach examines how agents or players might act when trying to optimize a utility function (Acquaviva 2017). The United States Department of Defense (DoD), for example, states that applying game theory techniques in cyberspace may assist in analyzing an adversary's preferred tactics (DoD 2011). Game theory can also guide resource allocations to defend against intelligent antagonists by explicitly taking into account the intelligent and adaptive nature of

**Table 1** Traditional protective and reactive measures in cyberspace

| Technique | Definition |
|---|---|
| Tamperproof | Automated methods of identification based on unique measurable physiological or behavioral characteristics such as voice, fingerprints, or iris patterns (Matyas and Riha 2002) |
| Cryptography | Techniques that merge words with images to hide data in transit or storage. They are used for authentication of user and data. |
| Detection/prevention | Techniques including antivirus software, firewalls, and intrusion detection systems (IDS)<br>Antivirus programs scan the communication mediums and the storage devices, detect signs of malware presence, and remove them. Firewalls limit access to private networks connected to the Internet. IDS algorithms detect suspected intrusions and alert the network administrator in real time (Gueye 2011; Roy et al. 2010) |
| Honeypot | A fake computer system used in network security to waste the attacker's time and resources. The network administrator can also use the captured data from the attacker's actions to better protect the network. (McCarty 2003; Rowe et al. 2007; Carroll and Grosu 2011; Pibil et al. 2012) |
| Attribution | Attribution is the determination of the identity or the location of an attacker or an attacker's intermediary (Robinson et al. 2015; Wheeler and Larsen 2003). The identity can be physical such as a geographical address or digital such as an Internet Protocol (IP) address (Guan and Zhang 2010). The information captured by attribution can be used to improve defensive techniques and prevent future attacks (Nicholson et al. 2012) |

the threat (Bier et al. 2009). The arguments put forward to justify this approach are numerous. They particularly include (but are not limited to) its ability to model the non-cooperative and cooperative strategic interactions between multiple decision-makers with conflicting goals. The analytical setting may be static or dynamic, discrete or continuous, deterministic or stochastic, and linear or non-linear.

A cooperative game model examines how players might be working together to optimize a collective utility function (Acquaviva 2017). Cooperative games describe at high level the structure, strategies, and payoffs of subsets of players or coalitions. They are generally characterized by a characteristic function describing the outcome of each coalition.

A typical cooperative game in cyber domain may include a number of organizations or countries exchanging vulnerability information and attack detection procedures. By exchanging information on vulnerabilities, each member of the coalition will build new weapons using the newly learned vulnerabilities (Podins and Czosseck 2012). The UK government, for example, has initiated a cybersecurity hub that enables the exchange of information on cybersecurity threats between the public and private sectors (van Vuuren et al. 2012).

In a non-cooperative game, players seek to optimize their individual utility functions regardless of the utilities of the other players involved (Acquaviva 2017). Non-cooperative games are more general than cooperative games. They describe in detail the individual strategies and payoffs of each player. They focus on analyzing Nash equilibrium that no player can do better by unilaterally deviating from it (Breton et al. 2008; Bachrach et al. 2013; Brandenburger 2007).

Interactions in cyberspace are generally adversarial and inherently selfish. A game between a system administrator and an attacker trying to compromise or destroy the system is a typical non-cooperative game in this domain. In this case, the time spent controlling the system or the reward for destroying it may be the utility function for the attacker. The reward for controlling the system may be the utility function for the defender (Acquaviva 2017).

The aim of this chapter is to discuss the suitability of game theory to adversarial interaction between attackers and defenders in cyberspace. The chapter also sheds light on the main challenging issues surrounding its applicability in this domain. A new game formulation combining simulation and game-theoretic approaches is proposed to solve the problem of uncertain observability in the payoff matrix.

This chapter is organized into six sections. Following the introduction, Sect. 2 provides a comprehensive review of literature on the application of game theory in the cyber domain. Section 3 presents a resource allocation problem to show how the new approach can be used in cyberspace. In Sect. 4, a case study is presented to illustrate the suggested approach. The main challenges associated with the applicability of game-theoretic methods in cyberspace are discussed in Sect. 5. Concluding remarks as well as future research directions are indicated in Sect. 6.

## 2 Literature Review

Game theory is a common formalized way to inspire the development of defense algorithms in the physical world (Moisan and Gonzalez 2017; Coniglio 2013; Tambe 2011; Roy et al. 2010). A growing body of literature recognizes game theory as a sound theoretical foundation for modeling the strategic interactions between selfish agents in the cyber world. This literature can be divided into three main categories: resource allocation, network security, and cooperation models.

### *2.1 Resource Allocation*

Game theory can guide resource allocations to defend against intelligent attacks by explicitly taking into account the adaptive nature of the threat. In this game, the defender seeks to find the optimal resource allocation that maximizes his payoffs. The attacker seeks to minimize the risk of being traced back and punished (Acquaviva 2017). This problem is known in the game-theoretic literature as the allocation game (Bier et al. 2009).

Fielder et al. (2014), for example, proposed a game-theoretic model to optimally allocate cybersecurity resources such as administrators' time across different tasks.

In this game, the defender's solution is optimal independently from the attacker's strategy. The authors also found that a particular Nash equilibrium provides the most effective defense strategy and used real-life statistics to validate their result. More recently, Sokri (2018) used an allocation game to analyze the problem of common knowledge in cyberspace. The author incorporated uncertainty on each imprecise variable by changing its static value to a range of values.

Game theory is also used to determine the optimal investment in critical infrastructures such as networked systems. In this case, defensive investment is used to increase the effort needed by an attacker to achieve a certain probability of success. It can also be used to reduce the success probability of an attack, rather than increasing its effort. The game-theoretic framework determines the optimal allocation of the total defensive budget over the various components of the system in order to minimize the success probability of a potential attack or to maximize its expected cost (Azaiez and Bier 2007).

Game theory can also be used to investigate the optimal strategies for managing a sensitive security resource in response to APTs. Depending on the setting being modeled, the resource may be a password or an entire infrastructure. FlipIt, for example, is a two-player dynamic game where players may take control of the resource at any time by executing a stealthy move (i.e., not immediately detected). This idea implies that each player is allowed to move at arbitrary points in time, and the timing of the moves may be kept hidden from the other player. The objective is to maximize the fraction of time the player controls the resource while minimizing the cumulative move cost. FlipIt is characterized by the idea of stealthy moves or stealthy takeover (Rasouli et al. 2014; Hobbs 2015).

## *2.2 Network Security*

Game theory has also been proposed by several studies to understand defense strategies in network security. It offers a sound theoretical foundation for managing information security, modeling the strategic interactions in intrusion detection, and analyzing network defense mechanism design. It is useful for generalization of problems, formalizing the existing ad-hoc schemes, and future research (Alpcan and Basar 2004).

Bloem et al. (2006), for example, developed a stochastic and dynamic game to examine intrusion detection in access control systems. The authors used a game-theoretic approach to model the interaction between an attacker and a distributed IDS. They introduced the sensor network as a third player with a fixed probability distribution representing the output of the sensor network during the attack. The authors discussed the properties of the resulting system analytically and numerically.

Roy et al. (2010) presented a taxonomy for classifying the existing game-theoretic solutions designed to enhance network security. The authors provided a systematic description of how games can be played and what the outcomes might be. This information is used to define games with relevant concepts for network security problems.

Jafarian et al. (2013) combined game theory and constraint satisfaction optimization to proactively defend against denial-of-service attacks. In this static game, Nash equilibrium is determined by players' strategies and the cost associated with them. The optimal strategy for attack deterrence is determined while satisfying security and performance requirements of the network. Results showed that the method improves the protection of flow packets from being attacked against persistent attackers without causing any disruption for flows.

More recently, Musman and Turner (2018) described a game-oriented approach to minimizing cybersecurity risks for a given investment level. The game formulation uses the defender strategies to minimize the maximum cyber risk. The interested reader is referred to Information Resources Management Association (2018) for further information on this topic.

Game theory has also been used for studying the effects of deception on the interactions between an attacker and a defender of a computer network (Baston and Bostock 1988). In this literature, the defender can employ camouflage by disguising, for example, a honeypot as a normal system. Deception increases the attackers' uncertainty and effort (e.g., time and money) to determine whether a system is true or fake. Even long before computers existed, deception was widely used for information protection (Cohen 1998; Rowe et al. 2007; Carroll and Grosu 2011). Rowe et al. (2007), for example, summarized some game-theoretic aspects of introducing honeypots. The authors developed a mathematical model of deception and counterdeception to see at what point people could detect deception. Results show that attacks on honeypots decreased over time.

Carroll and Grosu (2011) performed a game-theoretical investigation of deception in network security. The authors used a dynamic game of incomplete information to examine a scenario where a defender can disguise normal systems as honeypots or honeypots as normal systems. The attacker observes the system and decides whether or not to proceed compromising the system. The authors determined and characterized the perfect Bayesian equilibria of the game. At an equilibrium, the players do not have any incentives to unilaterally deviate by changing their strategies.

## *2.3 Agent Cooperation*

Cooperative game theory can determine how the collective reward can be shared between selfish agents. It can also provide a mechanism to sustain the cooperative solution which is not a self-enforcing contract (Breton et al. 2008). A typical cooperative game in the existing literature may include a number of selfish agents and a principal controlling a computer network. To allow a reliable connectivity between a certain set of critical servers, the principal can incentivize the agents to cooperate by offering them a certain reward (Bachrach et al. 2013). It can also consist of a multi-mode attack combining different types of warfare that are more effective in tandem than when employed alone (Browne 2000).

Liu et al. (2005), for example, developed a preliminary game-theoretic formalization to capture the interdependency between attacker and defender objectives and strategies. The authors showed that the concept of incentives and utilities can be used to model attacker objectives. Bachrach et al. (2013) modeled a communication network where a failure of one node may disturb communication between other nodes as a simple coalitional game. The authors showed how various game-theoretic solution concepts can be used to characterize the fair share of the revenues an agent is entitled to.

Shamshirband et al. (2014) combined a game-theoretic approach and a fuzzy Q-learning algorithm in Wireless Sensor Networks. The authors implemented cooperative defense counter-attack scenarios for the victim node and the base station to operate as rational decision-maker players through a game theory strategy. The proposed model's attack detection and defense accuracy yield a greater improvement than the existing machine learning methods.

A recent survey of the existing game-theoretic approaches for cybersecurity can be found in Do et al. (2017).

## 3 Resource Allocation Game

In this section, we will show how a game-theoretic model can be used to optimally allocate resources in the cyber domain. The main challenges and open research questions associated with this formulation will be presented and discussed in Sect. 5.

Consider a security game between an attacker $a$ and a defender $d$ in a cyberinfrastructure system. Following Korzhyk et al. (2011), let $A = \{t_1, t_2, \ldots, t_n\}$ be a set of $n$ targets that the attacker may choose to attack. The defender seeks to prevent attacks by covering targets using cybersecurity resources from the set $R = \{r_1, r_2, \ldots, r_m\}$. In the physical world, targets may be flights and resources may be air marshals. In the cyber world, targets may be software vulnerabilities and resources may be protective devices such as firewalls (Gueye 2011).

The set $A$ corresponds to pure strategies for the attacker where each pure strategy refers to a single target to attack. Let $D$ be the set of all the possible resource allocations over the set of targets. If at most one resource is assigned to a target, there will be $n$ *Choose* $m$ combinations to allocate $m$ resources to $n$ targets (Jain et al. 2010). The defender pure strategies are represented by these resource allocations. The two players are allowed to play mixed strategies by assigning a probability distribution over the set of pure strategies (Coniglio 2013; Jain et al. 2010). If a player adopts his mixed strategy, the outcome of the game will be expressed as an expected value.

Let $\delta$ be a leader's mixed strategy consisting of a vector of the defender's pure strategies. Denote by $\delta_i$ the proportion of times assigned to the pure strategy $i$ when the defender plays the mixed strategy $\delta$.

Similarly, we denote by $\rho$ a mixed strategy of the attacker (the follower) and by $\rho_j$ the probability of the pure strategy $j$ when he plays the mixed strategy $\rho$. Let $E(U_d(i,j))$ be the expected utility of the defender and $E(U_a(i,j))$ the expected utility of the attacker when the defender plays pure strategy $i$ and the attacker plays pure strategy $j$.

One of the main challenging issues in security games is the problem of common knowledge concept. It is generally assumed in these games that the players are able to exactly evaluate their own payoffs and the payoffs of their opponents. In most real-world cybersecurity problems, this assumption is not always true. Using deterministic values of payoffs may make the committed strategies ineffective (Coniglio 2013; Sokri 2018). In this paper, utilities are seen as random variables generated by a stochastic simulation. Uncertainty is incorporated in the theoretical framework using their expected values.

Fixing the policy of the defender to some mixed strategy $\delta$, the first problem to solve is to find the attacker's best response to $\delta$. This optimization problem can be formulated as a linear program where the follower maximizes his expected utility given $\delta$.

$$\text{Max}_{\rho} \sum\nolimits_{i \in D} \sum\nolimits_{j \in A} \delta_i \rho_j E\left(U_a\left(i, j\right)\right) \tag{1}$$

$$\text{s.t.} \sum\nolimits_{j \in A} \rho_j = 1 \tag{2}$$

$$\rho_j \geq 0, \forall j. \tag{3}$$

While the constraints define the set of feasible solutions $\rho$ as a probability distribution over the set of targets $A$, it is straightforward to see that the optimal strategy for the follower is a pure strategy $\rho_j = 1$ for a $j$ that maximizes $\sum_{j \in A} \delta_i E\left(U_a\left(i, j\right)\right)$. This result can also be obtained using the corresponding dual problem which has the same optimal solution value

$$\text{Min}_{v}\ v \tag{4}$$

$$\text{s.t.} v \geq \sum\nolimits_{i \in D} \delta_i E\left(U_a\left(i, j\right)\right), \quad j \in A. \tag{5}$$

The corresponding complementary slackness condition is given by

$$\rho_j \left(v - \sum\nolimits_{i \in D} \delta_i E\left(U_a\left(i, j\right)\right)\right) = 0, \quad j \in A. \tag{6}$$

This condition implies that the follower expected reward is maximal for any pure strategy with $\rho_j > 0$.

Denoting by $\rho(\delta)$ the follower's best response to $\delta$, the leader seeks to solve the following problem:

$$\text{Max}_{\rho}\sum_{i\in D}\sum_{j\in A}\delta_i\rho(\delta)_j E\left(U_d\left(i, j\right)\right) \tag{7}$$

$$\text{s.t.}\sum_{i\in D}\delta_i = 1 \tag{8}$$

$$\delta_i \in [0, 1], \quad \forall i \in D. \tag{9}$$

The two constraints enforce the leader's mixed strategy to be feasible.

If we complete the leader's problem by including the follower's optimality conditions, the two programs can be formulated as a single mixed-integer quadratic problem (MIQP).

$$\text{Max}_{\delta,\rho,v}\sum_{i\in D}\sum_{j\in A}\delta_i\rho_j E\left(U_d\left(i, j\right)\right) \tag{10}$$

$$\text{s.t.}\sum_{i\in D}\delta_i = 1 \tag{11}$$

$$\sum_{j\in A}\rho_j = 1 \tag{12}$$

$$0 \le \left(v - \sum_{i\in D}\delta_i U_a\left(i, j\right)\right) \le \left(1 - \rho_j\right) M, \quad \forall j \in A \tag{13}$$

$$\delta_i \in [0, 1], \quad \forall i \in D \tag{14}$$

$$\rho_j \in \{0, 1\}, \quad \forall j \in A \tag{15}$$

$$v \in R \tag{16}$$

To simplify the complementary slackness condition represented by the rightmost inequality in Eq. (13), the attacker plays only pure strategies. Equations (12) and (15) characterize a feasible pure strategy for this player. In this formulation, $v$ is the follower's maximum payoff value and M is a large number.

## 4 Illustration

To illustrate the approach suggested in Sect. 3, consider the game in compact form in Table 2 (Sokri 2018; Jain et al. 2010; An et al. 2011). In this example, there are three targets and two defender resources. Each of defender's resources can only cover one target at a time. For each target, there are two payoffs: the payoff of the attacker and the payoff of the defender. Each payoff consists of two parts: one when the attacked target is covered and one when it's uncovered.

Let $U_d^c(t)$ be the defender's payoff if the attacked target $t$ is covered and $U_d^u(t)$ his payoff if the target is uncovered. Similarly, denote by $U_a^u(t)$ the attacker's payoff if the attacked target $t$ is uncovered and by $U_a^c(t)$ the attacker's payoff if the attacked target $t$ is covered. For each target $t$, the expected utilities of the defender and the attacker are respectively given by

$$U_d(t) = \rho_t \left(\delta_t U_d^c(t) + (1 - \delta_t)\, U_d^u(t)\right) \tag{17}$$

$$U_a(t) = \rho_t \left((1 - \delta_t)\, U_a^u(t) + \delta_t U_a^c(t)\right) \tag{18}$$

The expected utilities in Eqs. (17) and (18) depend simply on the attacked targets and their coverage. Uncertainty can furthermore be placed on each payoff using three-point estimates instead of single values.

This game has multiple equilibria of the form

$$\langle \delta = (\delta_1, \delta_2, 1)\,, \quad \rho = (0, 0, 1) \rangle\,. \tag{19}$$

This standard solution indicates that the attacker would aim the most valuable target no matter how defended it might be (Sokri 2018; Jain et al. 2010; An et al. 2011). A solution for the defender–attacker Stackelberg game that satisfies the constraints and the numerical convergence criterion is given by

$$\langle \delta = (0.75, 0.25, 1)\,, \quad \rho = (0, 0, 1) \rangle\,. \tag{20}$$

To find a robust solution, further refinement is needed. The equilibrium refinement may be based on some utility dominance criteria such as Pareto dominance (An et al. 2011).

**Table 2** Payoff table

| | Defender | | Attacker | |
|---|---|---|---|---|
| | Covered | uncovered | uncovered | Covered |
| Target 1 | 5 | 2 | 7 | 5 |
| Target 2 | 2 | 1 | 4 | 4 |
| Target 3 | 5 | 5 | 12 | 9 |

# 5 Application of the Game in Cyberspace: Challenges and Opportunities

Game theory has already produced several notable successes in numerous physical security domains. It was applied, for example, to randomize checkpoints at the Los Angeles International Airport (LAX), to assign federal air marshals to protect flights (Jain et al. 2010; Kiekintveld et al. 2015; Acquaviva 2017). Researchers have also used game theory to understand security and defense strategies in the cyber world. The application of game theory to this domain presents at least three main challenges: (1) the complexity of the cyber domain, (2) the dynamic nature of the analyzed games, and (3) the validity of the adopted assumptions.

## *5.1 Complexity of the Cyber Domain*

Cybersecurity is more complex than in physical security domains. In the cyber domain, digital attacks are often sophisticated and imperceptible to the human senses. They are highly dynamic overstepping all geographic and political boundaries (Moisan and Gonzalez 2017). To interact appropriately in the cyber domain under dynamically changing real-world scenarios, it is important to understand the entire cyberinfrastructure system. To this end, the holistic game inspired defense architecture suggested by Shiva et al. (2012) would be a good starting point.

Shiva et al. (2012) proposed a four-layer decision-making framework inspired by game theory. As illustrated in Fig. 1, the security scheme is organized into four layers. The first and innermost layer in the framework contains self-checking hardware and software components. The second layer consists of secure built-in or bolt-on applications employing self-checking concepts and components. The

**Fig. 1** Game inspired decision model (Adapted from Shiva et al. 2012)

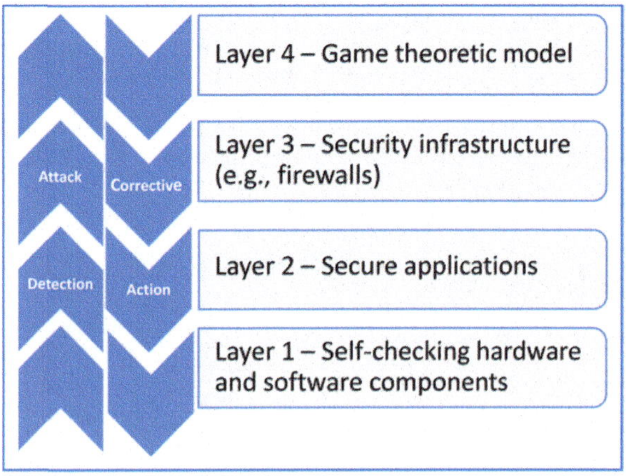

third layer is the security infrastructure consisting of intrusion detection system (IDS), firewalls, and antivirus software. The fourth and outermost layer uses game-theoretic analysis to provide the best action strategies. It receives input from the inner three layers, evaluates the committed or probable attack information, and elects the optimal decision for defense.

## *5.2 Static Versus Dynamic Perspectives*

A static model is a model where the system state is independent of time. It is an interaction where each player makes a single decision in isolation and under imperfect information. The well-known prisoner's dilemma falls under the category of static games. Decisions in static games can be seen as made simultaneously. Real-world security interactions are inherently dynamic where recent attacks are built upon previous attacks. A dynamic model is a model where the system state changes with time, and players are able to observe the outcome of previous moves before responding. Stealthy move games are examples of dynamic games. The dynamic perspective can be introduced to the suggested framework by playing the game within a finite or infinite time horizon. Factors that determine the objective function such as rewards and costs should be explicitly presented as functions of time. This addition can, however, result in a more complex and challenging problem.

## *5.3 Validity of Assumptions*

The game-theoretic framework in Sect. 3 relies on two main key assumptions. The game considers (1) two rational players with certain observability and (2) limited amount of homogeneous resources and targets with no explicit cost of moving. In real world, the defender may face multiple rational or irrational attackers, and the common knowledge on payoffs may be missing. The number of targets to be protected can be large and the attacker may aim more than a single target. The defender's resources may also be numerous and with explicit cost of moving. By making the formalism more realistic, the algorithm would not be able to find an optimal solution in a reasonable time. It is, therefore, necessary to combine game theory with other potential tools and techniques to enhance cyber conflict analysis. Table 3, adapted from DoD (2011), presents the potential techniques, their definitions, and their potential use in cyberspace.

Combining game theory with other techniques in cyberspace is still at its beginnings, and many open issues are still to be tackled. The future combined frameworks should be able:

**Table 3** Potential tools and techniques that may be combined in cyber conflict analysis

| Technique | Definition | Use in cyberspace |
|---|---|---|
| Game theory | The study of mathematical models of conflict and cooperation between intelligent rational decision-makers (Myerson 1991) | Investigate security decisions in a methodical manner |
| Computer simulation | Computer representations that model the real-world interactions | Process visualization<br>Variables and parameters randomization<br>War gaming |
| Genetic algorithms | A family of computational models inspired by evolution | Searching for a sequence of steps that will allow an adversary to achieve their objective |
| Graph theory | A graph is a set of nodes and links that models pairwise relationships between items | Network mapping<br>Bayesian network<br>Identification of strong and weak links and nodes in the adversary's critical requirements |
| Reliability modeling | The process of predicting the likelihood that a component or system will function prior to its implementation | Analyze the availability of a critical capability when resources and conditions are deficient or absent |
| Cyber forensic analysis | Methods to recover and analyze materials found in digital sources | Reconstructing events believed to be malicious |
| IDS | A device or software application that monitors a computer network or individual system for abnormal activity | Detect the step executed and initiate mitigation measures |

- To be dynamic where recent attacks are built upon previous ones;
- To model multiple self-interested agents (e.g., multiple unknown attackers from multiple locations);
- To handle multiple uncertainties in adversary payoffs and observations;
- To deal with bounded rationality of human adversaries by introducing stochastic actions.

## 6 Conclusion

The extensive use of ICT in military sector has changed the face of the battlefield and made cybersecurity an increasingly important concern. Cyber weapons are malicious software that exploit unknown vulnerabilities in the target's defense. The players in this new space can be individuals, devices, or software. Theirs interactions are generally non-cooperative and their objectives are inherently conflicting.

The game-theoretic reasoning has been recognized as well-suited to many problems in the cyber world.

The arguments put forward to justify its use are abundant. Game theory uses proven mathematics to investigate a large range of security decisions. It provides a sound theoretical foundation for understanding the strategic interactions between selfish agents and optimally allocating limited resources and sharing collective rewards.

Defense algorithms inspired by game theory have become very popular in the physical security world. Cyberinfrastructure systems are, however, more complex and the corresponding security threats are highly dynamic and sophisticated. Despite considerable effort from the research community, the application of game theory in cyber defense is still at its beginnings and needs further adaptation to deliver according to its potential.

Current cyber algorithms generally use static settings and rely on idealized assumptions such as common knowledge about the payoff matrix. They also assume that players are able to remember and process large amounts of information accurately. Applying game theory under these simplified conditions may make the resulting strategies ineffective. Scaling up the formalism to real-world-sized problems would make it very complex and intractable.

To be able to make the formalism more realistic and obtain sound and effective solutions in a reasonable time, we recommend combining game theory with other techniques and tools. The suggested techniques include computer simulation, genetic algorithms, graph theory, reliability modeling, and cyber forensic analysis. Tools may consist of IDS, firewalls, and antivirus software. Using these techniques and tools under a solid game-theoretic setting will provide huge potential to solve many cybersecurity standard problems.

## References

Acquaviva, J. R. (2017). *Optimal cyber-defence strategies for advanced persistent threats: A game theoretical analysis*. Master Thesis, The Pennsylvania State University.

Adams, A., Reich, P., & Weinstein, S. (2012). A non-militarised approach to cyber-security. In E. Filiol & R. Erra (Eds.), *Proceedings of the 11th European Conference on Information Warfare and Security* (pp. 1–8). Laval: Academic Conferences & Publishing International Ltd.

Alpcan, T., & Basar, T. A. (2004). Game theoretic approach to decision and analysis in network intrusion detection. In *Proceedings of the 42nd IEEE Conference on Decision and Control*. Hawaii: IEEE.

An, B., Tambe, M., Ordonez, F., Shieh, E., & Kiekintveld, C. (2011). Refinement of strong Stackelberg equilibria in security games. In *Proceedings of the 25th Conference on Artificial Intelligence* (pp. 587–593). Los Alamitos, CA: IEEE.

Aslanoglu, R., & Tekir, S. (2012). Recent cyberwar spectrum and its analysis. In *Proceedings of the 11th European Conference on Information Warfare and Security* (pp. 45–52). Laval: Academic Conferences & Publishing International Ltd..

Azaiez, N., & Bier, V. M. (2007). Optimal resource allocation for security in reliability systems. *European Journal of Operational Research, 181*(2), 773–786.

Bachrach, Y., Porat, E., & Rosenschein, J. S. (2013). Sharing rewards in cooperative connectivity games. *Journal of Artificial Intelligence Research, 47*, 281–311.

Baston, V. J., & Bostock, F. A. (1988). Deception games. *International Journal of Game Theory, 17*(2), 129–134.
Bernier, M., LeBlanc, S., & Morton, B. (2012). Metrics framework of cyber operations on command and control. In *Proceedings of the 11th European Conference on Information Warfare and Security* (pp. 53–62). Laval: Academic Conferences & Publishing International Ltd..
Bier, V. M., Cox, L. A., & Azaiez, M. N. (2009). Why both game theory and reliability theory are important in defending infrastructure against intelligent attacks (chapter 1). In V. M. Bier & M. N. Azaiez (Eds.), *Game theoretic risk analysis of security threats* (pp. 1–11). New York: Springer.
Bloem, M., Alpcan, T., & Basar, T. (2006). Intrusion response as a resource allocation problem. In *IEEE Conference on Decision and Control*. Piscataway, NJ: IEEE.
Bowen, P., Hash, J., & Wilson, M. (2006). *Information security handbook: A guide for managers*. Gaithersburg, MD: NIST Special Publication 800–100.
Brandenburger, A. (2007). *Cooperative game theory: Characteristic functions, allocations, marginal contribution*. New York: Stern School of Business, New York University.
Breton, M., Sokri, A., & Zaccour, G. (2008). Incentive equilibrium in an overlapping-generations environmental game. *European Journal of Operational Research, 185*(2), 687–699.
Browne, R. (2000). C4I defensive infrastructure for survivability against multi-mode attacks. In *Proceedings of 21st Century Military Communication-Architectures and Technologies for Information Superiority*. Piscataway, NJ: IEEE.
Carroll, T. E., & Grosu, D. (2011). A game theoretic investigation of deception in network security. *Security and Communication Networks, 4*(10), 1162–1172.
Cohen, F. (1998). A note on the role of deception in information protection. *Computers and Security, 17*(6), 483–506.
Coniglio, S. (2013). *Algorithms for finding leader-follower equilibrium with multiple followers*. Ph.D. Thesis, Politecnico di Milano.
Do, C. T., Tran, N. H., Hong, C., Kamhoua, C. A., Kwiat, K. A., Blasch, E., Ren, S., Pissinou, N., & Iyengar, S. S. (2017). Game theory for cyber security and privacy. *ACM Computing Surveys (CSUR), 50*(2), 30.
Fielder, A., Panaousis, E., Malacaria, P., Hankin, C., & Smeraldi, F. (2014). Game theory meets information security management. In *Information Security and Privacy Conference* (pp. 15–29). Berlin: Springer.
Guan, Y., & Zhang, L. (2010). Network forensics. In J. R. Vacca (Ed.), *Managing information security* (pp. 197–212). Rockland, MA: Syngress.
Gueye, A. (2011). *A game theoretical approach to communication security*. Ph.D. Dissertation, University of California.
Hobbs, J. (2015). *Dominion: A game of information exploitation*. Master Thesis, University of New Mexico.
Information Resources Management Association. (2018). *Game theory: Breakthroughs in research and practice* (1st ed.). Hershey PA: IGI Global.
Jafarian, J. H., Al-Shaer, E., & Duan, Q. (2013). Formal approach for route agility against persistent attackers. In *18th European Symposium on Research in Computer Security*. Egham: Springer.
Jain, M., Tsai, J., Pita, J., Kiekintveld, C., Rathi, S., Ordone, F., & Tambe, M. (2010). Software assistants for randomized patrol planning for the LAX airport police and the federal air marshals service. *Interfaces, 40*(4), 267–290.
Kiekintveld, C., Lisy, V., & Pibil, R. (2015). Game-theoretic foundations for the strategic use of honeypots in network security. In *Cyber warfare* (pp. 81–101). Berlin: Springer.
Korzhyk, D., Yin, Z., Kiekintveld, C., Conitzer, V., & Tambe, M. (2011). Stackelberg vs. nash in security games: An extended investigation of interchangeability, equivalence, and uniqueness. *Journal of Artificial Intelligence Research, 41*, 2011.
Liang, X., & Xiao, Y. (2013). Game theory for network security. *IEEE Communications Surveys and Tutorials, 15*(1), 472–486.

Liu, P., Zang, W., & Yu, M. (2005). Incentive-based modeling and inference of attacker intent, objectives, and strategies. *ACM Transactions on Information and System Security, 8*(1), 2005.

Matyas, V., & Riha, Z. (2002). Biometric authentication — security and usability. In B. Jerman-Blazic & T. Klobucar (Eds.), *Advanced communications and multimedia security. The International Federation for Information Processing (IFIP)* (Vol. 100). Boston, MA: Springer.

McCarty, B. (2003). The honeynet arms race. *IEEE Security Privacy, 1*(6), 79–82.

McDowell, M. (2009). *Understanding denial-of-service attacks. Security Tip (ST04–015).* Washington, DC: US-CERT.

Miyachi, T., Narita, H., Yamada, H., & Furuta, H. (2011). Myth and reality on control system security revealed by Stuxnet. In *The Society of Instrument and Control Engineers (SICE) Annual Conference* (pp. 1537–1540). Piscataway, NJ: IEEE.

Moisan, F., & Gonzalez, C. (2017). Security under uncertainty: Adaptive attackers are more challenging to human defenders than random attackers. *Frontiers in Psychology, 8*, 982.

Moore, T., Friedman, A., & Procaccia, A. D. (2010). Would a 'Cyber Warrior' protect us? Exploring trade-offs between attack and defense of information systems. In *Proceedings of the 2010 Workshop on New Security Paradigms* (pp. 85–94). New York: ACM.

Musman, S., & Turner, A. J. (2018). A game oriented approach to minimizing cybersecurity risk. *International Journal of Safety and Security Engineering, 8*(2), 212–222.

Myerson, R. B. (1991). *Game theory: Analysis of conflict*. Cumberland, MD: Harvard University Press.

Nicholson, A., Watson, T., Norris, P., Duffy, A., & Isbell, R. (2012). A taxonomy of technical attribution techniques for cyber attacks. In E. Filiol & R. Erra (Eds.), *Proceedings of the 11th European Conference on Information Warfare and Security* (pp. 188–197). Laval: Academic Conferences & Publishing International Ltd..

NIST. (2002). *Risk management guide for information technology systems* (pp. 800–830). Gaithersburg, MD: NIST Special Publication.

Ottis, R. (2008). Analysis of the 2007 cyber attacks against Estonia from the information warfare perspective. In *Proceedings of the 7th European Conference on Information Warfare* (pp. 163–168). Plymouth: Academic.

Pibil, R., Lisy, V., Kiekintveld, C., Bosansky, B., & Pechoucek, M. (2012). Game theoretic model of strategic honeypot selection in computer networks. In J. Grossklags & J. Walrand (Eds.), *Decision and Game Theory for Security. GameSec 2012. Lecture Notes in Computer Science* (pp. 201–220). Heidelberg: Springer.

Podins, K., & Czosseck, C. (2012). A vulnerability-based model of cyber weapons and its implications for cyber conflict. *International Journal of Cyber Warfare and Terrorism, 2*(1), 14–26.

Rasouli, M., Miehling, E., & Teneketzis, D. (2014). A supervisory control approach to dynamic cyber-security. In R. Poovendran & W. Saad (Eds.), *Decision and game theory for security* (pp. 99–117). New York: Springer International Publishing.

Robinson, M., Jones, K., & Janicke, H. (2015). Cyber warfare: Issues and challenges. *Computer and Security, 49*, 70–94.

Rowe, N. C., Custy, E. J., & Duong, B. T. (2007). Defending cyberspace with fake honeypots. *Journal of Computers, 2*(2), 25–36.

Roy, S., Ellis, C., Shiva, S., Dasgupta, D., Shandilya, V., & Wu, Q. (2010). A survey of game theory as applied to network security. *Proceedings of the 43rd Hawaii International Conference on System Sciences (HICSS), 43*(Part 1), 880–889.

Shamshirband, S., Patel, A., Anuar, N. B., Kiah, M. L. M., & Abraham, A. (2014). Cooperative game theoretic approach using fuzzy Q-learning for detecting and preventing intrusions in wireless sensor networks. *Engineering Applications of Artificial Intelligence, 32*, 228–241.

Shiva, S., Bedi, H., Simmons, C., Fisher, M., & Dharam, R. (2012). A holistic game inspired defense architecture. In *International Conference on Data Engineering and Internet Technology*. Los Alamitos, CA: IEEE.

Sokri, A. (2018). Optimal resource allocation in cyber-security: A game theoretic approach. *Procedia Computer Science, 134*, 283–288.

Tambe, M. (2011). *Security and game theory: Algorithms, deployed systems, lessons learned.* Cambridge: Cambridge University Press.

The American Department of Defence (DoD). (2011). *Cyber Intelligence Preparation of the Environment (CIPE).* Technical Task Order 11-0002, Version 1.

van Vuuren, J. J., Phahlamohlaka, J., & Leenen, L. (2012). Governance of Cybersecurity in South Africa. In *Proceedings of the 11th European Conference on Information Warfare and Security* (pp. 135–144). Laval: Academic Conferences & Publishing International Ltd..

Wheeler, D. A., & Larsen, G. N. (2003). *Techniques for cyber attack attribution.* Alexandria, VA: Institute for Defense Analysis. IDA Paper P-3792.

Zakrzewska, A., & Ferragut, E. (2011). Modeling cyber conflicts using an extended petri net formalism. In *Proceedings of IEEE Symposium on Computational Intelligence in Cyber Security* (pp. 60–67). Piscataway, NJ: IEEE.

Ziolkowski, K. (2010). Computer network operations and the law of armed conflict. *Military Law and Law of War Review, 49*(2), 47–94.

# A LQ Vaccine Communication Game

**Alessandra Buratto, Luca Grosset, and Bruno Viscolani**

**Abstract** The vaccination issue is a crucial problem nowadays. We see the presence of an anti-vaccination movement, which takes actions to spread the idea that vaccines are ineffective and even dangerous. We propose a model for this public health problem using the differential game framework and aspire to help understanding the effectiveness of communication policies. One player of the game is the health-care system, which aims to minimize the number of unvaccinated people at minimum cost. The second player is a pharmaceutical firm, which produces and sells a given type of vaccine, and wants to maximize its profit. To pursue their objectives, the two players run suitable vaccination advertising campaigns. We study the open-loop Nash equilibrium advertising strategies of the two players and observe that the communication policy of the pharmaceutical firm helps the health-care system to decrease the number of unvaccinated people.

**Keywords** Differential games · Vaccine communication policy · Advertising

## 1 Introduction

In recent times, the vaccination issue has become a crucial problem. This is a consequence of actions by the anti-vaccination movement, aimed to spread the idea that vaccines are ineffective and even dangerous (Carrillo-Santisteve and Lopalco 2012; Hotez 2017). The main claimed danger concerns the onset of severe neurological diseases caused by vaccinations, even if controlled studies have excluded such causal relations (Gasparini et al. 2015). The fear for the claimed bad consequences spreads easily and is difficult to contrast. Such a situation has caused a substantial reduction of the fraction of vaccinated people in Italy, as alleged by the "Istituto Superiore di Sanità" (Italian Higher Institute for Health) (Guerra et al.

A. Buratto (✉) · L. Grosset · B. Viscolani
Department of Mathematics Tullio Levi-Civita, University of Padova, Padova, Italy
e-mail: buratto@math.unipd.it; grosset@math.unipd.it; viscolani@math.unipd.it

 
P.-O. Pineau et al. (eds.), *Games in Management Science*,
International Series in Operations Research & Management Science 280,
https://doi.org/10.1007/978-3-030-19107-8_19

2017). One main spreading way of this vaccine policy resistance is the word-of-mouth communication and the phenomenon is amplified by the presence of fake information in social media. In fact, as observed by several scholars (see, e.g., El Ouardighi et al. 2016), word-of-mouth has become a relevant phenomenon to take into account.

Social benefits of vaccination should be easily recognized, but they are questioned in the case of diseases which have become rare. From a mere economic viewpoint, it has been estimated that societies undergo heavy costs from vaccine-preventable diseases, because of deaths, disabilities, and economic losses. The observed financial burden which can be attributed to unvaccinated people indicates the potential economic benefit of increasing people immunization (Ozawa et al. 2016).

We propose a model for this public health problem using the linear-quadratic (LQ) differential game frameworks (Dockner et al. 2000; Haurie et al. 2012), and aspire to help understanding the effectiveness of communication policies. To the best of our knowledge no differential game model is present in the relevant scientific literature, whereas the vaccination issue has been tackled through population-scale models (Reluga and Galvani 2011). Quite naturally, we must admit that communication is only one face of the vaccination problem. This requires also, for example, coordination and management efforts to provide vaccination coverage of populations moving through different geographical regions (Carrillo-Santisteve and Lopalco 2012). However, the word-of-mouth communication among the anti-vaccine people can only be contrasted through a fair communication campaign.

In Sect. 2 we present the model for the time evolution of the unvaccinated population under the effects of the word-of-mouth, health care system, and pharmaceutical firm communication. We treat the vaccine communication campaign as a dynamic advertising process (Huang et al. 2012). Objectives of the two players are discussed too. The health care system wants to minimize the number of unvaccinated people at minimum cost. The pharmaceutical firm, who produces and sells a given type of vaccine, aims at maximizing its profit, along with minimizing the number of unvaccinated people. In Sect. 3, in order to study the open-loop Nash equilibrium advertising strategies of the two players, we analyze the open-loop best response strategies using Pontryagin Maximum Principle. In Sect. 4 we illustrate numerically the results obtained. In Sect. 5 we conclude with some suggestions for further research.

## 2 Unvaccinated Subpopulation and Its Contrast

Let us denote by $x(t) > 0$ the number of unvaccinated people at time $t$; they spread some negative information on vaccines and vaccination through the social media (a word-of-mouth mechanism) with intensity proportional to their number. The effect is represented by the motion equation

$$\dot{x}(t) = \gamma x(t) , \tag{1}$$

where $\gamma > 0$ represents the word-of-mouth effectiveness.

Two actors are present on the scene, the *health-care system (S)* which is committed to keep/obtain the herd immunity of the whole population, and a *pharmaceutical firm (f)* which produces and sells the vaccines and looks for profit. Both the health-care system and the pharmaceutical firm want to contrast the growth of unvaccinated people, their vaccine hesitancy (World Health Organization 2016), and negative communication activity. The communication rates $u_s(t), u_f(t)$, of players $S$ and $f$, respectively, affect the number of unvaccinated people as represented by the Cauchy problem

$$\begin{aligned} \dot{x}(t) &= \gamma x(t) - \delta_s u_s(t) - \delta_f u_f(t), \\ x(0) &= x_0, \end{aligned} \tag{2}$$

where $\delta_s, \delta_f > 0$ are the effectiveness of the system and firm flows, while $x_0 > 0$ is the initial level of unvaccinated people. The terms in the r.h.s. of (2) represent the word-of-mouth effect of the unvaccinated subpopulation and the communication effects of the system and firm policies, respectively.

Let a time horizon $T > 0$ be given; we call $(x(t), u_s(t), u_f(t))$ an admissible solution if its state and control components are positive, $x(t), u_s(t), u_f(t) \geq 0$, $t \in [0, T]$, and solve the Cauchy problem (2). We choose a finite time horizon because $S$ (the health-care system) has a mandate which depends on the political vote. Moreover, $f$ (the pharmaceutical firm) has to present a financial report, which concerns also the vaccine communication activity. Both the political vote and the financial report have precise time definitions.

The model represented by Eq. (2) is symmetric to the one presented in Jørgensen and Zaccour (2004, p. 110) to explain the goodwill evolution depending on two simultaneous advertising efforts and on a "natural decay." Here we have a "natural growth" contrasted by two simultaneous vaccine communication efforts made by the health-care system and the pharmaceutical firm.

The two actors pursue different objectives, although they both want to minimize the number of unvaccinated people.

The health-care system aims to minimize both the costs induced by the unvaccinated people (which are assumed to be quadratic in the number of unvaccinated people) and the costs due to the communication campaign (which are assumed to be quadratic in the communication flow). Its objective may be represented by the functional

$$J_s = \int_0^T \left( \frac{\beta}{2} x^2(t) + \frac{\kappa_s}{2} u_s^2(t) \right) \mathrm{d}t + \frac{\eta}{2} x^2(T) \tag{3}$$

to be minimized over the finite programming interval $[0, T]$. The parameters $\beta > 0$ and $\kappa_s > 0$ are the marginal penalty rate of the unvaccinated subpopulation level and the marginal cost rate of the health-care system communication effort, respectively. The parameter $\eta > 0$ makes the residual value function a penalty which is as higher as the final unvaccinated subpopulation $x(T)$ is larger.

The pharmaceutical firm seeks to increase its revenue and minimize its loss of earnings. It also seeks to minimize its communication costs. Its objective may be represented by the functional

$$J_f = \int_0^T \left( \vartheta x(t) + \frac{\kappa_f}{2} u_f^2(t) \right) dt \tag{4}$$

to be minimized. The parameters $\vartheta > 0$ and $\kappa_f > 0$ are the marginal revenue loss rate of the unvaccinated subpopulation level and the marginal cost rate of the pharmaceutical firm communication effort, respectively. Quadratic advertising costs are widely used in the relevant literature, see, e.g., Jørgensen and Zaccour (2004).

Note that in the case of the firm objective functional (4), the cost rate for the firm is only linear in $x(t)$ because it is just a loss of income, whereas in the system functional (3) it is quadratic. Note also that the firm is not interested in the final value of $x(T)$ so that there is not any residual value in (4).

The model is an inhomogeneous linear-quadratic (LQ) differential game (Dockner et al. 2000, pp. 171–187; Haurie et al. 2012, pp. 271 and 197).

## 3 Equilibrium Vaccine Communication Campaigns

Even though the two players of the *vaccine communication game* have rather similar objectives, since they are both keen on diminishing the number of people who distrust vaccination as a useful medical treatment, their motivations are essentially different. The health-care system is moved mainly by social costs consideration, whereas the pharmaceutical firm simply wants to reduce its revenue loss.

We assume that the open loop information structure (Dockner et al. 2000, pp. 29–30) is available to both players, i.e., that the health-care system manager and the pharmaceutical firm do not observe the level of the anti-vaccine subpopulation after the initial time. We mean that both players $S$ and $f$ have to plan their communication campaigns at the beginning of the programming interval. We have several reasons for this choice. First, the number of unvaccinated people is not easy to measure, and it can be obtained only with some delay. Then, both players want to control the advertising expenditures and an open-loop solution allows to estimate the advertising costs in advance.

In order to compute the Nash equilibrium we look for the best response communication strategies of the system and the firm.

### 3.1 Pharmaceutical Firm Best Response

The pharmaceutical firm minimizes the objective functional (4), subject to the motion equation and initial condition (2) using the control $u_f(t) \geq 0$.

Here the function $u_s(t) \geq 0$ can be assumed as a guess of player $f$ on the communication strategy of player $S$. This is a case of linear quadratic control problem (Engwerda 2005, § 5.2).

The Hamiltonian function is

$$H_f(x, u_f, p_f, t) = -\vartheta x - \frac{\kappa_f}{2} u_f^2 + \left(\gamma x - \delta_s u_s(t) - \delta_f u_f\right) p_f ; \tag{5}$$

the Pontryagin Maximum Principle necessary conditions for an optimal open-loop solution (see, e.g., Seierstad and Sydsæter 1987, p. 85) are the control rule

$$u_f^*(t) = \arg\max_{u_f \geq 0} \left\{-\frac{\kappa_f}{2} u_f^2 - \delta_f p_f(t) u_f\right\} = \max\left\{0, -\frac{\delta_f}{\kappa_f} p_f(t)\right\} , \tag{6}$$

and the adjoint Cauchy problem

$$\dot{p_f}(t) = \vartheta - \gamma p_f(t) , \tag{7}$$

$$p_f(T) = 0 . \tag{8}$$

We obtain that

$$p_f(t) = \frac{\vartheta}{\gamma} \left(1 - e^{\gamma(T-t)}\right) \leq 0 , \qquad t \in [0, T] , \tag{9}$$

so that player $f$'s optimal communication rate is

$$u_f^*(t) = \frac{\vartheta \delta_f}{\gamma \kappa_f} \left(e^{\gamma(T-t)} - 1\right) . \tag{10}$$

It is important to observe that $u_f^*(t)$ is not affected by the information (or guess) on the communication strategy $u_s(t)$ of the health-care system.

### 3.2 Health-Care System Best Response

The health-care system managers minimize the objective functional (3), subject to the motion equation and initial condition (2) using the control $u_s(t) \geq 0$. Here the function $u_f(t) \geq 0$ can be assumed as a guess of the system on the firm communication strategy. Again, this is a case of linear quadratic control problem (Engwerda 2005, § 5.2).

The Hamiltonian function is

$$H_s(x, u_s, p_s, t) = -\frac{\beta}{2}x^2 - \frac{\kappa_s}{2}u_s^2 + \left(\gamma x - \delta_s u_s - \delta_f u_f(t)\right) p_s ; \tag{11}$$

the Pontryagin Maximum Principle necessary conditions for an optimal open-loop solution (see, e.g., Seierstad and Sydsæter 1987, p. 85) are the control rule

$$u_s^*(t) = \arg\max_{u_s \geq 0} \left\{-\frac{\kappa_s}{2}u_s^2 - \delta_s p_s(t) u_s\right\} = \max\left\{0, -\frac{\delta_s}{\kappa_s} p_s(t)\right\} , \tag{12}$$

and the adjoint Cauchy problem

$$\dot{p}_s(t) = \beta x(t) - \gamma p_s(t) , \tag{13}$$

$$p_s(T) = -\eta x(T) . \tag{14}$$

In the following claims we introduce two guesses on the adjoint function sign and structure. Hence, we will discuss existence and features of a solution compatible with them.

*Claim (Adjoint Function Sign)* The adjoint function is always negative, $p_s(t) < 0$, $t \in [0, T]$, therefore

$$u_s^*(t) = -\frac{\delta_s}{\kappa_s} p_s(t) . \tag{15}$$

*Claim (Adjoint Function Structure)* The adjoint function $p_s(t)$ can be expanded as

$$p_s(t) = \varphi(t) x(t) + \chi(t) , \tag{16}$$

for some continuously differentiable functions $\varphi(t)$ and $\chi(t)$.

After substituting (15) into the motion equation (2) and in view of the initial condition, we obtain

$$\dot{x}(t) = \gamma x(t) + \frac{\delta_s^2}{\kappa_s} p_s(t) - \delta_f u_f(t) , \tag{17}$$

$$x(0) = x_0 . \tag{18}$$

The candidate solutions are characterized by the two-point-boundary value problem for state and co-state functions $x(t)$ and $p(t)$ given by Eqs. (17), (18) and (13), (14).

Using the expansion (16) in Eqs. (17), (18) and (13), (14), we characterize the functions $\varphi(t)$ and $\chi(t)$. First, we obtain that function $\varphi(t)$ must satisfy the Cauchy problem

$$\begin{cases} \dot{\varphi}(t) + 2\gamma\varphi(t) + \dfrac{\delta_s^2}{\kappa_s}\varphi^2(t) - \beta = 0, \\ \varphi(T) = -\eta, \end{cases} \tag{19}$$

whose differential equation belongs to the Riccati type. Second, function $\chi(t)$ must satisfy the problem

$$\begin{cases} \dot{\chi}(t) = -\left(\dfrac{\delta_s^2}{\kappa_s}\varphi(t) + \gamma\right)\chi(t) + \delta_f\varphi(t)u_f(t), \\ \chi(T) = 0, \end{cases} \tag{20}$$

which involves a first order linear equation.

We observe that any solution $\varphi(t)$ to Cauchy problem (19) is invariant with respect to the function parameter $u_f(t)$, whereas any solution $\chi(t)$ to Cauchy problem (20) depends on it. We stress such dependence by calling $\chi\left(t; u_f\right)$ a solution to (20).

The above analysis can be synthesized by the statement that the unique open-loop optimal control is given (in feedback form) by

$$u_s(t) = -\frac{\delta_s}{\kappa_s}\left(\varphi(t)x(t) + \chi\left(t; u_f\right)\right), \tag{21}$$

where $\varphi(t)$ and $\chi\left(t, u_f\right)$ are continuously differentiable solutions of problems (19) and (20).

We notice that $u_f^*$ does not depend on $u_s^*$ because the effect of the two advertising flows on the state evolution is additive, and because the Hamiltonian (5) is linear in the state variable $x$. On the other hand, the Hamiltonian (11) is quadratic in $x$, thus making $u_s^*$ depend on $u_f^*$.

**Theorem 1** *If solutions to problems (19) and (20) exist and the control function $u_s(t)$ provided by (21) is positive, i.e., if $\varphi(t)x(t) + \chi\left(t; u_f\right) = p_s(t) < 0$, then both claims hold and the control $u_s(t)$ satisfies the necessary conditions to be a best response.*

The unique solution $u_s(t)$ is optimal, because the Hamiltonian (11) is concave in $(x, u)$, and the function involved in the residual value definition, $-\eta x^2/2$, is concave (see, e.g., Seierstad and Sydsæter 1987, The Mangasarian sufficiency theorem, p. 105).

Here we comment on the solutions to the two Cauchy problems (19) and (20).

In problem (19) we can verify that

$$\varphi(t) < 0, \qquad t \in [0, T], \tag{22}$$

in fact, the Riccati ordinary differential equation in (19) has the two constant solutions (equilibrium points)

$$\varphi_{1,2} = \frac{-\gamma \mp \sqrt{\gamma^2 + \beta\delta_s^2/\kappa_s}}{\delta_s^2/\kappa_s}, \tag{23}$$

which have opposite signs, $\varphi_1 < 0 < \varphi_2$. The equilibrium point $\varphi_1 < 0$ is unstable, whereas $\varphi_2 > 0$ is stable; it follows that the solution of the Cauchy problem with $\varphi(T) = -\eta < 0$ must necessarily satisfy

$$\varphi(t) \leq \max\{\varphi_1, -\eta\}, \qquad \text{for all } t \in [0, T],$$

which implies inequality (22). We observe in particular that $\varphi(t)$ is monotonically increasing if and only if $\eta \leq -\varphi_1$.

As far as problem (20) is considered, using the general formula (Alexéev et al. 1982, pp. 184–188) we determine the unique solution of problem (20),

$$\chi(t) = -e^{\omega(t)} \int_t^T \delta_f \varphi(s)\, u_f(s)\, e^{-\omega(s)}\, ds, \quad \omega(t) = -\int_t^T \left(\frac{\delta_s^2}{\kappa_s}\varphi(s) + \gamma\right) ds,$$

and, recalling that inequality (22) holds, while $u_f(t) \geq 0$ by assumption, we can verify that

$$\chi(t) \geq 0, \qquad t \in [0, T]. \tag{24}$$

Moreover, inequality (24) holds strictly for $t < T$, if $u_f(t) > 0$ in a left neighborhood of $T$.

Unfortunately, we cannot prove that $u_s^*(t) \geq 0$ for all $t \in [0, T]$ and for all choices of the model parameters. Here we provide a condition for the existence of a positive control over $[0, T]$, satisfying the necessary optimality conditions.

**Theorem 2** *Let $(u_s^*(t), x^*(t))$ be an optimal solution. If $x^*(t) > 0$ for all $t \in [0, T)$, then $u_s^*(t) > 0$ for all $t \in [0, T)$.*

*Proof* See Appendix.

## 3.3 Open-Loop Equilibrium

From the analysis developed so far we can conclude that player *f*'s best response communication policy is invariant with respect to player *S*'s policy, and this fact ensures that $u_f^*(t)$, as given by Eq. (10), is the unique player *f*'s component of an open-loop equilibrium, provided that an equilibrium exists. Therefore, we need to determine player *S*'s component to obtain an equilibrium. From Theorem 2 and the continuity of $x(t)$, we can easily prove the following existence result.

**Theorem 3** *If the time horizon* $T > 0$ *is small enough, then the pair* $(u_s^*(t), u_f^*(t)) = (u_s^*\left(t; u_f^*(t)\right), u_f^*(t))$, *whose second component is given by Eq. (10), while the first one is given by Eq. (21), after solving problems (19) and (20) with the use of* $u_f^*(t)$, *satisfies the necessary conditions to be an open-loop Nash equilibrium.*

In order to obtain a best response strategy $u_s^*(t; u_f^*(t))$ for player $S$ we need first to solve the Riccati equation in (19), which requires a numerical quadrature; then to integrate the linear equation in (20), which can be done through numerical quadrature; finally to use the results for the functions $\varphi(t)$ and $\chi(t)$ in Eq. (21). Solving the Cauchy problem (2) with the control functions $u_s^*(t; u_f^*(t))$ and $u_f^*(t)$ will provide the associated state function.

We can compare the optimal communication effort of the health-care system when it acts alone with its equilibrium communication effort in presence of the pharmaceutical firm.

**Corollary 1** *The communication activity of player f in equilibrium implies a lower communication effort by player S with respect to the one-actor policy.*

*Proof* As $u_f(t) > 0$, $t < T$, from (21) and (24) we obtain that the optimal control $u_s^*(t)$ is smaller than the one associated with $u_f(t) \equiv 0$, which (in feedback form) is

$$u_s^*(t) = -\frac{\delta_s}{\kappa_s}\varphi(t)\,x(t)\,. \tag{25}$$

□

## 4 Numerical Example

In this section, we analyze numerically some instances of the game in order to illustrate the results obtained.

Table 1 presents an overview of data common to all the game instances we consider, whereas we let the parameters $\gamma$ and $\delta_f$ take sundry values, so that

$$\gamma \in \{0.01\,, 0.05\,, 0.1\}\,, \qquad \delta_f \in \{0\,, 0.3\,, 0.6\}\,.$$

The parameter $\delta_f = 0$ sets a situation where player $f$'s communication campaign cannot have any effect; on the contrary, player $f$'s communication cam-

**Table 1** Fixed game parameters

| $\beta$ | $\delta_s$ | $\eta$ | $\vartheta$ | $\kappa_f$ | $\kappa_s$ | $x_0$ | $T$ |
|---|---|---|---|---|---|---|---|
| 0.3 | 0.6 | 0.9 | 0.1 | 0.8 | 4 | 50 | 10 |

paign is effective if $\delta_f > 0$. The different values of the parameter $\gamma$ represent low/intermediate/high effectiveness of the anti-vaccine word-of-mouth phenomenon.

**Sensitivity to Word-of-Mouth Effectiveness $\gamma$**

From Eq. (10) it follows that the pharmaceutical firm does some communication effort if and only if $\delta_f > 0$, namely when it can have an effect on the state evolution; for $\delta_f = 0.6$ we can observe the behavior as in Fig. 1. We can observe that the higher the word-of-mouth effectiveness $\gamma$, the stronger player $f$'s contrasting strategy. Moreover, the contribution of player $f$'s action in a neighborhood of $T$ is vanishing, in fact $u_f(T) = 0$.

In Fig. 2, player $S$'s optimal communication strategy $u_s^*$ is plotted, for different values of the parameter $\gamma$. Again, the higher the word-of-mouth effectiveness $\gamma$, the stronger player $S$'s optimal communication strategy.

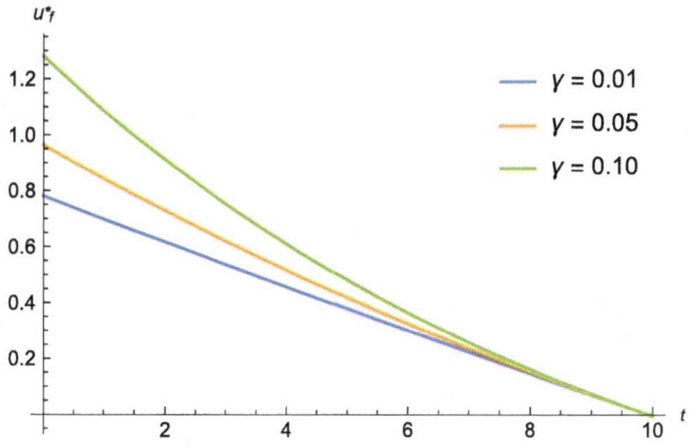


**Fig. 1** f strategy for $\delta_f = 0.6$

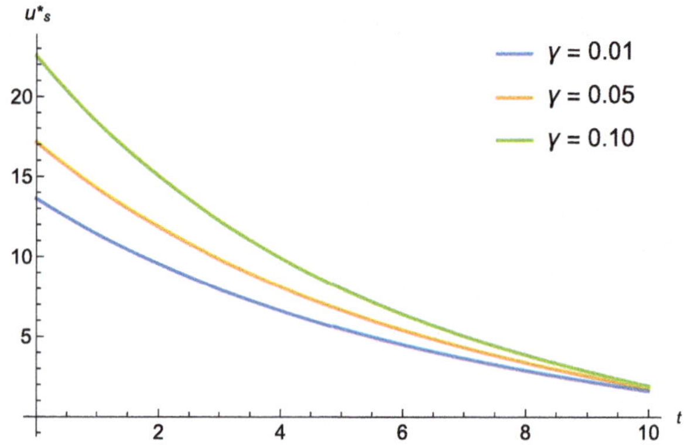


**Fig. 2** S strategy for $\delta_f = 0.6$

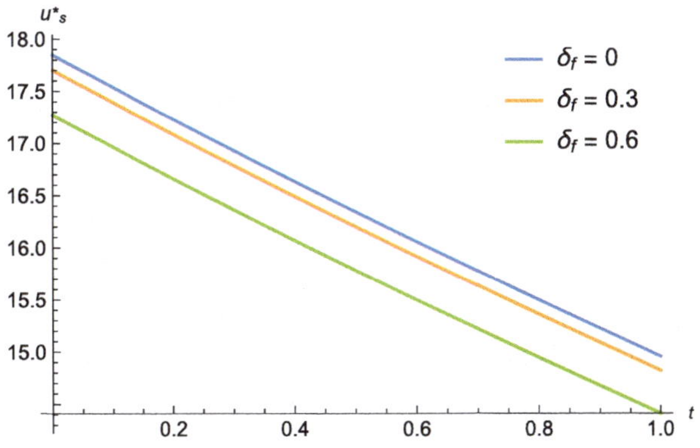


**Fig. 3** S strategy for $\gamma = 0.05$ and $t \in [0, 1]$

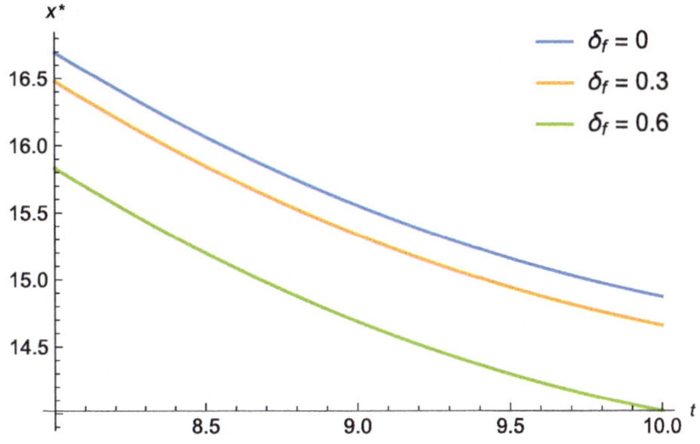


**Fig. 4** Unvaccinated people for $\gamma = 0.05$ and $t \in [8, 10]$

**Sensitivity to Communication Effectiveness $\delta_f$**

Player *S*'s strategy in equilibrium, given by Eq. (21), is shown in Fig. 3. We can see that the effective presence of player *f*'s communication effort ($\delta_f > 0$) makes player *S*'s communication effort weaker.

The evolution of the unvaccinated people is plotted in Fig. 4; we observe that when the communication campaign of the pharmaceutical firm is effective ($\delta_f > 0$) then the joint action of the two players causes a stronger reduction of the unvaccinated people.

**Equilibrium Costs and Final Level of Unvaccinated People**

In Table 2 we display the equilibrium payoffs of the two players, together with the associated level of the state function, for different values of the parameters $\gamma$ (word-of-mouth effectiveness) and $\delta_f$ (player *f*'s communication effectiveness).

**Table 2** Players' equilibrium costs and final state

| | $\delta_f = 0$ | | | $\delta_f = 0.3$ | | | $\delta_f = 0.6$ | | |
|---|---|---|---|---|---|---|---|---|---|
| | $J_f$ | $J_s$ | $x(T)$ | $J_f$ | $J_s$ | $x(T)$ | $J_f$ | $J_s$ | $x(T)$ |
| $\gamma$ | | | | | | | | | |
| 0.01 | 26.01 | 2358.75 | 13.27 | 26.06 | 2325.69 | 13.09 | 26.22 | 2228.42 | 12.53 |
| 0.05 | 26.27 | 2975.55 | 14.87 | 26.39 | 2927.55 | 14.66 | 26.75 | 2786.53 | 14.02 |
| 0.1 | 25.49 | 3924.96 | 15.80 | 25.75 | 3849.86 | 15.56 | 26.55 | 3629.74 | 14.84 |

From the results in Table 2, we observe in general that as $\gamma$ increases the final unvaccinated people level $x(T)$ and player $S$'s total cost increase. On the other hand, as $\delta_f$ increases the final unvaccinated people level $x(T)$ and player $S$'s total cost decrease. Finally, player $f$'s total cost increases as $\delta_f$ increases, whereas it is a quasi-concave function of $\gamma$.

# 5 Conclusion

The aim of this study was to shed some light on the importance of a pharmaceutical firm information activity to build awareness of the vaccines role in the fight against infectious diseases. We found relevant information on the health-care situation and viewpoint in recent documents of health organizations, WHO, and of groups of medical researchers. We have proposed a model with two actors, the health-care system, $S$, and a pharmaceutical firm, $f$, who direct their communication efforts to reduce the unvaccinated population, while such population tends to grow spontaneously.

The main results of our model are the following: first, the presence of the pharmaceutical firm is useful for the health-care system because the firm's communication campaign helps to decrease the number of unvaccinated people at the end of the programming interval. Second, the pharmaceutical firm invests to reduce the number of unvaccinated people and this investment decreases the total expenditure of the health-care system. Therefore, the presence of the pharmaceutical firm is useful to society, because it shares with the health-care system the objective of reducing the number of unvaccinated people. Finally, the real problem that emerges from our model analysis is the effectiveness of the word-of-mouth communication of the unvaccinated people: the stronger this phenomenon, the higher the expenditures of both pharmaceutical firm and health-care system. Moreover, as far as the unvaccinated population exists, its final value is strictly increasing in the effectiveness of the word-of-mouth phenomenon.

The study may be extended in different directions. On one hand, the definitions (3) and (4) of the players objective functions could be modified in order to consider semi-altruistic preferences of the agents (see Brekke et al. 2017, 2012). On the other hand, it could be interesting to analyze a stochastic evolution of the number of unvaccinated people. This goal could be obtained by introducing a stochastic effect in the communication campaigns and changing the ordinary differential equation (2) into a stochastic one (see Grosset and Viscolani 2004; Buratto and Grosset 2006). Moreover, a new definition of the objective functionals (3) and (4) should be provided.

**Acknowledgements** The authors thank two anonymous referees for their interesting comments and suggestions. The third author wants to thank D. Bonandini for instructive talks on the vaccination problem in Italy.

# Appendix

*Proof (Theorem 2)* Let $u_s(t) \geq 0$, $t \in [0, T]$, be an admissible control of player $S$, which determines the state function $x(t) > 0\,, t \in [0, T)$. We recall that the function $u_f(t)$ is a known parameter in the present context.

Let us assume that

$$u_s(t) = 0\,, \qquad t \in [\tau, \tau + h)\,,$$

for some $\tau \in [0, T - 2h]$, $h > 0$, and let us define the spike variation

$$u_s^\alpha(t) = \begin{cases} \alpha\,, & t \in [\tau, \tau + h)\,, \\ u_s(t)\,, & t \notin [\tau, \tau + h)\,, \end{cases}$$

of the control $u_s(t)$, where $\alpha > 0$. Let the admissible control $u_s^\alpha(t)$, jointly with $u_f(t)$, determine the state function $x^\alpha(t)\,, t \in [0, T]$, and assume that

$$x^\alpha(t) > 0\,, \qquad t \in [\tau, \tau + 2h)\,.$$

Remark that

$$x^\alpha(\tau) = x(\tau)\,,$$

$$x^\alpha(t) \leq x(t)\,, \qquad t \in (\tau, T)\,,$$

and

$$x^\alpha(t) < x(t) - \delta_s h\alpha\,, \qquad t \geq \tau + h\,.$$

In order to compare the values of the objective functional $J_s$ associated with the control pairs $(u_s, u_f)$ and $(u_s^\alpha, u_f)$, let us define

$$\Delta J_s = J_s[u_s^\alpha, u_f] - J_s[u_s, u_f]$$

and look for an upper bound of it. We observe that

$$\begin{aligned}\Delta J_s \leq\ & \Delta J_s\ |_{[\tau,\tau+h)}\ + \Delta J_s\ |_{[\tau+h,\tau+2h)} \\ =\ & \frac{\beta}{2}\int_{\tau}^{\tau+h}\Big((x^\alpha)^2(t) - x^2(t)\Big)\,\mathrm{d}t\ + \frac{\kappa_s}{2}\int_{\tau}^{\tau+h}\alpha^2\,\mathrm{d}t \\ & + \frac{\beta}{2}\int_{\tau+h}^{\tau+2h}\Big((x^\alpha)^2(t) - x^2(t)\Big)\,\mathrm{d}t\ ,\end{aligned}$$

because $\Delta J_s\ |_{[0,\tau)} =\ 0$, $\Delta J_s\ |_{[\tau+2h,T]} \leq\ 0$, and because of the residual value function negative variation. The upper bound just obtained is strictly less than

$$\frac{\kappa_s}{2}\int_{\tau}^{\tau+h}\alpha^2\,\mathrm{d}t\ + \frac{\beta}{2}\int_{\tau+h}^{\tau+2h}\Big((x^\alpha)^2(t) - x^2(t)\Big)\,\mathrm{d}t\ ,$$

because

$$\frac{\beta}{2}\int_{\tau}^{\tau+h}\Big((x^\alpha)^2(t) - x^2(t)\Big)\,\mathrm{d}t\ <\ 0\,.$$

The last upper bound is strictly less than

$$\begin{aligned}&\frac{\kappa_s h}{2}\alpha^2 + \frac{\beta}{2}\int_{\tau+h}^{\tau+2h}\Big((x(t) - \delta_s h\alpha)^2 - x^2(t)\Big)\,\mathrm{d}t \\ &\quad = \frac{h\alpha}{2}\left\{\Big(\kappa_s + \beta\delta_s^2 h^2\Big)\alpha\ -\ 2\beta\delta_s\int_{\tau+h}^{\tau+2h} x(t)\,\mathrm{d}t\right\}\,.\end{aligned}$$

Now, the last expression is negative if and only if

$$\alpha\ <\ \frac{2\beta\delta_s}{\kappa_s + \beta\delta_s^2 h^2}\int_{\tau+h}^{\tau+2h} x(t)\,dt\ ,$$

which is true for suitably small $\alpha\ >\ 0$, because the right-hand side of the above inequality is strictly positive. □

## References

Alexéev, V., Tikhomirov, V., & Fomine, S. (1982). *Commande optimale*. Moscou: Éditions MIR.

Brekke, K. R., Cellini, R., Siciliani, L., & Straume, O. R. (2017). Differential games in health-care markets: Models of quality competition with fixed prices. In J. Apaloo & B. Viscolani (Eds.), *Advances in dynamic and mean field games. Annals of the International Society of dynamic games* (Vol. 15, pp. 145–167). Cham: Birkhäuser.

Brekke, K. R., Siciliani, L., & Straume, O. R. (2012). Quality competition with profit constraints. *Journal of Economic Behavior and Organization, 84*, 642–659.
Buratto, A., & Grosset, L. (2006). A communication mix for an event planning: A linear quadratic approach. *Central European Journal of Operations Research, 14*(3), 247–259.
Carrillo-Santisteve, P., & Lopalco, P. L. (2012). Measles still spreads in Europe: Who is responsible for the failure to vaccinate? *Clinical Microbiology and Infection, 18*(Suppl. 5), 50–56.
Dockner, E., Jørgensen, S., Long, N. V., & Sorger, G. (2000) *Differential games in economics and management science*. Cambridge: Cambridge University Press.
El Ouardighi, F., Feichtinger, G., Grass, D., Hartl, R., & Kort, P. M. (2016). Autonomous and advertising-dependent 'word of mouth' under costly dynamic pricing. *European Journal of Operational Research, 251*(3), 860–872.
Engwerda, J. C. (2005) *LQ dynamic optimization and differential games*. Chichester: Wiley.
Gasparini, R., Panatto, D., Lai, P. L., & Amicizia, D. (2015). The "urban myth" of the association between neurological disorders and vaccinations. *Journal of Preventive Medicine and Hygiene, 56*, e1–e8.
Grosset, L., & Viscolani, B. (2004). Advertising for a new product introduction: A stochastic approach. *Top: Trabajos de Investigacion Operativa, 12*(1), 149–168.
Guerra, R., Ricciardi, W., & Siliquini, R. (2017). Vaccinazioni: Stato dell'arte, falsi miti e prospettive. *Quaderni del Ministero della Salute, 27*. Vaccinations: State of the art, false myths and perspectives, Report of the Italian Health Ministry.
Haurie, A., Krawczyk, J. B., Zaccour, G. (2012). *Games and dynamic games*. Singapore: World Scientific.
Hotez, P. J. (2017). How the Anti-Vaxxers are winning. *The New York Times*. The Opinion Pages, February 8 (p. A25).
Huang, J., Leng, M., & Liang, L. (2012). Recent developments in dynamic advertising research. *European Journal of Operational Research, 220*(3), 591–609.
Jørgensen, S., & Zaccour, G. (2004). *Differential games in marketing*. Boston: Kluwer Academic Publishers.
Ozawa, S., Portnoy, A., Getaneh, H., Clark, S., Knoll, M., Bishai, D., Yang, H. K., & Patwardhan, P. D. (2016). Modeling the economic burden of adult vaccine-preventable diseases in the United States. *Health Affairs*, published online October 12.
Reluga, T. C., & Galvani, A. P. (2011). A general approach for population games with application to vaccination. *Mathematical Biosciences, 230*, 67–78.
Seierstad, A., & Sydsæter, K. (1987). *Optimal control theory with economic applications*. Amsterdam: North-Holland.
World Health Organization. (2016, June 7). Addressing vaccine hesitancy. http://www.who.int/immunization/programmes_systems/vaccine_hesitancy/en/

# On the Stability of a Two-Player International Environmental Agreement with Intra-Industry Trade

**Sébastien Debia**

**Abstract** An international environmental agreement is an unstable coalition by nature, which results in a drive to design mechanisms to provide stability. This note numerically shows that when an international environmental agreement is coupled with intra-industry trade with complementary intermediate goods, then two-player cooperation is a Nash equilibrium. This result may be of interest given that over half the trade between developed countries is intra-industry.

**Keywords** International environmental agreement · Intra-industry trade · Cooperative game

## 1 Introduction

The solution to tackle anthropic global warming is well-known, as the main obstacle to its implementation: when contemplating whether to sign an environmental agreement, countries must consider that there is an incentive to free-ride. Pollution abatement is a public good, and every country will profit from the effort of one, resulting in a prisoner's dilemma: non-cooperation is the equilibrium, while cooperation is Pareto-dominant (Hardin 1982).

The game-theory literature about international environmental agreements bases its analysis on a simplified representation of the economy. Countries produce a pollution-emitting homogeneous good to increase their wealth (Dockner and Long 1993; Barrett 1994; Breton et al. 2010). Based on this assumption, the fundamental situation is a non-cooperative equilibrium: the public good not provided by a country can be supplied by another, thereby increasing the incentive to free-ride. The literature accordingly focuses on the mechanisms allowing the cooperative solution

S. Debia (✉)
HEC Montréal, Montréal, QC, Canada
e-mail: sebastien.debia@hec.ca

 
P.-O. Pineau et al. (eds.), *Games in Management Science*,
International Series in Operations Research & Management Science 280,
https://doi.org/10.1007/978-3-030-19107-8_20

to be stable. However, little attention has been given to trade, as the homogeneity assumption implies that the goods are perfect substitutes.

The relationship between trade and environmental policy is complex. Countries with homogeneous goods tend to soften their environmental policy as an indirect means of subsidizing their exports (Barrett 1994). While this result is not fully supported by empirical evidence, the literature generally agrees that when a country has a strong environmental regulation, this is detrimental to its volume of exports (Cherniwchan et al. 2017). Nevertheless, these authors emphasize the emergence of a new empirical literature based on heterogeneous-firm models à la Melitz (Melitz 2003) to analyze the link between intra-industry trade and environmental policy.[1] Because trade makes competition more intense, less-productive firms are pushed out of the market, which then raises the economy's overall productivity. Concomitantly, we may see a reduction in emissions, since input use is reduced.

This note links two strands of the literature by bringing the structure of intra-industry trade into the game-theoretic framework of coalition stability. Countries produce differentiated intermediate goods used for the production of final goods in each country, along a constant elasticity of substitution (CES) production function (Melitz 2003; Arkolakis et al. 2012; Edmond et al. 2015). The consumption of the final goods and the trading of the intermediate goods generate welfare for each country, but they also generate environmental damage.

The stylized model considered in this note is a very simple two-player-two-(intermediate)-good static game. The production of the intermediate goods generates transboundary pollution, while the production of the final goods does not. To reduce the model to its simplest form in order to capture the main feature of intra-industry trade, it is assumed that each country's economy is small relative to the rest of the world's, so that they consider the international price of intermediate goods as given. I compute the equilibrium of the economy, assuming that each country plays either cooperatively or non-cooperatively, and design a bi-matrix game to determine if cooperation is a Nash equilibrium (or not). I then determine the dominant strategy for each player.

This framework is interpretable as a non-binding agreement, such as the Paris COP 21. Countries agree on the principle of cooperation, but may deviate when it comes to applying the agreement.

The contribution of this note is to numerically provide conditions on the final-good-production function under which environmental cooperation is a Nash equilibrium. While the equilibrium is the usual prisoner's dilemma when the two intermediate goods are substitutes, cooperation becomes the Nash equilibrium when the two intermediate goods are relative complements. The latter case is far from

[1]Intra-industry trade refers to international trade between similar countries and within a particular sector, such as car or computer manufacturing. It accounted for over half the total trade of developed countries in the late 1990s (OECD 2002).

insignificant, as most final goods require a precise quantity of intermediate goods for their production.[2]

The main mechanism behind this result is that the increased cost of a stronger environmental policy is reflected through the world prices. When goods are relative complements, the demand for these goods is relatively inelastic because they are needed for the production process. If a country deviates from cooperation, it reduces the price of its domestic intermediate good, while the price of the imported good remains high. In other words, a deviating country drastically reduces its terms of trade, an effect that is detrimental to its wealth. This result provides support for advocating that intra-industry trade might be beneficial for the environment, in that it fosters environmental cooperation.

The note is organized as follows: Sect. 2 describes the economic structure of the model. Section 3 defines the meta-game and the optimality conditions associated with each subgame. Section 4 displays the numerical results. Section 5 briefly concludes.

## 2 The Economic Structure

I consider a partial-equilibrium model with two countries indexed by $i$, $i = 1, 2$. Each country produces a final good for local consumers and has an intermediate goods sector used as an input in the production of the final good in either countries. The intermediate good can be exported to the other country. Both countries are symmetric, except for their respective demand function in the differentiated intermediate goods.

**Final Goods** Each country produces a quantity of final goods $q_i$ for their local consumers. Their inverse-demand functions $P_i(q_i)$ are as follows:

$$P_i(q_i) = 100 - q_i.$$

The final goods are produced along a constant elasticity of substitution (CES) in both countries:

$$q_i = \left(y_i^{\rho_i} + m_i^{\rho_i}\right)^{\frac{1}{\rho_i}},$$

where $y_i$ is the quantity of intermediate goods produced by country $i$, and $m_i$ is the quantity of intermediate goods imported by country $i$ from the other country. The parameter $\rho_i$ measures whether the two goods $y_i$ and $m_i$ are substitutes or complements:

[2]For example, a computer needs a precise quantity of rare earth and other metal to be built. This case is so relevant that Cherniwchan et al. (2017) exemplify the Melitz model with a Leontief (perfect complement) function for the final goods' production.

$$\rho_i = \frac{s_i - 1}{s_i},$$

where $s_i$ is the elasticity of substitution between $y_i$ and $m_i$. As $\rho_i$ goes to $-\infty$, the two intermediate goods are perfect complements in the production of the final good: the CES function approximates a Leontief function. Conversely, as $\rho_i$ goes to 1, the two intermediate goods are perfect substitutes in the production of the final good. When $\rho_i$ tends to 0, the CES function is equivalent to a Cobb–Douglas function. Below this threshold, the two goods are considered relative complements, and above it they are considered to be relative substitutes.

**Intermediate Goods** Both countries produce a differentiated intermediate good that they can use locally and/or export. The variable $y_i$ is the quantity of intermediate good produced by country $i$ for local usage, and the variable $x_i$ is the quantity of the same intermediate good for export. The intermediate good in each country is costless to produce but generates a negative externality: transboundary pollution. For simplicity, we assume that the emissions rate is equal to one and that this pollution generates the same quadratic damage in both countries. Each intermediate-good producer does not internalize the environmental damage but has to pay the environmental tax set by the host country.

For each unit of exports, the transporter must be paid an iceberg transportation cost $\tau$, $\tau > 1$, in kind by the exporter. That is, each country's total production amounts to $y_i + \tau x_i$.

**Trading Prices** In this two-player setting, the quantity of goods imported by country $i$ must match the quantity of goods exported by country $j$. This equality defines the world price $w_j$ for the intermediate good produced in country $j$. Formally:

$$\mathbb{R} \ni w_j \perp x_j - m_i = 0, \qquad j = 1, 2, \ i \neq j.$$

The equilibrium condition is represented as a slackness condition, which states that a world price is nonzero only if exports and imports match.

The imported goods must be paid at a market price including the transport cost (CIF price). That is, country $i$ must pay $\tau w_j$ per unit of import $m_i$. The exporter only receives the market price, excluding the transport cost (FOB price), $w_j$ for $x_j$, while it has produced $\tau$ times the number of units sold.

To summarize, the exporter pays the transport cost in kind, while the importer supports the transport cost financially. While questionable, this double accountancy is usual in the international trade literature (Arkolakis et al. 2012; Edmond et al. 2015). In our framework, it is nevertheless interpretable as the pollution caused by transport. We consider that each country is a price taker with respect to these prices.

## 3 The Meta-Game

After a non-binding environmental agreement, such as the Paris COP 21, has been signed, the question remains as to whether each party is willing to cooperate. Each country's cooperation is modeled such that its implicit tax rate takes full account of the environmental marginal damage. On the other hand, a country is non-cooperative if it considers only the local damage in setting the tax. This approach to cooperation is relatively standard in the game theoretic literature about transboundary pollution (Dockner and Long 1993). The environmental agreement being non-binding, each country is free to decide whether it will play cooperatively or not. Hence, we may face cases where one country implements a low production policy in order to limit pollution, while the other country may want to free-ride and increase its production. These cases will be referred to as deviation equilibria.

**The Non-cooperative Subgame Equilibrium** In the non-cooperative scenario, each country maximizes its local welfare considering the other country's actions and the market price as given. Country $i$'s maximization program is written as follows, for $i \neq j$:

$$\max_{q_i, y_i, x_i, m_i \geq 0} q_i(100 - q_i/2) + w_i x_i - \tau w_j m_i - \frac{1}{2}\left(\sum_{k=1}^{2} y_k + \tau x_k\right)^2, \tag{1}$$

$$\text{s.t. } q_i = \left(y_i^{\rho_i} + m_i^{\rho_i}\right)^{\frac{1}{\rho_i}}, \qquad (p_i). \tag{2}$$

Country $i$'s welfare consists of the consumer surplus plus the export benefits minus the import costs minus the local environmental damage.

Note here that $q_i$ is only used to simplify the exposition, as is exactly defined by $y_i$ and $m_i$. The dual variable $p_i$ associated with the constraint (2) is interpretable as the price of the intermediate good produced by country $i$. The Karush–Kuhn–Tucker (KKT) conditions provide the necessary conditions for the game to be at equilibrium. Deriving for each $i$ and using some simple substitutions yields

$$\mathbb{R} \ni w_i \perp x_i - m_j = 0, \qquad i = 1, 2,\ i \neq j, \tag{3}$$

$$\mathbb{R} \ni p_i \perp q_i - \left(y_i^{\rho_i} + m_i^{\rho_i}\right)^{\frac{1}{\rho_i}} = 0, \qquad i = 1, 2, \tag{4}$$

$$0 \leq q_i \perp p_i - 100 + q_i \geq 0, \qquad i = 1, 2, \tag{5}$$

$$0 \leq y_i \perp \sum_{k=1}^{2} (y_k + \tau x_k) - \left(\frac{q_i}{y_i}\right)^{1-\rho_i} p_i \geq 0, \qquad i = 1, 2, \tag{6}$$

$$0 \leq x_i \perp \tau \sum_{k=1}^{2} (y_k + \tau x_k) - w_i \geq 0, \qquad i = 1, 2, \tag{7}$$

$$0 \leq m_i \perp \tau w_j - \left(\frac{q_i}{m_i}\right)^{1-\rho_i} p_i \geq 0, \qquad i = 1, 2, \ i \neq j, \tag{8}$$

As previously discussed, (3) and (4) refer to the formation of intermediate and final goods price by matching supply and demand. Despite the fact that these prices are not formally constrained to be non-negative, the equality between price and marginal utility described in (5), associated with a non-negative marginal cost, implicitly constrains the prices to be non-negative. Equation (6) implies that for the local sales of intermediate goods to be positive, its marginal environmental (local) damage must equal its marginal productivity, evaluated along the price of the final good. In the same vein, provided $m_i$ is nonzero, the association of (3), (7), and (8) implies the same principle, where the environmental damage is augmented by the transport cost:

$$0 < m_i \implies \tau^2 \sum_{k=1}^{2} (y_k + \tau x_k) = \left(\frac{q_i}{m_i}\right)^{1-\rho_i} p_i, \quad i = 1, 2, \ i \neq j.$$

**The Cooperative Subgame Equilibrium** In the cooperative game, players maximize their joint payoff to set their strategy. That is, they maximize the following objective function:

$$\max_{q_i, y_i, x_i, m_i \geq 0} \sum_{i=1}^{2} \left[ q_i(100 - q_i/2) + w_i x_i - \tau w_j m_i - \frac{1}{2} \left( \sum_{k=1}^{2} y_k + \tau x_k \right)^2 \right], \tag{9}$$

subject to the CES production functions (2) for each $i$. Compared to the non-cooperative game, only KKT conditions (6) and (7) change:

$$0 \leq y_i \perp 2 \sum_{k=1}^{2} (y_k + \tau x_k) - \left(\frac{q_i}{y_i}\right)^{1-\rho_i} p_i \geq 0, \qquad i = 1, 2, \tag{10}$$

$$0 \leq x_i \perp 2\tau \sum_{k=1}^{2} (y_k + \tau x_k) - w_i \geq 0, \qquad i = 1, 2, \tag{11}$$

As expected, the marginal environmental damage is augmented to take into account the global effect of transboundary pollution.

**The Deviation Subgames Equilibria** The situation where one player cooperates while the other does not is simply represented by each player adopting the corresponding KKT conditions. That is, the non-cooperative player maximizes its individual welfare, while the cooperative player maximizes the coalition welfare.

**Proposition 1** *The Cournot–Nash equilibrium exists and is unique for each subgame.*

*Proof* Using the optimality conditions of the supply for exported goods, $p_x$ and $p_y$ are replaced by the marginal environmental damage. The existence of an equilibrium is straightforward. For uniqueness, the concavity of the welfare function must be shown.

1. Let us first show that a CES function is concave, such that its Hessian matrix is negative semi-definite. Define

$$f(y,m) = \left(y^{\rho} + m^{\rho}\right)^{1/\rho} = g(y,m)^{1/\rho},$$

   with $y, m \geq 0$ and $\rho \leq 1$. Denote by $\mathcal{F}$ the Hessian matrix of $f(y,m)$, and by $\mathcal{G}$ the Hessian matrix of $g(y,m)$.

   (a) When $\rho \in (0, 1]$, $g(y,m)$ is a concave function since $\mathcal{G}$ is a diagonal matrix with negative terms.
   (b) When $\rho < 0$, $f$ is monotone decreasing in $g(y,m)$. It is thus a monotone increasing function of $-g(y,m)$. Thus, $-\mathcal{G}$ is negative semi-definite, such that $-g(y,m)$ is concave.
   (c) $f$ being a monotone increasing function of a concave function, it must be quasi-concave. A quasi-concave function that is homogeneous of degree 1 is concave. Hence $f(y,m,)$ is concave, and $\mathcal{F}$ is negative semi-definite.

2. At equilibrium, the cost of the imported good is equal to $\tau$ times the marginal damage, $\tau > 1$. The Hessian matrix associated with the non-cooperative game $\mathcal{H}$ is thus a negative translation of the CES Hessian matrix $\mathcal{F}$, that is,

$$\mathcal{H} = p \times \mathcal{F} - T = p \times \begin{pmatrix} f_{yy} & f_{ym} \\ f_{my} & f_{mm} \end{pmatrix} - \begin{pmatrix} f_y^2 + 1 & f_y f_m + \tau \\ f_m f_y + \tau & f_m^2 + \tau^2 \end{pmatrix},$$

   where $p$ is the price, $p \geq 0$. The second matrix $T$ is positive definite: the diagonal terms are positive and the determinant reduces to $(f_m - \tau f_y)^2 > 0$. Hence, $-T$ is negative definite, implying that $\mathcal{H}$ is negative definite and the welfare function is strictly concave.

Thus, the Cournot–Nash equilibrium of the non-cooperation game is unique. The equilibrium uniqueness of the other subgames is shown in a similar way. □

Because of the CES function, a closed-form solution cannot be derived from the necessary conditions of each subgame. It is however possible to characterize the intensity of imports $(y_i/m_i)$ at equilibrium for the different regimes. For interior solutions, marginal productivity equals marginal cost for both $y_i$ and $m_i$. Dividing one by the other, you obtain the intensity of imports $(y_i/m_i)$ for each $i$ as a function of its relative marginal cost. If both countries are cooperative or non-cooperative, then

$$\left(\frac{y_i}{m_i}\right)^{1-\rho_i} = \tau^2.$$

In other words, if both countries use the same strategy regime, cooperating or not will have the same impact on the intensity of trade. Then it is of course the case that two cooperative countries will produce less, and thus import less, because of higher environmental costs. However, if country $i$ is cooperative while country $j$ is non-cooperative, then

$$\left(\frac{y_i}{m_i}\right)^{1-\rho_i} = \frac{\tau^2}{2}, \qquad \text{and} \left(\frac{y_j}{m_j}\right)^{1-\rho_j} = 2\tau^2.$$

That is, the cooperating country imports more intensively, while the deviating country does the opposite. Indeed, the stronger environmental policy of the cooperative country decreases the volume of its intermediate goods. Such a scarcity always increases the price, even more so as $\rho_i$ moves away from 1. The effect of this deviation on price is the main driver of the next section's results.

## 4 Results

This section provides numerical solutions for the Nash equilibrium of the meta-game under various assumptions about the production function's elasticity parameters. The game is solved by plotting the different KKT conditions into GAMS and using the PATH solver. The transport cost is assumed to be $\tau = 1.1$.

Figure 1 represents the impact of the elasticity parameters $\rho_i$ on the Nash equilibrium of the meta-game. Note that $\rho_i$ can take values between $-\infty$ and 1. The range of parameter values in our experiment is chosen large enough to capture the behavior of the solution when $\rho$ tends to infinity.

When the two intermediate goods are relative complements in both final goods production functions, then environmental cooperation is always the Nash equilibrium of the meta-game. For high elasticity values in both production functions, the Nash equilibrium is the non-cooperative one and it is Pareto-dominated by cooperation. If goods are relative substitutes for country $i$'s final goods production but relative complements for country $j$'s, $i = 1, 2$, $j \neq i$, then:

1. If Country $i$'s CES production function has a low value of substitutability ($\rho_i \to 0$), and Country $j$'s has a high value of complementarity ($\rho_j \to -1$), then cooperation is the Nash equilibrium.
2. If Country $i$'s CES production function has an intermediate value of substitutability, then it tends to deviate from cooperation while Country $j$ sticks to cooperation at equilibrium.
3. If Country $i$'s CES production function has a high value of substitutability (close to 1), then non-cooperation is the Nash equilibrium.

Along the diagonal of symmetric players ($\rho_1 = \rho_2$), there exists cases where the elasticities are close to zero and there exists two Nash equilibria.

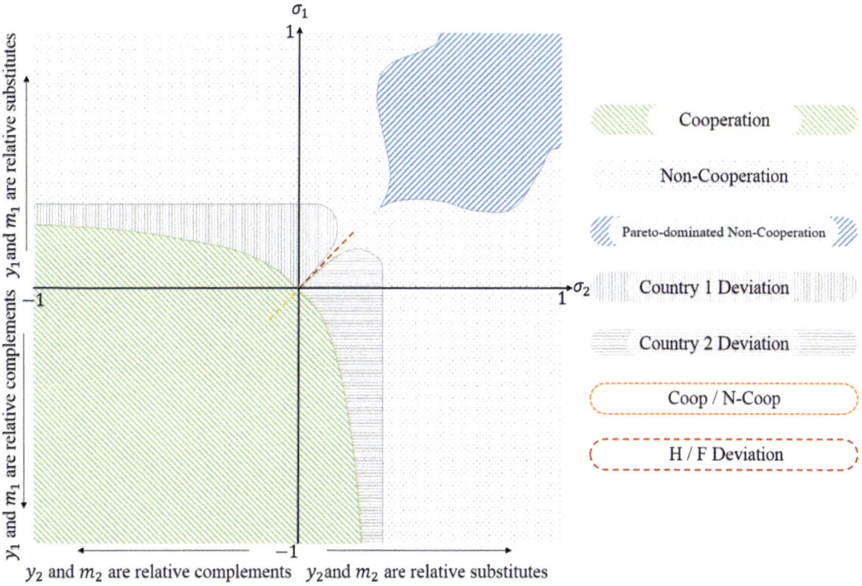


**Fig. 1** Nash equilibrium(-a) of the meta-game

(a) If $\rho_i \to 0^-$ for all $i$, both intermediate goods are weak complements, and both cooperation and non-cooperation are eligible to be a Nash equilibrium.
(b) If $\rho_i \to 0^+$ for all $i$, both intermediate goods are weak substitutes, and both deviation solutions are eligible to be a Nash equilibrium.

Figure 2 represents the pattern of trade for each equilibrium in its respective regime (cooperative, non-cooperative, and deviation equilibria). First, there is bilateral trade when the two players are symmetric, along the diagonal, except when the two intermediate goods are perfect substitutes. Intuitively, if country $i$'s CES function exhibits the perfect substitution property, it does not import. However, once this country's function is of imperfect substitution, it imports. When the two goods are complements ($\rho_i, \rho_j < 0,\ j \neq i$), the imports of country $i$ depend on its relative complementarity w.r.t. the other country ($\rho_i/\rho_j$). When the relative complementarity of country $i$ increases ($\rho_i/\rho_j > 1$), this country is exporting unilaterally.[3]

This effect is explained by the pattern of production, represented in Fig. 3, which corresponds almost exactly to the regime of trade. The only differences are explained by the transport cost, which makes importing more costly. If one country

[3]Note that a numerical model is ill-conditioned for very low quantity of inputs and negative elasticity parameters value $\rho_i$. To avoid numerical instability, quantities are restricted to be greater than $10^{-5}$. Only interior variables are considered non-null.

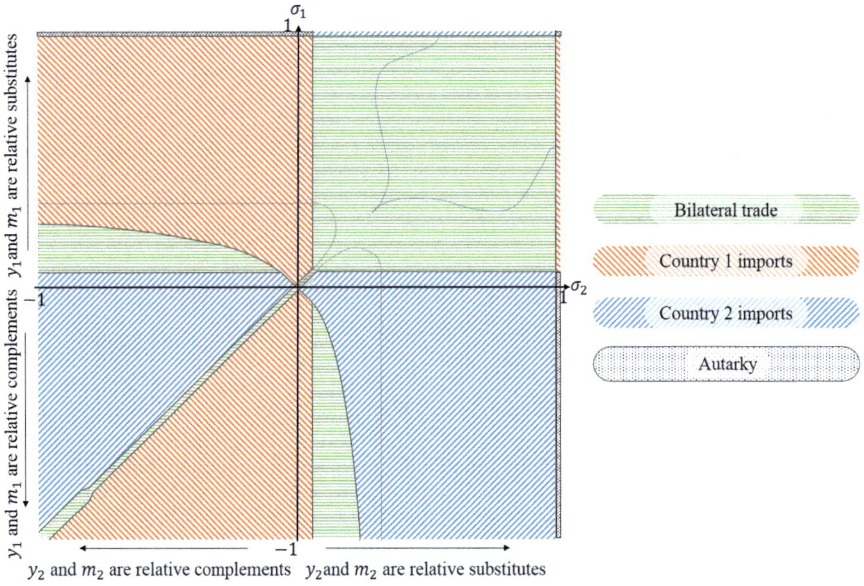


**Fig. 2** Pattern of trade

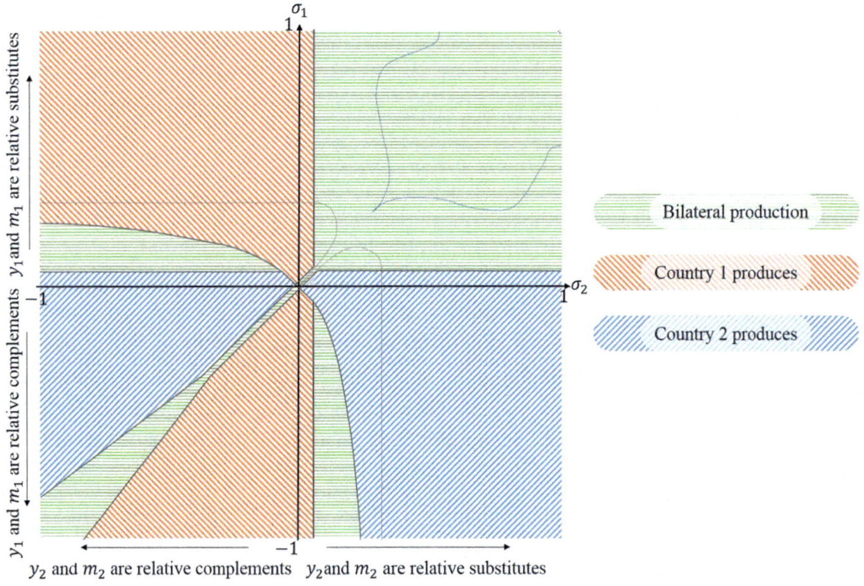


**Fig. 3** Pattern of production for local usage

uses perfect substitutes, it allocates all its effort to less-costly local production. If both goods are complements, the cases where there is production but no imports occur for close-to-the-lower-bounds production: since imports are more expensive, the volume of imports is below this lower bound and considered to be zero.

## 5 Conclusion

This note analyzes the stability of environmental cooperation in a simple two-player game with tradable goods. It models a situation of intra-industry trade where a final good is produced by combining two intermediate goods through a CES function. I test the impact of the elasticity of substitution of these CES functions and find that if the two intermediate goods are relative complements, then cooperation is a Nash equilibrium. If cooperation is the equilibrium of the static game, then it might also be the case for the dynamic game with pollution stock.

While an environmental agreement in a globalized economy is primarily a general equilibrium problem among $n$ players, this study is useful in providing insights about the properties of intra-industry trade. In the current context, where major players, such as the USA, disengage from the (non-binding) Paris agreement, it offers hope that agreements concerning this specific structure of trade are sustainable.

This study can be extended in many ways. In addition to the usual extensions to n-players and higher degrees of parametrization of the analyzed function, an extension in a dynamic framework seems achievable. Second, it would be interesting to test the results' robustness when the economy is subject to market power and/or when the countries are able to anticipate world prices. Last but not least, the frontier between cooperation and non-cooperation seems to be defined by a smooth function, so that it would be interesting to find its analytical form in terms of the elasticities of substitution.

**Acknowledgements** The author wishes to thank Michèle Breton, the two anonymous referees, and the co-editors for their comments. Any remaining errors are the author's own.

## References

Arkolakis, C., Costinot, A., & Rodríguez-Clare, A. (2012). New trade models, same old gains? *American Economic Review, 102*(1), 94–130.

Barrett, S. (1994). Strategic environmental policy and international trade. *Journal of Public Economics, 54*, 325–338.

Breton, M., Sbragia, L., & Zaccour, G. (2010). A dynamic model for international environmental agreements. *Environmental and Resource Economics, 45*(1), 25–48.

Cherniwchan, J., Copeland, B. R., & Taylor, M. S. (2017). Trade and the environment: New methods, measurements, and results. *Annual Review of Economics, 9*(1), 59–85.

Dockner, E. J., & Long, N. V. (1993). International pollution control: Cooperative versus noncooperative strategies. *Journal of Environmental Economics and Management, 25*(1), 13–29.

Edmond, C., Midrigan, V., & Xu, D. Y. (2015). Competition, markups, and the gains from international trade. *American Economic Review, 105*(10), 3183–3221.

Hardin, R. (1982). *Collective action*. Baltimore: The John Hopkins University Press.

Melitz, M. J. (2003). The impact on intra-industry reallocations and aggregate industry productivity. *Econometrica, 71*(6), 1695–1725.

OECD. (2002). Intra-Industry and Intra-Firm Trade and the Internationalisation of Production (Ch. 6), OECD Economic Outlook, 71 (pp. 159–170).

# Stable Coalition Structures in Dynamic Competitive Environment

**Elena Parilina and Artem Sedakov**

**Abstract** We consider a finite horizon dynamic competition model in discrete time in which firms are not restricted from cooperation with each other and can form coalitions of any size. For every coalition of firms, we determine profits of its members by two approaches: without the redistribution of profits inside the coalition and with such redistribution using a solution from cooperative game theory. Next, for each approach we examine the stability of a coalition structure in the game. When we find a stable coalition structure, we then verify whether it is dynamically stable, that is, stable over time with respect to the same profit distribution method chosen in the initial time period.

**Keywords** Dynamic competition · Coalition structure · Stability

## 1 Introduction

In the chapter we consider a dynamic competition model, in which firms choose their outputs in each time period. The market price is formed based on the decision of firms and on the price in the previous time period. We assume that the level of influence of the previous period price depends on the market state. Having this competitive model, we make an assumption that firms may cooperate in coalitions of any size forming a coalition structure. If the coalition structure is formed, each firm acts to maximize the profit of the coalition it belongs to. If the firms are supposed to have non-transferable profits, they are paid by initially given payoff functions. But if firms' profits are transferable, a cooperative point solution which redistributes the profits between firms is calculated. In both cases, a firm may have an interest in deviating from a coalition it belongs by joining another coalition or becoming a singleton. If no firm has a profitable deviation from its coalition, the coalition

E. Parilina (✉) · A. Sedakov
Saint Petersburg State University, Saint Petersburg, Russia
e-mail: e.parilina@spbu.ru; a.sedakov@spbu.ru



P.-O. Pineau et al. (eds.), *Games in Management Science*,
International Series in Operations Research & Management Science 280,
https://doi.org/10.1007/978-3-030-19107-8_21

structure is called stable (Parilina and Sedakov 2014; Sedakov et al. 2013). The stability of a coalition structure is determined with respect to a profit distribution method. In the definition of a stable coalition structure one may find similarity with the Nash equilibrium concept. The existence of a stable coalition structure with respect to the Shapley value in three-person games is proved in Sedakov et al. (2013). In four-person games the existence of a stable coalition structure is proved for special classes of transferable utility games (TU games). It is shown that a stable coalition structure may not exist in general (Sun and Parilina 2018).

The problem of stability of a coalition structure is actual in many applied problems. When the coalition structure is unstable, it is difficult to keep it the same over time and realize the game without changing the structure. There also exist other approaches to determine the stability of a coalition structure (e.g., see Carraro 1999). In the abovementioned paper, to be stable the coalition structure should be (i) internally stable, i.e., each player loses if he leaves his coalition becoming a singleton, (ii) externally stable, i.e., each player-singleton loses if he joins any coalition or another singleton, and, finally, (iii) intracoalitionally stable, i.e., each player from a coalition loses if he leaves his coalition and joins another one. In Parilina and Sedakov (2015) a process of changing coalition structures over time is considered. The authors introduce the concept of d-stability of a coalition structure, which players would never change once it is reached.

Even if the coalition structure is stable in the whole game, i.e., the stability conditions are satisfied in the initial time period, it may become unstable on the corresponding equilibrium state trajectory in some intermediate time period. Therefore, we define a *dynamically stable* coalition structure which is stable not only in the game but in any subgame starting from any intermediate time period and the corresponding state.

In this chapter we examine a competition model with finite time horizon and linear-quadratic profit functions of firms-competitors (see Carlson and Leitmann 2005). The firms are allowed to cooperate by forming a coalition structure. We determine the conditions for firms' strategies to form an open-loop coalition Nash equilibrium. By a coalition Nash equilibrium we mean a Nash equilibrium among players-coalitions in the given coalition structure. There are two options to determine firms' profits in the game. If the profits are non-transferable, the firms are paid according to their initially given payoff functions. If they are transferable, we determine the characteristic function according to the concepts in Chander and Tulkens (1997) and Rajan (1989). Based on the characteristic function, a cooperative point solution is defined using the Shapley value adopted for the games with a given coalition structure (Aumann and Dreze 1974; Shapley 1953). We determine a stable coalition structure for the initial time period and a dynamically stable coalition structure. As an example, a game with three firms is considered which admits five possible coalition structures. Interestingly, in the case of non-transferable payoffs, there are no stable coalition structures, but when firms redistribute their profits according to the Shapley value, there exists a unique stable coalition structure which is not the grand coalition. We then verify that this structure is also dynamically stable.

The chapter is organized as follows. Section 2 presents the theoretical model of the dynamic game and the conditions of a coalition equilibrium. In Sect. 3, we formulate the concept of stability for a coalition structure when profits are both non-transferable or transferable. In the latter case, the Shapley value is chosen as a cooperative point solution. We provide an illustrative example in Sect. 4, and briefly conclude in Sect. 5.

## 2 The Model

We consider a market of firms composing a finite set $N$ with $|N| = n \geqslant 2$. Producing and selling a product, firms compete in quantities over a finite set of periods $\mathscr{T} = \{0, 1, \ldots, T\}$ with the initial market price $p_0$ for the product. In each period $t \in \mathscr{T} \setminus T$, a firm $i \in N$ selects its quantity $q_i(p_0, t) \in \mathbb{R}_+$ to be produced for this period. A market price $p(t) \in \mathbb{R}_+$ satisfies the *state equation*

$$p(t+1) = sp(t) + (1-s)\left(a - b\sum_{i\in N} q_i(p_0, t)\right), \quad t \in \mathscr{T} \setminus T, \tag{1}$$

with the initial state $p(0) = p_0$. For a given $s \in [0, 1]$, the first summand in the r.h.s. of (1) represents the inertia in the market price while the second one reflects the price change as a reaction on produced output for some positive constants $a$ and $b$. Under an open-loop information structure (Haurie et al. 2012), an *open-loop strategy* of firm $i$ is a profile of quantities $q_i(p_0) = (q_i(p_0, 0), \ldots, q_i(p_0, T-1))$ which $i$ decides to produce during the planning horizon. Denote a *strategy profile* by $q(p_0) = (q_1(p_0), \ldots, q_n(p_0))$. Each firm $i$ aims to maximize its total discounted profit of the form

$$\pi_i(p_0, q(p_0)) = \sum_{t=0}^{T-1} \varrho^t \left[p(t)q_i(p_0, t) - \frac{c_i}{2} q_i^2(p_0, t)\right]$$

adopting its strategy $q_i(p_0)$, where $p(t)$ satisfies state equation (1) with initial state $p(0) = p_0$. A parameter $c_i > 0$ reflects firm $i$'s unit cots and $\varrho \in (0, 1]$ is a common discount factor. In period $T$ players have zero payoffs.

From now, we assume that firms are not restricted in cooperating with each other and can form any *coalition*, which is a nonempty subset of $N$. A partition $\mathscr{B} = \{B_1, \ldots, B_m\}$ of set $N$ is called a *coalition structure*. A strategy profile under the structure $\mathscr{B}$ will be denoted by $q^{\mathscr{B}}(p_0)$. A strategy of a coalition $B \in \mathscr{B}$ is a profile $q_B^{\mathscr{B}}(p_0) = \{q_i^B(p_0), i \in B\}$. Given a coalition structure $\mathscr{B}$, a strategy profile can then be written in terms of the structure, i.e., $q^{\mathscr{B}}(p_0) = \{q_B^{\mathscr{B}}(p_0), B \in \mathscr{B}\}$. Under the coalition structure, the aim of each firm is to maximize the profit of the coalition to which it belongs. More formally, jointly selecting a profile $q_B^{\mathscr{B}}(p_0)$, all firms from coalition $B \in \mathscr{B}$ maximize the sum $\pi_B^{\mathscr{B}}(p_0, q^{\mathscr{B}}(p_0)) = \sum_{i \in B} \pi_i(p_0, q^{\mathscr{B}}(p_0))$ subject to the state Eq. (1) with $p(0) = p_0$.

**Definition 1** A profile $\bar{q}^{\mathscr{B}}(p_0)$ is an *open-loop coalition Nash equilibrium* (or simply coalition Nash equilibrium) if

$$\pi_B^{\mathscr{B}}(p_0, \bar{q}^{\mathscr{B}}(p_0)) \geqslant \pi_B^{\mathscr{B}}(p_0, (q_B^{\mathscr{B}}(p_0), \bar{q}_{N \setminus B}^{\mathscr{B}}(p_0)))$$

for any coalition $B \in \mathscr{B}$ and its strategy $q_B^{\mathscr{B}}(p_0)$. Alternatively, $\bar{q}^{\mathscr{B}}(p_0)$ satisfies

$$\bar{q}_B^{\mathscr{B}}(p_0) = \arg \max_{q_B^{\mathscr{B}}(p_0)} \pi_B^{\mathscr{B}}(p_0, (q_B^{\mathscr{B}}(p_0), \bar{q}_{N \setminus B}^{\mathscr{B}}(p_0)))$$

for any $B \in \mathscr{B}$.

In particular, when $\mathscr{B} = \{\{1\}, \dots, \{n\}\}$, that is, all coalitions in coalition structure $\mathscr{B}$ are singletons, the coalition Nash equilibrium is a *Nash equilibrium*, while when $\mathscr{B} = \{N\}$, that is, all firms cooperate in one coalition, the coalition equilibrium is a *cooperative optimum*. A sequence of market prices $\bar{p}^{\mathscr{B}} = \{\bar{p}^{\mathscr{B}}(0) \equiv p_0, \bar{p}^{\mathscr{B}}(1), \dots, \bar{p}^{\mathscr{B}}(T)\}$ uniquely determined by coalition equilibrium $\bar{q}^{\mathscr{B}}(p_0)$ and state equation (1) is a *coalition equilibrium trajectory*. A coalition equilibrium trajectory determined by a cooperative optimum $\bar{q}^{\{N\}}(p_0)$ is a *cooperative trajectory* denoted by $\bar{p}^{\{N\}}$. Next, we can define the profit of firm $i$ under a coalition equilibrium $\bar{q}^{\mathscr{B}}(p_0)$, which is $\pi_i(p_0, \bar{q}^{\mathscr{B}}(p_0))$. Similarly, we define firm $i$'s *cooperative profit* $\pi_i(p_0, \bar{q}^{\{N\}}(p_0))$, i.e., its profit under a cooperative optimum $\bar{q}^{\{N\}}(p_0)$.

Let $\Gamma^{\mathscr{B}}(p_0)$ denote the dynamic game over the set of periods $\mathscr{T}$ with coalition structure $\mathscr{B}$ starting in state $p_0$. We now characterize an open-loop coalition Nash equilibrium in this game. A similar infinite-horizon two-person non-cooperative model is examined in Carlson and Leitmann (2005) for open-loop strategies. One can study this problem also by assuming a feedback information structure. However, to find the corresponding feedback coalition Nash equilibrium, one needs to assume the form of value functions.

**Theorem 1** *Under a coalition structure $\mathscr{B}$, an open-loop coalition Nash equilibrium $\bar{q}^{\mathscr{B}}$ is composed of the following strategies:*

$$\bar{q}_i^{\mathscr{B}}(p_0, t) = \frac{1}{c_i}\left[\bar{p}^{\mathscr{B}}(t) - \varrho b(1-s)\mu_B^{\mathscr{B}}(t+1)\right], \quad i \in B, \quad t \in \mathscr{T} \setminus T, \tag{2}$$

*where $\bar{p}^{\mathscr{B}}(t)$ and $\mu_B^{\mathscr{B}}(t)$, $B \in \mathscr{B}$, satisfy the recursive relations:*

$$\bar{p}^{\mathscr{B}}(t) = s\bar{p}^{\mathscr{B}}(t-1) + (1-s)\left(a - b\sum_{i \in N} \bar{q}_i^{\mathscr{B}}(p_0, t-1)\right), \quad t \in \mathscr{T} \setminus 0,$$

$$\mu_B^{\mathscr{B}}(t) = \sum_{i \in B} \bar{q}_i^{\mathscr{B}}(p_0, t) + \varrho s \mu_B^{\mathscr{B}}(t+1), \quad t \in \mathscr{T} \setminus \{0, T\},$$

*with* $\bar{p}^{\mathscr{B}}(0) = p_0$ *and* $\mu_B^{\mathscr{B}}(T) = 0$ *for any* $B \in \mathscr{B}$.

*Proof* For a coalition $B \in \mathscr{B}$, we define the Hamiltonian $\mathscr{H}_B^{\mathscr{B}}$:

$$\mathscr{H}_B^{\mathscr{B}} = \sum_{i \in B} \varrho^t \left[ p^{\mathscr{B}}(t) q_i^{\mathscr{B}}(p_0, t) - \frac{c_i}{2} (q_i^{\mathscr{B}}(p_0, t))^2 \right]$$
$$+ \lambda_B^{\mathscr{B}}(t+1) \left[ s p^{\mathscr{B}}(t) + (1-s) \left( a - b \sum_{i \in N} q_i^{\mathscr{B}}(p_0, t) \right) \right],$$

where $\lambda_B^{\mathscr{B}}(t+1)$ is a costate variable. From the maximum principle, for any coalition $B \in \mathscr{B}$, the following is true:

$$\frac{\partial \mathscr{H}_B^{\mathscr{B}}}{\partial q_i^{\mathscr{B}}(p_0, t)} = \varrho^t \left[ p^{\mathscr{B}}(t) - c_i q_i^{\mathscr{B}}(p_0, t) \right] - (1-s) b \lambda_B^{\mathscr{B}}(t+1)$$
$$= 0, \ i \in B, \ t \in \mathscr{T} \setminus T,$$
$$\frac{\partial \mathscr{H}_B^{\mathscr{B}}}{\partial p^{\mathscr{B}}(t)} = \varrho^t \sum_{i \in B} q_i^{\mathscr{B}}(p_0, t) + s \lambda_B^{\mathscr{B}}(t+1) = \lambda_B^{\mathscr{B}}(t), \quad t \in \mathscr{T} \setminus \{0, T\},$$
$$\lambda_B^{\mathscr{B}}(T) = 0.$$

Replacing costate variables $\lambda_B^{\mathscr{B}}(t)$ with scaled ones $\mu_B^{\mathscr{B}}(t)$ by $\mu_B^{\mathscr{B}}(t) = \varrho^{-t} \lambda_B^{\mathscr{B}}(t)$, $t \in \mathscr{T} \setminus 0$, and rewriting condition $\partial \mathscr{H}_B^{\mathscr{B}} / \partial q_i^{\mathscr{B}}(p_0, t) = 0$, we obtain the expressions from the statement of the theorem. □

## 3 Stability of a Coalition Structure

Assuming the firms are exogenously organized in a coalition structure $\mathscr{B}$, Theorem 1 provides equilibrium outputs $\bar{q}_i^{\mathscr{B}}(p_0)$ for each firm $i \in N$ under a coalition Nash equilibrium. Thus following the equilibrium profile $\bar{q}^{\mathscr{B}}(p_0)$, a firm $i$ can determine its profit $\pi_i(p_0, \bar{q}^{\mathscr{B}}(p_0))$ in the game. However under a different coalition structure $\mathscr{B}'$ resulting in a different coalition Nash equilibrium $\bar{q}^{\mathscr{B}'}(p_0)$, firm $i$'s profit $\pi_i(p_0, \bar{q}^{\mathscr{B}'}(p_0))$ will not necessarily coincide with $\pi_i(p_0, \bar{q}^{\mathscr{B}}(p_0))$. If firms were to create a coalition structure themselves, they would do it in a way that each firm would select the coalition which it does not want to leave, thus coming to a *stable* coalition structure. This approach is quite natural, and of course we are aware that there might be other reasons why firms should form a particular coalition structure. Although firms in a coalition focus on the total profit of this coalition, at the same time each firm also takes into account its individual profit in this coalition to measure its "satisfaction" from being a member.

In this section we consider two cases for determining a stable coalition structure: when firms' profits are either non-transferable or transferable.

For a given coalition structure $\mathscr{B}$, let $B(i)$ denote the coalition from $\mathscr{B}$ which contains firm $i$. Let also for some $B \in \mathscr{B}$ denote $\mathscr{B}_{-B} = \mathscr{B} \setminus B$.

## 3.1 Non-transferable Profits

We start with a case of non-transferable profits. This means that for a coalition structure $\mathscr{B}$, under the corresponding coalition Nash equilibrium $\bar{q}^{\mathscr{B}}$, a coalition $B \in \mathscr{B}$ receives its profit of $\pi_B^{\mathscr{B}}(p_0, \bar{q}^{\mathscr{B}}(p_0))$ while its member $i \in B$ gets $\pi_i(p_0, \bar{q}^{\mathscr{B}}(p_0))$.

**Definition 2** A coalition structure $\mathscr{B}$ is *stable* if for any firm $i \in N$ it holds that

$$\pi_i(p_0, \bar{q}^{\mathscr{B}}(p_0)) \geqslant \pi_i(p_0, \bar{q}^{\mathscr{B}'}(p_0)), \tag{3}$$

where $\mathscr{B}' = \{B(i) \setminus \{i\}, B \cup \{i\}, \mathscr{B}_{-B(i)\cup B}\}$ for any $B \in \mathscr{B} \cup \varnothing$ and $B \neq B(i)$. Otherwise, the coalition structure is *unstable*. Here we recall that $\bar{q}^{\mathscr{B}}(p_0)$ and $\bar{q}^{\mathscr{B}'}(p_0)$ are coalition Nash equilibria for coalition structures $\mathscr{B}$ and $\mathscr{B}'$, respectively.

The definition of the stable coalition structure assumes that a firm may leave a coalition and become a singleton; it may also join any other coalition in the structure. Moreover, if a firm $i$ leaves $B(i)$, the coalition $B(i) \setminus \{i\}$ does not break up and remains a part of the coalition structure. If firm $i$ decides to leave $B(i)$ in favor of some other coalition $B$, then the members of $B$ allow it to enter and form a coalition $B \cup \{i\}$ not blocking $B$ from the new member.

When a coalition structure, say $\mathscr{B}$, is stable, no firm wishes to change a coalition, i.e., each firm $i \in N$ prefers to be a member of $B(i) \in \mathscr{B}$. Here we stress the reader's attention that the proposed stability concept is related only to the initial game period $t = 0$ when firms are supposed to follow a prescribed equilibrium profile $\bar{q}^{\mathscr{B}}(p_0)$ in the whole game under $\mathscr{B}$. Indeed, for this coalition structure inequality (3) holds true. However in some game period $t \in \mathscr{T} \setminus 0$ under profile $\bar{q}^{\mathscr{B}}(p_0)$ on the coalition equilibrium trajectory $\bar{p}^{\mathscr{B}}$ in state $\bar{p}^{\mathscr{B}}(t)$, coalition structure $\mathscr{B}$ may become unstable. Let $\bar{q}^{\mathscr{B}}(\bar{p}^{\mathscr{B}}(t))$ denote a coalition Nash equilibrium in the subgame of game $\Gamma^{\mathscr{B}}(p_0)$ starting in period $t$ in state $\bar{p}^{\mathscr{B}}(t)$. We denote this subgame by $\Gamma^{\mathscr{B}}(\bar{p}^{\mathscr{B}}(t))$. Thus firms' profits in $\Gamma^{\mathscr{B}}(\bar{p}^{\mathscr{B}}(t))$ are of the form:

$$\pi_i(\bar{p}^{\mathscr{B}}(t), \bar{q}^{\mathscr{B}}(\bar{p}^{\mathscr{B}}(t))) = \sum_{\tau=t}^{T-1} \varrho^{\tau-t} \left[\bar{p}^{\mathscr{B}}(\tau)\bar{q}_i^{\mathscr{B}}(\bar{p}^{\mathscr{B}}(t), \tau) - \frac{c_i}{2}\left(\bar{q}_i^{\mathscr{B}}(\bar{p}^{\mathscr{B}}(t), \tau)\right)^2\right]$$

with

$$\bar{p}^{\mathscr{B}}(\tau+1) = s\bar{p}^{\mathscr{B}}(\tau)+(1-s)\left(a - b\sum_{i\in N}\bar{q}_i^{\mathscr{B}}(\bar{p}^{\mathscr{B}}(t),\tau)\right), \quad \tau \in \{t,\ldots,T-1\}.$$

If firm $i$ in this state leaves $B(i)$, that is, changes current coalition structure $\mathscr{B}$ to some other $\mathscr{B}'$, this will lead to another coalition Nash equilibrium $\bar{q}^{\mathscr{B}'}(\bar{p}^{\mathscr{B}}(t))$. We notice that the equilibrium in the subgame with new coalition structure $\mathscr{B}'$ will depend upon the state $\bar{p}^{\mathscr{B}}(t)$ in which $\mathscr{B}$ has been changed. Let further

$$\pi_i(\bar{p}^{\mathscr{B}}(t),\bar{q}^{\mathscr{B}'}(\bar{p}^{\mathscr{B}}(t))) = \sum_{\tau=t}^{T-1}\varrho^{\tau-t}\left[\bar{p}^{\mathscr{B}'}(\tau)\bar{q}_i^{\mathscr{B}'}(\bar{p}^{\mathscr{B}}(t),\tau) - \frac{c_i}{2}\left(\bar{q}_i^{\mathscr{B}'}(\bar{p}^{\mathscr{B}}(t),\tau)\right)^2\right]$$

with

$$\bar{p}^{\mathscr{B}'}(\tau+1) = s\bar{p}^{\mathscr{B}'}(\tau)+(1-s)\left(a - b\sum_{i\in N}\bar{q}_i^{\mathscr{B}'}(\bar{p}^{\mathscr{B}}(t),\tau)\right), \quad \tau \in \{t,\ldots,T-1\},$$

and $\bar{p}^{\mathscr{B}'}(t) \equiv \bar{p}^{\mathscr{B}}(t)$ denote firm $i$'s profit in the subgame starting in period $t$ in state $\bar{p}^{\mathscr{B}}(t)$ in the new coalition structure $\mathscr{B}'$ under coalition Nash equilibrium $\bar{q}^{\mathscr{B}'}(\bar{p}^{\mathscr{B}}(t))$. Thus we come to the definition.

**Definition 3** A coalition structure $\mathscr{B}$ is *dynamically stable* if for any firm $i \in N$ and any game period $t \in \mathscr{T}$ it holds that

$$\pi_i(\bar{p}^{\mathscr{B}}(t),\bar{q}^{\mathscr{B}}(\bar{p}^{\mathscr{B}}(t))) \geqslant \pi_i(\bar{p}^{\mathscr{B}}(t),\bar{q}^{\mathscr{B}'}(\bar{p}^{\mathscr{B}}(t))), \tag{4}$$

where $\mathscr{B}' = \{B(i)\setminus\{i\}, B\cup\{i\}, \mathscr{B}_{-B(i)\cup B}\}$ for any $B \in \mathscr{B}\cup\varnothing$ and $B \neq B(i)$.

The dynamic stability of $\mathscr{B}$ means its stability in *any* game period along the coalition equilibrium trajectory $\bar{p}^{\mathscr{B}}$.

## 3.2 Transferable Profits

Now we move to the case of transferable profits. Here we assume that for a coalition structure $\mathscr{B}$ under the corresponding coalition Nash equilibrium $\bar{q}^{\mathscr{B}}(p_0)$ a coalition $B \in \mathscr{B}$ receives its profit of $\pi_B^{\mathscr{B}}(p_0,\bar{q}^{\mathscr{B}}(p_0))$ while the profit of its members from cooperation has to be determined by redistributing $\pi_B^{\mathscr{B}}(p_0,\bar{q}^{\mathscr{B}}(p_0))$ among them. We will determine the profits of firms by a *cooperative solution* of a corresponding TU game with a coalition structure. A TU game with a coalition structure is a triple $(N, v_0, \mathscr{B})$, where $N$ is a player set (the set of firms), $v_0$ is a characteristic function measuring a *worth* of any coalition, and finally $\mathscr{B}$ is a coalition structure. When transiting from a normal-form game to the corresponding TU game, there is no unique way in determining the characteristic function. We define it in two steps. Given a coalition structure $\mathscr{B}$, at the first step we define the value $v_0(B, \mathscr{B})$ as the profit of the coalition of firms $B \in \mathscr{B}$ under a coalition Nash equilibrium $\bar{q}^{\mathscr{B}}(p_0)$ in the dynamic game between players-coalitions from $\mathscr{B}$. Thus $v_0(B, \mathscr{B}) = \pi_B^{\mathscr{B}}(p_0, \bar{q}^{\mathscr{B}}(p_0))$. Next, for a coalition $S \subset B$, we define the value $v_0(S, \mathscr{B})$ as the total profit of its members under a coalition Nash equilibrium $\hat{q}^{S,\mathscr{B}}(p_0)$ in a dynamic game between players-firms from $\mathscr{B}$ when (i) firms from coalition $S$ jointly maximize the total profit of this coalition, (ii) each firm from $B \setminus S$ maximizes its own profit, and (iii) each firm $i \in N \setminus B$ maximizes the total profit of coalitions they belong to, given the state equation (1). In other words, $\hat{q}^{S,\mathscr{B}}(p_0)$ is of the form:

$$\hat{q}_i^{S,\mathscr{B}}(p_0) = \begin{cases} \arg\max\limits_{q_S^{S,\mathscr{B}}(p_0)} \pi_S^{\mathscr{B}}(p_0, (q_S^{S,\mathscr{B}}(p_0), \hat{q}_{N\setminus S}^{S,\mathscr{B}}(p_0))), \ i = S, \\ \arg\max\limits_{q_i^{S,\mathscr{B}}(p_0)} \pi_i^{\mathscr{B}}(p_0, (q_i^{S,\mathscr{B}}(p_0), \hat{q}_{N\setminus\{i\}}^{S,\mathscr{B}}(p_0))), \ i \in B \setminus S, \\ \arg\max\limits_{q_{B'}^{S,\mathscr{B}}(p_0)} \pi_{B'}^{\mathscr{B}}(p_0, (q_{B'}^{S,\mathscr{B}}(p_0), \hat{q}_{N\setminus B'}^{S,\mathscr{B}}(p_0))), \ i = B', \ B' \in \mathscr{B}_{-B}. \end{cases}$$

Therefore, the characteristic function is given by

$$v_0(S, \mathscr{B}) = \begin{cases} \pi_B^{\mathscr{B}}(p_0, \bar{q}^{\mathscr{B}}(p_0)), & S = B, \ B \in \mathscr{B}, \\ \pi_S^{\mathscr{B}}(p_0, \hat{q}^{S,\mathscr{B}}(p_0)), & S \subset B, \ B \in \mathscr{B}, \\ 0, & S = \varnothing, \\ \sum\limits_{\substack{B \in \mathscr{B}, \\ B \subseteq S}} \pi_B^{\mathscr{B}}(p_0, \bar{q}^{\mathscr{B}}(p_0)) + \sum\limits_{\substack{B \in \mathscr{B}, \\ B \not\subseteq S, B \cap S \neq \varnothing}} \pi_{B\cap S}^{\mathscr{B}}(p_0, \hat{q}^{B\cap S,\mathscr{B}}(p_0)), & \text{otherwise.} \end{cases}$$

**Theorem 2** *Under a coalition structure $\mathscr{B}$, for any $S \subset B$, an open-loop coalition Nash equilibrium $\hat{q}^{S,\mathscr{B}}(p_0)$ is composed of the following strategies:*

$$\hat{q}_i^{S,\mathscr{B}}(p_0, t) = \begin{cases} \frac{1}{c_i}\left[\hat{p}^{S,\mathscr{B}}(t) - \varrho b(1-s)\mu_S^{S,\mathscr{B}}(t+1)\right], \ i \in S, \\ \frac{1}{c_i}\left[\hat{p}^{S,\mathscr{B}}(t) - \varrho b(1-s)\mu_i^{S,\mathscr{B}}(t+1)\right], \ i \in B \setminus S, \\ \frac{1}{c_i}\left[\hat{p}^{S,\mathscr{B}}(t) - \varrho b(1-s)\mu_{B'}^{S,\mathscr{B}}(t+1)\right], \ i \in B', \ B' \in \mathscr{B}_{-B}, \end{cases} \quad t \in \mathscr{T} \setminus T,$$

*where $\hat{p}^{S,\mathscr{B}}(t)$ and $\mu_S^{S,\mathscr{B}}(t)$, $\mu_i^{S,\mathscr{B}}(t)$, $i \in B \setminus S$, $\mu_{B'}^{S,\mathscr{B}}(t)$, $B' \in \mathscr{B}_{-B}$, satisfy the recursive relations:*

$$\hat{p}^{S,\mathscr{B}}(t) = s\hat{p}^{S,\mathscr{B}}(t-1) + (1-s)\left(a - b\sum_{i\in N} \hat{q}_i^{S,\mathscr{B}}(p_0, t-1)\right), \quad t \in \mathscr{T} \setminus 0,$$

$$\mu_S^{S,\mathscr{B}}(t) = \sum_{i \in S} \hat{q}_i^{S,\mathscr{B}}(p_0, t) + \varrho s \mu_S^{S,\mathscr{B}}(t+1), \quad t \in \mathscr{T} \setminus \{0, T\},$$

$$\mu_i^{S,\mathscr{B}}(t) = \hat{q}_i^{S,\mathscr{B}}(t) + \varrho s \mu_i^{S,\mathscr{B}}(t+1), \quad i \in B \setminus S, \quad t \in \mathscr{T} \setminus \{0, T\},$$

$$\mu_{B'}^{S,\mathscr{B}}(t) = \sum_{i \in B'} \hat{q}_i^{S,\mathscr{B}}(p_0, t) + \varrho s \mu_{B'}^{S,\mathscr{B}}(t+1), \quad B' \in \mathscr{B}_{-B}, \quad t \in \mathscr{T} \setminus \{0, T\},$$

*with* $\hat{p}^{S,\mathscr{B}}(0) = p_0$, $\mu_S^{S,\mathscr{B}}(T) = 0$, $\mu_i^{S,\mathscr{B}}(T) = 0$, $i \in B \setminus S$, *and* $\mu_{B'}^{S,\mathscr{B}}(T) = 0$, $B' \in \mathscr{B}_{-B}$.

*Proof* For a coalition $S \subset B$, each firm $i \in B \setminus S$, and each coalition $B' \in \mathscr{B}_{-B}$ we define the Hamiltonians $\mathscr{H}_S^{S,\mathscr{B}}$, $\mathscr{H}_i^{S,\mathscr{B}}$, and $\mathscr{H}_{B'}^{S,\mathscr{B}}$, respectively:

$$\mathscr{H}_S^{S,\mathscr{B}} = \sum_{i \in S} \varrho^t \left[ p^{S,\mathscr{B}}(t) q_i^{S,\mathscr{B}}(p_0, t) - \frac{c_i}{2} (q_i^{S,\mathscr{B}}(p_0, t))^2 \right]$$
$$+ \lambda_S^{S,\mathscr{B}}(t+1) \left[ s p^{S,\mathscr{B}}(t) + (1-s) \left( a - b \sum_{i \in N} q_i^{S,\mathscr{B}}(p_0, t) \right) \right],$$

$$\mathscr{H}_i^{S,\mathscr{B}} = \varrho^t \left[ p^{S,\mathscr{B}}(t) q_i^{S,\mathscr{B}}(p_0, t) - \frac{c_i}{2} (q_i^{S,\mathscr{B}}(p_0, t))^2 \right]$$
$$+ \lambda_i^{S,\mathscr{B}}(t+1) \left[ s p^{S,\mathscr{B}}(t) + (1-s) \left( a - b \sum_{i \in N} q_i^{S,\mathscr{B}}(p_0, t) \right) \right],$$

$$\mathscr{H}_{B'}^{S,\mathscr{B}} = \sum_{i \in B'} \varrho^t \left[ p^{S,\mathscr{B}}(t) q_i^{S,\mathscr{B}}(p_0, t) - \frac{c_i}{2} (q_i^{S,\mathscr{B}}(p_0, t))^2 \right]$$
$$+ \lambda_{B'}^{S,\mathscr{B}}(t+1) \left[ s p^{S,\mathscr{B}}(t) + (1-s) \left( a - b \sum_{i \in N} q_i^{S,\mathscr{B}}(p_0, t) \right) \right],$$

where $\lambda_S^{S,\mathscr{B}}(t)$, $\lambda_i^{S,\mathscr{B}}(t)$, $i \in B \setminus S$, and $\lambda_{B'}^{S,\mathscr{B}}(t)$, $B' \in \mathscr{B}_{-B}$, are costate variables. From the maximum principle the following is true:

$$\frac{\partial \mathscr{H}_S^{S,\mathscr{B}}}{\partial q_i^{S,\mathscr{B}}(p_0, t)} = \varrho^t \left[ p^{S,\mathscr{B}}(t) - c_i q_i^{S,\mathscr{B}}(p_0, t) \right] - (1-s) b \lambda_S^{S,\mathscr{B}}(t+1) = 0,$$
$$i \in S, \quad t \in \mathscr{T} \setminus T,$$

$$\frac{\partial \mathscr{H}_S^{S,\mathscr{B}}}{\partial p^{S,\mathscr{B}}(t)} = \varrho^t \sum_{i \in S} q_i^{S,\mathscr{B}}(p_0, t) + s \lambda_S^{S,\mathscr{B}}(t+1) = \lambda_S^{S,\mathscr{B}}(t), \quad t \in \mathscr{T} \setminus \{0, T\},$$

$$\lambda_S^{S,\mathscr{B}}(T) = 0,$$

$$\frac{\partial \mathscr{H}_i^{S,\mathscr{B}}}{\partial q_i^{S,\mathscr{B}}(p_0,t)} = \varrho^t \left[p^{S,\mathscr{B}}(t) - c_i q_i^{S,\mathscr{B}}(p_0,t)\right] - (1-s)b\lambda_i^{S,\mathscr{B}}(t+1) = 0,$$

$$i \in B \setminus S,\ t \in \mathscr{T} \setminus T,$$

$$\frac{\partial \mathscr{H}_i^{S,\mathscr{B}}}{\partial p^{S,\mathscr{B}}(t)} = \varrho^t q_i^{S,\mathscr{B}}(p_0,t) + s\lambda_i^{S,\mathscr{B}}(t+1) = \lambda_i^{S,\mathscr{B}}(t), \quad t \in \mathscr{T} \setminus \{0, T\},$$

$$\lambda_i^{S,\mathscr{B}}(T) = 0,$$

$$\frac{\partial \mathscr{H}_{B'}^{S,\mathscr{B}}}{\partial q_i^{S,\mathscr{B}}(p_0,t)} = \varrho^t \left[p^{S,\mathscr{B}}(t) - c_i q_i^{S,\mathscr{B}}(p_0,t)\right] - (1-s)b\lambda_{B'}^{S,\mathscr{B}}(t+1) = 0,$$

$$i \in B',\ t \in \mathscr{T} \setminus T,$$

$$\frac{\partial \mathscr{H}_{B'}^{S,\mathscr{B}}}{\partial p^{S,\mathscr{B}}(t)} = \varrho^t \sum_{i \in B'} q_i^{S,\mathscr{B}}(p_0,t) + s\lambda_{B'}^{S,\mathscr{B}}(t+1) = \lambda_{B'}^{S,\mathscr{B}}(t), \quad t \in \mathscr{T} \setminus \{0, T\},$$

$$\lambda_{B'}^{S,\mathscr{B}}(T) = 0.$$

First we replace costate variables $\lambda_S^{S,\mathscr{B}}(t)$, $\lambda_i^{S,\mathscr{B}}(t)$, $i \in B \setminus S$, and $\lambda_{B'}^{S,\mathscr{B}}(t)$, $B' \in \mathscr{B}_{-B}$, with scaled ones by $\mu_S^{S,\mathscr{B}}(t) = \varrho^{-t}\lambda_S^{S,\mathscr{B}}(t)$, $\mu_i^{S,\mathscr{B}}(t) = \varrho^{-t}\lambda_i^{S,\mathscr{B}}(t)$, $i \in B \setminus S$, and $\mu_{B'}^{S,\mathscr{B}}(t) = \varrho^{-t}\lambda_{B'}^{S,\mathscr{B}}(t)$. Next, rewriting conditions $\partial \mathscr{H}_S^{S,\mathscr{B}}/\partial q_i^{S,\mathscr{B}}(p_0,t) = 0$, $i \in S$, $\partial \mathscr{H}_i^{S,\mathscr{B}}/\partial q_i^{S,\mathscr{B}}(p_0,t) = 0$, $i \in B \setminus S$, and $\partial \mathscr{H}_{B'}^{S,\mathscr{B}}/\partial q_i^{S,\mathscr{B}}(p_0,t) = 0$, $i \in B'$, $B' \in \mathscr{B}_{-B}$, we obtain the expressions from the statement of the theorem.

□

A *cooperative point solution* to the game $(N, v_0, \mathscr{B})$ with a coalition structure $\mathscr{B}$ is a map that assigns a profile $\xi[v_0, \mathscr{B}] \in \mathbb{R}^n$ to the TU game such that $\sum_{i \in B} \xi_i[v_0, \mathscr{B}] = v_0(B, \mathscr{B})$ for all $B \in \mathscr{B}$. In this definition we relax the individual rationality condition as the characteristic function may not be superadditive by its construction. As cooperative point solutions we may consider different ones, e.g., the Shapley value, the nucleolus, etc. (see Aumann and Dreze (1974) for cooperative solutions of a TU game with a coalition structure).

**Definition 4** A coalition structure $\mathscr{B}$ is *stable* with respect to a cooperative point solution if for any firm $i \in N$ it holds that $\xi_i[v_0, \mathscr{B}] \geqslant \xi_i[v_0, \mathscr{B}']$ where $\mathscr{B}' = \{B(i) \setminus \{i\}, B \cup \{i\}, \mathscr{B}_{-B(i) \cup B}\}$ for any $B \in \mathscr{B} \cup \varnothing$ and $B \neq B(i)$. Otherwise the coalition structure is *unstable*.

In a similar way, we can determine a dynamically stable coalition structure. For this reason we have to determine the cooperative point solution in each subgame starting in state $\bar{p}^{\mathscr{B}}(t)$, $t \in \mathscr{T} \setminus 0$ on coalition equilibrium trajectory $\bar{p}^{\mathscr{B}}$. To do this, we first define a TU subgame $(N, v_t, \mathscr{B})$ with coalition structure $\mathscr{B}$ where $v_t$ is the characteristic function in this subgame. We let $v_t(B, \mathscr{B}) = \pi_B^{\mathscr{B}}(\bar{p}^{\mathscr{B}}(t), \bar{q}^{\mathscr{B}}(\bar{p}^{\mathscr{B}}(t)))$ for any $B \in \mathscr{B}$. And for a coalition $S \subset B$, we define the value $v_t(S, \mathscr{B})$ as the profit of the coalition of firms $S$ under a coalition Nash equilibrium $\hat{q}^{S,\mathscr{B}}(\bar{p}^{\mathscr{B}}(t))$ in a dynamic subgame similarly, i.e., when (i) firms from coalition $S$ jointly maximize the total profit of this coalition, (ii) each firm from $B \setminus S$ maximizes its own profit, and (iii) each firm $i \in N \setminus B$ maximizes the total profit of coalitions they belong to. In other words, $\hat{q}^{S,\mathscr{B}}(\bar{p}^{\mathscr{B}}(t))$ is of the form:

$$\hat{q}_i^{S,\mathscr{B}}(\bar{p}^{\mathscr{B}}(t)) = \begin{cases} \arg \max\limits_{q_S^{S,\mathscr{B}}(\bar{p}^{\mathscr{B}}(t))} \pi_S^{\mathscr{B}}(\bar{p}^{\mathscr{B}}(t), (q_S^{S,\mathscr{B}}(\bar{p}^{\mathscr{B}}(t)), \hat{q}_{N\setminus S}^{S,\mathscr{B}}(\bar{p}^{\mathscr{B}}(t)))), & i = S, \\ \arg \max\limits_{q_i^{S,\mathscr{B}}(\bar{p}^{\mathscr{B}}(t))} \pi_i^{\mathscr{B}}(\bar{p}^{\mathscr{B}}(t), (q_i^{S,\mathscr{B}}(\bar{p}^{\mathscr{B}}(t)), \hat{q}_{N\setminus\{i\}}^{S,\mathscr{B}}(\bar{p}^{\mathscr{B}}(t)))), & i \in B \setminus S, \\ \arg \max\limits_{q_{B'}^{S,\mathscr{B}}(\bar{p}^{\mathscr{B}}(t))} \pi_{B'}^{\mathscr{B}}(\bar{p}^{\mathscr{B}}(t), (q_{B'}^{S,\mathscr{B}}(\bar{p}^{\mathscr{B}}(t)), \hat{q}_{N\setminus B'}^{S,\mathscr{B}}(\bar{p}^{\mathscr{B}}(t)))), & i = B', \\ & B' \in \mathscr{B}_{-B}. \end{cases}$$

Therefore the characteristic function is given by

$$v_t(S, \mathscr{B}) = \begin{cases} \pi_B^{\mathscr{B}}(\bar{p}^{\mathscr{B}}(t), \bar{q}^{\mathscr{B}}(\bar{p}^{\mathscr{B}}(t))), & S = B,\ B \in \mathscr{B}, \\ \pi_S^{\mathscr{B}}(\bar{p}^{\mathscr{B}}(t), \hat{q}^{S,\mathscr{B}}(\bar{p}^{\mathscr{B}}(t))), & S \subset B,\ B \in \mathscr{B}, \\ 0, & S = \varnothing, \\ \sum\limits_{B\in\mathscr{B}, B\subseteq S} \pi_B^{\mathscr{B}}(\bar{p}^{\mathscr{B}}(t), \bar{q}^{\mathscr{B}}(\bar{p}^{\mathscr{B}}(t))) & \\ \quad + \sum\limits_{B\in\mathscr{B}, B\nsubseteq S, B\cap S\neq\varnothing} \pi_{B\cap S}^{\mathscr{B}}(\bar{p}^{\mathscr{B}}(t), \hat{q}^{B\cap S,\mathscr{B}}(\bar{p}^{\mathscr{B}}(t))), & \text{otherwise.} \end{cases}$$

Using the same cooperative point solution in the subgame $(N, v_t, \mathscr{B})$, we get a profile $\xi[v_t, \mathscr{B}]$ of firms' cooperative profits.

**Definition 5** We call a coalition structure $\mathscr{B}$ *dynamically stable* with respect to a cooperative point solution in case of transferable profits if for any firm $i \in N$ and any game period $t \in \mathscr{T}$ it holds that

$$\xi_i[v_t, \mathscr{B}] \geqslant \xi_i[v_t, \mathscr{B}'], \tag{5}$$

where $\mathscr{B}' = \{B(i) \setminus \{i\}, B \cup \{i\}, \mathscr{B}_{-B(i)\cup B}\}$ for any $B \in \mathscr{B} \cup \varnothing$ and $B \neq B(i)$, meaning that $\mathscr{B}$ is stable at *any* time period along the coalition equilibrium trajectory $\bar{p}^{\mathscr{B}}$.

*Remark 1* In Rajan (1989), the author proposes an alternative scheme of determining the characteristic function in TU oligopoly games. Following this approach, (i) firms from coalition $S$ jointly maximize the total profit of this coalition, (ii) firms from $B \setminus S$ jointly maximize the total profit of $B \setminus S$, and (iii) each firm $i \in N \setminus B$ maximizes the total profit of coalition $B(i)$. In other words, the coalition Nash equilibrium $\check{q}^{S,\mathscr{B}}(p_0)$ is given by:

$$\check{q}_R^{S,\mathscr{B}}(p_0) = \begin{cases} \arg\max\limits_{q_S^{S,\mathscr{B}}(p_0)} \pi_S^{\mathscr{B}}(p_0, (q_S^{S,\mathscr{B}}(p_0), \hat{q}_{N\setminus S}^{S,\mathscr{B}}(p_0))), & R = S, \\ \arg\max\limits_{q_{B\setminus S}^{S,\mathscr{B}}(p_0)} \pi_{B\setminus S}^{\mathscr{B}}(p_0, (q_{B\setminus S}^{S,\mathscr{B}}(p_0), \hat{q}_{N\setminus(B\setminus S)}^{S,\mathscr{B}}(p_0))), & R = B \setminus S, \\ \arg\max\limits_{q_{B'}^{S,\mathscr{B}}(p_0)} \pi_{B'}^{\mathscr{B}}(p_0, (q_{B'}^{S,\mathscr{B}}(p_0), \hat{q}_{N\setminus B'}^{S,\mathscr{B}}(p_0))), & R = B', \ B' \in \mathscr{B}_{-B}, \end{cases}$$

while the coalition Nash equilibrium $\check{q}^{S,\mathscr{B}}(\bar{p}^{\mathscr{B}}(t))$ in subgame, starting in state $\bar{p}^{\mathscr{B}}(t), t \in \mathscr{T} \setminus 0$, on coalition equilibrium trajectory $\bar{p}^{\mathscr{B}}$, takes the following form:

$$\check{q}_R^{S,\mathscr{B}}(\bar{p}^{\mathscr{B}}(t)) = \begin{cases} \arg\max\limits_{q_S^{S,\mathscr{B}}(\bar{p}^{\mathscr{B}}(t))} \pi_S^{\mathscr{B}}(\bar{p}^{\mathscr{B}}(t), (q_S^{S,\mathscr{B}}(\bar{p}^{\mathscr{B}}(t)), \hat{q}_{N\setminus S}^{S,\mathscr{B}}(\bar{p}^{\mathscr{B}}(t)))), & R = S, \\ \arg\max\limits_{q_{B\setminus S}^{S,\mathscr{B}}(\bar{p}^{\mathscr{B}}(t))} \pi_{B\setminus S}^{\mathscr{B}}(\bar{p}^{\mathscr{B}}(t), (q_{B\setminus S}^{S,\mathscr{B}}(\bar{p}^{\mathscr{B}}(t)), \hat{q}_{N\setminus(B\setminus S)}^{S,\mathscr{B}}(\bar{p}^{\mathscr{B}}(t)))), & R = B \setminus S, \\ \arg\max\limits_{q_{B'}^{S,\mathscr{B}}(\bar{p}^{\mathscr{B}}(t))} \pi_{B'}^{\mathscr{B}}(\bar{p}^{\mathscr{B}}(t), (q_{B'}^{S,\mathscr{B}}(\bar{p}^{\mathscr{B}}(t)), \hat{q}_{N\setminus B'}^{S,\mathscr{B}}(\bar{p}^{\mathscr{B}}(t)))), & R = B', \\ & B' \in \mathscr{B}_{-B}. \end{cases}$$

Given the above coalition Nash equilibria, one can determine characteristic functions $v_t(S, \mathscr{B})$, $S \subseteq N, t \in \mathscr{T}$, under this approach:

$$\check{v}_t(S, \mathscr{B}) = \begin{cases} \pi_B^{\mathscr{B}}(\bar{p}^{\mathscr{B}}(t), \bar{q}^{\mathscr{B}}(\bar{p}^{\mathscr{B}}(t))), & S = B, \ B \in \mathscr{B}, \\ \pi_S^{\mathscr{B}}(\bar{p}^{\mathscr{B}}(t), \check{q}^{S,\mathscr{B}}(\bar{p}^{\mathscr{B}}(t))), & S \subset B, \ B \in \mathscr{B}, \\ 0, & S = \varnothing, \\ \sum\limits_{B\in\mathscr{B}, B\subseteq S} \pi_B^{\mathscr{B}}(\bar{p}^{\mathscr{B}}(t), \bar{q}^{\mathscr{B}}(\bar{p}^{\mathscr{B}}(t))) & \\ \quad + \sum\limits_{B\in\mathscr{B}, B\nsubseteq S, B\cap S\neq\varnothing} \pi_{B\cap S}^{\mathscr{B}}(\bar{p}^{\mathscr{B}}(t), \check{q}^{B\cap S,\mathscr{B}}(\bar{p}^{\mathscr{B}}(t))), & \text{otherwise}, \end{cases}$$

with $\bar{p}^{\mathscr{B}}(0) \equiv p_0$. Then we are able to determine the corresponding cooperative solutions $\xi[\check{v}_t, \mathscr{B}]$, $t \in \mathscr{T}$, and verify whether the coalition structure $\mathscr{B}$ is (dynamically) stable.

## 4 An Example

We consider a market of three firms, $N = \{1, 2, 3\}$ competing in quantities over a finite set of periods $\mathscr{T} = \{0, 1, \ldots, 10\}$ with parameters: $s = 0.8$, $a = 30$, $b = 1$, $c_1 = 1$, $c_2 = 2$, $c_3 = 3$, a discount factor $\varrho = 0.9$ and the initial market price $p(0) = 10$.

Five coalition structures can be formed by three firms: $\mathscr{B}_1 = \{\{1\}, \{2\}, \{3\}\}$, $\mathscr{B}_2 = \{\{1, 2\}, \{3\}\}$, $\mathscr{B}_3 = \{\{1, 3\}, \{2\}\}$, $\mathscr{B}_4 = \{\{1\}, \{2, 3\}\}$, $\mathscr{B}_5 = \{\{1, 2, 3\}\}$.

First, for each coalition structure we calculate a coalition Nash equilibrium, the corresponding price trajectory, and the profits of the firms under this equilibrium for two cases: when the profits are non-transferable and when they are the components of the Shapley value in the game with a given coalition structure. Non-transferable profits are represented in Table 1. The analysis of these profits shows that all five coalition structures are unstable if the firms are paid by the initially given payoff functions:

- for $\mathscr{B}_1$, firm 1 benefits if it joins firm 2 which results in coalition structure $\mathscr{B}_2$;
- for $\mathscr{B}_2$, firm 2 has an incentive to become a singleton thus forming structure $\mathscr{B}_1$;
- for $\mathscr{B}_3$, firm 1 will benefit by joining firm 2;
- for $\mathscr{B}_4$, firm 1 will benefit by joining coalition $\{2, 3\}$;
- and finally for $\mathscr{B}_5$, firm 2 has an incentive to deviate becoming a singleton.

Since in the case of non-transferable profits there is no stable coalition structure, then there cannot be any dynamically stable coalition structure.

Now consider the case of transferable profits. We use the Shapley value $\text{Sh}[v_t, \mathscr{B}] = (\text{Sh}_1[v_t, \mathscr{B}], \dots, \text{Sh}_n[v_t, \mathscr{B}])$, $t \in \mathscr{T}$, as a cooperative point solution in the game and any subgame. Its components are given by

$$\text{Sh}_i[v_t, \mathscr{B}] = \sum_{S \subseteq B(i),\ i \in S} \frac{(|B(i)| - |S|)!(|S| - 1)!}{|B(i)|!} \big(v_t(S, \mathscr{B}) - v_t(S \setminus \{i\}, \mathscr{B})\big), \quad i \in N.$$

We note that the Shapley value for a TU game with a coalition structure (or the Aumann–Dreze value (Aumann and Dreze 1974)) is defined by a so-called restricted characteristic function. For any coalition $B \in \mathscr{B}$ and subcoalition $S \subseteq B$, the value of the restricted characteristic function coincides with $v_t(S, \mathscr{B})$. The Shapley values for all possible coalition structures and all subgames are represented in Table 2. The analysis of firms' profits in the transferable case shows that $\mathscr{B}_2$ is the only stable coalition structure with respect to the Shapley value at $t = 0$ because there are no profitable deviations for any firm. Other four coalition structures are unstable with respect to the Shapley value. Indeed,

**Table 1** Firms' profits

| $\mathscr{B}$ | $\pi_1^{\mathscr{B}}(p_0, \bar{q}^{\mathscr{B}}(p_0))$ | $\pi_2^{\mathscr{B}}(p_0, \bar{q}^{\mathscr{B}}(p_0))$ | $\pi_3^{\mathscr{B}}(p_0, \bar{q}^{\mathscr{B}}(p_0))$ |
|---|---|---|---|
| $\mathscr{B}_1$ | 422.344 | 231.706 | 158.167 |
| $\mathscr{B}_2$ | 446.836 | 223.418 | 186.940 |
| $\mathscr{B}_3$ | 442.399 | 261.002 | 147.466 |
| $\mathscr{B}_4$ | 454.596 | 235.079 | 156.719 |
| $\mathscr{B}_5$ | 486.875 | 243.438 | 162.292 |

Non-transferable case

- for $\mathscr{B}_1$, firm 1 will benefit if it joins firm 2;
- for $\mathscr{B}_3$, firm 1 can make a profitable deviation by joining firm 2 and therefore forming a structure $\mathscr{B}_2$;
- for $\mathscr{B}_4$, firm 2 will benefit by joining firm 1;
- and finally for $\mathscr{B}_5$, firm 1 has an incentive to become a singleton.

Coalition equilibrium trajectories (equilibrium prices) for different coalition structures are depicted in Fig. 1. For any $t = 1, \ldots, 10$, the price $\bar{p}^{\mathscr{B}_1}(t)$ for the case when firms do not cooperate is the smallest, and the price $\bar{p}^{\mathscr{B}_5}(t)$ for the case of full cooperation is the largest as expected.

Moreover, the analysis of Table 2 shows that the structure $\mathscr{B}_2$ is also dynamically stable, so it satisfies Definition 5, i.e., there are no profitable deviations of any firm in any time period $t = 0, \ldots, 9$ when the game is realized along the coalition equilibrium trajectory $\bar{p}^{\mathscr{B}_2}$ calculated for the game with coalition structure $\mathscr{B}_2$.

**Table 2** Firms' profits (the Shapley values)

| $t$ | $\mathscr{B}$ | $\text{Sh}_1[v_t, \mathscr{B}]$ | $\text{Sh}_2[v_t, \mathscr{B}]$ | $\text{Sh}_3[v_t, \mathscr{B}]$ | $t$ | $\mathscr{B}$ | $\text{Sh}_1[v_t, \mathscr{B}]$ | $\text{Sh}_2[v_t, \mathscr{B}]$ | $\text{Sh}_3[v_t, \mathscr{B}]$ |
|---|---|---|---|---|---|---|---|---|---|
| 0 | $\mathscr{B}_1$ | 422.344 | 231.706 | 158.167 | 5 | $\mathscr{B}_1$ | 336.968 | 177.409 | 119.639 |
| | $\mathscr{B}_2$ | 430.446 | 239.808 | 186.940 | | $\mathscr{B}_2$ | 342.448 | 182.890 | 133.982 |
| | $\mathscr{B}_3$ | 427.021 | 261.002 | 162.844 | | $\mathscr{B}_3$ | 340.182 | 191.721 | 122.853 |
| | $\mathscr{B}_4$ | 454.596 | 232.669 | 159.130 | | $\mathscr{B}_4$ | 351.803 | 178.421 | 120.651 |
| | $\mathscr{B}_5$ | 452.758 | 258.405 | 181.441 | | $\mathscr{B}_5$ | 353.934 | 192.172 | 132.136 |
| 1 | $\mathscr{B}_1$ | 440.274 | 239.261 | 162.871 | 6 | $\mathscr{B}_1$ | 288.484 | 150.155 | 100.959 |
| | $\mathscr{B}_2$ | 448.325 | 247.312 | 190.187 | | $\mathscr{B}_2$ | 292.709 | 154.381 | 110.950 |
| | $\mathscr{B}_3$ | 444.94 | 266.987 | 167.537 | | $\mathscr{B}_3$ | 290.968 | 160.062 | 103.444 |
| | $\mathscr{B}_4$ | 470.542 | 240.335 | 163.945 | | $\mathscr{B}_4$ | 298.540 | 150.973 | 101.777 |
| | $\mathscr{B}_5$ | 469.588 | 264.982 | 185.207 | | $\mathscr{B}_5$ | 300.822 | 160.827 | 109.890 |
| 2 | $\mathscr{B}_1$ | 433.687 | 233.943 | 158.893 | 7 | $\mathscr{B}_1$ | 231.874 | 119.054 | 79.790 |
| | $\mathscr{B}_2$ | 441.417 | 241.673 | 183.845 | | $\mathscr{B}_2$ | 234.562 | 121.742 | 85.409 |
| | $\mathscr{B}_3$ | 438.182 | 259.186 | 163.388 | | $\mathscr{B}_3$ | 233.457 | 124.586 | 81.374 |
| | $\mathscr{B}_4$ | 460.982 | 235.080 | 160.030 | | $\mathscr{B}_4$ | 237.343 | 119.588 | 80.324 |
| | $\mathscr{B}_5$ | 460.929 | 257.827 | 179.542 | | $\mathscr{B}_5$ | 239.214 | 125.345 | 84.976 |
| 3 | $\mathscr{B}_1$ | 411.162 | 220.230 | 149.256 | 8 | $\mathscr{B}_1$ | 165.154 | 83.545 | 55.816 |
| | $\mathscr{B}_2$ | 418.359 | 227.426 | 171.183 | | $\mathscr{B}_2$ | 166.215 | 84.607 | 57.727 |
| | $\mathscr{B}_3$ | 415.359 | 242.328 | 153.452 | | $\mathscr{B}_3$ | 165.780 | 85.409 | 56.442 |
| | $\mathscr{B}_4$ | 434.787 | 221.384 | 150.409 | | $\mathscr{B}_4$ | 166.932 | 83.757 | 56.028 |
| | $\mathscr{B}_5$ | 435.606 | 241.631 | 167.657 | | $\mathscr{B}_5$ | 167.842 | 85.818 | 57.654 |
| 4 | $\mathscr{B}_1$ | 378.076 | 200.926 | 135.856 | 9 | $\mathscr{B}_1$ | 86.073 | 43.036 | 28.691 |
| | $\mathscr{B}_2$ | 384.533 | 207.383 | 154.221 | | $\mathscr{B}_2$ | 86.073 | 43.036 | 28.691 |
| | $\mathscr{B}_3$ | 381.852 | 219.350 | 139.632 | | $\mathscr{B}_3$ | 86.073 | 43.036 | 28.691 |
| | $\mathscr{B}_4$ | 397.501 | 202.044 | 136.973 | | $\mathscr{B}_4$ | 86.073 | 43.036 | 28.691 |
| | $\mathscr{B}_5$ | 399.088 | 219.280 | 151.529 | | $\mathscr{B}_5$ | 86.073 | 43.036 | 28.691 |

Transferable case

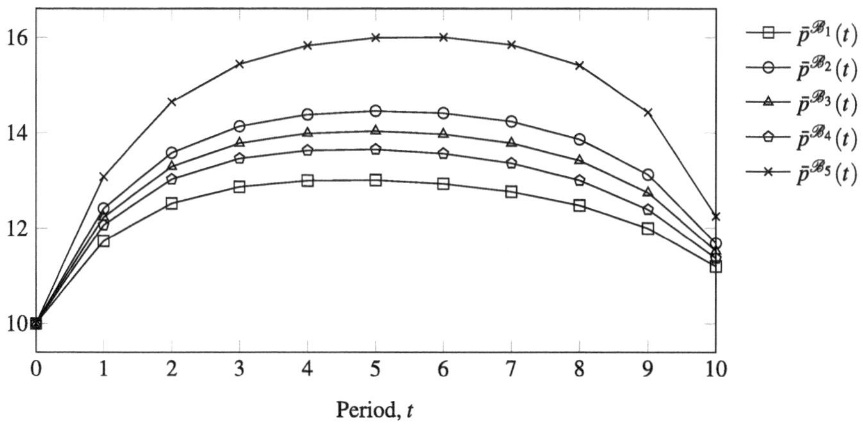


**Fig. 1** Price under coalition Nash equilibrium for all possible coalition structures (coalition equilibrium trajectories)

**Table 3** The Shapley values based on characteristic functions given in Remark 1 for the game with coalition structure $\mathscr{B}_5$

| $t$ | $\text{Sh}_1[\check{v}_t, \mathscr{B}_5]$ | $\text{Sh}_2[\check{v}_t, \mathscr{B}_5]$ | $\text{Sh}_3[\check{v}_t, \mathscr{B}_5]$ |
|---|---|---|---|
| 0 | 453.830 | 257.999 | 180.775 |
| 1 | 470.503 | 264.627 | 184.647 |
| 2 | 461.662 | 257.534 | 179.103 |
| 3 | 436.144 | 241.405 | 167.346 |
| 4 | 399.432 | 219.123 | 151.343 |
| 5 | 354.103 | 192.08 | 132.059 |
| 6 | 300.858 | 160.788 | 109.893 |
| 7 | 239.179 | 125.341 | 85.0157 |
| 8 | 167.805 | 85.8249 | 57.6838 |
| 9 | 86.0727 | 43.0364 | 28.6909 |

This motivates firms to keep this coalition structure the same in the game and not to change it in any intermediate game period once the game has been started.

Following Remark 1, we may calculate the characteristic functions under another approach (see the definition of $\check{v}_t(S, \mathscr{B})$). The values of these functions and the corresponding Shapley values for the three-person game differ only for coalition structure $\mathscr{B}_5$. The Shapley values $\text{Sh}[\check{v}_t, \mathscr{B}_5]$ for $t = 0, \ldots, 9$ are presented in Table 3. Analyzing the values in Tables 2 and 3, we observe that the coalition structure $\mathscr{B}_2$ is also dynamically stable under this approach.

## 5 Conclusion

We have considered a linear-quadratic dynamic game in which firms, competing in a market, may cooperate and form not only the grand coalition but also smaller coalitions being components of a coalition structure. The firms in the coalitions obtain their profits according to a cooperative point solution (e.g., the Shapley value, the nucleolus). The conditions for coalition Nash equilibrium strategies of firms have been obtained. We examined the stability of the coalition structure meaning its Nash stability according to which no firm has an incentive to individual deviation from the coalition it belongs to. We have considered an example for which the grand coalition is unstable, but there exists another coalition structure which is stable not only for the whole game but also along the state equilibrium trajectory corresponding to this coalition structure, that is, dynamically stable. It is interesting to find the general conditions under which a coalition structure is stable (or dynamically stable) for the class of dynamic games considered in the chapter. One can also develop stronger stability conditions which would protect a coalition structure against deviations of any group of players. These developments are left for future research.

**Acknowledgements** This research was supported by the Russian Science Foundation under grant No. 17-11-01079.

## References

Aumann, R. J., & Dreze, J. H. (1974). Cooperative games with coalition structures. *International Journal of Game Theory, 3*, 217–237.

Carlson, D. A., & Leitmann, G. (2005). The direct method for a class of infinite horizon dynamic games. In C. Deissenberg & R. F. Hartl (Eds.), *Optimal control and dynamic games. Advances in computational management science* (Vol. 7, pp. 319–334). Boston, MA: Springer.

Carraro, C. (1999). The structure of international environmental agreements. In C. Carraro (Ed.), *International environmental agreements on climate change. Fondazione Eni Enrico Mattei (Feem) series on economics, energy and environment* (Vol. 13, pp. 9–26). Dordrecht: Springer.

Chander, P., & Tulkens, H. (1997). A core of an economy with multilateral environmental externalities. *International Journal of Game Theory, 26*, 379–401.

Haurie, A., Krawczyk, J. B., & Zaccour, G. (2012). *Games and dynamic games*. Singapore: Scientific World.

Parilina, E., & Sedakov, A. (2014). Stable bank cooperation for cost reduction problem. *The Czech Economic Review, 8*(1), 7–25

Parilina, E., & Sedakov, A. (2015). Stochastic approach for determining stable coalition structure. *International Game Theory Review, 17*(4), 155009

Rajan, R. (1989). Endogenous coalition formation in cooperative oligopolies. *International Economic Review, 30*(4), 863–876

Sedakov, A., Parilina, E., Volobuev, Yu., & Klimuk, D. (2013). Existence of stable coalition structures in three-person games. *Contributions to Game Theory and Management, 6*, 407–422.

Shapley, L. S. (1953). A value for *n*-person games. In H. W. Kuhn & A. W. Tucker (Eds.), *Contributions to the theory of games* (Vol. II, pp. 307–317). Princeton: Princeton University Press.

Sun, F., & Parilina, E. (2018). Existence of stable coalition structures in four-person games. *Contributions to Game Theory and Management, 11*, 224–248.

Printed by Printforce, the Netherlands